ENVIRONMENTAL IMPACT OF GENETICALLY MODIFIED CROPS

ENVIRONMENTAL IMPACT OF GENETICALLY MODIFIED CROPS

Edited by

Natalie Ferry and Angharad M.R. Gatehouse

School of Biology
Institute for Research on Environment and Sustainability
Newcastle University
UK

www.cabi.org

CABI is a trading name of CAB International

CABI Head Office
Nosworthy Way
Wallingford
Oxfordshire OX10 8DE
UK

CABI North American Office
875 Massachusetts Avenue
7th Floor
Cambridge, MA 02139
USA

Tel: + 44 (0)1491 832111
Fax: + 44 (0)1491 833508
E-mail: cabi@cabi.org
Website: www.cabi.org

Tel: + 1 617 395 4056
Fax: + 1 617 354 6875
E-mail: cabi-nao@cabi.org

A catalogue record for this book is available from the British Library, London, UK.

Library of Congress Cataloging-in-Publication Data

Environmental impact of genetically modified crops/edited by Natalie Ferry and Angharad
M.R. Gatehouse.
        p. cm.
    Includes bibliographical references and index.
    ISBN 978-1-84593-409-5 (alk. paper)
1. Transgenic plants--Environmental aspects. 2. Food--Safety measures. 3.
Agricultural biotechnology. I. Ferry, Natalie. II. Gatehouse, A. M. R. III. Title.

SB123.57.E584 2009
577.5'5--dc22

                                                                    2008028686

ISBN: 978 1 84593 409 5

Typeset by SPi, Pondicherry, India.
Printed and bound in the UK by MPG Books Group.

The paper used for the text pages in this book is FSC certified. The FSC (Forest
Stewardship Council) is an international network to promote responsible management
of the world's forests.

Contents

Contributors

Professor Ramón Albajes, *Centre UdL-IRTA, Universitat de Lleida, Rovira Roure 191, Lleida 25198, Spain; E-mail: Ramon.Albajes@irta.cat*

Em Professor Klaus Ammann, *Delft University of Technology, Julianalaan, NL-2628 BC, Delft, The Netherlands; E-mail: Klaus.ammann@ips.unibe.ch*

Elisabeth P.J. Burgess, *Horticulture and Food Research Institute of New Zealand Ltd, Mt Albert Research Centre, Private Bag 92169, Auckland, New Zealand; E-mail: eburgess@hortresearch.co.nz*

Dr Teresa Capell, *Univeritat de Lleida, Department de Produccio Vegetal i Ciencia Forestal, Av. Alcalde Rovira Roure, 191, E-25198 Lleida, Spain; E-mail: Teresa.Capell@pvcf.udl.cat*

Professor Yves Carrière, *Department of Entomology, University of Arizona, Tucson, AZ 85721, USA; E-mail: ycarrier@ag.arizona.edu*

Professor Pedro Castañera, *Departamento de Biología de Plantas, Centro de Investigaciones Biológicas, CSIC, Ramiro de Maeztu 9, Madrid 28040, Spain; E-mail: castan@cib.csic.es*

Professor Paul Christou, *ICREA and Univeritat de Lleida, Department de Produccio Vegetal i Ciencia Forestal, Av. Alcalde Rovira Roure, 191, E-25198 Lleida, Spain; E-mail: christou@pvcf.udl.es*

Dr Martin G. Edwards, *School of Biology, Institute for Research on Environment and Sustainability, Newcastle University, Newcastle upon Tyne NE1 7RU, UK; E-mail: martin.edwards@ncl.ac.uk*

Dr Natalie Ferry, *School of Biology, Institute for Research on Environment and Sustainability, Newcastle University, Newcastle upon Tyne NE1 7RU, UK; E-mail: Natalie.ferry@ncl.ac.uk*

Dr Monica Garcia-Alonso, *Syngenta, Jealott's Hill International Research Centre, Bracknell RG42 6EY, UK; E-mail: monica.garcia-alonso@syngenta.com*

Professor Angharad M. R. Gatehouse, *School of Biology, Institute for Research on Environment and Sustainability, Newcastle University, Newcastle upon Tyne NE1 7RU, UK; E-mail: a.m.r.gatehouse@ncl.ac.uk*

Mr Derick George, *School of Biology, Institute for Research on Environment and Sustainability, Devonshire Building, Newcastle University, Newcastle upon Tyne, NE1 7RU, UK; E-mail: derick.george @ncl.ac.uk; Ministry of Agriculture, Department of Agricultural Research, PO Box 0033, Gaborone, Botswana; E-mail: dgeorge@gov.bw*

Professor Kanglai He, *The State Key Laboratory for Biology of Plant Disease and Insect Pests, Institute of Plant Protection, Chinese Academy of Agricultural Sciences, 2 West Yuanmingyuan Road, Beijing 100193, China; E-mail: klhe@ippcaas.cn*

Dr Richard L. Hellmich, *USDA–ARS, Corn Insects and Crop Genetics Research Unit and Department of Entomology, Iowa State University, Genetics Laboratory c/o Insectary, Ames, Iowa 50011-3140, USA; E-mail: richard.hellmich@ars.usda.gov*

Ms Udhaya Kannan, *School of Biology, Institute for Research on Environment and Sustainability, Newcastle University, Newcastle upon Tyne NE1 7RU, UK; E-mail: udhayabiotech@gmail.com*

Dr Ajay Kohli, *School of Biology, Institute for Research on Environment and Sustainability, Newcastle University, Newcastle upon Tyne NE1 7RU, UK; E-mail: ajay.kohli@ncl.ac.uk; Plant Breeding Genetics and Biotechnology, International Rice Research Institute, DAPO Box 7777, Metro Manila, The Phillipines.*

Dr Louise A. Malone, *Horticulture and Food Research Institute of New Zealand Ltd, Mt Albert Research Centre, Private Bag 92169, Auckland, New Zealand; E-mail: lmalone@hortresearch.co.nz*

Mr Michael Meissle, *Agroscope Reckenholz-Tänikon Research Station ART, Reckenholzstr, 191, 8046 Zurich, Switzerland; E-mail: michael.meissle@art.admin.ch*

Dr Thomas E. Nickson, *Monsanto Company, 800 N. Lindbergh Blvd, St Louis, MO 63141, USA; E-mail: thomas.nickson@monsanto.com*

Professor Tony G. O'Donnell, *Dean, Faculty of Natural and Agricultural Sciences, The University of Western Australia, 35 Stirling Highway, Crawley, Perth, WA 6009, Australia; E-mail: tony.odonnell@uwa.edu.au*

Dr Félix Ortego, *Departamento de Biología de Plantas, Centro de Investigaciones Biológicas, CSIC, Ramiro de Maeztu 9, Madrid 28040, Spain; E-mail: ortego@cib.csic.es*

Professor Micheal D.K. Owen, *Associate Department Chair, Professor of Agronomy and Extension Weed Science, 2104 Agronomy Hall, Ames, Iowa 50011, USA; E-mail: mdowen@iastate.edu*

Professor Maurizio G. Paoletti, *Dipartimento di Biologia, Via Ugo Bassi 58/B, Università di Padova, Padova, Italy; E-mail: maurizio.paoletti@unipd.it*

Professor Richard H. Phipps, *Deputy Director Centre for Dairy Research and Professor of Animal Science and Applied Biotechnology, School of Agriculture, Policy and Development, University of Reading, Reading RG6 6AR, UK; E-mail: r.h.phipps@reading.ac.uk*

Professor David Pimentel, *Department of Entomology, Systematics and Ecology, Cornell University, 5126 Comstock Hall, Ithaca, New York 14853-2601, USA; E-mail: pd18@cornell.edu*

Professor Xavier Pons, *Centre UdL-IRTA, Universitat de Lleida, Rovira Roure 191, Lleida 25198, Spain; E-mail: Xavier.Pons@irta.cat*

Siam Popluechai, *School of Biology, Institute for Research on Environment and Sustainability, Newcastle University, Newcastle upon Tyne NE1 7RU, UK; E-mail: siam.popluechai@newcastle.ac.uk; School of Science, Mae Fah Luang University, 333 Muang, Chiang Rai 57100, Thailand*

Professor Guy M. Poppy, *School of Biological Sciences, University of Southampton, Southampton SO16 7PX, UK; E-mail: g.m.poppy@soton.ac.uk*

Manish Raorane, *School of Biology, Institute for Research on Environment and Sustainability, Newcastle University, Newcastle upon Tyne NE1 7RU, UK; E-mail: manish.raorane@newcastle.ac.uk*

Dr Alan Raybould, *Syngenta, Jealott's Hill International Research Centre, Bracknell, Berkshire RG42 6EY, UK; E-mail: alan.raybould@syngenta.com*

Dr Ben Raymond, *Department of Zoology, University of Oxford, South Parks Road, Oxford OX1 3PS, UK; E-mail: benjamin.raymond@zoo.ox.ac.uk*

Dr Jörg Romeis, *Agroscope Reckenholz-Tänikon Research Station ART, Reckenholzstr. 191, 8046 Zurich, Switzerland; E-mail: joerg.romeis@art.admin.ch*

Professor J. Keith Syers, *School of Science, Mae Fah Luang University, 333 Muang, Chiang Rai 57100, Thailand; E-mail: keith@mfu.ac.th*

Professor Bruce E. Tabashnik, *Department of Entomology, University of Arizona, Tucson, AZ 85721, USA; E-mail: brucet@ag.arizona.edu*

Dr Francesca G. Tencalla, *Monsanto Europe SA, Avenue de Tervueren 270–272, B-1150 Brussels; ToxMinds BVBA, Waversesteenweg 138, B-1560 Hoeilaart, Belgium; E-mail: Francesca.tencalla@toxminds.com*

Professor Zhenying Wang, *The State Key Laboratory for Biology of Plant Disease and Insect Pests, Institute of Plant Protection, Chinese Academy of Agricultural Sciences, 2 West Yuanmingyuan Road, Beijing 100193, China; E-mail: zywang@ippcaas.cn*

Dr Ron Wheatley, *Scottish Crop Research Institute, Dundee, Scotland, UK; E-mail: ron.wheatley@scri.ac.uk*

Dr Mark J. Whittingham, *School of Biology, Ridley Building, Newcastle University, Newcastle upon Tyne NE1 7RU, UK; E-mail: m.j.whittingham@ncl.ac.uk*

Professor Denis J. Wright, *Department of Life Sciences, Sir Alexander Fleming Building, Imperial College London, London SW7 2AZ, UK; E-mail: d.wright@imperial.ac.uk*

Dr Yongjun Zhang, *The State Key Laboratory for Biology of Plant Disease and Insect Pests, Institute of Plant Protection, Chinese Academy of Agricultural Sciences, 2 West Yuanmingyuan Road, Beijing 100193, China; E-mail: yjzhang@ippcaas.cn*

Preface

Agriculture is a vital development tool for achieving the Millennium Development Goal that calls for halving by 2015 the share of people suffering from extreme poverty and hunger.

(World Bank, 2008)

World population increased fourfold during the last century, with current estimates placing it at 9.2 billion by 2050. We are now facing a situation where food demand is beginning to outstrip supply, a scenario predicted some two centuries ago by the Reverend Robert Malthus (Jesus College Cambridge; 1766–1834). This situation is compounded by the fact that we may be at the limit of the existing genetic resources available in our major crops (Gressel, 2008). Thus, new genetic resources must be found and only new technologies will enable this. Lord Robert May (President of the Royal Society, March 2002) stated:

We couldn't feed today's world with yesterday's agriculture and we won't be able to feed tomorrow's world with today's.

One such route is through the use of recombinant DNA technology to produce transgenic crops expressing desirable agronomic traits such as enhanced resistance to insect pests and herbicide tolerance; such crops were first commercialized in the mid-1990s. By 2007, approximately 12 million farmers in 23 countries (12 developing and 11 industrialized countries) grew biotech (genetically modified (GM)) crops, with an additional 29 countries having granted regulatory approvals since 1996. In 2007, the global market value of such crops was an estimated US$6.9 billion, representing 16% of the US$42.2 billion global crop protection market and 20% of the estimated US$34 billion global seed market. This biotech crop market comprised US$3.2 billion for biotech maize (equivalent to 47% of global biotech crop market), US$2.6 billion for biotech soybean (37%), US$0.9 billion for biotech cotton (13%) and US$0.2 billion for biotech canola (3%); of this market, US$5.2 billion (76%) was in the industrial countries and

US$1.6 billion (24%) was in the developing countries. The accumulated global value for the 11-year period, since biotech crops were first commercialized in 1996, is estimated at US$42.4 billion and this figure is projected to increase to approximately US$7.5 billion for 2008 (James, 2008).

Despite increased yields, accompanied by a decrease in pesticide use, this technology remains one of the most controversial agricultural issues of current times. Many consumer and environmental lobby groups believe that GM crops will bring very little benefit to growers and to the general public, and that they will have a deleterious effect on the environment. This is at a time when food and fuel are competing for land, and climate change threatens to compromise current resources. Thus, population growth, changing diets, higher transport costs and a drive towards bio-fuels are forcing food prices up (see Chapter 18, this volume). During the 12 months from March 2007 to March 2008, the price of wheat and rice, two major food staples, increased by 130% and 74%, respectively, while soya and maize, important components of feeds, increased by 87% and 31%, respectively (Bloomberg, FAO/Jackson Son & Co). The UN's Food and Agriculture Organization (FAO) stated that the food crisis had thrown an additional 75 million people into hunger and poverty in 2007 alone.

It is, and will continue to be, a priority for agriculture to produce more crops on less land. The minimization of losses to biotic and abiotic stresses would go some way to optimizing the yield on land currently under cultivation. Traditionally agricultural production has kept pace, and even outstripped, human population growth; however, we currently face a set of unique challenges. One of the greatest dangers to agriculture is its vulnerability to global climate change; the expected impacts are for more frequent and severe drought and flooding and shorter growing seasons. The performance of crops under stress will depend on their inherent genetic capacity and on the whole agroecosystem in which they are managed. It is for this reason that any efforts to increase the resilience of agriculture to climate change must involve the adoption of stress-tolerant plants as well as more prudent management of crops, animals and the natural resources that sustain their production.

This book seeks to present a balanced view on the environmental impact of GM crops. The first part, 'Genetically Modified Crops: Three Chapters, Three Views', sets the scene. Chapter 1 addresses the role of biotechnology in sustainable agriculture, describing the 'state-of-the-art' technology used to produce such crops. It also discusses the status of first-generation GM crops and current advances being made in molecular farming (use of plants to produce value-added molecules) concluding with the role that politics plays in delivering, or not, this technology. The theme of GM crops and sustainability is further developed in Chapter 2, which focuses on the benefits of the technology to meet demand in the face of an ever-increasing global population and current constraints on agricultural production, including societal demands for reduced pesticide usage. The final chapter in this part, Chapter 3, examines the need to increase and make food production more rational, to conserve natural resources and to reduce food (crop) losses to pests. Further, it addresses the possibility, and benefits, of converting annual grains into perennial grains. In addition to its benefits, the chapter critically evaluates perceived risks of the technology and its failure to prevent the continued decline in production and availability of world cereal grains. Finally, it considers the impact of attitudes and practices on food biodiversity.

The second part, 'Agricultural Biotechnology: Risks, Benefits and Potential Ecological Impact', considers in detail the role and contribution of both insect-resistant transgenic crops and herbicide-tolerant crops to agriculture and global food security. However, authorization for commercialization, whether for import or cultivation, must take into account the outcome of the environmental risk assessment, a formal assessment of food and feed safety, and in certain cases also consider political, economic and societal factors. Although the details of the risk assessment frameworks for GM crops vary from country to country, the general principles, upon which they are based, are comparable. These main principles and regulatory aspects of this process are discussed at length in Chapter 4. Chapter 5 specifically focuses on the use of crops genetically modified to produce crystal (Cry) toxins from *Bacillus thuringiensis* (*Bt*) for enhanced resistance to insect pests. It describes the mode of action of these Cry toxins and discusses the potential for evolution of resistance to these toxins in insect populations. Further, it considers the monitoring of resistance to *Bt* crops and the required/recommended deployment strategies for such crops, a theme further developed in the following chapter. Chapter 6 also considers the influence of three broad ecological factors on the evolution and management of pest resistance to insect-resistant transgenic crops: population-level processes; bottom-up effects of the host plant; and top-down effects deriving from natural enemies and pathogens. The exploitation of such ecological factors in resistance management strategies, recent developments in the area and a novel genetic method for resistance management are also discussed. The impact of *Bt*-expressing crops on non-target and organisms and, in particular, beneficial insects including natural enemies (such as predators and parasitoids) and pollinators are discussed in Chapters 8 and 9, respectively. These two chapters also consider the potential impact of transgenic crops expressing other insecticidal proteins, e.g. protease inhibitors and lectins and even plants which have been metabolically engineered. The potential role of pollinators in transgene flow is also considered. Moving on from insect-resistant transgenic crops, Chapter 7 discusses in detail the use of herbicide-tolerant GM crops and the value that they provide, both economically and environmentally. This chapter also addresses some of the key questions about the benefits and risks of transgenic crops, focusing specifically on glyphosate-resistant crops. Further, it considers their impact on biological diversity, their coexistence with non-transgenic crops and the impact that they, and supporting agronomic management systems, have on weed communities, specifically population shifts, evolved resistance and transgene flow to near-relative weeds. Soil ecosystem functioning is vital to the sustained functioning of the biosphere, with plant inputs as the major drivers. Thus, the potential impact of transgenic crops on soil and water ecology is essential. Chapter 10 considers the impact of such crops, together with changes in agronomic practice, such as cultivation technique and timing, planting and sowing rates, pesticide usage, etc. as drivers of community structure in the rhizosphere.

A major concern regarding the large-scale growing of transgenic crops has been their potential impact on biodiversity, including crop biodiversity. Chapter

11 thus considers the needs for biodiversity and the consequences of its loss. The author presents data to suggest that in tropical environments with a naturally high biodiversity the interactions between potentially invasive hybrids of transgenic crops and their wild relatives should be buffered through the complexity of the surrounding ecosystems. This theme is further developed in Chapter 12, which considers the impact of GM crops on birds since they are important biodiversity indicators as they depend on a range of invertebrates and plants for food. Farmland bird populations have declined dramatically, especially in Europe, in the latter half of the 20th century, due to intensification in agricultural practice. While there is little direct evidence of effects of GM crops on birds, there is considerable evidence for potential indirect effects from the recent UK Farm-Scale Evaluation (FSE) trials (see Chapter 2, this volume).

Another major concern relates to the safety of such crops for human consumption. Although the World Health Organization stated that the consumption of DNA from all sources (including plants improved through biotechnology) is safe and does not produce a risk to human health, concern was expressed that 'transgenic' DNA and the novel protein produced by the inserted gene might accumulate in milk, meat and eggs. Chapter 13 provides a synthesis of the major studies carried out to date of first-generation transgenic crops as well as the fate of transgenic DNA and encoded proteins in GM feed. It concludes that there is no evidence to suggest that food derived from animals fed GM products is anything other than as safe and as nutritious as that produced from conventional feed ingredients.

In response to the negative environmental impacts, limited sources and rising prices of fossil fuels, all of which pose significant environmental and socioeconomic challenges, there has been a focus on using plants as a source of biofuels. The final chapter in this part, Chapter 14, reviews the potential of Jatropha as a model, non-edible, oilseed plant and the research needed to realize its potential as a bioenergy crop, including transgenic-based approaches.

Currently the USA is the major grower of GM crops, and in 2007 accounted for approximately 50% of the market share, equivalent to 57.7 million ha. Understandably, most studies have focused on the USA. In the third part, a more global perspective is taken with chapters focusing on Europe, China and Africa. Chapter 15 presents the situation with regard to European commercial plantings of GM crops, of which Spain represents the major grower. GM maize crops expressing the Cry1Ab toxin from *Bacillus thuringiensis* have been cultivated in Spain on a commercial scale since 1998, reaching an area of about 75,000 ha (around 21% of total maize-growing area) in 2007. This chapter addresses case-specific monitoring for pest resistance as well as gene-flow and environmental concerns within a European context. China was in fact the first country to commercialize biotech crops with the commercialization of tobacco in the early 1990s. The last few years have seen a significant increase in the area planted to GM crops, primarily as a consequence of the growing of *Bt* cotton, first commercialized there in 1997. Chapter 16 provides a comprehensive overview of the status of *Bt* cotton in China which, in 2007, accounted for 66% of the cotton crop. It also discusses in detail programmes throughout the country to monitor the evolution of *Bt* resistance within pest populations. Finally, it addresses some socio-economic aspects of growing *Bt* cotton and the substantial contribution it has made to the alleviation of poverty of some 8 million smallholder, resource-poor cotton farmers. The final chapter in this part, Chapter 17, considers many of the critical challenges that Africa is currently facing, exacerbated by the fact that the level of crop production has not kept pace with population growth. The author reflects that while it is not suggested that GM crops are a panacea for Africa's problems, the technology may still bring immense benefits to its people. The constraints to biotechnological development in Africa that need to be resolved are many, including a lack of resources, political instability, lack of networks, intellectual property right law, trade imbalances, the current legislative framework, the actual crops chosen for modification, as well as issues relating to biosafety.

The final part of the book, 'The Future of Agriculture', contains only Chapter 18, and asks the reader to consider what is an acceptable environmental impact in relation to GM crops. It frames the question in the context of the evolution of agriculture over time from the first crop domestication events to the current highly input-dependent, artificial environments in which we grow crops modified by artificial selection to be very different from their wild progenitors. It is broadly agreed that crop production must be more sustainable. This chapter also poses the question as to whether agricultural innovations such as GM crops can contribute to sustainability and meet the needs, at least in part, of an increasing global population.

References

Gressel, J. (2008) *Genetic Glass Ceilings. Transgenics for Crop Biodiversity*. The John Hopkins University Press, Baltimore, Maryland.

James, C.A. (2008) Global status of commercialized biotech/GM crops: 2008. ISAAA Briefs. Brief 38.

World Bank (2008) *The World Development Report, Agriculture for Development*. ISBN-13:978-0-8213-807-7 World Bank, Washington, DC, 365 pp.

I Genetically Modified Crops: Three Chapters, Three Views

1

Transgenic Crops and Their Applications for Sustainable Agriculture and Food Security

P. Christou[1,2] and T. Capell[2]

[1]ICREA; [2]Univeritat de Lieida, Department de Produccio Vegetal i Cliencia Forestal, Spain

Keywords: Genetic transformation, biotic and abiotic stresses, value-added traits, political dimension of transgenic crops

Summary

The potential of transgenic crops to make major contributions to food security and agricultural sustainability worldwide is indisputable. One of the major advantages of transgenic technology is the fact that genes from any source can be accessed and introduced into target crops, facilitating the creation of improved varieties. Here, we review briefly the methodology available for generating transgenic crops and we discuss a number of target traits in the context of food security. Major objectives include: resistance to insect pests and tolerance to weeds; resistance to fungal, bacterial and viral diseases; tolerance to a range of abiotic stresses including drought, salinity, cold, hypoxia; improvement in yield and nutritional content; utilizing the plant cell's machinery as a factory to produce valuable recombinant proteins and metabolites, to name a few. The political dimension of transgenic crops is also discussed as this constitutes an inseparable element of their further deployment and utilization.

Introduction

Sustainable agriculture and food security are critical foundations that underpin human society. Without them, we face the inevitable collapse of our farming communities, irreversible environmental damage, food shortages and eventual economic failure. Sustainable agriculture refers to the ability of a farm to produce crops indefinitely and profitably, without damaging the ecosystem, e.g. through soil erosion, nutrient depletion and overuse of water (Altieri, 1995). Unfortunately, the need to balance profitability and environmental stewardship is a significant economic and scientific challenge, since agriculture by its very nature is one of the most expensive and environmentally harmful practices carried out by humans. Since the dawn of the agricultural age, excessive tillage, irrigation and, more recently, monoculture and the use of agrichemicals have exerted strong environmental pressures, leading to land erosion, destruction of natural habitats and pollution of soil water and groundwater. In addition to the economic and scientific challenges, political considerations and cultural diversity in different geographical locations complicate agricultural development, particularly in the face of world trade agreements and subsidies.

One of the most intractable issues to address is food security, which is defined by the Food and Agriculture Organization (FAO) as the situation that exists when all people, at all times, have access to sufficient, safe and nutritious food that meets their dietary needs and food preferences for an active and healthy life (FAO, 2001). In much of the developing world, food security is something that cannot be taken for granted. Indeed, more than 840 million people in developing countries are chronically undernourished, surviving on fewer than 2000 calories per day (Pinstrup-Andersen et al., 1999; FAO, 2001). Many more people, perhaps half of the world's population in total, suffer from diseases caused by dietary deficiencies and inadequate supplies of vitamins and minerals (Graham et al., 2001). Faced with this immediate problem, most governments are willing to support short-term, non-sustainable agriculture to increase the availability of food, ignoring the fact that continuing food security depends on the sustainable production of food crops in the long term.

There is no unique solution to the problem of sustainable agriculture, but the development of improved plant varieties with enhanced performance and reduced environmental impact is one beneficial strategy. Crop varieties with improved agronomic performance can be generated using a number of methods, some based on conventional breeding, others on more recent developments in biotechnology and perhaps some by combining both conventional and biotechnology strategies (Huang et al., 2002a). The use of transgenic plants offers great promise for the integration of improved varieties into traditional cropping systems because improved plant lines can be generated quickly and with relative precision once suitable genes for transfer have been identified (Christou and Twyman, 2004). It is recognized that biotechnology is not a magic wand that can achieve sustainable agriculture and free the world from poverty, hunger and malnutrition, but the use of transgenic plants as one component of a wider strategy including conventional breeding and other forms of agricultural research can contribute substantially towards the achievement of these goals, both now and in the future.

In this chapter, we outline a number of different routes by which genetically enhanced plants can promote sustainable agriculture and food security. After outlining the methods used to produce transgenic plants, we discuss different ways in which crop yields can be maximized while reducing reliance on chemical inputs, thereby protecting the environment and reducing input costs. We then look at ways in which transgenic plants can be used to produce value-added products to increase the diversity of wealth generation strategies in agriculture, help reduce poverty, improve health and thus maintain economic stability.

Transgenic Methodology

Transgenic plants contain additional genetic material called the *transgene*, which may comprise one or more heterologous genes that provide the plant with novel properties or the ability to synthesize novel products (Twyman *et al.*, 2002). Since plants generally display a high degree of developmental plasticity, the normal route to transgenesis is through the introduction of the transgene into cultured plant cells or tissues, followed by subjecting those cells or tissues to selection for a *marker gene* contained in the introduced DNA, and then regenerating from the surviving (transgenic) cells a whole, fertile plant known as the *primary transformant* or *primary event*. Seeds from the primary transformant then yield *transgenic lines*, wherein each cell in each plant contains the same introduced DNA in the same position within the plant genome. Depending on the species, isolated stem segments, leaf discs or seed-derived callus tissue may be able to regenerate an entire new plant under appropriate culture conditions. For most plant species, some form of tissue culture step is therefore necessary for the successful production of transgenic plants, although in the case of *Arabidopsis* whole plant (*in planta*) transformation strategies are possible which minimize or eliminate the need for subsequent tissue culture (Feldmann and Marks, 1987). However, such *in planta* transformation methods are not widely applicable.

There are two major strategies for gene transfer to plants: one based on *Agrobacterium tumefaciens*, a soil bacterium which causes plant tumours, and another based on the bombardment of plant tissue with microprojectiles, typically DNA-coated gold particles. The *Agrobacterium* method exploits a naturally occurring plasmid called the tumour-inducing (Ti) plasmid, which has the ability to transfer a small piece of DNA (the transferred DNA, usually called T-DNA) into the genome of wounded plant cells. The T-DNA is bracketed by direct repeats which are the only sequences required for transfer. They are recognized by proteins encoded by the *vir* region of the same plasmid, and these proteins orchestrate the transfer process. In nature, the T-DNA contains genes encoding plant hormones so that the transformation of plant cells results in the growth of callus tissue, forming a tumour known as a gall. The T-DNA also contains genes that give plant cells the ability to synthesize novel amino acid derivatives, known as opines, which the bacterium uses as a food source. In this way, the bacterium creates its own ideal habitat in a living plant. For plant transformation in the laboratory, the T-DNA is disarmed by removing the tumour-causing genes and replacing them with the transgene. In modern transformation procedures, a binary vector system is used, where the bacterium is transformed with two plasmids, one small vector containing the transgene within T-DNA repeats, and a helper plasmid containing the *vir* functions (Tzfira and Citofsky, 2003; Gelvin, 2003). Examples include the pGreen series of binary vectors (Hellens *et al.*, 2000) and the GATEWAY vectors (Karimi *et al.*, 2002). The former plasmid system allows any arrangement of selectable marker and reporter gene at the right and left T-DNA borders without compromising the choice of restriction sites for cloning. The GATEWAY system is based on site-specific recombination mediated by phage λ and provides a rapid and highly efficient way to move DNA sequences into multiple vector systems for functional analysis and protein expression. The *Agrobacterium* method is widely used for the transformation of a number of dicot species and many transformation protocols are variants of the leaf-disc method originally developed by Horsch *et al.* (1985). Essentially, small discs cut from leaves are incubated with the bacterium, leading to the infection and transformation of the peripheral cells. These are regenerated under selection to yield primary events.

Particle bombardment is a conceptually simple transformation method in which the transgene DNA is precipitated on to small metal particles that are in turn loaded on to a larger projectile. This is accelerated towards a retaining screen under the force of pressured gas or an electrical discharge, the screen acting as a stopping plate for the larger projectile but containing apertures large enough for the microprojectiles to continue their journey towards the plant tissue. The microprojectiles penetrate the plant cells and some particles become lodged in the nucleus, wherein the DNA diffuses from the particle and integrates into the genomic DNA (Twyman and Christou, 2004). Since bombardment is an entirely physical process (in that no genes or gene products are required to effect the transfer process), it circumvents several major limitations of the *Agrobacterium* system (Altpeter *et al.*, 2005). First, it is possible to achieve the transformation of any species and cultivar by this method because there are no 'host range' limitations, meaning that the range of plants transformable by particle bombardment is restricted only by the competence of cells for regeneration. Second, there are no intrinsic vector requirements, so transgenes of any size and arrangement can be introduced, and multiple-gene co-transformation is straightforward, all without the need for vector backbone sequences. Therefore, particle bombardment can be used to transform species and cultivars currently outside the *Agrobacterium* host range (including most elite cultivars of cereals) and can be used for 'clean DNA' transformation, where only the transgene, not the entire vector, is used for the transformation process (Fu *et al.*, 2000; Agrawal *et al.*, 2005). Particle bombardment has been used most widely for generating commercial transgenic crops, and the delivery of transgenes into embryonic tissues by particle bombardment remains the principal direct DNA transfer technique in plant biotechnology (James, 2006).

It has been suggested that particle bombardment will be supplanted by the *Agrobacterium* method, at least for the production of commercial genetically enhanced crops (Tzfira and Citofsky, 2002; Gelvin, 2003; Valentine, 2003). There is a widely held belief that *Agrobacterium*-mediated transformation is more precise, more controllable and therefore 'cleaner' than particle bombardment, but there have been many reports of vector backbone co-transfer by *A. tumefaciens* (reviewed by Kohli *et al.*, 2003) and it is clear, in the light of recent innovations, that particle bombardment allows much more precise control over transgene structure (Altpeter *et al.*, 2005). It therefore seems unlikely that particle bombardment will be displaced as a transformation method in the foreseeable future.

Countering Biotic Constraints: Weeds, Pests and Diseases

Pests and pathogens are collectively described as *biotic stresses* and are defined as biological constraints that reduce yields either by adversely affecting plant growth and development, or by consuming and/or spoiling the products of food crops in the field or in storage. In terms of food production, the most significant biotic constraints include weeds, insect pests, viral and microbial diseases, and nematodes. Together, these factors are thought to reduce crop yields worldwide by up to 30%. In developing countries, this figure may be much higher because the climatic conditions favour the survival and breeding of insect pests, which not only feed on plants but also act as vectors for many viral diseases. Weeds and insect pests have been identified as primary targets for transgenic technology, and the vast majority of commercially grown transgenic plants are modified for herbicide tolerance, insect resistance or both (James, 2006). Making plants resistant to pests and diseases removes the requirement for expensive and hazardous chemicals, and allows crops to be grown productively on smaller areas of land.

Weeds

Weeds compete with crops for resources, in some cases by parasitism. The cost of weeds, measured in terms of reduced yields, the application of herbicides and the mechanical and manual labour required to remove them, is probably the largest single input into agriculture. For this reason, weed control has been identified as the primary target for first-generation genetically modified (GM) technology and most of the transgenic plants grown in the world today have been modified for herbicide tolerance, allowing the use of safe, broad-spectrum herbicides such as glyphosate. Although herbicide-resistant crops encourage the use of chemical inputs, the overall environmental impact is positive, since total herbicide use is lowered in conjunction with herbicide-tolerant crops. Glyphosate, for example, is active against a wide range of plants but has a very low toxicity to wildlife, farm animals and humans; in soil, it is rapidly bound to soil particles and inactivated by bacteria (EPA, 1993). The use of glyphosate, instead of more specific herbicides which persist longer in the environment, are more toxic to wildlife and also more expensive, is a significant economic and environmental benefit which contributes to sustainable agriculture and food security (Gressel, 2002).

Because GM technology in the West is driven predominantly by the potential for commercial gain, research has focused on the weed problems facing farmers in the industrialized nations. There has been little interest in developing crops with resistance to the weed species that plague subsistence farmers in the developing world, even though this would have an immediate impact on food security. For example, *Striga* is a genus of parasitic flowering plants that infests cereal and legume crops throughout Africa. It is particularly difficult to control in maize crops and causes yield losses estimated at US$7 billion every year (Berner *et al.*, 1995). Genes allowing the selective control of this weed by herbicide application have been identified (Joel *et al.*, 1995), but thus far no transgenic varieties have been produced. Progress towards the selective control of *Striga* in Africa has been made solely through mutation and conventional breeding for imazapyr resistance in maize (Kanampiu *et al.*, 2001, 2002).

Insect pests

Many of our crops are also food for insect pests, and about one-half of all crop production in the developing world is thought to be lost to insects, 15% of these losses occurring due to postharvest consumption and spoilage (Gatehouse *et al.*, 1993). Insects not only cause direct yield losses by damaging and consuming plants, but they also act as vectors for many viral diseases and the damage they inflict often facilitates secondary microbial infections.

In the West, pest control is heavily dependent on chemical inputs, which are both expensive and damaging to the environment. The chemicals are non-selective, killing harmless and beneficial insects as well as pests, and they accumulate in water and soil. Constitutive exposure can lead to the evolution of resistance in insect populations, which reduces their efficiency. Generally, chemical pesticides are too expensive for farmers in the developing world and in any case are often ineffective against sap-sucking pests such as the rice brown planthopper (*Nilaparvata lugens*).

Insect pests are therefore an important target for GM technology (Christou *et al.*, 2006). The genetic modification of plants to express insect-resistance genes offers the potential to overcome all of the shortcomings listed above, since genes that show exquisite specificity towards particular pest species have been isolated from bacteria and other sources, thus minimizing the threat towards non-target organisms. Furthermore, the expression of such proteins within plants allows the effective control of insects that feed or shelter within the plant. Finally, the probability of insects becoming resistant to transgenic plants can be reduced by a number of strategies, such as pyramiding resistance genes affecting different receptors in the target insect, conditional expression and the provision of 'safe-havens' or refuges, to reduce selection pressure. Several different types of genes have been exploited to control insect pests, including bacterial toxins, lectins and protease inhibitors. We use toxins from the spore-forming bacterium *Bacillus thuringiensis* (*Bt*) as an example.

Bt toxins are also known as crystal (Cry) proteins or δ-endotoxins. They are expressed as inert protoxins that are activated by proteinases within the midgut; in Lepidoptera, this is highly alkaline. This provides an important safety barrier, since the environment in which the toxins are activated is unique to insects, and thus it is safe for other animals, and humans, to ingest plants expressing *Bt* toxins. Once activated, the toxins interact with receptors on the midgut epithelium cells creating pores in the plasma membrane by disrupting osmotic balance. This results in paralysis and ultimately the death of the insect.

Many excellent accounts of the economic, environmental and health benefits of *Bt* crops in the West have been published (e.g. de Maagd *et al.*, 1999; Hilder and Boulter, 1999; Llewellyn and Higgins, 2002), but their greatest impact is felt in developing countries which are the worst affected by pest infestations. Farm surveys of randomly selected households cultivating insect-resistant GM rice varieties demonstrate that when compared with households cultivating non-GM rice, small and poor farm households benefit from adopting GM rice by both higher crop yields and reduced use of pesticides, which also contribute to improved health. For rice, the development and implementation of appropriate resistance management strategies and resolution of trade policy barriers are key constraints that have delayed earlier widespread cultivation of the crop (Huang *et al.*, 2005). For cotton, key documented benefits are a 70% reduction in insecticide applications in *Bt* cotton fields in India, resulting in a saving of up to US$30/ha in insecticide costs, with an increase of 80–87% in yield of harvested cotton (Qaim and Zilberman, 2003) and a dramatic reduction in pesticide applications in *Bt* cotton fields in China. The same survey revealed that the percentage of farmers with pesticide poisoning was reduced from 22% to 4.7% (Huang *et al.*, 2002b).

A field evaluation has been carried out to assess potential hazards of growing Compa, a transgenic *Bt* maize variety (Eizaguirre *et al.*, 2006). Two categories of potential hazards were investigated: the potential of the target corn borer, *Sesamia nonagrioides*, to evolve resistance to *Bt* maize, and effects on non-target species (herbivores and predators). Pest larvae collected in *Bt* fields at later growth stages, in which event 176 *Bt* maize expresses the toxin at sublethal concentrations, had longer diapause and post-diapause development than larvae collected in non-*Bt* fields, a feature that might lead to a certain isolation between populations in both type of fields and accelerate *Bt*-resistance evolution. Transgenic maize did not have a negative impact on non-target pests in the field or on natural predators; more aphids and leafhoppers but similar numbers of cutworms and wireworms were counted in *Bt* versus non-*Bt* fields.

Recent developments in *Bt* transgenic technology include the stacking of multiple *Bt* genes and the use of fusion genes to provide enhanced resistance against a range of insect pests. An example of the first strategy is the simultaneous introduction of three insecticidal genes, *cry1Ac*, *cry2A* and *gna* (the latter encoding a lectin which is toxic to sap-sucking homopteran pests), into indica rice to control three major pests: rice leaf folder (*Cnaphalocrocis medinalis*), yellow stem borer (*Scirpophaga incertulas*) and brown planthopper (Bano-Maqbool *et al.*, 2001). The triple transgenic plants were more resistant to pests compared to their binary transgenic counterparts. This is one of the few examples where transgene pyramiding was used in a crop plant to create durable resistance against multiple insect pests with different feeding modes.

The fusion gene strategy has also been shown to provide enhanced resistance. For example, the efficacy of *Bt* toxins was increased by creating fusion between domain III of Cry1Ac and domains I and II of various other Cry1 family proteins (Karlova *et al.*, 2005). Similarly, a hybrid toxin was developed against *Spodoptera litura*, a polyphagous pest, that is tolerant of most *Bt* toxins (Singh *et al.*, 2004). In this case, a domain with weak toxicity in the naturally occurring Cry1Ea toxin was replaced with the homologous 70 amino acid region of Cry1Ca. The synthetic gene was further optimized for high-level expression in plants and was introduced into tobacco and cotton, resulting in extreme toxicity to *Spodoptera litura* at all stages of larval development. Mehlo *et al.* (2005) engineered plants with a fusion protein combining Cry1Ac with the galactose-binding domain of the non-toxic ricin B-chain. This fusion increased the potential number of interactions at the molecular level in target insects, so that transgenic rice and maize plants engineered to express the fusion protein were significantly more toxic in insect bioassays than those containing the *Bt* gene alone. They were also resistant to a wider range of insects, including important pests that are not normally susceptible to *Bt* toxins.

Crop diseases

Many plants have evolved resistance mechanisms that protect them either generally from pathogens or against particular pathogen species. The transfer of genes from resistant to susceptible species is one GM strategy that can be used to accelerate conventional breeding for disease resistance (Salmeron and Vernooij, 1998; Stuiver and Custers, 2001).

   One of the most prevalent bacterial diseases in our food crops is bacterial blight of rice, which causes losses >US$250 million every year in Asia alone. This disease has received a great deal of attention due to the discovery of a resistance gene complex in the related wild species *Oryza longistaminata*. The trait was introgressed into cultivated rice line IR-24 and was shown to confer resistance to all known isolates of the blight pathogen *Xanthomonas oryzae* pv. *oryzae* in India and the Philippines (Khush *et al.*, 1990).

   Further investigation of the resistance complex resulted in the isolation of a gene, named *Xa21*, encoding a receptor tyrosine kinase (Song *et al.*, 1995). The transfer of this gene to susceptible rice varieties resulted in plant lines showing strong resistance to a range of isolates of the pathogen (Wang *et al.*, 1996; Tu *et al.*, 1998; Zhang *et al.*, 1998). The *Xa21* gene has also been stacked with two genes for insect resistance to generate a rice line with resistance to bacterial blight and a range of insect pests, and this is on the verge of commercial release in China (Huang *et al.*, 2002a). As with insect-resistance genes, the widespread use of transgenic plants carrying a single resistance factor could prompt the evolution of new pathogen strains with counteradaptive properties. Therefore, other blight-resistance transgenes are being tested for possible deployment either alone or in combination with *Xa21*. For example, Tang *et al.* (2001) have produced rice plants expressing a ferredoxin-like protein that had previously been shown to delay the hypersensitive response to the pathogen *Pseudomonas syringae* pv. *syringae*. In inoculation tests with *X. oryzae* pv. *oryzae*, all the transgenic plants showed enhanced resistance against the pathogen.

   Viral diseases also cause significant losses, and viruses are another major target of GM technology. Virus resistance can be achieved in a variety of ways, but one of the most effective is to introduce into the transgenic plant one or more genes from the virus itself, a strategy known as *pathogen-derived resistance*. One way to achieve pathogen-derived virus resistance is to express a coat protein gene, since this can help to block virus replication. Several examples of this strategy have been demonstrated in rice, which is host to more than ten disease-causing viruses. In South-east Asia, for example, tungro is the most damaging disease. This is caused by the combined action of two viruses: rice tungro bacilliform virus (RTBV) and rice tungro spherical virus (RTSV). In Central and South America, rice hoja blanca virus (RHBV) has been known to cause up to 100% losses. Pathogen-derived resistance to these diseases has been achieved in experimental plants by expressing coat protein genes from RTBV, RTSV and RHBV (Kloti *et al.*, 1996; Lentini *et al.*, 1996; Sivamani *et al.*, 1999).

In Africa, the major cause of disease in lowland rice ecosystems is rice yellow mottle virus (RYMV). The only naturally occurring resistance genes to RYMV are found in African landraces, which are difficult to cross with the cultivated varieties. Therefore, transgenes were constructed from the RNA polymerase gene of RYMV, which encodes a highly conserved component of the virus replicative machinery (Pinto *et al.*, 1999). The gene was transferred to three West African cultivated rice varieties that are grown in regions with the worst records of viral disease and where the yield gaps are the highest. All three varieties were shown to be resistant to RYMV, and one of the varieties was resistant to isolates of the virus from several different locations in Africa. In the best-performing lines, viral replication was completely blocked over several generations. The resistance appeared to be RNA-mediated.

Countering Abiotic Constraints: Resistance to Stress

After pests and diseases, unfavourable environmental conditions (*abiotic stresses*) represent the next major productivity constraint in the developing world. The most significant abiotic factors affecting food production are drought/salinity, cold and (in Asia) flooding. The development of crops with an inbuilt capacity to withstand these effects could help to stabilize crop production and hence significantly contribute to sustainable food security (Holmberg and Bulow, 1998; Bray *et al.*, 2002; Zhang *et al.*, 2000). Furthermore, since only 35% of the world's potential arable land is currently in use, the modification of plants to prosper in inhospitable environments such as saline soils could help to expand agricultural production to ensure continuing food security in the coming decades.

Drought and salinity stress

Many plants respond to drought (prolonged dehydration) and increased salinity by synthesizing small, very soluble molecules such as betaines, sugars, amino acids and polyamines. These are collectively termed *compatible solutes*, and they increase the osmotic potential within the plant, therefore preventing water loss in the short term and helping to maintain a normal physiological ion balance in the longer term (Yancey et al., 1982). Compatible solutes are nontoxic even at high concentrations, so transgenic approaches have been used to make them accumulate in crop plants in order to improve drought and salinity tolerance (Chen and Murata, 2002; Serraj and Sinclair, 2002).

Several species have been engineered to produce higher levels of glycine betaine but in most cases the levels achieved have fallen short of the 5–40 µmol g^{-1} fresh weight observed in plants that naturally accumulate this molecule under salt stress conditions (Sakamoto and Murata, 2000, 2001). However, transgenic rice plants expressing betaine aldehyde dehdrogenase (BADH), one of the key enzymes in the glycine betaine synthesis pathway, accumulated the molecule to levels in excess of 5 µmol g^{-1} fresh weight (Sakamoto et al., 1998). In China, transgenic rice plants expressing BADH are likely to be the first commercially released GM plants developed for abiotic stress tolerance, and will be available for small-scale subsistence farmers as well as large producers (Huang et al., 2002b). In our laboratory, we have studied the effects of polyamine accumulation on drought tolerance in rice plants. We have generated transgenic rice plants expressing the *Datura stramonium* arginine decarboxylase (*adc*) gene, and these plants produced much higher levels of putrescine under stress, enhancing spermidine and spermine synthesis and ultimately protecting the plants from drought. Wild-type plants, however, are not able to raise their spermidine and spermine levels after 6 days of drought stress and consequently exhibit the classical drought-stress response (Capell et al., 2004).

Flooding stress (hypoxia)

While the absence of water is detrimental to plants, too much water can also be a problem particularly in rain-fed areas where the level of precipitation can be excessive. The main consequence of flooding is oxygen deficit in the roots, which induces ethylene synthesis in the aerial parts of the plant, resulting in chlorosis, senescence and eventually death (Stearns and Glick, 2003). The increase in ethylene synthesis in flooded plants is due to the induction of 1-aminocyclopropane-1-carboxylate (ACC) synthase. The ACC produced in the roots is transported to the aerial parts of the plant, where it is converted into ethylene by another enzyme, ACC oxidase (reviewed by Grichko and Glick, 2001b).

Transgenic tomato plants expressing ACC deaminase, a catabolic enzyme that draws ACC away from the ethylene synthesis pathway, showed increased tolerance to flooding stress and were less subject to the deleterious effects of root hypoxia on plant growth than non-transformed plants. The most significant improvements were achieved by expressing the transgene under the control of the A. *rhizogenes* root-specific *rol*D promoter (Grichko and Glick, 2001a). Transgenic tomato plants have also been produced in which the endogenous ACC synthase or ACC oxidase genes have been suppressed by antisense RNA. In plants transformed with antisense ACC synthase, ethylene production was lowered to less than 1% of normal levels (John, 1997). Antisense ACC oxidase plants showed lower ethylene levels following root submergence (English *et al.*, 1995).

Yield and Nutritional Improvement

As well as addressing constraints that increase the *yield gap* (the gap between the maximum potential yield of a crop, which is known as the *yield ceiling*, and the actual yield), transgenic technologies can be employed to lift the yield ceiling itself, e.g. by increasing the efficiency of photosynthesis, increasing the efficiency of nutrient uptake and accumulation or increasing the efficiency of primary metabolism. Further strategies include modifying the developmental potential of the plant either to change the plant's architecture (e.g. increase the number of seeds produced by grain crops) or artificially extend the growing season (e.g. induce early flowering or multiple flowering seasons per year).

In terms of yield enhancement, photosynthesis is perhaps the most obvious target for genetic intervention because it determines the rate of carbon fixation, and therefore the overall size of the organic carbon pool. Attempts to modify the major enzymes responsible for photosynthate assimilation, i.e. Rubisco and the enzymes of the Calvin cycle, have been restricted mainly to tobacco (e.g. Miyagawa *et al.*, 2001; Whitney and Andrews, 2001). However, progress has been made in crop species by attempting to introduce components of the energy-efficient C_4 photosynthetic pathway into C_3 plants, which lose a proportion of their fixed carbon through photorespiration. The key step in C_4 photosynthesis is the conversion of CO_2 into C_4 organic acids by the enzyme phosphoenolpyruvate carboxylase (PEPC) in mesophyll cells. The maize gene encoding PEPC has been transferred into several C_3 crops, including potato (Ishimaru *et al.*, 1998) and rice (Matsuoka *et al.*, 1998; Ku *et al.*, 1999), in order to increase the overall level of carbon fixation. Transgenic rice plants were also produced expressing pyruvate orthophosphate dikinase (PPDK) and nicotinamide adenine dinucleotide phosphate (NADP)-malic enzyme (Ku *et al.*, 1999). Preliminary field trials in China and Korea demonstrated 10–30% and 30–35% yield increases for PEPC and PPDK transgenic rice plants, respectively, which was quite unexpected since only one C_4 enzyme was expressed in each case. In the PEPC transgenic plants, there was also an unanticipated secondary effect in which Rubisco showed reduced inhibition by oxygen (Ku *et al.*, 1999). As an extension of this strategy, there is now great interest in increasing the amount of fixed carbon in plants in order to produce ethanol for fuel. This would contribute to sustainable agriculture by producing a whole new generation of cash crops and reducing our reliance on fossil fuels (e.g. Wu and Birch, 2007; see Chapter 14, this volume).

There are also many examples of transgenic technology being employed to increase the nutritional value of plants, either through direct interference with nutrient accumulation or through the modification of primary or secondary metabolism. An example of the former strategy is the expression of seed storage proteins or developmental regulators to increase the protein content of food. Specific reports include the expression of the *AmA1* seed albumin gene from *Amaranthus hypochondriacus* in potato, which has been shown to double the protein content and increase the content of essential amino acids (Chakraborty *et al.*, 2000), and the expression of the *Gpc-B1* gene in wheat, which improved grain protein, zinc and iron content by accelerating senescence and the rate of grain filling (Uauy *et al.*, 2006).

The creation of 'Golden Rice' with enhanced vitamin A content is an example of the metabolic engineering approach to nutritional improvement. Vitamin A deficiency is prevalent in the developing world, and is probably responsible for the death of 2 million children every year. In surviving children, vitamin A deficiency is a leading, but avoidable, cause of blindness (WHO, 2001). Humans can synthesize vitamin A if provided with the precursor molecule β-carotene (also known as provitamin A), a pigment found in many plants but not cereal grains. Therefore, a strategy was devised to introduce the correct metabolic steps into rice endosperm to facilitate β-carotene synthesis. The synthesis of carotenes in plants is a branch of the isoprenoid pathway and the first committed step is the joining of two geranylgeranyl diphosphate (GGPP) molecules to form the precursor phytoene. The conversion of phytoene into β-carotene requires three additional enzyme activities: phytoene desaturase, β-carotene desaturase and lycopene β-cyclase. Rice and other cereal grains accumulate GGPP but lack the subsequent enzymes in the pathway, so the genes for all three enzymes are required.

The first major breakthrough was the development of rice grains accumulating phytoene. Burkhardt *et al.* (1997) described rice plants transformed with the phytoene synthase gene from the daffodil (*Narcissus pseudonarcissus*) that accumulated high levels of this metabolic intermediate. The same group then produced rice plants expressing two daffodil genes and one bacterial gene, which recapitulated the entire heterologous pathway and produced golden-coloured rice grains containing more than 1.5 µg g^{-1} of β-carotene (Ye *et al.*, 2000). Further work saw the development of 'Golden Rice II' in which the daffodil phytoene synthase gene was replaced with its more efficient maize homologue, resulting in a 23-fold improvement in β-carotene content (up to 37 µg g^{-1}) compared to the first generation of Golden Rice (Paine *et al.*, 2005).

Molecular Pharming

Molecular pharming refers to the use of plants to produce value-added molecules, typically recombinant pharmaceutical proteins (plant-made pharmaceuticals (PMPs)) and industrial enzymes (plant-made industrial proteins (PMIs); Twyman *et al.*, 2005). The impact of molecular pharming on sustainable agriculture and food security is indirect, and is brought about by converting traditionally low-value food and feed crops into cash crops, which bring a considerable premium to the farmer. This could reflect the production of high-value pharmaceutical products on a small scale, or lower-value industrial enzymes such as phytase or amylase on a larger scale. In this way, molecular pharming increases the spectrum of wealth-generating strategies in agriculture and allows farmers to diversify. Second, many of the products currently considered for molecular pharming are either medical or veterinary products, which could reduce the cost of pharmaceutical production, particularly in developing countries. In this way, molecular pharming could bring inexpensive medicines to those most in need, helping to improve health and well-being.

Medically relevant proteins made in plants are often subdivided into three convenient categories: vaccines, recombinant antibodies and 'all others', the latter group including blood products, growth factors, cytokines, enzymes and structural proteins. Examples of the most advanced products in each of these categories are considered below.

Plant-derived vaccines

Plant-derived vaccines can be divided into two categories: those designed for veterinary use and those designed for medical use. A veterinary vaccine was the first PMP to be approved for commercial use, in February 2006, and there is a large body of both immunogenicity and challenge data to support the efficacy of such vaccines. In a number of reports, plant-derived recombinant subunit vaccines have protected animals against (in some cases lethal) challenges with the pathogen (reviewed by Twyman *et al.*, 2005). Clinical trials showed that an oral vaccine expressed in plants gives protection against a virulent viral pathogen in livestock. The trials were conducted on swine using an edible form of a vaccine for transmissible gastroenteritis virus (TGEV; Lamphear *et al.*, 2002). Another PMP veterinary vaccine has been produced by Guardian Bioscience for the prevention of coccidiosis in poultry. This subunit is produced in transgenic canola and Canadian Food Inspection Agency (CFIA) phase II trials are ongoing.

There have been several human clinical trials involving plant-derived, recombinant oral vaccines, all of which have been successful in that they produced serum and/or secretory antibody responses against the antigen in the test subjects. Trials with transgenic potato and maize expressing the enterotoxigenic *Escherichia coli* (ETEC) labile toxin B-subunit (LTB), one of the most potent known oral immunogens, induced at least fourfold increase in serum immunoglobulin G (IgG) against LTB (Tacket *et al.*, 1998, 2004). The same group also described the results of a clinical trial performed using transgenic potato tubers expressing the Norwalk virus capsid protein (NVCP; Tacket *et al.*, 2000), with nearly all of the volunteers showing significant increases in the numbers of immunoglobulin A (IgA)-antibody forming cells (AFCs). A clinical trial has also been carried out using orally delivered hepatitis B virus (HBV) surface antigen produced in lettuce (Kapusta *et al.*, 1999) and potatoes (Richter *et al.*, 2000). Of 33 participants given either two or three 1 mg doses of the antigen, about half showed increased serum IgG titres against the virus.

The production of anti-idiotype antibodies recognizing malignant B cells is a useful approach for vaccination against diseases such as non-Hodgkin's lymphoma. McCormick *et al.* (1999) produced a plant-derived single-chain variable fragment (scFv) antibody based on the well-characterized mouse lymphoma cell line 38C13. When administered to mice, the scFv stimulated the production of anti-idiotype antibodies capable of recognizing 38C13 cells. This provided immunity against lethal challenge with the lymphoma. It is envisaged that this strategy could be used as a rapid production system for tumour-specific vaccines customized for each patient and capable of recognizing unique markers on the surface of any malignant B cells. The rapid derivation of such prophylactic antibodies can be facilitated by the use of virus-infected plants. Twelve or more such products have been taken through phase I and phase II trials by the former Large Scale Biology Company Inc. The same company also took a tobacco-derived vaccine against feline parvovirus through advanced (phase III) efficacy trials in concert with Schering Plough Animal Health (SPAH), for the prevention of panleukopenia in kittens.

Finally, Yusibov *et al.* (2002) have carried out a trial involving 14 volunteers given spinach infected with alfalfa mosaic virus vectors expressing the rabies virus glycoprotein and nucleoprotein. Five of these individuals had previously received a conventional rabies vaccine. Three of those five and all nine of the initially naive subjects produced antibodies against rabies virus while no such response was seen in those given normal spinach.

Plant-derived antibodies

Three categories of plant-derived antibodies have been tested in phase II clinical trials. The first is the idiotypic scFv molecules discussed above, which are used for protection against lymphoma. The second is a full-length IgG specific for EpCAM (a marker of colorectal cancer) produced in maize and developed as the drug Avicidin by NeoRx and Monsanto. Although Avicidin demonstrated some anti-cancer activity in patients with advanced colon and prostate cancers, it was withdrawn from phase II trials in 1998 because it also resulted in a high incidence of diarrhoea (Gavilondo and Larrick, 2000).

The most advanced plant-derived antibody is CaroRx, a chimeric secretory IgA/G produced in transgenic tobacco plants which have completed phase II trials sponsored by Planet Biotechnology Inc. (Ma *et al.*, 1998). As stated earlier, secretory antibody production requires the expression of four separate components, which in this case were initially expressed in four different tobacco lines that were crossed over two generations to stack all the transgenes in one line. Nicholson *et al.* (2005) describe the same achievement in rice following simultaneous bombardment of callus tissue with four separate constructs. In the latter case, all transgenes were co-introduced into the same plants at the same time and this resulted in a time saving of at least 18 months compared to the tobacco experiments. The antibody is specific for the major adhesin (SA I/II) of *Streptococcus mutans*, the organism responsible for tooth decay in humans. Topical application following elimination of bacteria from the mouth helped to prevent recolonization by

S. mutans and led to the replacement of this pathogenic organism with harmless endogenous flora.

Most recently, the Cuban regulatory agencies have approved the production and use of a plant-derived antibody against hepatitis B virus. Although this is not used directly as a pharmaceutical product, it is used in the purification process to isolate the cognate antigen, i.e. the HBV surface antigen. Because it is used in a biopharmaceutical manufacturing process, the antibody has to meet all the same purity and homogeneity criteria imposed on the product itself.

A Caveat: Promise Versus Politics

Advances in plant transformation and gene expression technology allow the introduction of novel traits into our crop plants. Genetically enhanced crops have the potential to address some of the causes of hunger, both directly (by increasing the availability of food) and indirectly (by reducing poverty and increasing health in developing countries). Crop failure, due to pests and diseases, could be averted by the adoption of plants that are resistant to such biotic stresses. The development of plants that are tolerant of extreme environments could allow marginal soils to be brought into agricultural use, and could allow plants to survive periods of drought or flooding. These measures, in combination with conventional breeding and developments in other agricultural practices, may produce the estimated 50% increase in grain yields required over the next 50 years to cope with the anticipated increase in population.

Transgenic strategies can also be used to modify the nutritional properties of plants and address the widespread problem of malnutrition in developing countries. The harvestable products of plants can be improved by promoting the uptake, accumulation or synthesis of bioavailable minerals and vitamins, by increasing or modifying the content of amino acids and fats/oils and by eliminating antinutritional factors. Finally, plants can be modified to produce valuable molecules, thus extending the economic potential of agriculture in new directions. All these transgenic approaches provide useful strategies to make agriculture more sustainable, and increase wealth, food security, health and well-being.

Despite these anticipated benefits of transgenic crops, politics has a way of becoming embroiled with agricultural production and distribution, often with negative effects. GM crops have become a political football between the European Union (EU) and the USA, with the playing field being shifted to southern African states on the verge of starvation. Recent controversy centred on the refusal of some African states to accept American aid in the form of GM food, for no other reason than political pressure from certain EU quarters. This is not only inexcusable and hypocritical but also unethical. We hope that in time the value of GM crops, as a component of a serious drive focusing on sustainability, will contribute significantly to improving food security in the developing world.

References

Agrawal, P.K., Kohli, A., Twyman, R.M. and Christou, P. (2005) Transformation of plants with multiple cassettes generates simple transgene integration patterns and high expression levels. *Molecular Breeding* 16, 247–260.

Altieri, M.A. (1995) *Agroecology: The Science of Sustainable Agriculture.* Westview Press, Boulder, Colorado.

Altpeter, F., Baisakh, N., Beachy, R., Bock, R., Capell, T., Christou, P., Daniell, H., Datta, K., Datta, S., Dix, P.J., Fauquet, C., Huang, N., Kohli, A., Mooibroek, H., Nicholson, L., Nguyen, T.T., Nugent, G., Raemakers, C.J.J.M., Romano, A., Somers, D.A., Stoger, E., Taylor, N. and Visser, R.G.F. (2005) Particle bombardment and the genetic enhancement of crops: myths and realities. *Molecular Breeding* 15, 305–327.

Bano-Maqbool, S., Riazuddin, S., Loc, N.T., Gatehouse, A.M.R., Gatehouse, J.A. and Christou, P. (2001) Expression of multiple insecticidal genes confers broad resistance against a range of different rice pests. *Molecular Breeding* 7, 85–93.

Berner, D.K., Kling, J.G. and Singh, B.B. (1995) *Striga* research and control: a perspective from Africa. *Plant Disease* 79, 652–660.

Bray, E.A., Bailey-Serres, J. and Weretilnyk, E. (2002) Responses to abiotic stress. In: Buchanan, B.B., Gruissem, W. and Jones, R.L. (eds) *Biochemistry and Molecular Biology of Plants.* American Society of Plant Pathologists, Rockville, Massachusetts, pp. 1158–1203.

Burkhardt, P.K., Beyer, P., Wunn, J., Kloti, A., Armstrong, G.A., Schledz, M., vonLintig, J. and Potrykus, I. (1997) Transgenic rice (*Oryza sativa*) endosperm expressing daffodil (*Narcissus pseudonarcissus*) phytoene synthase accumulates phytoene, a key intermediate of provitamin A biosynthesis. *The Plant Journal* 11, 1071–1078.

Capell, T., Bassie, L. and Christou, P. (2004) Modulation of the polyamine biosynthetic pathway in transgenic rice confers tolerance to drought stress. *Proceedings of the National Academy of Sciences of the USA* 101, 9909–9914.

Chakraborty, S., Chakraborty, N. and Datta, A. (2000) Increased nutritive value of transgenic potato by expressing a nonallergenic seed albumin gene from *Amaranthus hypochondriacus. Proceedings of the National Academy of Sciences of the USA* 97, 3724–3729.

Chen, T.H.H. and Murata, N. (2002) Enhancement of tolerance of abiotic stress by metabolic engineering of betaines and other compatible solutes. *Current Opinion in Plant Biology* 5, 250–257.

Christou, P. and Twyman, R.M. (2004) The potential of genetically enhanced plants to address food insecurity. *Nutrition Research Reviews* 17, 23–42.

Christou, P., Capell, T., Kohli, A., Gatehouse, J.A. and Gatehouse, A.M.R. (2006) Recent developments and future prospects in insect pest control in transgenic crops. *Trends in Plant Science* 11, 302–308.

De Maagd, R.A., Bosch, D. and Stiekema, W. (1999) *Bacillus thuringiensis* toxin-mediated insect resistance in plants. *Trends in Plant Science* 4, 9–13.

Eizaguirre, M., Albajes, R., Lopez, C., Eras, J., Lumbierres, B. and Pons, X. (2006) Six years after the commercial introduction of Bt maize in Spain: field evaluation, impact and future prospects. *Transgenic Research* 15, 1–12.

English, P.J., Lycett, G.W., Roberts, J.A. and Jackson, M.B. (1995) Increased 1-aminocyclopropane-1-carboxylic acid oxidase activity in shoots of flooded tomato plants raises ethylene production to physiologically active levels. *Plant Physiology* 109, 1435–1440.

EPA (1993) *Re-registration Eligibility Decision: Glyphosate.* Office of Prevention, Pesticides and Toxic Substances. US Environmental Protection Agency, Washington, DC.

FAO (2001) *The State of Food Insecurity in the World.* FAO, Rome.

Feldmann, K.A. and Marks, M.D. (1987) *Agrobacterium*-mediated transformation of germinating seeds of *Arabidopsis thaliana* – a non-tissue culture approach. *Molecular and General Genetics* 208, 1–9.

Fu, X.D., Duc, L.T., Fontana, S., Bong, B.B., Tinjuangjun, P., Sudhakar, D., Twyman, R.M., Christou, P. and Kohli, A. (2000) Linear transgene constructs lacking vector backbone sequences generate low-copy-number transgenic plants with simple integration patterns. *Transgenic Research* 9, 11–19.

Gatehouse, A.M.R., Shi, Y., Powell, K.S., Brough, C., Hilder, V.A., Hamilton, W.D.O., Newell, C.A., Merryweather, A., Boulter, D. and Gatehouse, J.A. (1993) Approaches to insect resistance using transgenic plants. *Philosophical Transactions of the Royal Society of London Series B – Biological Sciences* 342, 279–286.

Gavilondo, J.V. and Larrick, J.W. (2000) Antibody production technology in the mil-len-nium. *Biotechniques* 29, 128–145.

Gelvin, S.B. (2003) *Agrobacterium*-mediated plant transformation: the biology behind the 'gene-jockeying' tool. *Microbiology and Molecular Biology Reviews* 67, 16–37.

Graham, R.D., Welch, R.M. and Bouis, H.E. (2001) Addressing micronutrient malnutrition through enhancing the nutritional quality of staple foods: principles, perspectives and knowledge gaps. *Advances in Agronomy* 70, 77–142.

Gressel, J. (2002) Transgenic herbicide-resistant crops – advantages, drawbacks and fail-safes. In: Oksman-Caldentey, K.-M. and Barz, W.H. (eds) *Plant Biotechnology and Transgenic Plants*. Marcel Dekker, New York, pp. 597–633.

Grichko, V.P. and Glick, B.R. (2001a) Amelioration of flooding stress by ACC deaminase-containing plant growth-promoting bacteria. *Plant Physiology and Biochemistry* 39, 11–17.

Grichko, V.P. and Glick, B.R. (2001b) Ethylene and flooding stress in plants. *Plant Physiology and Biochemistry* 39, 1–9.

Hellens, R.P., Edwards, E.A., Leyland, N.R., Bean, S. and Mullineaux, P.M. (2000) PGreen: a versatile and flexible binary Ti vector for *Agrobacterium*-mediated transformation. *Plant Molecular Biology* 42, 819–832.

Hilder, V.A. and Boulter, D. (1999) Genetic engineering of crop plants for insect resist-ance – a critical review. *Crop Protection* 18, 177–191.

Holmberg, N. and Bulow, L. (1998) Improving stress tolerance in plants by gene transfer. *Trends in Plant Science* 3, 61–66.

Horsch, R.B., Fry, J.E., Hoffmann, N.L., Eichholtz, D., Rogers, S.G. and Fraley, R.T. (1985) A simple and general method for transferring genes into plants. *Science* 227, 1229–1231.

Huang, J., Pray, C. and Rozelle, S. (2002a) Enhancing the crops to feed the poor. *Nature* 418, 678–683.

Huang, J., Rozelle, S.D., Pray, C.E. and Wang, Q. (2002b) Plant biotechnology in China. *Science* 295, 674–677.

Huang, J., Hu, R., Rozelle, S. and Pray, C. (2005) Insect-resistant GM rice in farmer's fields: assessing productivity and health effects in China. *Science* 308, 688–690.

Ishimaru, K., Okhawa, Y., Ishige, T., Tobias, D.J. and Ohsugi, R. (1998) Elevated pyruvate orthophosphate dikinase (PPDK) activity alters carbon metabolism in C3 transgenic potatoes with a C4 maize PPDK gene. *Physiologia Plantarum* 103, 340–346.

James, C. (2006) *Global Status of Commercialized Transgenic Crops: 2006*. ISAAA Briefs No 35, ISAAA, Ithaca, New York.

Joel, D.M., Kleifeld, Y., Losner-Goshen, D., Herzlinger, G. and Gressel, J. (1995) Transgenic crops against parasites. *Nature* 374, 220–221.

John, P. (1997) Ethylene biosynthesis: the role of 1-aminocyclopropane-1-carboxylate (ACC) oxidase, and its possible evolutionary origin. *Physiologia Plantarum* 100, 583–592.

Kanampiu, F.K., Ransom, J.K. and Gressel, J. (2001) Imazapyr seed dressings for *Striga* control on acetolactate synthase target-site resistant maize. *Crop Protection* 20, 885–895.

Kanampiu, F.K., Ranson, J.K., Gressel, J., Jewell, D., Friesen, D., Grimanelli, D. and Hoisington, D. (2002) Appropriateness of biotechnology to African agriculture: *Striga* and maize as paradigms. *Plant Cell Tissue and Organ Culture* 69, 105–110.

Kapusta, J., Modelska, A., Figlerowicz, M., Pniewski, T., Letellier, M., Lisowa, O., Yusibov, V., Koprowski, H., Plucienniczak, A. and Legocki, A.B. (1999) A plant-derived edible vaccine against hepatitis B virus. *FASEB Journal* 13, 1796–1799.

Karimi, M., Inze, D. and Depicker, A. (2002) GATEWAY vectors for *Agrobacterium*-mediated plant transformation. *Trends in Plant Science* 7, 193–195.

Karlova, R., Weemen-Hendriks, M., Naimov, S., Ceron, J., Dukiandjiev, S. and de Maagd, R.A. (2005) *Bacillus thuringiensis* delta endotoxin Cry1Ac domain III enhances activity against *Heliothis virescens* in some, but not all Cry1-Cry1Ac hybrids. *Journal of Invertebrate Pathology* 88, 169–172.

Khush, G.S., Bacalangco, E. and Ogawa, T. (1990) A new gene for resistance to bacterial blight from *O. longistaminata*. *Rice Genetics Newsletter* 7, 121–122.

Kloti, A., Futterer, J., Terada, R., Bieris, D., Wunn, J., Burkhardt, P.K., Chen, G., Hohn, T.H., Biswas, G.C. and Potrykus, I. (1996) Towards genetically engineered resistance to tungro virus. In: Khush, G.S. (ed.) *Rice Genetics III*. IRRI, Los Banos, The Philippines, pp. 763–767.

Kohli, A., Twyman, R.M., Abranches, A., Wegel, E., Shaw, P., Christou, P. and Stoger, E. (2003) Transgene integration, organization and interaction in plants. *Plant Molecular Biology* 52, 247–258.

Ku, M.S.B., Agarie, S., Nomura, M., Fukayama, H., Tsuchida, H., Ono, K., Hirose, S., Toki, S., Miyao, M. and Matsuoka, M. (1999) High level expression of maize phosphoenol pyruvate carboxylase in transgenic rice plants. *Nature Biotechnology* 17, 76–80.

Lamphear, B.J., Streatfield, S.J., Jilka, J.M., Brooks, C.A., Barker, D.K., Turner, D.D., Delaney, D.E., Garcia, M., Wiggins, B., Woodard, S.L., Hood, E.E., Tizard, I.R., Lawhorn, B. and Howard, J.A. (2002) Delivery of subunit vaccines in maize seed. *Journal of Controlled Release* 85, 169–180.

Lentini, Z., Calvert, L., Tabares, E., Lozano, I., Ramirez, B.C. and Roca, W. (1996) Genetic transformation of rice with viral genes for novel resistance to rice hoja blanca virus. In: Khush, G.S. (ed.) *Rice Genetics III*.

IRRI, Los Banos, The Philippines, pp. 780–783.

Llewellyn, D.J. and Higgins, T.J.V. (2002) Transgenic crop plants with increased tolerance to insect pests. In: Oksman-Caldentey, K.-M. and Barz, W.H. (eds) *Plant Biotechnology and Transgenic Plants*. Marcel Dekker, New York, pp. 571–595.

Ma, J.K., Hikmat, B.Y., Wycoff, K., Vine, N.D., Chargelegue, D., Yu, L., Hein, M.B. and Lehner, T. (1998) Characterization of a recombinant plant monoclonal secretory antibody and preventive immunotherapy in humans. *Nature Medicine* 4, 601–606.

Matsuoka, M., Nomura, M., Agarie, S., Tokutomi, M. and Ku, M.S.B. (1998) Evolution of C4 photosynthetic genes and overexpression of maize C4 genes in rice. *Journal of Plant Research* 111, 333–337.

McCormick, A.A., Kumagai, M.H., Hanley, K., Turpen, T.H., Hakim, I., Grill, L.K., Tuse, D., Levy, S. and Levy, R. (1999) Rapid production of specific vaccines for lymphoma by expression of the tumor-derived single-chain Fv epitopes in tobacco plants. *Proceedings of the National Academy of Sciences of the USA* 96, 703–708.

Mehlo, L., Gahakwa, D., Nghia, P.T., Loc, N.T., Capell, T., Gatehouse, J.A., Gatehouse, A.M.R. and Christou, P. (2005) An alternative strategy for sustainable pest resistance in genetically enhanced crops. *Proceedings of the National Academy of Sciences of the USA* 102, 7812–7816.

Miyagawa, Y., Tamoi, M. and Shigeoka, S. (2001) Overexpression of a cyanobacterial fructose-1,6-sedoheptulose-1,7-bisphosphatase in tobacco enhances photosynthesis and growth. *Nature Biotechnology* 19, 965–969.

Nicholson, L., Gonzalez-Melendi, P., van Dolleweerd, C., Tuck, H., Perrin, Y., Ma, J.K., Fischer, R., Christou, P. and Stoger, E. (2005) A recombinant multimeric immunoglobulin expressed in rice shows assembly-dependent subcellular localization in endo-sperm cells. *Plant Biotechnology Journal* 3, 115–127.

Paine, J.A., Shipton, C.A., Chaggar, S., Howells, R.M., Kennedy, M.J., Vernon, G., Wright, S.Y., Hinchliffe, E., Adams, J.L., Silverstone, A.L. and Drake, R. (2005) Improving the nutritional value of Golden

Rice through increased pro-vitamin A content. *Nature Biotechnology* 23, 482–487.

Pinstrup-Andersen, P., Pandra-Lorch, R. and Rosegrant, M.W. (1999) *World Food Prospects: Critical Issues for the Early Twenty-First Century.* Food Policy Report, International Food Policy Research Institute, Washington, DC.

Pinto, Y.M., Kok, R.A. and Baulcombe, D.C. (1999) Resistance to rice yellow mottle virus (RYMV) in cultivated African rice varieties containing RYMV transgene. *Nature Biotechnology* 17, 702–707.

Qaim, M. and Zilberman, D. (2003) Yield effects of genetically modified crops in developing countries. *Science* 299, 900–902.

Richter, L.J., Thanavala, Y., Arntzen, C.J. and Mason, H.S. (2000) Production of hepatitis B surface antigen in transgenic plants for oral immunization. *Nature Biotechnology* 18, 1167–1171.

Sakamoto, A. and Murata, N. (2000) Genetic engineering of glycine betaine synthesis in plants: current status and implications for enhancement of stress tolerance. *Journal of Experimental Botany* 51, 81–88.

Sakamoto, A. and Murata, N. (2001) The use of bacterial choline oxidase, a glycine betaine-synthesizing enzyme, to create stress-resistant transgenic plants. *Plant Physiology* 125, 180–188.

Sakamoto, A., Murata, A. and Murata, N. (1998) Metabolic engineering of rice leading to biosynthesis of glycine betaine and tolerance to salt and cold. *Plant Molecular Biology* 38, 1011–1019.

Salmeron, J.M. and Vernooij, B. (1998) Transgenic approaches to microbial disease resistance in crop plants. *Current Opinion in Plant Biology* 1, 347–352.

Serraj, R. and Sinclair, T.R. (2002) Osmolyte accumulation: can it really help increase crop yield under drought conditions? *Plant Cell and Environment* 25, 333–341.

Singh, P.K., Kumar, M., Chaturvedi, C.P., Yadav, D. and Tuli, R. (2004) Development of hybrid delta endotoxin and its expression in tobacco and cotton for control of a polyphagous pest *Spodoptera litura*. *Transgenic Research* 13, 397–410.

Sivamani, E., Huet, H., Shen, P., Ong, C.A., de Kochko, A., Fauquet, C. and Beachy, R.N. (1999) Rice plant (*Oryza sativa* L.) containing rice tungro spherical virus (RTSV) coat protein transgenes are resistant to virus infection. *Molecular Breeding* 5, 177–185.

Song, W.Y., Wang, G., Chen, L., Kim, H., Pi, L.Y., Holsten, T., Gardner, J., Wang, B., Zha, W., Zhu, L., Fauquet, C. and Ronald, P. (1995) A receptor kinase-like protein encoded by the rice disease resistance gene, *Xa-21*. *Science* 270, 1804–1806.

Stearns, J.C. and Glick, B.R. (2003) Transgenic plants with altered ethylene biosynthesis or perception. *Biotechnology Advances* 21, 193–210.

Stuiver, M.H. and Custers, J.H.H.V. (2001) Engineering disease resistant plants. *Nature* 411, 865–868.

Tacket, C.O., Mason, H.S., Losonsky, G., Clements, J.D., Levine, M.M. and Arntzen, C.J. (1998) Immunogenicity in humans of a recombinant bacterial-antigen delivered in transgenic potato. *Nature Medicine* 4, 607–609.

Tacket, C.O., Mason, H.S., Losonsky, G., Estes, M.K., Levine, M.M. and Arntzen, C.J. (2000) Human immune responses to a novel Norwalk virus vaccine delivered in transgenic potatoes. *Journal of Infectious Disease* 182, 302–305.

Tacket, C.O., Pasetti, M.F., Edelman, R., Howard, J.A. and Streatfield, S. (2004) Immunogenicity of recombinant LT-B delivered orally to humans in transgenic corn. *Vaccine* 22, 4385–4389.

Tang, K.X., Sun, X.F., Hu, Q.N., Wu, A.Z., Lin, C.H., Lin, H.J., Twyman, R.M., Christou, P. and Feng, T.Y. (2001) Transgenic rice plants expressing the ferredoxin-like protein (API) from sweet pepper show enhanced resistance to *Xanthomonas oryzae* pv. *oryzae*. *Plant Science* 160, 1035–1042.

Tu, J., Ona, I., Zhang, Q., Mew, T.W., Khush, G.S. and Datta, S.K. (1998) Transgenic rice variety 'IR72' with *Xa21* is resistant to bacterial blight. *Theoretical and Applied Genetics* 97, 31–36.

Twyman, R.M. and Christou, P. (2004) Plant transformation technology – particle bombardment. In: Christou, P. and Klee, H. (eds) *Handbook of Plant Biotechnology.* Wiley, New York, pp. 263–289.

Twyman, R.M., Christou, P. and Stoger, E. (2002) Genetic transformation of plants and their cells. In: Oksman-Caldentey, K.M. and Barz, W. (eds) *Plant Biotechnology and Transgenic Plants.* Marcel Dekker, New York, pp. 111–141.

Twyman, R.M., Schillberg, S. and Fischer, R. (2005) Transgenic plants in the biopharmaceutical market. *Expert Opinion on Emerging Drugs* 10, 185–218.

Tzfira, T. and Citovsky, V. (2003) The *Agrobacterium*–plant cell interaction. Taking biology lessons from a bug. *Plant Physiology* 133, 943–947.

Uauy, C., Distelfeld, A., Fahima, T., Blechl, A. and Dubcovsky, J. (2006) A NAC gene regulating senescence improves grain protein, zinc, and iron content in wheat. *Science* 314, 1298–1301.

Valentine, L. (2003) *Agrobacterium tumefaciens* and the plant: the David and Goliath of modern genetics. *Plant Physiology* 133, 948–955.

Wang, G.L., Song, W.Y., Ruan, D.L., Sideris, S. and Ronald, P.C. (1996) The cloned gene, *Xa-21*, confers resistance to multiple *Xanthomonas oryzae* pv. *oryzae* isolates in transgenic plants. *Molecular Plant–Microbe Interactions* 9, 850–855.

Whitney, S.M. and Andrews, T.J. (2001) Plastome-encoded bacterial ribulose-1, 5-bisphosphate carboxylase/oxygenase (Rubisco) supports photosynthesis and growth of tobacco. *Proceedings of the National Academy of Sciences of the USA* 98, 14738–14743.

WHO (2001) Combating vitamin A deficiency. Available at: http://www.who.int/nut/vad.htm

Wu, L. and Birch, R.G. (2007) Doubled sugar content in sugarcane plants modified to produce a sucrose isomer. *Plant Biotechnology Journal* 5, 109–117.

Yancey, P.H., Clark, M.E., Hand, S.C., Bowlus, R.D. and Somero, G.N. (1982) Living with water stress: evolution of osmolyte systems. *Science* 217, 1214–1222.

Ye, X.D., Al-Babili, S., Kloti, A., Zhang, J., Lucca, P., Beyer, P. and Potrykus, I. (2000) Engine-ering the provitamin A (beta-carotene) biosynthetic pathway into (carotenoid-free) rice endosperm. *Science* 287, 303–305.

Yeo, E.T., Kwon, H.B., Han, S.E., Lee, J.T., Ryu, J.C. and Byun, M.O. (2000) Genetic engineering of drought-resistant potato plants by introduction of the trehalose-6-phosphate synthase (*TPS1*) gene from *Saccharomyces cerevisiae. Molecules and Cells* 10, 263–268.

Yusibov, V., Hooper, D.C., Spitsin, S.V., Fleysh, N., Kean, R.B., Mikheeva, T., Deka, D., Karasev, A., Cox, S., Randall, J. and Koprowski, H. (2002) Expression in plants and immunogenicity of plant virus-based experimental rabies vaccine. *Vaccine* 20, 3155–3164.

Zhang, J., Kueva, N.Y., Wang, Z., Wu, R., Ho, T.-H. and Nguyen, H.T. (2000) Genetic engineering for abiotic stress tolerance in crop plants. *In Vitro Cellular and Developmental Biology – Plant* 36, 108–114.

Zhang, S., Song, W.Y., Chen, L., Ruan, D., Taylor, N., Ronald, P., Beachy, R. and Fauquet, C. (1998) Transgenic elite indica rice varieties, resistant to *Xanthomonas* pv. *oryzae. Molecular Breeding* 4, 551–558.

2 Environmental Benefits of Genetically Modified Crops

M.G. Edwards[1] and G.M. Poppy[2]

[1]School of Biology, Institute for Research on Environment and Sustainability, Newcastle University, Newcastle upon Tyne, UK; [2]School of Biological Sciences, University of Southampton, Southampton, UK

Keywords: Environmental benefit, agricultural biotechnology, herbicide reduction, agri-environmental conservation, Farm-scale Evaluations (UK FSEs)

Summary

Modern agriculture is challenged with an extremely difficult task where it must produce more food to support the expansion of the global population, with less synthetic inputs and without increasing its global footprint. A crucial factor in the expansion of the global population to its current levels was the advances initiated by the development of improved varieties and increased use of fertilizer, irrigation and synthetic pesticides during the Green Revolution. While the ability of pesticides (insecticides and pesticides) to reduce crop losses must be recognized, their potential negative effects on public health, with particular emphasis in developing countries, and the environment cannot be ignored nor allowed to continue unchecked. The response of the agricultural industry in bringing forward new precision farming technology, such as reduced application rates of targeted pesticides with lower toxicity and persistency, has some limited benefit. However, with an increasing world population, a slowing of the rate of crop improvement through conventional breeding and a declining area of land available for food production, there is a need for new technologies to produce more food of improved nutritional value in an environmentally acceptable and sustainable manner. In addition, there is also pressure to ensure that adoption of new techniques will actually benefit our agricultural environment. One answer to these challenges is through the use of genetically modified crops, and it is with a rational and thoughtful introduction of this technology that our agricultural systems can continue to support the global communities for generations to come.

Introduction

A crucial factor in the expansion of the global population to its current level was the advances initiated by the development of improved varieties and increased use of fertilizer, irrigation and synthetic pesticides during the Green

Revolution. While the ability of pesticides (insecticides and herbicides) to reduce crop losses must be recognized, their potential negative effects on public health, with particular emphasis in developing countries, and the environment must also be addressed. The response of the agricultural industry in bringing forward new precision farming technology such as reduced application rates of targeted pesticides with lower toxicity and persistency is noted. However, with an increasing world population, a slowing of the rate of crop improvement through conventional breeding and a declining area of land available for food production, there is a need for new technologies to produce more food of improved nutritional value in an environmentally acceptable and sustainable manner. While the authors recognize that the introduction of genetically modified (GM) crops to the UK and Europe still remains controversial, the benefits of these crops, including their effect on pesticide use, are now being realized by the global community. Published data are used to estimate what effect GM crops have had on pesticide use first on a global basis, and then to predict what effect they would have if widely grown in the European Union (EU). On a global basis, GM technology has reduced pesticide use, with the size of the reduction varying between crops and the introduced trait. It is estimated that the use of GM soybean, oilseed rape (canola), cotton and maize varieties modified for herbicide tolerance and insect-protected GM varieties of cotton reduced pesticide use by a total of 22.3 million kg of formulated product in the year 2000. Estimates indicate that if 50% of the maize, oilseed rape, sugarbeet and cotton grown in the EU were GM varieties, pesticide used in the EU per annum would decrease by 14.5 million kg of formulated product (equating to 4.4 million kg active ingredient). In addition, there would be a reduction of 7.5 million ha sprayed which would save 20.5 million l of diesel and result in a reduction of approximately 73,000 t of carbon dioxide being released into the atmosphere.

An Historical Perspective

The dawn of agriculture occurred some 10,000 years ago with the domestication of cereals, soon to be followed by other crops. In the Western world, the evolution of agriculture has been divided into four discrete periods – prehistoric, Roman, feudal and scientific – with each being associated with specific advancements or developments. The prehistoric, or Neolithic era (10,000 BCE), was thus recognized as the era of crop domestication originating in the regions of low to middle latitude. The Roman era (1000 BCE–500 CE) saw the introduction of metal tools, the use of animals for farm work and the development of the manipulation of watercourses for irrigation, while the feudal era (height, 1100 CE) saw the beginning of international trade based on exportation of crops. Interestingly, the era known as the scientific era started as early as the 16th century and although there is documentary evidence for the use of pest control from ancient times, its adoption is primarily attributed to this era. While mineral-based pesticides (arsenates and copper salts) had been used previously, major advances in the development of synthetic insecticides did not occur until the end of the Second World War and was accompanied by the intensification of farming. The Green Revolution, fathered in the 1960s by Norman Borlaug, heralded one of the major agricultural developments of the last century. The production of new cereal varieties, coupled with increased use of fertilizers, irrigation and pesticides, provided many of the technological inputs required to feed an expanding world population. The introduction of synthetic pesticides in 1947 has led to a reduction of crop losses due to insects, diseases and weeds. Even so, these losses for eight of the world's major crops are estimated at US$244 billion per annum, representing 43% of world production (Oerke, 1999, 2006) and postharvest losses contribute a further 10%. Paoletti and Pimentel (2000) estimated that, if it were not for synthetic pesticide use, the situation would be exacerbated and losses might well increase by a further 30%. Reports emerging from China quote officials saying that pesticide use saves China millions of tonnes of food and fibre every year (Huang, 2003; Pray *et al.*, 2001, 2002). Thus, the combined effects of improved varieties, increased fertilizer use and irrigation coupled with increased pesticide use have been instrumental in allowing world food production to double in the 35 years to 1996, and this was accompanied respectively by 6.87- and 3.48-fold increases in the global annual rate of nitrogen and phosphorus fertilization (Tilman, 1999). Nevertheless, pesticides have been associated with a number of negative events that were unforeseen at the time of their adoption but must be considered now before their impact on the environment becomes unmanageable.

Some Concerns Associated with Conventional Crop Production Practices

Effects of pesticide use on public health

In the 1970s, the World Health Organization (WHO) estimated that there were globally 500,000 pesticide poisonings per year, resulting in some 5000 deaths (Farah, 1994; Gunnell *et al.*, 2007). The Environmental Protection Agency (EPA) estimates that between 10,000 and 20,000 cases of pesticide poisoning occur in agricultural workers each year in the USA. The problem of pesticide toxicity may be worse in developing countries due to less education towards, and lack of, awareness of the inherent dangers of pesticides, inadequate protective clothing and lack of appropriate training. This was demonstrated by studies with women working on smallholder cotton farms in southern Africa (Rother, 1998, 2006). Rother reported that while the women appreciated that pesticides were poisons and had to be kept under lock and key, they were seen mixing pesticides with the same water that supplied the household. Women also collected edible weeds and grew vegetables for domestic use among the sprayed cotton. In earlier field studies with rice growers in the Philippines, over half the farmers claimed sickness due to pesticide use (Cuyno *et al.*, 2001). These examples and others (Repetto and Baliga, 1996; London *et al.*, 2002; Eddleston *et al.*, 2008; Thundiyil *et al.*, 2008) show the inherent risks to pesticide users, particularly in developing countries.

Environmental effects of pesticide use

Fortunately, the regulations covering the use of synthetic pesticides since their introduction in the mid-1940s have been improved and controlled by tighter legislation. In the late 1960s, Rachel Carson in her controversial book *Silent Spring* expressed the view that the increasing use of synthetic pesticides would have a serious negative effect on the environment. To the detriment of our natural environment, her hypothesis has been supported by numerous examples since its publication. For example, the Royal Society for the Protection of Birds has linked the dramatic decline in UK farmland bird life to a number of factors such as the intensification of agriculture which includes increased pesticide use (Krebs *et al.*, 1999; Chapter 12, this volume). Paoletti and Pimentel (2000) cite numerous examples of well-documented cases where pesticide application has been directly responsible for specific incidents in which large numbers of birds have been killed. The EPA (Ellenberger *et al.*, 1989) has estimated that carbofuron kills 1–2 million birds per year, and Paoletti and Pimentel (2000) have argued that based on a conservative estimate of 10% mortality, close to 70 million birds are killed annually in the USA as a direct effect of pesticide use. Thankfully, we are now seeing evidence of a turnaround in the situation due to the introduction of stricter legislation and new technology such as reduced application rates of targeted pesticides of lower toxicity and persistency. Precautionary methods of using buffer zones and low drift technology are also used to try to decrease potential negative environmental effects of agrochemical use.

Do we need a new technology?

The human population is ever increasing, with conservative estimates predicting that the population will rise to approximately 10 billion by 2050. Thus, the major challenges facing the world are to feed and provide shelter for a world population that is increasing at an exponential rate. Furthermore, it is essential to protect human health, and ensure social and economic conditions that are conducive to the fulfilment of the human potential. Agriculture must play a major role in achieving these goals by providing both ever-increasing food yields and ever-increasing supply of natural products required by industry. The recent excitement and drive behind the development and use of crops as biofuels only serves to exacerbate the constraints on agriculture. Thus, the challenge in the forthcoming decades is to achieve maximum production of food and other products without further irreversible depletion or destruction of the natural environment, against a backdrop of climate change which not only is predicted to result in the loss of agricultural land as a consequence of rising sea levels, but is also likely to have a major impact on the dynamics of pest populations. Agriculture must become an integral part of a sustainable global society. Agricultural sustainability integrates three major goals: environmental health, economic profitability and social and economic equity. It thus rests on the principle that we must meet the needs of the present without compromising the ability of future generations to meet their own needs. Current figures suggest that to feed a world population of 10 billion in 2050 without allowing for additional imports of food, Africa will have to increase its food production by 300%, Latin America by 80% and Asia by 70%. Even North America, which is not usually associated with food shortages, would have to increase its food production by 30% to feed its own projected population of 348 million. Given the current scenario of some 800 million people going hungry on a daily basis and an estimated 30,000 (half of them children) dying every day due to hunger and malnutrition, it is clear that society has many major challenges to address. One step towards achieving sustainability is to identify current major constraints on crop productivity. Tilman (1999) noted that this major challenge to decrease the environmental impact of agriculture, while maintaining or improving its productivity and sustainability, would have no single easy solution. Simply putting more land into agricultural use, thereby increasing the 'agricultural foot-print', is not a viable option in the long term.

Biotechnology offers many opportunities for agriculture and provides the means to address many of the constraints placed on productivity outlined above. It uses the conceptual framework and technical approaches of molecular biology and plant cell culture systems to develop commercial processes and products. With the rapid development of biotechnology, agriculture has moved from a resource-based to a science-based industry, with plant breeding being dramatically augmented by the introduction of recombinant DNA technology based on knowledge of gene structure and function. The concept of utilizing a transgenic approach to host-plant resistance was realized in the mid-1990s with the commercial introduction of transgenic maize, potato and cotton plants expressing genes encoding the insecticidal δ-endotoxin from *Bacillus thuringiensis* (*Bt*). Similarly, the role of herbicides in agriculture entered a new era with the introduction of glyphosate-resistant soybeans in 1995 (Cerdeira and Duke, 2006, see Chapter 7, this volume). Despite the increasing disquiet over the growing of such crops in Europe and Africa (at least by the media and certain non-government organizations (NGOs) in recent years, the figures demonstrate that the market is increasing, year-on-year (James, 2007). It is evident that to meet the demands of food production in the coming years we cannot ignore the importance of effective pesticide usage; however, this must be used in environmentally sound practices. The use of GM technology and non-persistent contact pesticides is proving to satisfy these two requirements leading agriculture into a scenario where we are now seeing both direct and indirect benefits to the environment through the use of biotechnology in agriculture.

Effect of GM Crops on Pesticide Use

Herbicide-tolerant soybean

Herbicide-tolerant (HT) soybean, with 58.3 million ha grown globally (James, 2007), is currently the dominant transgenic crop. Heimlich *et al.* (2000) noted that when comparing 1997 to 1998, the overall rate of herbicide use in GM soybeans declined by nearly 10%. Also, based on regression analysis, the authors estimated that 2.5 million kg of glyphosate replaced 3.3 million kg of formulated products of other synthetic herbicides such as imazethapyr, pendimethalin and trifluralin. Further work by Carpenter (2001), Beckie (2006) and later by Bonny (2008) supports this finding. The Dutch Centre for Agriculture and Environment (Hin *et al.*, 2001) report concluded that in the USA the overall difference in pesticide use between GM and conventional soybeans ranged from +7% to −40% (1995–1998) with an average reduction of 10%. It should be noted, however, that the report said that the reduction might be associated with a number of other factors including soil type and climate. The report also concluded that as a result of adopting HT soybeans, glyphosate was replacing other herbicides with less-favourable environmental profiles. Nelson and Bullock (2003) used data from 431 farms in 20 locations in the USA to model the effect of introducing HT soybeans on herbicide use. Their preliminary results indicate that, while the GM crop made the use of 16 herbicides redundant, it increased glyphosate use by fivefold. They also noted that glyphosate has a number of desirable characteristics when compared with other pesticides and noted that the EPA has given glyphosate its lowest toxicity rating. While there is evidence to indicate that the introduction of HT soybean will reduce herbicide use by up to 10%, it should be noted that some authors have concluded that their use had little net effect on total herbicide used (Gianessi and Carpenter, 2000). However, even a modest reduction in pesticide use applied to 25 million ha would be highly significant in reducing overall pesticide use.

Herbicide-tolerant oilseed rape (canola)

While the area of HT canola is small in comparison to soybean, there are still nearly 3 million ha grown, mainly in Canada. In 2000, over 80% of the canola growers in Western Canada adopted transgenic varieties and grew them on 55% of the 5 million ha of canola. The Canola Council of Canada (2001) reported that in addition to finding that the introduction of HT canola increased yields by about 10%, transgenic crops required less herbicide than conventional crops. The total amount of herbicide used was reduced by 1.5 million kg in 1997 and by 6.0 million kg of formulated product in 2000. Furthermore, growers planting transgenic crops used less fuel due to fewer field operations and fuel savings increased from 9.5 million l in 1997 to 31.2 million l of diesel in 2000. This saving equated to CAN$13.1 million, and clearly contributed to improved profitability and enhanced competitiveness of the Canadian canola growers. The decrease in diesel use would also reduce emissions of the green house gas carbon dioxide by approximately 110,000 t.

GM maize

The area of GM maize grown globally is 35.4 million ha (James, 2007) and the use of HT maize has on average reduced herbicide use by 30% (0.69 kg ha^{-1}). This is equivalent to a reduction of 3.5 million kg formulated product per year.

The area of insect-protected maize currently grown is around 16 million ha (James, 2006). While the European corn borer (ECB) is a serious insect pest of maize grain crops causing losses ranging from 0.75 to 7.5 million t of grain per year in the USA, only 5% of the crop was sprayed against ECB due to the problems of assessing the correct time to spray. As a result of the previously low levels of pesticides sprayed, the introduction of *Bt* maize has only resulted in a modest decrease in insecticide usage. It is generally considered that the main reason for growing *Bt* maize is for the increased yields which occur when infestation of ECB is controlled by this technology. However, Munkvold *et al.* (1999) have established that the use of *Bt* maize has the added advantage of reducing mycotoxin contamination, thus producing safer grain for both human and animals.

GM cotton

The herbicide use in cotton is expected to decline with the adoption of HT cotton varieties. Application rates for conventional varieties vary from 4.9 to 8.0 kg formulated product per hectare compared with 2.5–4.0 kg ha^{-1} for glyphosate-tolerant GM varieties. Thus, with 9.8 million ha grown the decrease in pesticide use associated with the introduction of HT cotton was estimated at over 20 million kg formulated product in 2005. In addition, the introduction of HT varieties has been associated with a reduction in spray applications of 1.8 million ha.

In addition to being grown in the USA, transgenic cotton is also grown in China, Mexico, Australia, Argentina and South Africa. Cotton is highly susceptible to a number of serious insect pests such as tobacco budworm, cotton bollworm and pink bollworm, and requires a sustained insecticide spray programme. Gianessi and Carpenter (2000) calculated that between 1995, the year before *Bt* varieties were introduced, and 1999, the amount of insecticide used decreased by 1.2 million kg of formulated product, which represents 14% of all insecticides. In addition the number of spray applications per hectare was reduced by 15 million, which represented a 22% reduction. Agnew and Baker (2001) have stated that *Bt* cotton has helped to reduce insecticide use in Arizona cotton to the lowest levels in the past 20 years. Edge *et al.* (2001) reported that studies in the USA, Australia, China, Mexico and Spain all demonstrated an overall reduction in insecticide sprays. While the introduction of *Bt* cotton has markedly reduced the amount of pesticide used and the number of spray applications required per hectare, Gianessi and Silvers (2000) noted that many of the traditional pesticides used in cotton production also had poor environmental characteristics. China is a major producer of cotton and their growers are among the largest users of pesticides. Data by Huang *et al.* (2003) examined the effect of biotechnology on pesticide use in cotton crops in China. They found that the effect of introducing *Bt* cotton on pesticide use was dramatic. In a survey conducted in 1999 and 2000, they reported that on average growers that used *Bt* cotton reduced pesticide use from 55 to 16 kg formulated product per hectare and the number of times the crop was sprayed from 20 to seven. In addition to a reduction in pesticide use of 70%, the authors also noted that the use of the highly toxic organochlorines and organophosphates was all but eliminated. *Bt* cotton was first adopted in India as hybrids in 2002. India grew approximately 50,000 ha of officially approved *Bt* cotton hybrids for the first time in 2002, and doubled its *Bt* cotton area to approximately 100,000 ha in 2003. The *Bt* cotton area increased again fourfold in 2004 to reach over 0.5 million ha. In 2005, the area planted with *Bt* cotton in India continued to increase, reaching 1.3 million ha, an increase of 160% over 2004.

In 2006, the record increases of adoption in India continued with almost a tripling of area of *Bt* cotton from 1.3 to 3.8 million ha. In 2006, this tripling in area was the highest year-on-year growth for any country in the world. Of the 6.3 million ha of hybrid cotton in India in 2006, which represent 70% of all the cotton area in India, 60% or 3.8 million ha was *Bt* cotton – a remarkably high proportion in a fairly short period of 5 years. With this dramatic increase one might expect the use of *Bt* cotton to provide many of the same benefits as those noted in China. The reduction in pesticide could amount to 11.4 million kg per year of formulated product.

Examples of other GM crops in which pesticide use can be reduced

The 500,000 ha of potatoes grown in the USA are currently treated with 1.2 million kg of pesticide. Both Colorado potato beetle (CPB) and aphids (which transmit a number of viral diseases, including potato leaf roll virus (PLRV) and potato virus Y (PVY)) present major problems of control for growers. In 1996, GM potatoes that were protected against CPB were made available to growers. Initial results showed that their use decreased the number of insecticide applications from 2.78 to 1.58 ha^{-1} and the amount of insecticide used from 2.17 to 1.74 kg ha^{-1} (Gianessi and Carpenter, 2000). A further development in which these potatoes were also protected against PLRV and PVY produced further reductions in insecticide use and applications required.

A European Perspective

Between 1990 and 1995, the annual amount of pesticide active ingredients used in the EU declined from 307,000 to 253,000 t, which represents an 18% reduction. This was due to a number of factors including lower dose rates, better application technology, changes in farm management practices, national mandatory reduction schemes, as well as payment for agri-environmental schemes. The EU 6th Environmental Action Plan has continued to focus on pesticide reduction as a priority in relation to environmental degradation. It is against this background of reducing pesticide input that the potential of GM crops to further reduce pesticide use in the EU will be estimated.

The estimates indicate that if 50% of the maize, oilseed rape, sugarbeet and cotton were grown in the EU as HT or *Bt* varieties, the amount of pesticide used would fall by 14.5 million kg formulated product per annum, which represents a decrease of 4.4 million kg of active ingredient. In addition, there would be a reduction of 7.5 million ha sprayed (Brookes and Barfoot, 2006).

Effect of GM Crops on the Agricultural Environment

In addition to the benefits of reduced exposure achieved through a simple reduction to the total amount of formulated product needed per annum, the use of GM crops has now been demonstrated to directly benefit agricultural environmental systems. In the largest ever field trials of GM crops in the world, GM and conventional varieties of four crops were compared in terms of their impact to the UK agri-environment. The crops were winter-sown oilseed rape, spring-sown oilseed rape, beet and maize. The GM crops had been genetically modified to make them resistant to specific herbicides. While the transgenic crops currently available to the market had been assessed as safe in terms of human health and direct impacts upon the environment, there had been insufficient research to determine whether there might be any significant effects on farmland wildlife resulting from the way that the crops would be managed (Advisory Committee on Releases to the Environment, 2000). The Farm-scale Evaluations (FSEs) of these genetically modified herbicide-tolerant (GMHT) crops were established to bridge this important gap in our knowledge (Firbank *et al.*, 1999; Squire *et al.*, 2003).

The FSEs of spring-sown GMHT crops were conducted in the UK from 2000 to 2002 (Firbank *et al.*, 1999, 2003a). The effects of the management regimes associated with conventional and GM beet (*Beta vulgaris* L.), maize (*Zea mays* L.) and oilseed rape (*Brassica napus* L.) crops on weed plant and invertebrate indicators within fields and in field margins were compared. The first results were published in October 2003 for vegetation (Heard *et al.*, 2003), soil-surface-active invertebrates (Brooks *et al.*, 2003), epigeal and aerial arthropods (Haughton *et al.*, 2003), field boundary invertebrates and vegetation (Roy *et al.*, 2003), and plant and invertebrate trophic groups (Hawes *et al.*, 2003). Briefly, each experiment comprised a randomized block design, with whole fields as blocks and with the treatment (conventional or GMHT) replicated once on half-field units in each field. The FSEs were unusual as they were highly controversial and attracted intense examination because of the public concern over genetic modification (Perry *et al.*, 2003). As a response, the research proposed by the contractors was overseen by a Scientific Steering Committee that scrutinized closely the planned design and analysis which became the subject of considerable discussion and further research (Perry *et al.*, 2003). Additionally, Firbank *et al.* (2003b) and others (Lawton, 2003; May, 2003; Webb, 2003; Pollock, 2004) have emphasized the prime importance of the FSE database as a source of baseline measurements for the abundance of biodiversity to inform changes in policy for British agriculture. Research is now showing how biodiversity can be enhanced in arable landscapes by the manipulation of farming systems (Dewar *et al.*, 2003) and their adjacent field margins (Sotherton, 1991), and there is a perceived need to restore the balance between agricultural production and wildlife (see Chapters 11 and 12, this volume).

Herbicide-tolerant winter-sown oilseed rape

Winter-sown oilseed rape is the most widely cultivated of all four crops studied. About 330,000 ha are grown all over the UK, but mostly in the south of the country. The results of the FSE trials found that there were similar total weed densities in both GM and conventional winter rape fields. However, significant differences were observed in the abundance of different types of weed. It was found that there were more grass weeds, but fewer broad-leaved, flowering weeds in the GM crops than in the conventional crops. This marked imbalance between grass and broad-leaved weeds was only found in winter rape. The source of this imbalance was probably due to the reduced efficacy of glyphosate against grasses and that the conventional herbicide is applied at a much later stage of weed growth. By October of the trials, farmers had treated half of the conventional winter rape crop with pre-emergence herbicides. At this time, broad-leaved weed densities were two times higher and numbers of grass weeds were three times higher in the GM crop than in the conventional crop. Subsequent to the herbicide application, broad-leaved weed densities in the GM crop fell to one-third of those found in the conventional crop. However, herbicide application did not make any difference to the grass weeds; their densities remained three times higher until harvest. These increased densities manifested in a five times greater differences in seed numbers from grass weeds in the GM crop and this difference persisted in the seedbank in the following year. However, broad-leaved weeds produced three times more seeds in the conventional crop, resulting in differences in weed numbers that lasted to the next year. The increased densities of broad-leaved weeds resulted in more butterflies and bees in conventional winter rape because there were more flowering broad-leaved weeds. The numbers of springtails were higher in the GM crop. They feed on rotting broad-leaved weeds, so they benefited from weeds killed later in the year when they were larger and more abundant. By comparison with all other crops, though, few differences were found in the numbers of insects and other small animals between the GM and conventional winter rape.

Herbicide-tolerant spring-sown oilseed rape

In any one year, spring rape can cover 60,000 ha. Typically, it is sown in April and harvested in September. In the FSE trials, farmers treated almost half of the conventional spring rape crops with herbicides applied before the weeds emerged in March. This meant that weed density was, on average, higher in the GM fields until farmers applied the GM herbicide. Then, weed density, particularly of broad-leaved plants, fell drastically. As seen in the winter-sown varieties, the sharp reduction of these broad-leaved weeds was due to the fact that herbicide used in these trials acts more effectively against these types of weeds and is less effective against grass weeds. Although the numbers of surviving broad-leaved weeds were similar in conventional and GM crops, the plants had a 70% lower biomass in the GM crops. It was also reported that seed rain was lower, with 80% fewer broad-leaved weed seeds. Overall, the weed seedbank was smaller following GM crops. After 1 year in the conventional spring rape fields, the seedbank of broad-leaved weeds doubled but it only increased slightly in the GM equivalent. Butterfly numbers were higher in the fields and field margins of conventional spring rape crops, attracted mainly by the greater numbers of flowering weeds in and around the crop. Most other insect groups, including bees, were found in similar numbers in the GM and conventional fields, although there were some seasonal differences. Springtails and spiders were significantly more abundant in GM crops in July and August, respectively, just before harvest. This was probably because springtails feed on rotting weeds, which were more abundant in the GM crops late in the year. The GM herbicides are used later in the year so the weeds are larger when they are treated, providing more food for springtails. The spiders were probably feeding off the springtails. One particular type of seed-eating ground beetle also appeared more frequently in the conventional rape fields because there were more seeds for them to eat.

Herbicide-tolerant beet

About 170,000 ha of sugarbeet are typically grown in the UK (including cultivated field margins). In spring, the density of weed seedlings growing in the GMHT beet fields was four times that in the conventional beet fields because many farmers used pre-emergence herbicides on the conventional halves of the field. However, applying the broad-spectrum glyphosate herbicide to the GM crops in May resulted in the weed density being reduced by a half when compared with the conventional crops. After this, the biomass of the remaining weeds was six times lower and the seed rain was three times lower compared with the conventional crops. The seedbank of weed grasses remained the same at the end of the trials as it was at the beginning 2 years earlier. The seedbank of broad-leaved weeds remained constant in the GM fields but increased in the conventional fields. Although the numbers of bees and butterflies present in beet crops are typically never very large, there were even fewer in the GMHT beet crops than in the conventional crops; it was assumed that this was due to fewer suitable flowering weeds to attract them. Bee numbers, while generally low everywhere, were even lower in the GM crops, falling to their lowest in August in the crops and in July in the field margins. The populations of two insects, springtails and true bugs, also showed some differences between crops. There were more springtails and some of their predators, such as one species of ground beetle (*Bembidion* spp.), were present in the GM crop in August than in the conventional crop. This was probably because springtails feed on rotting weeds, which were more abundant in the GM crops later in the year. As mentioned above, the herbicides used on the GM crops are employed later in the year so the weeds are bigger when they are killed, providing more food for springtails. On the other hand, populations of true bugs were much smaller than those found in conventional crops, probably because they could not find enough weeds and seeds to eat in the GM fields. One particular type of seed-eating ground beetle was also more frequent in the conventional beet fields because there were more seeds for them to eat.

Herbicide-tolerant maize

British farmers typically grow more than 100,000 ha of forage maize annually to make into silage to feed to cattle. Maize, unlike beet or rape, can be grown continuously in the same field. Both the density and biomass of broad-leaved weeds were three times higher in the GMHT maize fields than in the conventional maize fields. Taken together, the weeds in the GM crops produced twice as many seeds as the weeds in the conventional crops. There was no effect on the seedbank. Over the growing season from May to September, butterflies were attracted to the GMHT maize fields and field margins in the same numbers as the conventional fields. However, significantly more butterflies visited the GM crops in July, while August saw nearly three times as many honeybees in the GM field boundaries as in the conventional fields, probably because there were more plants in flower in the field margins at this time. It must, however, be stressed that even in the GM fields, numbers of bees and butterflies were still low. Most groups of insects were found in similar numbers in both crops. The main differences were a consistently greater number of springtails in the GM crops, especially in August, for reasons already given above. There were also more butterflies in the GM crops in July, and more honeybees in August. There were fewer spiders in the GM crop margins, reflecting the lower abundance of plants to aid web-spinning. Just before publication of the results for spring crops in October 2003, the EU announced a ban on three herbicides used extensively in conventional maize, including the persistent chemical atrazine. This meant that comparisons the researchers had made between conventional and GM maize crops might not hold for the future, because most farmers in the study had used one of these chemicals on their conventional maize fields. This point was subsequently addressed by Perry et al. (2004) who concluded that growing GMHT maize would still benefit wildlife, but the effects would be reduced by about one-third. This is because agricultural experts agree that farmers are likely to use equally persistent chemicals on conventional maize fields.

Indirect Environmental Benefits of GM

While it is clear that the major benefit of the introduction of GM crops is a direct reduction in the number of pesticides applied and also the reduction in the use of the most environmentally unsound chemical pesticides, there are now signs that this shift in agricultural practices is benefiting the environment through many indirect routes.

Soil conservation

HT crops may lead to environmental benefits by facilitating a shift to conservation tillage practices. Specifically, these crops may allow farmers to eliminate pre-emergent herbicides that are incorporated into the soil and rely on post-emergent herbicides, such as glyphosate. The shift to postemergent control of weeds may promote no-till and conservation tillage practices that can decrease soil erosion and water loss and increase soil organic matter (Cannell and Hawes, 1994; Soon and Clayton, 2002; Rieger *et al.*, 2008). Studies are needed to address whether soils are improving as a result of growing crops genetically engineered for herbicide tolerance (see Chapter 10, this volume).

Phytoremediation

The genetic modification of plant or microorganisms may provide *in situ* remediation of polluted soils, sediments, surface waters and aquifers. Transgenic plants can increase removal of toxic heavy metals from polluted soils and waters and sequester these into plant tissue available for harvest (Gleba *et al.*, 1999; Zhu *et al.*, 1999; Wolfenbarger and Phifer, 2000; Peuke and Rennenberg, 2005; Arshad *et al.*, 2007), or can transform pollutants into less toxic forms (Bizily *et al.*, 2000; Watanabe, 2001; Meagher, 2006). Environmental remediation through transgenic plants has not yet been used widely, so net environmental benefits have not been measured.

Mitigation of direct effects

The FSE trials of GMHT sugarbeet showed that there was potential to have an adverse impact on food for farmland birds if a 'weed free' management approach was adopted. Previous work at Broom's Barn Research Station demonstrated that innovative crop management practices deploying GMHT beet had the potential to deliver food for farmland birds in spring or autumn. Pidgeon *et al.* (2007) demonstrate an extremely cheap and simple mitigation approach to avoid any adverse impacts on bird populations. This is achieved by simply leaving two crop rows in every 100 unsprayed. Pidgeon states: 'The economic benefits for a hard pressed farming sector are large. This demonstrates beyond reasonable doubt that GMHT beet can be economically and environmentally beneficial.'

Reduction in herbicide losses in surface runoff

Through the introduction of GMHT soybean and maize and replacing the traditionally used residual herbicides with short half-life, strongly sorbed contact herbicides, the loss of herbicide through surface runoff has been significantly reduced. Previously the concentration of residual herbicides in surface runoff had often been detected at concentrations exceeding their maximum contamination levels (MCL). Shipitalo *et al.* (2008) were able to demonstrate that the losses of glyphosate when used with GMHT soybean were approximately one-seventh that of metribuzin and about one-half that of alachlor, the residual herbicides it replaces. Similar benefits were seen when using glufosinate where runoff was one-fourth of atrizine. Additionally, both glyphosate and glufosinate concentrations in runoff were significantly below their MCL.

Discussion and Conclusions

The authors recognize that the debate surrounding genetic modification is complex and that it is a technology to which many individuals and organizations are opposed. However, it is a technology that has the potential to deliver many of the solutions needed in modern sustainable agriculture, and therefore, we cannot afford to ignore its place. In countries where GM crops are at present widely grown, published data show that the adoption of GM technology can lead to a marked reduction in pesticide use. However, the size of the reduction varies between crops and the introduced trait. For example, only a modest reduction in pesticide use of 10% is associated with the introduction HT soybeans but a large and highly significant reduction of 60% in pesticide use is recorded for *Bt* varieties of cotton. Although the total reduction in pesticide use of 2.9 million kg associated with HT soybeans is important, the most valuable contribution to environmental benefits of GM soybeans may be that they encourage farmers to use conservation tillage techniques. It is estimated that the use of HT soybean, oilseed rape, cotton and maize varieties reduced pesticide use by a total of 22.3 million kg of formulated product in the year 2000. It is important that further studies are conducted to quantify the benefits to the environment that can occur from such a large reduction in pesticide use. Further data have been presented for the likely impact in terms of pesticide use if GM crops were introduced into the EU. The estimate indicates that if 50% of the maize, oilseed rape, sugarbeet and cotton were grown as HT or insect-protected GM varieties the amount of pesticide used in the EU per annum would fall by 14.5 million kg of formulated product. In addition, the amount of active ingredient applied would decrease by 4.4 million kg and there would be a reduction of 7.5 million ha sprayed, which would save 20.5 million l of diesel and result in a reduction of approximately 73,000 t of carbon dioxide being released into the atmosphere (Taylor *et al.*, 1993). These values could increase markedly as countries such as Turkey growing 700,000 ha of cotton enter the EU. Despite the limitations in the analysis and the overall complexities of the debate, the authors believe that GM technology has the potential to markedly reduce overall pesticide use. Further, if less chemical is used and the number of spray applications is reduced there will be a considerable saving in support energy required for crop production. While large-scale commercial plantings of GM crops have not yet occurred in the EU, a 50% planting of maize, oilseed rape, sugarbeet and cotton to GM varieties could result in the saving of 7.60×10^{11} GJ of energy per year or the equivalent of 20.5 million l of diesel fuel. These calculations assume an energy cost of 115 MJ ha^{-1} for spray application and an energy value of diesel of 37 MJ l^{-1} (Bailey *et al.*, 2003). Further savings on energy would be made through including not only the different energy costs of pesticide productions but also the fact that the use of less pesticide will require less raw ingredients and inerts, less diesel fuel in the manufacturing process, less fuel for shipment and storage, less water and fuel during spraying, and of course, less packaging for their containment and distribution to and within the agricultural sector. Further research is also required to investigate and estimate the impacts of the use of GM crops on the frequency and severity of pollution incidents relating to pesticides and watercourses. Looking to the future, a recent study by Kline and Company, a New Jersey-based consulting firm, analysed the future trends in pesticide use in the USA by 2009. Their analyses of the market indicated that by 2009, HT and insect-protected crops would contribute to an annual reduction of 20 million and 6 million kg of herbicide and insecticide active ingredient, respectively. The authors feel that if the reductions indicated and those envisaged could be achieved, then there should be a flow of positive environmental benefits to society at large. While it is important that the rigorous investigation of the impact of the introduction of GM crops in the EU continues, it is surprising that some of the positive aspects of their introduction appear to have been ignored. For instance, we would agree with Carpenter *et al.* (2001) who suggests that while scientists continue to debate risks, such as the effects of genetically engineered maize pollen on butterfly populations, dramatic reductions in pesticide use achieved through the introduction of GM crops continues to be largely ignored. Additionally, we find ourselves in a position where we can manipulate our agricultural practices to gain the greatest benefits that are available when GM crops become widespread in the EU. Studies are already showing that by identifying any potential risks of the use of biotechnology, agricultural practices can be altered thus allowing effective mitigation of any undesirable effects.

References

Advisory Committee on Releases to the Environment 2000 ACRE Annual Report no. 6. London: Defra. Available at http://www.defra.gov.uk/environment/acre/pubs.htm#annrpt

Agnew, G.K. and Baker, P.B. (2001) Pest and pesticide usage patterns in Arizona cotton. Proc. 2001 Beltwide Cotton Conferences. National Cotton Council, Memphis, TN. pp. 1046–1054.

Arshad, M., Saleem, M. and Hussain, S. (2007) Perspectives of bacterial ACC deaminase in phytoremediation. *Trends in Biotechnology* 25, 356–362.

Bailey, A.P., Basford, W.D., Penlington, N., Park, J.R., Keatinge, J.D.H., Rehman, T., Tranter, R.B. and Yates, C.M. (2003) A comparison of energy use in conventional and integrated arable farming systems in the UK. *Agriculture, Ecosystems and Environment* 97, 241–253.

Beckie, H.J. (2006) Herbicide-resistant weeds: management tactics and practices. *Weed Technology* 20, 793–814.

Bizily, S.P., Rugh, C.L. and Meagher, R.B. (2000) Phytodetoxification of hazardous organomercurials by genetically engineered plants. *Nature Biotechnology* 18, 213–217.

Bonny, S. (2008) Genetically modified glyphosate-tolerant soybean in the USA: adoption factors, impacts and prospects. A review. *Agronomy for Sustainable Development* 28, 21–32.

Brookes, G. and Barfoot, P. (2006) *GM Crops: The First Ten Years – Global Socio-Economic and Environmental Impacts.* ISAAA, Ithaca, New York.

Brooks, D.R., Bohan, D.A., Champion, G.T., Haughton, A.J., Hawes, C., Heard, M.S., Clark, S.J., Dewar, A.M., Firbank, L.G., Perry, J.N., Rothery, P., Scott, R.J., Woiwod, I.P., Birchall, C., Skellern, M.P., Walker, J.H., Baker, P., Bell, D., Browne, E.L., Dewar, A.J.G., Fairfax, C.M., Garner, B.H., Haylock, L.A., Horne, S.L., Hulmes, S.E., Mason, N.S., Norton, L.R., Nuttall, P., Randle, Z., Rossall, M.J., Sands, R.J.N., Singer, E.J. and Walker, M.J. (2003) Invertebrate responses to the management of genetically modified herbicide-tolerant and conventional spring crops. I. Soil-surface-active invertebrates. *Philosophical Transactions of the Royal Society of London Series B-Biological Sciences* 358, 1847–1862.

Cannell, R.Q. and Hawes, J.D. (1994) Trends in tillage practices in relation to sustainable crop production with special reference to temperate climates. *Soil and Tillage Research* 30, 245–282.

Canola Council of Canada (2001) Impact of transgenic canola on growers, industry and environment. Available at: http://www.canola-council.org

Carpenter, J.E. (2001) Case studies in benefits and risks of agricultural biotechnology: roundup ready soybeans and Bt field corn. *National Centre for Food and Agricultural Policy,* Washington, DC.

Carpenter, J.E., Gianessi, L.P. and Silvers, C.S. (2001) Insecticidal Bt plants versus chemical insecticides. *Abstracts of Papers of the American Chemical Society* 222, U69.

Cerdeira, A.L. and Duke, S.O. (2006). The current status and environmental impacts of glyphosate-resistant crops: a review. *Journal of Environmental Quality* 35, 1633–1658.

Cuyno, L.C.M., Norton, G.W. and Rola, A. (2001) Economic analysis of environmen-tal benefits of integrated pest manage-ment: a Philippine case study. *Agricultural Economics* 25, 227–233.

Dewar, A.M., May, M.J., Woiwod, I.P., Haylock, L.A., Champion, G.T., Garner, B.H., Sands, R.J.N., Qi, A.M. and Pidgeon, J.D. (2003) A novel approach to the use of genetically modified herbicide tolerant crops for environmental benefit. *Proceedings of the Royal Society of London Series B-Biological Sciences* 270, 335–340.

Eddleston, M., Buckley, N.A., Eyer, P. and Dawson, A.H. (2008) Management of acute organophosphorus pesticide poison-ing. *Lancet* 371, 597–607.

Edge, J.M., Benedict, J.H., Carroll, J.P. and Reding, H.K. (2001) Bollgard cotton: an assessment of global, economic, environ-mental and social benefits. *Journal of Cotton Science* 5, 1–8.

Ellenberger, J., Bascietto, J. and Barrett, M. (1989) *Carbofuran: A Special Review Technical Support Document.* US Environ-mental Agency, Office of Pesticides and Toxic Substances, Washington, DC.

Farah, J. (1994) *Pesticide Policies in Developing Countries: Do They Encourage Excessive Use?* World Bank, Washington, DC.

Firbank, L.G. (2003) The farm scale evaluations of springsown genetically modified crops – introduction. *Philosophical Transactions of the Royal Society of London Series B-Biological Sciences* 358, 1777–1778.

Firbank, L.G., Dewar, A.M., Hill, M.O., May, M.J., Perry, J.N., Rothery, P., Squire, G.R. and Woiwod, I.P. (1999) Farm-scale evalu-ation of GM crops explained. *Nature* 399, 727–728.

Firbank, L.G., Heard, M.S., Woiwod, I.P., Hawes, C., Haughton, A.J., Champion, G.T., Scott, R.J., Hill, M.O., Dewar, A.M., Squire, G.R., May, M.J., Brooks, D.R., Bohan, D.A., Daniels, R.E., Osborne, J.L., Roy, D.B., Black, H.I.J., Rothery, P. and Perry, J.N. (2003a) An introduction to the farm-scale evaluations of genetically modi-fied herbicide-tolerant crops. *Journal of Applied Ecology* 40, 2–16.

Firbank, L.G., Perry, J.N., Squire, G.R., Bohan, D.A., Brooks, D.R., Champion, G.T., Clark, S.J., Daniels, R.E., Dewar, A.M., Haughton,

A.J., Hawes, C., Heard, M.S., Hill, M.O., May, M.J., Osborne, J.L., Rothery, P., Roy, D.B., Scott, R.J. and Woiwod, I.P. (2003b) The implications of spring-sown genetically modified herbicide-tolerant crops for farmland biodiversity: a commentary on the farm scale evaluations of spring sown crops. Available at: http://www.defra.gov.uk/environment/gm/fse/results/fse-commentary.pdf

Gianessi, L.P. and Carpenter, J.E. (2000) *Agricultural Biotechnology: Benefits of Transgenic Soybeans*. National Center for Food and Agricultural Policy, Washington, DC.

Gleba, D., Borisjuk, N.V., Borisjuk, L.G., Kneer, R., Poulev, A., Sarzhinskaya, M., Dushenkov, S., Logendra, S., Gleba, Y.Y. and Raskin, I. (1999) Use of plant roots for phytoremediation and molecular farming. *Proceedings of the National Academy of Sciences of the USA* 96, 5973–5977.

Gunnell, D., Eddleston, M., Phillips, M.R. and Konradsen, F. (2007) The global distribution of fatal pesticide self-poisoning: systematic review. *BMC Public Health*, 7, 357.

Haughton, A.J., Champion, G.T., Hawes, C., Heard, M.S., Brooks, D.R., Bohan, D.A., Clark, S.J., Dewar, A.M., Firbank, L.G., Osborne, J.L., Perry, J.N., Rothery, P., Roy, D.B., Scott, R.J., Woiwod, I.P., Birchall, C., Skellern, M.P., Walker, J.H., Baker, P., Browne, E.L., Dewar, A.J.G., Garner, B.H., Haylock, L.A., Horne, S.L., Mason, N.S., Sands, R.J.N. and Walker, M.J. (2003) Invertebrate responses to the management of genetically modified herbicide-tolerant and conventional spring crops. II. Within-field epigeal and aerial arthropods. *Philosophical Transactions of the Royal Society of London Series B-Biological Sciences* 358, 1863–1877.

Hawes, C., Haughton, A.J., Osborne, J.L., Roy, D.B., Clark, S.J., Perry, J.N., Rothery, P., Bohan, D.A., Brooks, D.R., Champion, G.T., Dewar, A.M., Heard, M.S., Woiwod, I.P., Daniels, R.E., Young, M.W., Parish, A.M., Scott, R.J., Firbank, L.G. and Squire, G.R. (2003) Responses of plants and invertebrate trophic groups to contrasting herbicide regimes in the farm scale evaluations of genetically modified herbicide-tolerant crops. *Philosophical Transactions of the Royal Society of London Series B-Biological Sciences* 358, 1899–1913.

Heard, M.S., Hawes, C., Champion, G.T., Clark, S.J., Firbank, L.G., Haughton, A.J., Parish, A.M., Perry, J.N., Rothery, P., Scott, R.J., Skellern, M.P., Squire, G.R. and Hill, M.I. (2003) Weeds in fields with contrasting conventional and genetically modified herbicide-tolerant crops. I. Effects on abundance and diversity. *Philosophical Transactions of the Royal Society of London Series B-Biological Sciences* 358, 1819–1832.

Heimlich, R.E., Fernandez-Cornejo, J., McBride, W., Koltz-Ingram, C., Jans, S. and Brooks, N. (2000) *Genetically Engineered Crops: Has Adoption Reduced Pesticide Use?* USDA Economic Research Service, Washington, DC.

Hin, C.J.A., Schenkelaars, P. and Pak, G.A. (2001) Agronomic and environmental impacts of the commercial cultivation of glyphosate tolerant soybean in the USA. Centre for Agriculture and Environment Utrecht, June 2001. Available at: http://www.bothends.org/strategic/soy21.pdf

Huang, J.K., Hu, R.F., Pray, C., Qiao, F.B. and Rozelle, S. (2003) Biotechnology as an alternative to chemical pesticides: a case study of Bt cotton in China. *Agricultural Economics* 29, 55–67.

James, C. (2006) *Global Status of Commercialized Biotech/GM Crops: 2006.* ISAAA, Ithaca, New York.

James, C. (2007) *Global Status of Commercialized Biotech/GM Crops: 2007.* ISAAA, Ithaca, New York.

Krebs, J.R., Wilson, J.D., Bradbury, R.B. and Siriwardena, G.M. (1999) The second silent spring? *Nature* 400, 611–612.

Lawton, J.H. (2003) *The Guardian*, 17 October 2003.

London, L., De Grosbois, S., Wesseling, C., Kisting, S., Rother, H.A. and Mergler, D. (2002) Pesticide usage and health consequences for women in developing countries: out of sight, out of mind? *International Journal of Occupational and Environmental Health* 8, 46–59.

May, R.M. (2003) Royal Society submission to ACRS consultation on GM farm-scale evaluations. Available at: http://royalsociety.org/document.asp?id=1358

Meagher, R.B. (2006) Plants tackle explosive contamination. *Nature Biotechnology* 24, 161–163.

Munkvold, G.P., Hellmich, R.L. and Rice, L.G. (1999) Comparison of fumonisin concentrations in kernels of transgenic Bt maize hybrids and nontransgenic hybrids. *Plant Disease* 83, 130–138.

Nelson, G.C. and Bullock, D.S. (2003) Simulating a relative environmental effect of glyphosate-resistant soybeans. *Ecological Economics* 45, 189–202.

Oerke, E.C. (1999) The importance of disease control in modern plant production. *Modern Fungicides and Antifungal Compounds II*, 11–17.

Oerke, E.C. (2006) Crop losses to pests. *Journal of Agricultural Science* 144, 31–43.

Paoletti, M.G. and Pimentel, D. (2000) Environmental risks of pesticides versus genetic engineering for agricultural pest control. *Journal of Agricultural and Environmental Ethics* 12, 279–303.

Perry, J.N., Rothery, P., Clark, S.J., Heard, M.S. and Hawes, C. (2003) Design, analysis and statistical power of the farm-scale evaluations of genetically modified herbicide-tolerant crops. *Journal of Applied Ecology* 40, 17–31.

Perry, J.N., Firbank, L.G., Champion, G.T., Clark, S.J., Heard, M.S., May, M.J., Hawes, C., Squire, G.R., Rothery, P., Wolwod, I.P. and Pidgeon, J.D. (2004) Ban on triazine herbicides likely to reduce but not negate relative benefits of GMHT maize cropping. *Nature* 428, 313–316.

Peuke, A.D. and Rennenberg, H. (2005) Phytoremediation – molecular biology, requirements for application, environmental protection, public attention and feasibility. *Embo Reports* 6, 497–501.

Pidgeon, J.D., May, M.J., Perry, J.N. and Poppy, G.M. (2007) Mitigation of indirect environmental effects of GM crops. *Proceedings of the Royal Society B-Biological Sciences* 274, 1475–1479.

Pollock, C.J. (2004) Why it's time for GM Britian. *The Guardian*, 26 February 2004.

Pray, C., Ma, D.M., Huang, J.K. and Qiao, F.B. (2001) Impact of Bt cotton in China. *World Development* 29, 813–825.

Pray, C.E., Huang, J.K., Hu, R.F. and Rozelle, S. (2002) Five years of Bt cotton in China – the benefits continue. *Plant Journal* 31, 423–430.

Repetto, R. and Baliga, S.S. (1996) Pesticides and the immune system: the public health risks. Executive summary. *Central European Journal of Public Health* 4, 263–265.

Rieger, S., Richner, W., Streit, B., Frossard, E. and Liedgens, M. (2008) Growth, yield, and yield components of winter wheat and the effects of tillage intensity, preceding crops, and N fertilisation. *European Journal of Agronomy* 28, 405–411.

Rother, H.A. (1998) Influences of pesticide risk perception on the health of rural South African women and children. *International Conference on Pesticide Use in Developing Countries – Impact on Health and Environment*. San Jose, Costa Rica, 23–29 February 1998.

Rother, H.A. (2006) Pesticide health risks: a socially constructed perception. *Epidemiology* 17, S196.

Roy, D.B., Bohan, D.A., Haughton, A.J., Hill, M.O., Osborne, J.L., Clark, S.J., Perry, J.N., Rothery, P., Scott, R.J., Brooks, D.R., Champion, G.T., Hawes, C., Heard, M.S. and Firbank, L.G. (2003) Invertebrates and vegetation of field margins adjacent to crops subject to contrasting herbicide regimes in the farm scale evaluations of genetically modified herbicide-tolerant crops. *Philosophical Transactions of the Royal Society of London Series B-Biological Sciences* 358, 1879–1898.

Shipitalo, M.J., Malone, R.W. and Owens, L.B. (2008) Impact of glyphosate-tolerant soybean and glufosinate-tolerant corn production on herbicide losses in surface runoff. *Journal of Environmental Quality* 37, 401–408.

Soon, Y.K. and Clayton, G.W. (2002) Eight years of crop rotation and tillage effects on crop production and N fertilizer use. *Canadian Journal of Soil Science* 82, 165–172.

Sotherton, N.W. (1991) Conservation head-lands, a practical combination of intensive cereal farming and conservation. In: Firbank, L.G., Carter, N., Darbyshire, J.F. and Potts, G.F. (eds) *Ecology of Temperate Cereal Fields*. Blackwell Scientific Publications, Oxford, pp. 373–397.

Squire, G.R., Brooks, D.R., Bohan, D.A., Champion, G.T., Daniels, R.E., Haughton, A.J., Hawes, C., Heard, M.S., Hill, M.O., May, M.J., Osborne, J.L., Perry, J.N., Roy, D.B., Woiwod, I.P. and Firbank, L.G. (2003) On the rationale and interpretation of the farm scale evaluations of genetically modified herbicide-tolerant crops. *Philosophical Transactions of the Royal Society of London Series B-Biological Sciences* 358, 1779–1799.

Taylor, A.E.B., O'Callaghan, P.W. and Probert, S.D. (1993) Energy audit of an English farm. *Applied Energy* 44, 315–335.

Thundiyil, J.G., Stober, J., Besbelli, N. and Pronczuk, J. (2008) Acute pesticide poi-soning: a proposed classification tool. *Bulletin of the World Health Organization* 86, 205–209.

Tilman, D. (1999) Global environmental impacts of agricultural expansion: the need for sustainable and efficient practices. *Proceedings of the National Academy of Sciences of the USA* 96, 5995–6000.

Watanabe, M.E. (2001) Can bioremediation bounce back? *Nature Biotechnology* 19, 1111–1115.

Webb, J. (2003) Editorial: a victory for reason. *New Scientist* 180, (2418.).

Wolfenbarger, L.L. and Phifer, P.R. (2000) Biotechnology and ecology – the ecologi-cal risks and benefits of genetically engi-neered plants. *Science* 290, 2088–2093.

Zhu, Y.L., Pilon-Smits, E.A.H., Jouanin, L. and Terry, N. (1999) Overexpression of glutath-ione synthetase in Indian mustard enhances cadmium accumulation and tol-erance. *Plant Physiology* 119, 73–79.

3 Developing a 21st Century View of Agriculture and the Environment

D. Pimentel[1] and M.G. Paoletti[2]

[1]Department of Entomology, Systematics and Ecology, Cornell University, Ithaca, New York, USA; [2]Dipartimento di Biologia, Università di Padova, Padova, Italy

Keywords: Genetically engineered crops, environment, pesticides, annual grain, minilivestock, perennial grain, food diversity, world food security

Summary

Both poverty and malnutrition are serious problems in the world and both are interrelated. Food security for the poor depends on an adequate supply of food and/or the ability to purchase food. Unfortunately in the world today, more than 3.7 billion people are malnourished because of shortages of calories, protein, several vitamins, iron and iodine (World Health Organization, 2005; Rhodes, 2005). People can die because of shortage of any one of these nutrients or a combination of them. The total of 3.7 billion malnourished people is the largest number ever in history.

In the world today, there are more than 6.5 billion humans (PRB, 2006). Based on current rates of increase, the world population is projected to double to more than 13 billion in about 58 years (PRB, 2006). At a time when the world population continues to expand at a rate of 1.2% per year, adding more than a quarter million people daily, providing adequate food becomes an increasingly difficult problem. Conceivably, the numbers of the malnourished will reach 5 billion in a few decades. Reports from the Food and Agriculture Organization of the United Nations and the US Department of Agriculture, as well as numerous other international organizations, further confirm the serious nature of the global food supply. For example, the per capita availability of world cereal grains, which make up 80% of the world's food supply, has been declining for more than two decades. This decline is taking place despite all the current agricultural and biotechnological facilities available.

Malnourished people are more susceptible to numerous diseases, like malaria, tuberculosis, schistosomiasis and AIDS. The World Health Organization reports that there are more than 2.4 billion people infected with malaria, 2 billion infected with tuberculosis, 600 million infected with schistosomiasis and 40 million infected with AIDS (Pimentel et al., 2004).

In this chapter, we will examine the need to increase and make more rational food production, to conserve natural resources, to reduce food (crop) losses to pests, to consider the possibility and benefits of converting annual grains into perennial grains and to consider new crops and innovative minilivestock.

World Energy Resources and Food

Humankind relies on various sources of power for food production, housing, clean water and a productive environment. These sources range from human, animal, wind and water energy, to wood, coal, natural gas and oil sources. Of these, fossil-fuel resources have been most effective in increasing food production (Pimentel and Pimentel, 1996).

It is estimated that about 445 quads (1 quad = 10^{15} BTU; 1 British thermal unit (BTU) = 1055.05585 J) from fossil fuel and renewable energy sources are used worldwide each year for all human needs. In addition, about 50% of all solar energy captured by photosynthesis is incorporated into biomass worldwide and is used by humans. Although this amount of biomass energy is very large (600 quads year^{-1}), it is inadequate to meet the food needs of a rapidly increasing population (Pimentel *et al.*, 1999). To compensate, about 384 quads of fossil energy (oil, gas and coal) are utilized each year worldwide (Pimentel *et al.*, 1999). Of this amount about 100 quads of fossil energy are utilized in the USA, with an estimated 19% for just the food system (Pimentel *et al.*, 2008). Annually, the US population utilizes more than twice as much fossil energy as all the solar energy captured by all harvested crops, forest production and all other vegetation (100 quads consumed versus about 50 quads collected by biomass).

The current high rate of energy expenditure throughout the world is directly related to many factors, including rapid population growth, urbanization and high resource-consumption rates. Indeed, fossil energy use has been increasing at a rate faster than the rate of growth in world population. From 1970 to 1995, energy use has been doubling every 30 years, whereas the world population has doubled every 40–50 years. Future energy use is projected to double every 32 years, while the population is projected to double in 58 years.

The overall projections of the availability of fossil energy resources for agricultural production are discouraging because of the limits of fossil energy. The world supply of oil and natural gas is projected to last 40–50 years (Duncan, 2001). Coal supplies could last 50–100 years. These estimates are based on current consumption rates and current population numbers.

Vital Cropland and Freshwater Resources

More than 99.9% of the human food supply (calories) comes from the land, and less than 0.01% from oceans and other aquatic ecosystems (FAO, 2004). At a time when food production should be increasing to meet human nutrition needs, the per capita availability of world cropland has declined 20% during the past decade (Worldwatch Institute, 2001). Annually, more than 10 million ha of valuable cropland are degraded and lost because of wind and water erosion of soil (Preiser, 2005), plus an additional 10 million ha are abandoned due to severe salinization (Thomas and Middleton, 1993).

Furthermore, approximately 75 billion t of topsoil is being washed and blown away each year (Wilkinson and McElroy, 2007). Soil is being eroded from cropland at rates that range from $10 t\ ha^{-1}\ year^{-1}$ in the USA and Europe to about $30 t\ ha^{-1}\ year^{-1}$ in Africa, South America and Asia (Pimentel, 2006). Loss of soil is insidious, for example, during a single night one rain or wind storm can remove 1 mm of soil, which is equivalent to nearly 14 t of soil and if due to sheet erosion, this loss of soil may not readily be apparent.

The world's valued forests are being removed to replace lost cropland (Pimentel *et al.*, 2006). Globally, an average of only 0.23 ha of cropland per capita is available. In contrast, 0.5 ha per capita is available to support the diverse food system of the USA and Europe (Pimentel and Wilson, 2004). In the USA, cropland now occupies 17% of the total land area, but little additional cropland is suitable for future expansion of US agriculture (USDA, 2003).

Adequate freshwater, which supports the very survival of every plant and animal on earth, is in short supply globally. A human requires slightly more than 1 l of drinking water each day. In contrast, in the USA, to produce the food to feed a human each day requires more than 1600 l of water (Pimentel *et al.*, 2004). It is perhaps not surprising that more than 70% of all available freshwater is used in world agriculture (UNESCO, 2001). For example, to produce 1 ha of maize requires 5 million l of water (more than 500,000 gallons $acre^{-1}$ (1 gallon = 3.7853 l; 1 acre = 0.4047 ha; Pimentel *et al.*, 2004). As populations continue to increase, more water will be consumed and water conflicts within, and between, countries will escalate (Pimentel *et al.*, 2004).

Crop Losses to Pests and the Development of Genetically Engineered Crops

Crop losses to pests are one of the most serious problems facing world food security today. Pests are now destroying more than 40% of potential food crop production, despite the application of 3 billion kg of pesticides each year. Major emphasis has been placed on the use of genetically engineered/modified organisms (GMOs) to help reduce losses of food crops to insects, plant pathogens and weeds (see Chapters 2 and 7, this volume). However, while attempting to reduce crop losses to pests, we must exert great care to avoid releasing species that may be invasive pest species.

Genetic engineering of crops started approximately 20–25 years ago and initially focused on the control of pests, with the priority on insects and weeds (NAS, 1984; Malcolm, 2006). This was an appropriate focus because there already existed some genetic systems, such as the insecticidal proteins from *Bacillus thuringiensis* (*Bt*), whose encoding genes could be inserted into crops with minimal change and effort. Genetic engineering differs from regular plant breeding because genetic material from organisms unrelated to the crop plant can be inserted into the genetic make-up of the crop (Biotechnology, 2006). Plant breeding, on the other hand, alters the genetics of the target crop through a regular selection process. This regular selection of crops has worked extremely well in the past, providing effective control of hundreds of pests, for example, control of cabbage yellows and the Hessian fly, an insect pest of wheat.

As with all technologies, including genetic engineering, some have proven highly effective, while others have had negative impacts on public heath and the environment. These various beneficial and negative impacts of genetic engineering technologies will be examined in detail below. However, before discussing the benefits and costs of genetic engineering, it would be appropriate to discuss world pest control and associated problems.

World pest control

Approximately 70,000 species of pests exist in the world, but of these, only 10% are considered serious pests (Pimentel, 1997). Included in this total are approximately 10,000 species of insects and mites, 50,000 species of plant pathogens and about 10,000 species of weeds. In the USA, about 60% of the insect and mite species present have moved from feeding on native vegetation to feeding on introduced crops. Concerning plant pathogens and weeds, most of these pests have been introduced as invasive species.

Despite the use of more than 3.0 billion kg of pesticides applied worldwide (Table 3.1) at an annual cost of US$40 billion, pest insects, plant pathogens and weeds continue to destroy more than 40% of the potential world food production (Pimentel, 1997). Pre-harvest crop losses are approximately 14% for pest insects, 13% for crop diseases and about 13% for weeds. After harvest, another 20% of the food is lost to another group of pests, including insects and microbes (Pimentel, 1997). In the USA, the postharvest losses are estimated to be about 10%.

The opportunities for improved pest management are apparent when we note that insecticide use in the USA grew more than tenfold from 1945 to

Table 3.1. Estimated annual pesticide use.

Country/region	Pesticide use (10^6 t)
USA	0.5
Canada	0.2
Europe	1.0
Other, developed	0.5
Asia, developing	0.3
China	0.2
Latin America	0.2
Africa	0.1
Total	3.0

date, while losses to insects in US crops increased from 7% in 1945 to 13% to date (Pimentel, 1997). For some crops, like maize, losses to insects increased from 3.5% in 1945 to 12% in 2000, despite a more than 1000-fold increase in insecticide use (Pimentel *et al.*, 2000). Maize production is now the largest user of insecticides in the USA. These increased insect problems (primarily the corn-rootworm complex) are due to the planting of more than half of the maize crop without crop rotations (Pimentel, 2000a,b).

The applications of insecticides, herbicides and fungicides are estimated to reduce pest losses by approximately 20–57%, if no alternatives were available (NAS, 2000). If non-chemical alternatives were employed, the losses would range from 10% to 30%. Various non-chemical controls, such as plant breeding for host-plant resistance and biological pest control, are estimated to provide approximately 10–30% reduction in losses of crops to pests worldwide (D. Pimentel, Ithaca, 1991, unpublished data).

Total costs of world crop losses to pests are estimated to be US$500 billion per year (Jefferson and Porceddu, 2004). The funds spent for pesticide application are estimated to be US$40 billion per year worldwide (Pan-UK, 2003). Thus pesticides, when properly applied, return approximately US$4 per dollar invested (Pimentel, 1997). This does not take into account the negative impacts of pesticides on public health and the environment that total about US$12 billion per year in the USA alone (Pimentel, 2005). In the USA, for example, for every dollar of benefits in pest control with pesticides, there is an associated and additional public health and environmental cost of 30%.

Public health and environmental impacts of pesticides

Human poisonings and their related illnesses are clearly the highest price paid for pesticide use. Worldwide, an estimated 26 million suffer from pesticide poisonings each year; approximately 3 million are poisoned seriously enough to be hospitalized and about 220,000 severely enough to prove fatal (Richter, 2002).

The situation is especially serious in developing countries, even though these nations only utilize an estimated 20% of the total pesticides applied in the world (Pimentel and Lehman, 1993). A high pesticide poisoning to death ratio occurs in developing countries, where there tend to be inadequate occupational safety standards, protective clothing and washing facilities; insufficient enforcement of safety regulations; poor labelling of pesticides; illiteracy; and insufficient knowledge of pesticide hazards.

Questionable genetic engineering technologies

Genetically engineering microbial genes in crops for insect biocontrol may not be as useful in some cases as using the microbe directly as a biological control organism. For example, the nuclear polyhedrosis virus is highly pathogenic to the cabbage looper and need not be genetically engineered for use. The cabbage looper pest population can be controlled simply by placing five infected loopers in 400 l of water and spraying this concoction over a hectare of crop plants (D. Pimentel, Ithaca, 1991, unpublished data).

Long-term human consumption and various other data have demonstrated that consuming this natural virus on human foods, which is highly specific for the cabbage looper, is clearly not a threat to humans and other mammals (D. Pimentel, Ithaca, 1991 unpublished data).

Insecticidal Molecules for Crop Protection

Various toxic chemicals already exist in some food crops, for example cyanide in cassava and clover, alkaloids in potatoes and animal blood thinners in clovers, all of which can kill animals if sufficient quantities of the plant are consumed. Some of these toxins can be used for insect pest and plant pathogen control in trees and shrubs, but not in human foods or forages utilized by domestic livestock (Culliney *et al.*, 1992).

Use of *Bt* genes in grain maize for control of the corn borer

The corn borer causes about 10% losses in maize and is generally not treated with insecticides (Pimentel and Raven, 2000). The reasons for seldom treating maize for control of this insect pest are first, that seldom does the corn borer cause significant losses in maize (Pimentel, 2000a,b), and second, insecticide treatments have to be perfectly timed to control the corn borer. Once the corn-borer eggs hatch and the young larvae bore into the maize, then it is extremely difficult to control the pest because they cannot be reached by an insecticide application.

Expression of insecticidal genes from *B. thuringiensis* in grain maize has conferred advantages because the maize plant is now resistant to corn-borer caterpillar larvae. However, such material is costly for the farmer to plant because the grain cannot be used for human consumption but can only be used as livestock feed (Biotechnology Issues, 2005). A few years ago, by mistake, *Bt* maize was mixed in with maize targeted for use as human food. This resulted in US$6 billion in maize that could not be used for human food and thus could only be used as livestock feed. About 78% of maize is fed to livestock in the USA. It is perhaps noteworthy that *Bt* sweetcorn (developed by Syngenta and commercialized in the USA in 1998), as opposed to grain maize, is consumed by humans.

Another problem with *Bt* maize is the potential for pollen drift to organic maize; this has resulted in the contamination of maize that could not subsequently be sold as organic maize (Pimentel and Raven, 2000). In addition, there is also the potential for *Bt* maize pollen to drift to non-target plants so exposing beneficial insects and protected butterflies, like the Monarch butterfly and the blue karela butterfly (Pimentel and Raven, 2000; see also Chapters 8 and 9, this volume). Various claims and counter claims were made as to the seriousness of the impacts of *Bt* maize pollen on the Monarch butterfly and other endangered butterflies and moths (Gatehouse *et al.*, 2002). However, no one argued that killing Monarch or karela butterflies offers any advantages.

The difficulty in assessing the impacts of *Bt* maize on the environment rested with the fact that some investigators were using different strains of *Bt* maize, and that some of these maize genotypes were more toxic than others (Pimentel and Raven, 2000). For example, some of the *Bt* maize strains were highly toxic, whereas others were nearly non-toxic. In addition, different butterfly species, like the Monarch and swallow-tail butterflies, were used in the tests (Pimentel and Raven, 2000). It was generally concluded that the use of insecticides was having a greater impact on the Monarch butterflies than the *Bt*-contaminated maize pollen. However, the Monarch butterflies are suffering from pesticides and loss of natural habitat in Mexico and the USA, and the potential for losses due to *Bt*-contaminated maize pollen would only add to these additional negative impacts on the Monarch butterflies. *Bt* maize residues have shown effect on some detritivores that normally live on agroecosystems such as Enchytraeidae and earthworms *Lumbricus terrestris* (Zwahlen *et al.*, 2003; Frouz *et al.*, 2004). Carbon and nitrogen isotopes in engineered *Bt* maize and isogenic maize have been shown to have different nuclides ratios in vegetal materials and isopods fed on corn residues (Rossi *et al.*, 2006).

There have been claims about the benefits of *Bt* maize and other genetically modified (GM) crops reducing the need for pesticides. This is true for GM cotton. However, some studies in China have documented that pesticide use with GM cotton was just as large as with non-GM cotton (Connor, 2006; Wang and Just, 2006; for a comprehensive review of GM crops in China, see Chapter 16, this volume). In addition, farmers were paying three times the price of conventional cotton seed to purchase GM cotton seed (Connor, 2006; Wang and Just, 2006). An additional problem could be the cotton seed oil produced in places such as Australia that is used for human consumption.

Biotechnology and Crops

An estimated 100 million ha in the world are planted to transgenic crops (James, 2007). The USA accounts for 57.7 million ha (50% of global biotech area), spurred by a growing market for ethanol. Although the acreage of GM crops in developing/transition countries has dramatically increased since 2004, this has been mainly due to the cultivation of GM cotton. Currently, the developing countries that need help in increasing food production are benefiting very little from GMOs (Altieri, 2001).

In addition, the developed nations are not benefiting in terms of crop yields because 75% of the GM crops being grown are herbicide-resistant crops (Pew, 2004). Herbicide-tolerant crops, in general, do not increase crop yields but increase the use of herbicides (Benbrook, 2003). The reason that the chemical companies that own most of the seed companies are pushing herbicide-resistant crops is to sell more herbicide to benefit the chemical companies.

Perceived Risks of GMOs

Genetic engineering and herbicide-resistant crops

Most of the research dealing with GMOs focuses on herbicide resistance in crops, with an estimated 54% of the research involved with herbicide resistance (Paoletti and Pimentel, 1996, 2000). Herbicide resistance generally does not increase yields, it generally increases herbicide (pesticide) use and pollution of the environment (Institute of Science in Society, 2003). It would be of greater help to farmers if the focus were on the development of pest-resistant crops rather than pesticide (herbicide)-resistant crops (Paoletti and Pimentel, 1996).

Several crops, including soybeans and maize, were selected for herbicide tolerance and these two crops occupy most (70%) of the land currently cultivated with GM crops (Reuters, 2002; Knezevic, 2006). The advantage of this technology is that the crop area can be heavily treated with some toxic herbicide to destroy all weed plants in the crop field. The resulting 'clean culture' has benefits to the crop but negative impacts of increasing soil erosion and rapid water runoff (Pimentel, 2006). In some instances, an estimated 25% of the farmers use this technology along with no-till conservation tillage practices (Council for Biotechnology Information, 2006). No-till can significantly reduce soil erosion, but 'clean culture' using herbicide-resistant crops tends to significantly increase soil erosion.

Herbicide tolerance in crops results in the increased use of herbicides and increased pollution of surface water and groundwater resources (Institute of Science in Society, 2005). About 74% of the 500 million kg of pesticides used in the USA are herbicides (About Pesticides, 2001). Herbicides were detected more frequently and at higher concentrations than insecticides in the US Geological Survey (USGS; Larson et al., 1999). Overall, 11 herbicides and three insecticides were detected in more than 10% of the water samples. Herbicides are now the number one pesticide pollutant in surface water and groundwater resources (Larson et al., 1999). Pesticide concentrations in drinking water rarely exceeded the acceptable standards for drinking water, but some of the pesticides frequently exceeded the criteria established for the protection of aquatic organisms.

Herbicide pollution is also a threat to public health. Herbicides in dust and breathed by humans have been documented by Mary Ward of the National Cancer Institute (Raloff, 2006). Herbicides are not only a threat to animals, but also to non-target crops and natural vegetation. The estimate is that herbicide damage to non-target crops is more than US$1.4 billion per year in the USA (Pimentel, 2005).

Herbicide impacts on various crops also include increasing insect pests and plant pathogen pests. For instance, when the herbicide 2,4-D was applied to maize at recommended dosages, the corn leaf aphid and corn borer populations both increased twofold to threefold higher than on untreated maize (Oka and Pimentel, 1976). In addition, corn smut disease increased nearly fivefold on the treated maize, and maize resistance to the southern leaf blight was lost in maize treated with 2,4-D (Oka and Pimentel, 1976). There are numerous other examples of where recommended dosages of herbicides actually increased insect pests and plant pathogens (Pimentel, 1994).

Redesigning crops to harvest fuel

Ethanol has become a major push in agriculture (Pimentel and Patzek, 2005). One of the priorities is to modify crops so that they have reduced amounts of lignin (Pollack, 2006). This approach might increase the amount of starches and sugars present in various plants but may well turn out to be a disaster because the lignin is required by the plant to help prevent lodging.

Health risks of GMOs

The insertion of novel genes into crops, such as a gene from the Brazil nut to enhance the nutritional content of the soybean, can create various health problems in people who are allergic to various plant proteins, including those of nuts (Embar, 2006). Several investigators have documented that when genetically engineered soybeans were fed to mice and rabbits their cells and enzymatic activities were altered (Malatesta et al., 2002; Vecchio et al., 2004; Tudisco et al., 2006). These studies emphasize the need for careful consideration of what traits are to be expressed and what are the consequences, if any, for human/animal health (see also Chapter 13, this volume).

Release of genetically engineered native organisms

Just because the GMO might be a native organism does not mean that it could not hybridize with other native organisms and develop new races of plants and animals. Adding or deleting a gene from a native species may significantly alter the organism's ecology, including its potential for increased pathogenicity (Pimentel et al., 1989). The safest procedure may be to use an organism from a region with very different climatic conditions, for example, engineering an organism from the tropics for subsequent use in the northern part of the USA, as it would have a low probability of surviving the winter and upsetting the ecosystem.

Genetic engineers have experimentally produced drugs and other chemicals by manipulating the genetic make-up of various plants (Caruso, 2007). There are several risks associated with this technology including contaminating the food supply. The best approach would be to produce these possibly hazardous materials in secured greenhouses where insects and pollen cannot invade or escape.

Risks of introducing genes into crops that might subsequently become pests

An estimated 50,000 species of plants, animals and microbes have been introduced into the USA and some of these species have become serious pests (Pimentel et al., 2000). Some of the serious pests that were introduced include the pathogenic fungus that has all but destroyed the American chestnut tree, the gypsy moth and Johnson grass weed (Pimentel *et al.*, 2000). It has been documented that 128 species of crop plants that were intentionally introduced into the USA have subsequently become serious weed pests (Pimentel, 1995).

Herbicide resistance in weeds

When large quantities of pesticides are used, for example when herbicides are used with herbicide-resistant crops, there is the potential for weeds to evolve resistance to such herbicides.

Pesticide resistance is a serious problem in the USA today; more than 500 species of pests have evolved resistance to pesticides in the USA, including weeds (Pimentel, 2005). When pests develop resistance, it results in greater application rates as well as newer, often more toxic and more expensive, pesticides being applied. Some unintentional results may include increased crop losses, destruction of beneficial organisms, destruction of non-target crops and increased pollution of the environment. The total environmental costs per year in the USA totals US$1.5 billion (Pimentel, 2005).

Perennial Grains and GMOs

As mentioned, more than 99% of all food comes from agriculture and less than 1% from the oceans and other aquatic ecosystems (FAO, 2004). An estimated 99% of all grains that make up 80% of all food comes from annual grains. There are serious environmental and agricultural problems associated with current annual grain production, as well as major benefits (Pimentel *et al.*, 1986). The prime environmental concerns with annual grains are: large quantities of energy required in tilling the soil; planting large quantities of grains each year; significant soil erosion and water runoff problems; and the timing of planting to obtain the benefits of seasonal rainfall.

Humans rely on a very narrow species base for their food. About 80% of the foods utilized by humans comes from only 15 species, nine plant species and six livestock types. The crops are rice, wheat, maize, sorghum, millet, rye, barley, common bean, soybean, groundnut, cassava, potato, sweet potato, coconut and banana, and livestock types are cattle, buffalo, sheep, goats, swine and chicken. When one considers that there are perhaps 15 million species of plants, animals and microbes in the world, human primary dependence on only 24 species of organisms for food is surprising.

Early agriculture first started in about 11,000 BC in the Mediterranean Region and the first crops were grains (McCorriston, 2006). These first grains were mostly annuals. Before examining the potential of new crops that might be utilized for food, we should inquire why nine grain crops – rice, wheat, maize, sorghum, millet rye, barley, common bean, soybean and groundnut – now provide about 80% of the plant nutrients consumed by humans. A major advantage of grains is that they are capable of producing large quantities of carbohydrate and protein per hectare. Grains can yield from 1000 to 14,000 kg of grain per hectare in a 3- to 4-month period. A yield of 1500 kg ha^{-1} is sufficient to feed five persons as vegetarians for a year. When cereal grains and legume grains are combined in the correct proportions, not only are human food energy needs well taken care of, but the protein, calcium, iron, vitamins, thiamin and riboflavin needs of adults are also met.

Advantages of perennial grains

There are many energy, cropping and environmental advantages of perennial grains. Genetic engineering offers the opportunity to convert most of the annual grain crops into perennial grain crops. First, perennialism is a problem for some crops, like barley and soybeans (Pimentel et al., 1986), and another major problem is winter hardiness. For example, a maize variety exists that is perennial (H. Ilitis, University of Wisconsin, 2000, personal communication), but maize is a tropical plant. Converting the various grain crops into perennials will be a major challenge, even for genetic engineering; however, it can be accomplished. When perennial grains are mentioned, we suggest perennial crops that could be planted once every 5 or 6 years; a planting programme similar to the programme for the perennial lucerne crop could be followed.

Energy benefits

Just to till the soil and get the land ready for annual grain planting requires about 45 l of diesel fuel or nearly 500,000 kcal ha^{-1} (Pimentel and Pimentel, 1996). For the resource-poor farmer in a developing nation who is required to till the land by hand, perennial crops would offer a tremendous advantage. About 400 h of labour or 2.5 months of labour per hectare are required to hand till the land with a hoe or similar instrument (Pimentel and Pimentel, 1996). This amounts to about 240,000 kcal or about half the energy required to carry out the same tillage using a tractor. However, the drudgery of tilling the soil strictly by hand cannot be emphasized enough.

Cropping benefits

Whether the land is prepared by hand or by mechanization, timing of preparation of the land and planting the crop is critical to rainfall and freezing conditions. The farmer would prefer relatively dry soil conditions for tillage and for planting the seeds of the grain crop. Once planted, the farmer hopes for rainfall to germinate the crop and stimulate its growth. If the rainfall comes early, it is a problem for the farmers because it will be difficult to till the wet soil. Timing of tillage and the planting of the crop are critical in regards to rainfall; a successful grain yield depends on this timing.

   With a perennial grain crop, rainfall and freezing are no longer threats to the farmer's crop because the crop is already in the ground and ready to grow. Another major advantage of perennial grains is that they can begin collecting solar energy immediately in the spring and can also collect and store some solar energy in the fall (autumn) months. This added exposure of the perennial crop to sunlight can nearly double the amount of solar energy captured by the crop during its growing season. This has been illustrated with maize and the planting of cover crops after the maize crop has been harvested. The maize collects the equivalent of 1.0 unit of solar energy and the cover crops collect about 0.8 equivalent in solar energy (Pimentel, 2006). Therefore, nearly double the solar energy can be captured per year using a cover crop, compared with the solar energy captured only by an annual maize crop (Pimentel, 2006).

Soil and water conservation

After the serious problem of the rapid growth in the world's human population, probably the second most significant environmental problem is soil erosion, for several reasons: (i) more than 99% of human food supply comes from the land that is now being degraded by soil erosion (Pimentel, 2006); (ii) approximately 75 billion t of topsoil is being washed and blown away each year (Wilkinson and McElroy, 2007); (iii) soil is being eroded from cropland at rates that range from $10\,t\ ha^{-1}\ year^{-1}$ in the USA and Europe to about $30\,t\ ha^{-1}\ year^{-1}$ in Africa, South America and Asia (Pimentel, 2006); (iv) loss of soil is insidious; for example, during one night a rain or windstorm can remove 1 mm of soil and nearly 14 t of soil. If uniform sheet erosion has occurred, this loss of soil may not be apparent. The presence of perennial grains on the land would serve to conserve vital soil and water resources (Pimentel *et al.*, 1986).

Food and Grain Use in Western Countries

In most developed countries, grains are not directed as human food but provide feed for the meat, milk and egg industries. In the USA, for instance, only 22% of maize is directly consumed by humans and in other countries under rapid growth, like China, the traditional, highly vegetarian diet is soon going to change and, especially in urban areas, imitate the Western diet that emphasizes meat consumption. Changes of food preference imply increased environmental impacts and contamination (Paoletti *et al.*, 1999).

Livestock, especially beef, pork and chickens, need grain, fodder or pasture if intensively grown, and yields are poor. By promoting minor consumption of animal products such as meat, egg and milk, more grains could be available to humans.

Minilivestock

In traditional communities, and especially in the tropics, small, non-conventional animals have largely been adopted as human food; these will be referred to as minilivestock. For instance, in addition to small vertebrates, up to 2000 terrestrial edible invertebrates have been listed (Paoletti and Bukkens, 1997; De Foliart, 2005; Paoletti, 2005b).

In most cases, minilivestock are hunted, trapped and either collected in the wild or directly from crops on which they might be pests (for instance some grasshoppers and caterpillars). There are possibilities to develop small-scale production of such minilivestock which would offer a better rate of transformation of feed energy and an additional source of food to grains (Collavo *et al.*, 1995; Cerda *et al.*, 2005; Paoletti, 2005a,b). For instance, housecrickets need only 1.7 kg feed to produce 1 kg of meat (Collavo *et al.*, 2005) and to produce this amount only a few square metres of land are needed. Crickets, palmworms and caterpillars are among the promising and more sustainable, complementary resources (De Foliart, 2005).

Food Biodiversity

Traditional knowledge of biodiversity

The limited number of crop (10–15) and animal (5–8) species that provide the food base for humans on the planet has been suggested as a major limitation to food security. This, along with the large-scale cultivation of monocultures and inappropriate technologies, threatens natural resources including soil, biodiversity, energy and the environment. In different parts of the world there exist diversified models of agriculture and subsistence horticulture that emphasize diversity and plurality of organisms to be targeted as resources (Paoletti, 2005a). In some cases these models are more sustainable than just the monoculture of a few crops that rely on high inputs of energy. Different sustainable farming systems, including organic, integrated and precision farming, try to reduce unwanted inputs and negative environmental effects; however, much work and research is needed to achieve these goals. It is a priority to improve farming systems and explore all existing options.

Loss of biodiversity and knowledge

A reduction in the diversity of food favoured by consumers has been a consequence of urbanization in many countries, especially in the developed world (Paoletti, 1999). Most consumers do not regard availability of foods as a problem, and are not interested in how it is produced. A reduction in food diversity is not viewed as a problem by consumers, who prefer a constant supply of a limited range of products over a diverse range of foodstuffs variable in quality and supply. The high cost of production of foodstuffs favoured by consumers, which involve transportation, greenhouse maintenance and heating out of season, and chemical inputs to maintain monocultures, is not considered by the consumer. Greater education is needed to make consumers aware of the value of food diversity, and to improve its acceptability globally, in order to achieve more sustainable food production practices.

Acknowledgement

David Primentel is grateful that this research was supported in part by the Podell Emeriti Award at Cornell University.

References

About Pesticides (2001) *US Environmental Protection Agency 2000–2001 Pesticide Marketing Estimates: Usage.* EPA, Washington, DC.

Altieri, M.A. (2001) The ecological impacts of agricultural biotechnology. Available at: http://www.actionbioscience.org/biotech/altieri.html

Benbrook, C.M. (2003) Impacts of genetically engineered crops on pesticide use in the United State: the first eight years. Bio Tech InfoNet, Technical Paper No. 6.

Biotechnology (2006) Learning center. Available at: http://www.childrensmuseum.org/biotech/faqus.htm

Biotechnology Issues (2005) Maize. Available at: http://stockholm.usembassy.gov/biotech/us-gov3.html

Cerda, H., Araujo, Y., Glew, R.H. and Paoletti, M.G. (2005) Palm worm (coleoptera,*Curcu lionidae*: *Rhynchophorus palmarum*). Traditional food. Examples from Alto Orinoco, Venezuela. In: Paletti, M.G. (ed.) *Ecological Implications of Minilivestock.* Science Publishers, Enfield, New Hampshire, pp. 353–366.

Collavo, A., Glew, R.H., Yunk-Sheng, H., Lu-Te, C., Bosse, R. and Paletti, M.G. (2005) Housecricket small scale farming. In: Paoletti, M.G. (ed.) *Ecological Implications of Minilivestock.* Science Publishers, Enfield, New Hampshire, pp. 519–544.

Connor, S. (2006) Farmers use as much pesticide with GM crops, US study finds. *Independent, UK.* 27 July 2006. Available at: http://www.commondreams.org/cgi-bin/print.cgi?file = headlines06/0727-06.htm

Council for Biotechnology Information (2006) Plant biotechnology improves wildlife habitat, water quality. Available at: http://www.whybiotech.com/index.asp?id = 2185

Culliney, T.W., Pimentel, D. and Pimentel, M. (1992) Pesticides and natural toxicants in foods. *Agriculture, Ecosystems and Environment* 41, 297–320.

De Foliart, G.R. (2005) Overview of the role of edible insects in preserving biodiversity. In: Paoletti, M.G. (ed.) *Ecological Implications of Minilivestock.* Science Publishers, Enfield, New Hampshire, pp. 123–140.

Duncan, R.C. (2001) World energy production, population growth, and the road to the Olduvai Gorge. *Population and Environment* 22(5), 503–522.

Embar, W. (2006) Genetic modification. Vegan Peace. Available at: http://www.veganpeace.com/organic/gmo.htm

FAO (2004) Food balance sheets. Food and Agricultural Organization. United Nations, Geneva.

Frouz, J., Elhottova, D., Ourkova, M. and Kocourek, F. (2004) The effect of Bt-corn on soil invertebrates and decomposition rates of corn post-harvest residues under field and laboratory conditions. *Journal of Sustainable Agriculture* 32(4), 645–655.

Gatehouse, A.M.R., Ferry, N. and Raemaekers, R.J.M. (2002) The case of the Monarch Butterfly: a verdict is returned. *TRENDS in Genetics* 18(5), 249–251.

Institute of Science in Society (2003) GM crops increase pesticide use. Available at: http://www.i-sis.org.uk/GMCIPU.php

Institute of Science in Society (2005) GM crops increase pesticide use. Available at: http://www.i-sis.org.uk/GMOIPU.php

James, C.A. (2007) Global status of commercialized biotech/GM crops: 2007. ISAAA Briefs. Brief 37.

Jefferson, R. and Porceddu, M.C. (2004) Fostering democratic innovation as a means of reducing the 10/90 gap in health.

Available at: http://www.globalforumhealth.
org/Forum8-CDROM/OralPresentations/
Connett-Poceddu%20M.doc

Knezevic, S.Z. (2006) Use of herbicide tolerant crops as a component of an Integrated
Weed Management Program. University
of Nebraska (Lincoln), Institute of Agriculture and Natural Resources. Lincoln,
Nebraska.

Larson, S.J., Gilliom, R.J. and Capel, P.D.
(1999) Pesticides in streams of the
United States – initial results from
national water-quality assessment program. US Geological Survey. Water-
Resources Investigation Report 98-4222,
Sacramento, California.

Malatesta, M., Caporaloni, C., Gavaudan, S.,
Rocchi, M.B.L., Tiberi, C. and Gazzanelli,
G. (2002) Ultrastructural morphometrical
and immunocytochemical analyses of
hepatocyte nuclei from mice fed on genetically modified soybean. *Cell Structure and
Function* 27, 173–180.

Malcolm, L. (2006) The story of rice. Available
at: http://www/abc.net.au/science/slab/rice/
story.htm

McCorriston, J. (2006) Breaking the rain bar-
rier and the tropical spread of Near Eastern
agriculture into Southern Arabia. In:
Kennett, D.J. and Winterhalder, B. (eds)
*Behavioral Ecology and the Transition to
Agriculture*. University of California Press,
Berkeley, California, pp. 217–236.

NAS (1984) *Genetic Engineering of Plants:
Agricultural Research Opportunities and
Policy Concerns*. National Academies of
Science, National Academies Press,
Washington, DC.

NAS (2000) *Benefits, Costs and Contemporary
Use Patterns: Benefits of Pesticides*.
National Academies of Science, National
Academies Press, Washington, DC.

Oka, I.N. and Pimentel, D. (1976) Herbicide
(2,4-D) increases insect and pathogen
pests on corn. *Science* 193, 239–240.

PAN-UK (2003) Current pesticide spectrum,
global use and major concerns. Available
at: http://www.pan-uk.org/briefing/SIDA_Fil/
Chap1.htm

Paoletti, M.G. (1999) Using bioindicators
based on biodiversity to assess landscape
sustainability. *Agriculture, Ecosystems and
Environment* 74, 1–18.

Paoletti, M.G. (2005a) State of Amazon.
Biodiversity management and loss of tradi-
tional knowledge in the largest forest. In:
De Dapper, M. (ed.) *Tropical Forests in a
Changing Global Context*. The Royal
Academy of Overseas Sciences, Brussels,
pp. 93–111.

Paoletti, M.G. (ed.) (2005b) *Ecological
Implications of the Use of Minilivestock.
Insects, Rodents, Frogs and Snails*. Science Publishers, Enfield, New Hampshire,
648 pp.

Paoletti, M.G. and Bukkens, S. (eds) (1997)
Minilivestock. *Ecology of Food and Nutrition*
36(2–4), 95–346.

Paoletti, M.G. and Pimentel, D. (1996) Genetic
engineering in agriculture and the environment. *BioScience* 46(9), 665–673.

Paoletti, M.G., Giampietro, M., Han Chunru,
G., Pastore, S., Bukkens, G.F. and Baudry,
J. (eds) (1999) Agricultural intensification
and sustainability in PR China. *Critical
Reviews of Plant Sciences* 18(3), 257–487.

Pew (2004) Genetically modified crops in the
United States. Pew Initiative on Food and Biotechnology. Available at: http://pewagbiotech.
org/resources/factsheets/crops/

Pimentel, D. (1994) Insect population
responses to environmental stress and pollutants. *Environmental Reviews* 2(1), 1–15.

Pimentel, D. (1995) Biotechnology: environmental impacts of introducing crops and
biocontrol agents in North American agriculture. In: Hokkanen, H.M.T. and Lynch,
J.M. (eds) *Biological Control: Benefits and
Risks*. Cambridge University Press,
Cambridge, pp. 13–29.

Pimentel, D. (1997) *Techniques for Reducing
Pesticides: Environmental and Economic
Benefits*. Wiley, Chichester, UK.

Pimentel, D. (2000a) Genetically modified
crops and the agroecosystem: comments
of 'Genetically modified crops: risks and
promise' by Gordon Conway. *Conservation
Ecology* 4(1). Available at: http://www/
consecol.org/vol4/iss1/art10

Pimentel, D. (2000b) Economic and ecological perspectives for crops in cool wet
regions of the world. In: Parente, G. and

Frame, J. (eds) *Crop Development for the Cool and Wet Regions of Europe: Achievements and Future Prospects. COST Action 814. Proceedings of the Final Conference, Pordenone, Italy 10–13 May 2000.* European Communities, Belgium, pp. 31–41.

Pimentel, D. (2005) Environmental and economic costs of the application of pesticides primarily in the United States. *Environment, Development and Sustainability* 7, 229–252.

Pimentel, D. (2006) Soil erosion: a food and environmental threat. *Environment, Development and Sustainability* 8(1), 119–137.

Pimentel, D. and Lehman, H. (1993) *The Pesticide Question: Environment, Economics and Ethics.* Chapman & Hall, New York.

Pimentel, D. and Patzek, T. (2005) Ethanol production using corn, switchgrass, and wood: biodiesel production using soybean and sunflower. *Natural Resources Research* 14(1), 65–76.

Pimentel, D. and Pimentel, M. (1996) *Food, Energy and Society.* Colorado University Press, Niwot, Colorado.

Pimentel, D. and Raven, P.H. (2000) Bt corn pollen impacts on nontarget Lepidoptera: assessment of effects in nature. *Proceedings of the National Academy of Sciences* 97(15), 8198–8199.

Pimentel, D. and Wilson, A. (2004) World population, agriculture, and malnutrition. *World Watch*, 22–25 September/October.

Pimentel, D., Jackson, W., Bender, M. and Pickett, W. (1986) Perennial grains: an ecology of new crops. *Interdisciplinary Science Reviews* 11, 42–49.

Pimentel, D., Hunter, M.S., LaGro, J.A., Efroymson, R.A., Landers, J.C., Mervis, F.T., McCarthy, C.A. and Boyd, A.E. (1989) Benefits and risks of genetic engineering in agriculture. *BioScience* 39, 606–614.

Pimentel, D., Bailey, O., Kim, P., Mullaney, E., Calabrese, J., Walman, L., Nelson, F. and Yao, X. (1999) Will limits of the Earth's resources control human numbers? *Environment, Development, and Sustainability* 1(1), 19–39.

Pimentel, D., Lach, L., Zuniga, R. and Morrison, D. (2000) Environmental and economic costs of nonindigenous species in the United States. *BioScience* 50(1), 53–65.

Pimentel, D., Berger, B., Filiberto, D., Newton, M., Wolfe, B., Karabinakis, E., Clark, S., Poon, E., Abbett, E. and Nandaopal, S. (2004) Water resources: agricultural and environment. *BioScience* 54, 909–918.

Pimentel, D., Petrova, T., Riley, M., Jacquet, J., Ng, V., Honigman, J. and Valero, E. (2006) Conservation of biological diversity in agricultural, forestry, and marine systems. In Burk, A.R. (ed.) *Focus on Ecology Research.* Nova Science Publishers, New York, pp. 151–173.

Pimentel, D., Gardner, J.B., Bonniefield, A.J., Garcia, X., Grufferman, J.B., Horan C.M., Rochon, E.T., Schlenker, J.L. and Walling, E.E. (2008) Energy efficiency and conservation for individual Americans. *Environment Development and Sustainability.* DOI 10.1007/s10668-007-9128-x

Pollack, A. (2006) Redesigning crops to harvest fuel: scientists as custom tailors of genetics. *New York Times.* Friday, 8 September 2006.

PRB (2006) 2006 World Population Data Sheet. *Population Reference Bureau,* Washington, DC.

Preiser, R.F. (2005) Living within our environmental means: natural resources and an optimum human population. Available at: http://www.dieoff.org/page50.htm

Raloff, J. (2006) Farm fresh pesticides. *Science News.* Available at: http://sciencenews.org/scripts/printthis.asp?clip=%2F20060708%5Ffood%2Easp

Reuters, C.S. (2002) US Farmers look sold on gene-altered crops.

Richter, E.D. (2002) Acute human pesticide poisonings. In: Pimentel, D. (ed.) *Encyclopedia of Pest Management.* Marcel Dekker, New York, pp. 3–5.

Rhodes, F.H.T. (2005) On coming home: reunion with an elderly parent. Presentation 10 June 2005. Cornell University, Ithaca, New York.

Rossi, L., Costantini, M.L. and Brilli, M. (2006) Does stable isotope analysis separate transgenic and traditional corn (*Zea mays* L.) detritus and their consumers? *Applied Soil Ecology.* 35, 449–453.

Thomas, D.S.G. and Middleton, N.J. (1993) Salinization: new perspectives on a major desertification issue. *Journal of Arid Environments* 24, 95–105.

Tudisco, L., Lombardi, P., Bovera, F., d'Angelo, D., Cutrignelli, M.I., Mastellone, V., Terzi, V., Avallone, L. and Infascelli, F. (2006) Genetically modified soya bean in rabbit feeding: detection of DNA fragments and evaluation of metabolic effects by enzymatic analysis. *Animal Science* 82, 193–199.

UNESCO (2001) *Sharing Water Resources.* United Nations Education, Scientific and Cultural Organization, Paris.

USDA (2003) *Agricultural Statistics.* US Department of Agriculture, Washington, DC.

Vecchio, L., Cisterna, B., Malatesta, M., Martin, T.E. and Biggiogera, M. (2004) Ultrastructural analysis of testes from mice fed genetically modified soybean. *European Journal of Histochemistry* 48(4), 449–454.

Wang, S. and Just, D. (2006) Tarnishing silver bullets: Bt technology adoption, bounded rationality, and the outbreak of secondary pest infestations in China. Available at: http://www.grain.org/research/btcotton.cfm?id = 374

Wilkinson, B.H. and McElroy, J. (2007) The impact of humans on continental erosion and sedimentation. *Geological Sciences A Bulletin, January 1* 119(1–2), 140–156.

Worldwatch Institute (2001) *Vital Signs 2001.* Worldwatch Institute, Washington, DC.

Zwahlen, C., Hilbeck, A., Howald, R. and Nentwig, W. (2003) Effects of transgenic Bt corn litter on the earthworm Lumbricus terrestris. *Molecular Ecology* 12, 1077–1086.

II Agricultural Biotechnology: Risks, Benefits and Potential Ecological Impact

4 Environmental Risk Assessment

F.G. Tencalla,[1] T.E. Nickson[2] and M. Garcia-Alonso[3]

[1]Monsanto Europe SA, Brussels; [2]Monsanto Company, St Louis, Missouri, USA; [3]Syngenta, Jealott's Hill International Research Centre, Bracknell, UK

Keywords: Environmental risk assessment, genetically modified (GM) crops, comparative approach, agronomic/phenotypic and compositional/nutritional equivalence, tiered approach

Summary

In many world regions, regulatory frameworks are in place to ensure that all pre-commercial genetically modified (GM) crops are evaluated for potential impacts on human health, animal health and the environment according to established standards of risk assessment and current scientific knowledge before authorizations for import or planting are granted. The environmental risk assessment for GM crops follows the same fundamental principles as other risk assessment schemes, i.e. risk is a function of hazard and exposure. However, one of the main differences that sets GM crop risk assessment apart is that it is highly dependent on the crop and the introduced trait; hence, a case-by-case approach is required. For many crop/trait combinations, the assessment is based on a comparison with an appropriate conventional non-GM crop. If agronomic/phenotypic and compositional/nutritional equivalence between the GM crop and its non-GM counterpart is demonstrated, the environmental risk assessment can focus on what is different. For products with no appropriate comparator, further testing or a non-comparative-based evaluation may be required. The goal of the environmental risk assessment is to systematically collect information to support decision making. This is achieved by focusing on end points that are clearly defined and aligned with environmental management goals defined by public policy. A well-constructed risk assessment should follow a logical progression or 'tiered approach', where all information available at a given time is gathered and assessed to determine what, if any, additional data must be collected to reach satisfactory risk conclusions. The risk assessment provides regulators with information that allows them to make knowledge-based decisions about the GM crop. Final authorizations for commercialization, whether for import or cultivation, take into account the outcome of the environmental risk assessment, a formal assessment of food and feed safety, and in certain cases also consider political, economic and societal factors. Although the details of the risk assessment frameworks for GM crops vary from country to country, the general principles upon which they are based are comparable. Since 1996, over 100 GM crop/trait combinations have been placed on the market without negative environmental impacts, demonstrating the robustness of existing frameworks. This chapter reviews the main principles and regulatory aspects of the environmental risk assessment of GM crops.

> *The grand aim of all science is to cover the greatest number of empirical facts by logical deduction from the smallest number of hypotheses or axioms*
>
> Albert Einstein

Agriculture and Environmental Impact, Setting the Stage for Environmental Assessment of Biotechnology-derived Crops

Agriculture emerged as the major form of food production in the 'fertile crescent' of the Near East around 10,500 years ago. Prior to the 20th century, land seemed limitless and the proportion dedicated to agriculture was relatively small. When land was exhausted and no longer useful for farming, old fields were abandoned and new fields were created without much concern about issues such as biodiversity, conservation or erosion. However, as the human population exploded and life expectancies increased during the 19th century in parallel with revolutionary developments in technology and public health, the amount of land converted to food production rapidly grew and food production per unit area struggled to keep up with demand.

Agricultural land today covers approximately 38% of all available land surfaces (Ammann, 2003). Whether practised extensively or intensively, agriculture has a profound impact on the environment (ACRE, 2007; Sanvido et al., 2007). These impacts became the subject of much concern after the Second World War through growing awareness and a heightened environmental ethic. Concepts such as low-input and 'sustainable' agriculture became the focus of much public policy around the world. The goal now was to improve the productivity of agricultural systems, and at the same time markedly reduce adverse ecological implications. Older technologies and practices needed to be replaced with newer, more environmentally benign ones, but with regard for the impacts on economics and societies at large.

From the very beginning of plant domestication, humans have been selecting plants with desirable traits and using hybridization to improve crop performance. Mendel's revolutionary studies in plant genetics in the mid-1800s paved the way for modern plant breeding. In the 1960s, major advances were made to secure more sustainable food supplies, as progress in cellular genetics and cell biology contributed to what is known as the 'green revolution', resulting in significantly increased varieties of food crops containing traits for higher yield and pest or disease resistance worldwide. The emergence of molecular biology in the 1970s allowed the analysis of genetic sequences and the identification of genetic markers for desired traits. This was the birth of the next revolution in crop improvement, molecular-based (e.g. marker-assisted) breeding, which is the basis of many current conventional breeding strategies. The application of recombinant DNA techniques (e.g. genetic modification) to plants in the 1980s provided another tool to complement traditional methods by allowing faster and more precise breeding with specific genes and traits, thereby enabling plant breeders to make new improvements using modern biotechnology that were not possible using traditional methods.

Today, genetically modified (GM) crops, also known as 'biotech' crops, afford great potential to address world food demand through increased yield per unit area in a more environmentally sound manner (Borlaug, 2000). GM crops however are engulfed in controversy, which has led policy makers to regulate these products more strictly than crops developed with other methods. Prior to introduction into the environment, each GM crop must undergo a comprehensive risk assessment designed to inform decision makers about potential risks for humans, animals and the environment before authorizations for commercialization are granted.

This chapter will present some of the basic principles underlying the environmental risk assessment of GM crops. While risk assessment usually comprises both a human/animal health and an environmental aspect, here we will focus on issues relating to the impact of such crops on the environment.

Regulatory guidance related to the environmental risk assessment for GM crops

Generally, newly developed commodity crop varieties and the foods derived from them are not subject to a specific environmental risk assessment before commercialization[1]; in many countries, only variety registration (which focuses on agronomic characteristics) is required. However, as a result of discussions following the production of the first recombinant DNA in the 1970s (Fredrickson, 1979; Talbot, 1983), plants obtained through genetic modification are treated differently.

The first regulations and draft guidance for the risk assessment of GM organisms, including GM crops and the foods derived from them, were drawn up in the mid-1980s, approximately a decade before GM crops were initially commercialized (OECD, 1986; US OSTP, 1986). Individual world regions have since established specific pre-market regulatory systems, including an environmental risk assessment, for the import or cultivation of GM crops (Nickson and Fuchs, 1994; Nap *et al.*, 2003; Jaffe, 2004). Currently, guidance on the conduct of the environmental risk assessment exists in the regulations and guidelines of several of these world regions, including Australia (OGTR, 2000), Canada (CFIA, 2004) and the European Union (EU; EFSA, 2004). Furthermore, independent organizations such as Agbios in Canada have developed broadly applicable (non-country-specific) environmental risk assessment guidance for

---

[1] The Canadian regulatory system may be an exception since it is based on 'novelty' rather than the process used to produce the plant.

use in capacity-building projects around the world. This information is readily available on the Internet (http://www.agbios.com/main.php).

At an international level, the Food and Agriculture Organization (FAO) and the World Health Organization (WHO) built upon their work in developing standards, guidance and recommendations for food safety through the Codex process. As a result, the 'Principles of risk analysis for foods derived from GM crops' were published in 2003 (CAC, 2003). Another internationally recognized instrument, the Cartagena Protocol on Biosafety (CPB), also came into force in 2003. While the CPB is not standard-setting, it is the result of negotiations to develop an international convention aimed at the conservation and sustainable use of biodiversity. The focus of the CPB is to ensure the safe transboundary movement of living modified organisms (LMOs) including GM crops (CBD, 2000). The CPB provides only high-level guidance on risk assessment. It requires, among other things, that an environmental risk assessment be conducted prior to the first transboundary movement of LMOs and that the results be made available to all countries involved through a common mechanism, the Biosafety Clearing House.

Environmental risk assessment has both a human/animal health component and an environmental component. The methods used for food and feed risk assessment have been reviewed extensively by several authors, including König *et al.* (2004), Kuiper and Kleter (2003) and WHO (2005), and will not be further developed in this chapter, where the focus is on the environmental aspects.

Principles of environmental risk assessment for GM crops

Risk assessment for GM crops is based on well-established concepts used in other risk assessment schemes (Jaffe, 2004; König *et al.*, 2004), namely that risk is a function of hazard and the likelihood that this hazard will be realized (exposure). As in any risk assessment, an environmental risk assessment for GM crops identifies potential hazards, estimates the likelihood of the hazards being realized (exposure), analyses the potential severity of the consequences of the hazards being realized and ultimately characterizes risk[2] according to the well-known formulation:

Risk = function(hazard, exposure)

Many environmental risk assessment schemes, e.g. for chemicals, biocides or plant protection products, rely on widely accepted methods for hazard and exposure characterization and the ultimate evaluation of risk. The results obtained can be compared with 'safety' thresholds determined by regulatory authorities; risk values below or above given thresholds are considered acceptable, acceptable under specific conditions of management or unacceptable.

---

[2] EFSA (2004) defines hazard as 'the potential of an identified source to cause an adverse effect'. Exposure is the extent of contact of an individual or population to a given hazard source.

For GM crops, establishing uniform methods for the determination of: (i) hazard and exposure; and (ii) 'safety' thresholds have proven to be challenging. The environmental risks potentially posed by a given crop/trait combination are strongly linked to the crop type, the introduced trait and other factors such as the likely receiving environment, the extent of the release and the interactions among these elements. As a result, GM crops are generally evaluated on a case-by-case basis. The data collected provide a 'weight of evidence' allowing a determination of the magnitude of risk.

Early debates on how to structure the risk assessment of GM crops gave rise to additional principles that now help direct the process. Initially, it was unclear where to focus the assessment, but today there is wide scientific consensus that GM crops are very often plants well known to mankind through a history of long use in food, feed and fibre production. The only meaningful difference for many of these plants is that they have been modified using a new process to express specific traits/characteristics. As such, most risk assessment approaches start by comparing the GM crop to an appropriate conventional non-GM counterpart. Experience with the comparator crop serves as a baseline for the environmental risk assessment allowing the evaluation of 'familiarity' (OECD, 1993a; Hokanson *et al.*, 1999; Nickson and Horak, 2006) and 'substantial equivalence', as described by OECD (OECD, 1993b) and FAO/WHO (FAO/WHO, 2000).

The concept of familiarity is based on the fact that, as mentioned above, most GM plants are developed from organisms such as crop plants with a well-known biology and that have been cultivated for centuries. It is therefore appropriate to draw on this previous knowledge and experience and to use the non-GM plant as a comparator in the risk assessment in order to highlight any differences due to the genetic modification (EFSA, 2004; Nickson and Horak, 2006). For most of the major crop species, there are ample literature and data available to provide a context for assessing familiarity. The Organization for Economic Cooperation and Development (OECD, 2006), Canadian Food Inspection Agency (http://www.inspection.gc.ca/english/plaveg/bio/dir/biodoce.shtml) and US Department of Agriculture (http://www.aphis.usda.gov/brs/biology.html) have, for example, developed biology documents on many crops species.

The concept of substantial equivalence is based on the idea that an existing organism used as food or feed with a history of safe use can serve as a comparator when assessing the safety of a GM plant (EFSA, 2004). The GM crop and its non-GM counterpart are compared with regard to molecular, agronomic and morphological characteristics, as well as chemical composition. If no differences other than the intended modification (e.g. a newly produced protein) are found, the GM crop and its comparator can be considered 'substantially equivalent'. The risk assessment can then focus on the intended differences. If consistent differences between the GM plant and its comparator are found which go beyond the primary expected effects of introducing the new gene(s), this may indicate the occurrence of unintended effects (EFSA, 2004). Unintended effects are then assessed with respect to their safety, nutritional impact and environmental implications.

One should be cautious to avoid premature and scientifically unfounded risk conclusions about unintended effects as they relate to the safety of GM crops. While experts acknowledge that there is a potential for unintended effects to occur, these are possible both in the case of conventionally bred and biotechnology-derived crops. As such, they do not necessarily represent hazards (Kuiper et al., 2000; US EPA, 2004). The important distinction is that GM crops undergo a thorough safety assessment prior to approval by regulatory authorities, while conventional commodity crops generally do not.

Structuring an environmental risk assessment

The goal of an environmental risk assessment is to systematically collect information for the purpose of assessing the potential for a GM crop to be hazardous, the likelihood for the hazard to be realized (exposure) and the consequences, should this occur.[3] When planning a risk assessment, it is crucial to remember that the information to be gathered is not for the purpose of basic research but should contribute concretely to decision making. Basic research, as such, is conducted as part of the pre-market development of a new crop/trait combination and in some cases post-commercialization as well. It can also complement risk assessment when it targets important activities such as method validation or relevant experiments that will reduce the uncertainty in the risk assessment.

A well-constructed environmental risk assessment should follow a logical progression, in what is sometimes called a 'tiered approach' (variations of this approach have been described for chemical substances and GM crops by several authors, including Van Leuven, 1996; US EPA, 1998, 1999; Schuler et al., 2000; Nickson and McKee, 2002; Dutton et al., 2003; Wilkinson et al., 2003; Andow and Zwahlen, 2006; Garcia-Alonso et al., 2006; Rose, 2006). In a first step, referred to as 'Tier 0' by Garcia-Alonso et al. (2006) and more commonly known as 'problem formulation' (Raybould, 2006), all the information available on the GM crop is gathered, including data generated by the applicant to fulfil generic country-specific requirements and from published peer-reviewed papers. The information is used to conduct a first evaluation of the potential hazard of the GM crop and the extent of exposure under the proposed conditions of use, focusing the risk assessment on relevant risk areas. Risk conclusions are drawn for those areas where there is sufficient evidence to determine that: (i) the GM crop under evaluation will not pose meaningful hazard; or (ii) there is only negligible exposure. For those areas where the Tier 0

---

[3] In the EU for example, the complementary guidance notes laid down in Commission Decision 2002/623/EC suggest that, when drawing conclusions regarding potential environmental risks, the evaluation should be presented in six distinct steps: identification of the characteristics of the GM plant which may cause potential adverse effects (Step 1), evaluation of the likelihood of occurrence for each adverse effect (Steps 2 and 3), estimation of the risks posed by the GM plant (Step 4), application of risk management strategies (Step 5) and determination of the overall risk of the GM plant (Step 6).

or problem formulation step indicates a need for further investigation, i.e. there is both potential hazard and potential exposure, an experimental programme is designed with specific testing hypotheses and appropriate end points. These must be representative of the management goals or protection objectives defined by public policy (Wolt and Peterson, 2000; Raybould, 2006; Johnson *et al.*, 2007). Based on the testing hypotheses, experiments of increasing complexity (also referred to as increasing 'tiers'), ranging from simple well-controlled laboratory studies under worst-case conditions to larger-scale field trials, are conducted. After each tier of experimentation, an evaluation is made as to whether or not the results are sufficient to draw a risk conclusion and if so the risk assessment can stop.

Determining when enough information is available to come to a risk conclusion with sufficient certainty is one of the most common challenges in the risk assessment of GM crops. The decision depends on the crop/trait combination, the intended use and the amount of information already available. Because baseline information on various aspects of the ecology of agricultural systems is lacking, some scientists are of the opinion that larger-scale field testing is always necessary to understand the ecology and thus predict longer-term risks. However, field testing does not necessarily link to the purpose and objectives of risk assessment. In general, field testing should only be required when data from earlier tiers of the risk assessment do not allow one to draw an acceptable conclusion on the potential risk. Because of the inherent complexity of ecological interactions in the environment, field studies should be planned based on well-defined hypotheses and with clear measurable end points in mind. The results obtained from even the best planned tests are influenced by uncontrollable external factors such as weather or location and therefore may complement information from earlier tiers but cannot provide an answer alone. An example of tiered environmental risk assessment is presented in Chapter 8 (this volume).

Environmental risk assessment end points

Defining measurable and meaningful end points, and the methods to evaluate them, is key to the environmental risk assessment's goal of contributing to decision making. There are two distinct types of end points needed to conduct an environmental risk assessment: assessment end points and measurement end points. The US Environmental Protection Agency defines assessment end points as 'explicit expressions of the actual environmental value that is to be protected, operationally defined by an ecological entity and its attributes' (US EPA, 1998). They are linked to the environmental management goals defined at the start of the risk assessment. Measurement end points are the experimental outcomes, i.e. the results of tests which must be linked to the chosen assessment end points. Selecting representative measurement end points may be challenging in the case of only broadly defined assessment end points, for example 'impact on biodiversity' or 'impact of crop management practices'.

An appropriate risk assessment gathers data that clearly link the measurement end points with the environmental management goals through the assessment end points (Raybould, 2006). The specific experimental details will vary depending on the crop/trait combination and the geographical region being considered. For GM crops, measurement end points may include increased competitiveness, increased fitness, adverse effects on non-target organisms including threatened and endangered species, other adverse effects on biodiversity, increased pathogenicity and adverse effects on biogeochemical cycles/processes (Dale *et al.*, 2002; Conner *et al.*, 2003). In the EU for example, selection of appropriate measurement end points is guided by Directive 2001/18/EC, which requires consideration of:

1. The likelihood of the GM higher plant (GMHP) to become more persistent than the recipient or parental plants in agricultural habitats or more invasive in natural habitats;
2. Any selective advantage or disadvantage conferred to the GMHP;
3. The potential for gene transfer to the same or other sexually compatible plant species under conditions of planting the GMHP and any selective advantage or disadvantage conferred to those plant species;
4. Potential immediate and/or delayed environmental impact resulting from direct and indirect interactions between the GMHP and target organisms, such as predators, parasitoids and pathogens (if applicable);
5. Possible immediate and/or delayed environmental impact resulting from direct and indirect interactions of the GMHP with non-target organisms (also taking into account organisms which interact with target organisms), including impact on population levels of competitors, herbivores, symbionts (where applicable), parasites and pathogens.

Clearly, the challenge confronting risk assessors is to develop reasonable and testable hypotheses that will yield information relevant to assessing the impact of the GM crop compared to the non-GM crop in the context of the relevant end points. Tests are then designed to provide clear answers to these hypotheses (Fig. 4.1).

Data interpretation and appropriate comparators: the importance of a baseline for environmental risk assessment of GM Crops

When evaluating the impact of a given GM crop on any environmental compartment, it is important to consider normal background variation for the parameters measured in an agricultural context (i.e. the 'baseline'). This can be challenging because the agricultural environment has not been widely tested.

In many cases, the impact of factors such as location, meteorological conditions, crop rotation or soil quality on measurement end points is not well understood. From 2000 to 2002, a large-scale trial known as the 'Farm-scale Evaluations' was conducted throughout the UK to analyse the effects of

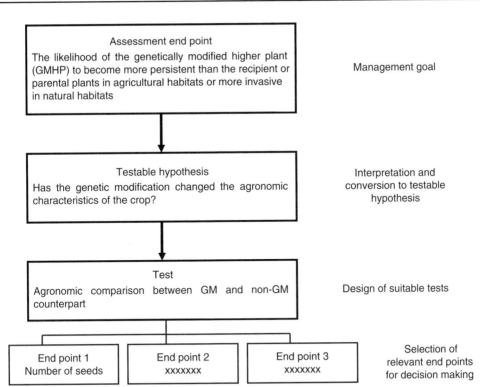

Fig. 4.1. Development of reasonable and testable hypothesis for the risk assessment of GM crops. Example of the end point 'persistence of the GM plant in the environment'.

GM herbicide-tolerant sugarbeet, maize and spring oilseed rape on farmland biodiversity with an emphasis on particular weeds and invertebrates (see Chapter 2, this volume; Squire *et al.*, 2003). This study, which included 273 fields around England, Wales and Scotland, showed that the differences in biodiversity between GM and non-GM fields of the same crop were much smaller than the differences due to the geographical location, the climate, the type of crop (sugarbeet, maize or rape), the cultivation method and the crop rotation strategies (Burke, 2003). Comparable results were found in the frame of the Ecology, Community Organization and Gender (ECOGEN) project which analysed the impact of insect-resistant *Bt* maize on soil organisms over 3 years in Europe (Debeljak *et al.*, 2007; Griffiths *et al.*, 2007b; Krogh *et al.*, 2007).

Often, it is not sufficient to conduct one-on-one comparisons between the GM crop and its non-GM counterpart. Information on a wide range of other non-GM varieties of the same crop should also be collected so that the full spectrum of responses from conventional plants can be taken into account. A systematic means to interpret experimental risk assessment data is critical (Nickson and Horak, 2006), as numerous publications have shown that many of the differences observed in GM crop studies are linked to genetic diversity in the crop variety and not to the newly introduced GM trait (e.g. Reynolds *et al.*, 2005; Griffiths *et al.*, 2007a). Because of the controversy over the proposed introduction of GM crops, much more baseline research is being undertaken and broader knowledge about the ecology of agricultural systems is being developed.

Other factors to consider for environmental risk assessment of GM crops

In most world areas, current agricultural practices including cultivation operations and pesticide applications are taken into account in the environmental risk assessment (Conner *et al.*, 2003). For example, some insect-resistant GM crops are highly specific to the target pests and a limited scope of closely related species. In evaluating the ecological significance of this information, it should be considered in the context of current practice, which may consist of repeated spraying of wide-spectrum insecticides that affect a broad range of target and non-target organisms. Although a certain level of risk to given non-target organisms occurs, this risk could be judged acceptable if it is lower than the risk incurred by the current methods (Candolfi, 2004; Marvier *et al.*, 2007).

Conclusions

With increasing demands on natural resources and food production, the use of new technologies such as GM crops in the agricultural sector has become essential in many world areas. Ensuring the sustainability of agricultural systems includes assessing the potential impact of new agricultural technologies on the environment before commercialization. Environmental risk assessment is a key tool to evaluate potential effects and allow regulators to make informed decisions before granting authorizations for import or cultivation.

Around the world, different countries have established various regulations for import or cultivation of GM crops. These regulations are generally science-based but may in some cases also take into account political, economic and societal factors. Since 1996, over 100 GM crop/trait combinations (http://www.agbios.com/) have been placed on the market without negative environmental impacts, reinforcing the robustness of existing frameworks.

Acknowledgements

The authors wish to acknowledge the contributions of several reviewers: Eric Sachs (Monsanto Company), Alan Raybould (Syngenta) and Patricia Ahl Goy (Syngenta).

References

ACRE (Advisory Committee on Releases to the Environment) (2007) Managing the footprint of agriculture: towards a comparative assessment of risks and benefits for novel agricultural systems. Report of the ACRE sub-group on wider issues raised by the Farm Scale Evaluations of herbicide tolerant GM crops. Available at: http://www.defra.gov.uk/environment/acre/fsewiderissues/pdf/acre-wi-final.pdf

Ammann, K. (2003) Biodiversity and agricultural biotechnology. A review of the impact of agricultural biotechnology on biodiversity. Available at: http://www.biosicherheit.de/pdf/dokumente/ammann_biodiversity.pdf

Andow, D. and Zwahlen, C. (2006) Assessing environmental risks of transgenic plants. *Ecology Letters* 9, 196–214.

Borlaug, N. (2000) Ending world hunger. The promise of biotechnology and the threat of antiscience zealotry. *Plant Physiology* 124, 487–490.

Burke, M. (2003) Summary of the scientific papers published in the Philosophical Transactions of the Royal Society (Biological Sciences). *Philosophical Transactions of the Royal Society (Biological Sciences)* 558, 1775–1889.

CAC (Codex Alimentarius Commission) (2003) Principles for the risk analysis of foods derived from modern biotechnology. CAC/GL 44–2003. Available at: http://www.who.int/foodsafety/biotech/en/codex_biotech_principles.pdf

Candolfi, M. (2004) A faunistic approach to assess potential side effects of genetically modified Bt corn on non-target arthropods under field conditions. *Biocontrol Science and Technology* 14, 129–170.

CBD (Cartagena Protocol on Biosafety) (2000) Cartagena protocol on biosafety to the convention on biological diversity. Available at: http://www.biodiv.org/doc/legal/cartagena-protocol-en.pdf

CFIA (Canadian Food Inspection Agency) (2004) Directive 94-08 (Dir94-08) Assessment criteria for determining environmental safety of plants with novel traits. Available at: http://www.inspection.gc.ca/english/plaveg/bio/dir/dir9408e.shtml

Conner, A.J., Glare, T.R. and Nap, J.P. (2003) The release of genetically modified crops into the environment. Part II. Overview of ecological risk assessment. *The Plant Journal* 33, 19–46.

Dale, P.J., Clarke, B. and Fontes, E.M.G. (2002) Potential for the environmental impact of transgenic crops. *Nature Biotechnology* 20, 567–574.

Debeljak, M., Cortet, J., Demšar, D., Krogh, P.H. and Džeroski, S. (2007) Hierarchical classification of environmental factors and agricultural practices affecting soil fauna under cropping systems using Bt maize. *Pedobiologia,* corrected proof available online 4 June 2007.

Dutton, A., Romeis, J. and Bigler, F. (2003) Assessing the risks of insect resistant transgenic plants on entomophagous arthropods: Bt maize expressing Cry1Ab as a case study. *BioControl* 48, 611–636.

EFSA (European Food Safety Authority) (2004) Guidance document of the Scientific Panel on Genetically Modified Organisms for the risk assessment of genetically modified plants and derived food and feed. *The EFSA Journal* 99, 1–94.

FAO/WHO (Food and Agriculture Organization/World Health Organization) (2000) Safety aspects of genetically modified foods of plant origin. Report of a joint FAO/WHO expert consultation on foods derived from biotechnology, 29 May–2 June 2000. Available at: http://www.fao.org/ag/agn/food/pdf/gmreport.pdf

Fredrickson, D.S. (1979) A history of recombinant DNA guidelines in the United States. *Recombinant DNA Technical Bulletin* 2, 87–90

Garcia-Alonso, M., Jacobs, E., Raybould, A., Nickson, T.E., Sowig, P., Willekens, H., Van der Kouwe, P., Layton, R., Amijee, F., Fuentes, A.M. and Tencalla, F. (2006) Assessing the risk of genetically modified plants to non-target organisms. *Environmental Biosafety Research* 5, 57–65.

Griffiths, B.S., Heckman, L.-H., Caul, S., Thompson, J., Scrimgeour, C. and Krogh, P.H. (2007a) Varietal effects of eight paired lines of transgenic Bt maize and near-isogenic non-Bt maize on soil microbial and nematode community structure. *Plant Biotechnology Journal* 5, 60–68.

Griffiths, B.S., Caul, S., Thompson, J., Birch, A.N.E., Cortet, J., Andersen, M.N. and Krogh, P.H. (2007b) Microbial and microfaunal community structure in cropping systems with genetically modified plants. *Pedobiologia* 51, 195–206.

Hokansen, K., Heron, D., Gupta, S., Koehler, S., Roseland, C., Shantaram, S., Turner, J., White, J., Schechtman, M., McCammon, S. and Bech, R. (1999) The concept of familiarity and pest-resistant plants. *Proceedings of a USDA-APHIS conference on 'Ecological Effects of Pest Resistance Genes in Managed Ecosystems', Bethesda, Maryland, 31 January–3 February 1999.*

Jaffe, G. (2004) Regulating transgenic crops: a comparative analysis of different regulatory processes. *Transgenic Research* 13, 5–19.

Johnson, K.L., Raybould, A.L., Hudson, M.D. and Poppy, G.M. (2007) How does scientific risk assessment of GM crops fit within the wider risk analysis? *Trends in Plant Science* 12, 1–5.

König, A., Cockburn, A., Crevel, R.W.R., Debruyne, E., Grafstroem, R., Hammerling, U., Kimber, I., Knudsen, I., Kuiper, H.A., Peijnenburg, A.A.C.M., Pennincks, A.H., Poulsen, M., Schauzu, M. and Wal, J.M. (2004) Assessment of foods derived from genetically modified (GM) crops. *Food and Chemical Toxicology* 42, 1047–1088.

Krogh, P.H., Griffiths, B., Demšar, D., Bohanec, M., Debeljak, M., Andersen, M.N., Sausse, C., Birch, A.N.E., Caul, S., Holmstrup, M., Heckmann, L.-H. and Cortet, J. (2007) Responses by earthworms to reduced tillage in herbicide tolerant maize and Bt maize cropping systems. *Pedobiologia* 51, 219–227.

Kuiper, H. and Kleter, G.A. (2003) The scientific basis for risk assessment and regulation of GM foods. *Trends in Food Science and Technology* 14, 277–293.

Kuiper, H.A., Kok, E.J. and Noteborn, H.J.P.M. (2000) Profiling techniques to identify differences between foods derived from biotechnology and their counterparts. Joint FAO/WHO expert consultation on foods derived from biotechnology. 29 May–2 June 2000. Available at: http://www.osservaogm.it/pdf/profiling.pdf

Marvier, M., McCreedy, C., Regetz, P. and Kareiva, P. (2007) A meta-analysis of effects of Bt cotton and maize on non-target invertebrates. *Science* 316, 1475–1477.

Nap, J.P., Metz, P.L.J., Escaler, M. and Conner, A. (2003) The release of genetically modified crops into the environment. Part I. Overview of current status and regulations. *The Plant Journal* 33, 1–18.

Nickson, T.E. and Fuchs, R.L. (1994) Environmental and regulatory aspects of using genetically modified plants in the field. In: Marshall, G. and Walters, D. (eds) *Molecular Biology in Crop Protection.* Chapman & Hall, London, pp. 246–262.

Nickson, T.E. and Horak, M.J. (2006) Assessing familiarity: the role of plant characterization. *Proceedings of the Ninth International Symposium on the Biosafety of Genetically Modified Organism.* Jeju Island, Korea, 24–29 September 2006, pp. 74–78.

Nickson, T.E. and McKee, M.J. (2002) Ecological assessment of crops derived through biotechnology. In: Thomas, J.A. and Fuchs, R.L. (eds) *Biotechnology and Safety Assessment,* 3rd edition. Academic Press, Amsterdam, The Netherlands, pp. 233–252.

OECD (Organization for Economic Cooperation and Development) (1986) Recombinant DNA safety considerations. Available at: http://www.oecd.org/dataoecd/45/54/1943773.pdf

OECD (Organization for Economic Cooperation and Development) (1993a) Safety considerations for biotechnology: scale-up of crop plants. Available at: http://www.oecd.org/LongAbstract/0,3425,en_2649_201185_1958517_1_1_1_1,00.html

OECD (Organization for Economic Cooperation and Development) (1993b) Safety evaluation of foods derived by modern bio-technology: concepts and principles. Available at: http://www.oecd.org/dataoecd/57/3/1946129.pdf

OECD (Organization for Economic Cooperation and Development) (2006) Safety assessment of transgenic organisms: OECD consensus documents (1 and 2), 823 pp. Available at: http://www.oecd.org/document/15/0,3343,en_2649_34387_37336335_1_1_1_1,00.html

OGTR (Office of the Gene Technology Regulator, Australia) (2000) Gene technology Act 2000. Available at: http://www.frli.gov.au/ComLaw/Legislation/ActCompilation1.nsf/0/0A2F 6253DBF1CBE7CA257313000E 8674/$file/GeneTechnology2000_WD02.pdf

Raybould, A. (2006) Problem formulation and hypothesis testing for environmental risk assessment of genetically modified crops. *Environmental Biosafety Research* 5, 119–125.

Reynolds, T.L., Nemeth, M.A., Glenn, K.C., Ridley, W.P. and Astwood, J.D. (2005) Natural variability of metabolites in maize grain: differences due to genetic background. *Journal of Agricultural and Food Chemistry* 53, 10061–10067.

Rose, R.I. (2006) Tier-based testing for potential effects of proteinaceous insecticidal plant-incorporated protectants on non-target arthropods in the context of regulatory risk assessments. *IOBC WPRS Bulletin* 29, 145–152.

Sanvido, O., Romeis, J. and Bigler, F. (2007) Ecological impacts of genetically modified crops: ten years of field research and commercial cultivation. *Advances in Biochemical Engineering/Biotechnology* 107, 235–278.

Schuler, T., Poppy, G.M. and Denholm, I. (2000) Recommendations for assessing effects of GM crops on non-target organisms. *British Crop Protection Council Conference: Pest and Diseases* 3, 1221–1228.

Squire, G.R., Brooks, D.R., Bohan, D.A., Champion, G.T., Daniels, R.E., Haughton, A.J., Hawes, C., Heard, M.S., Hill, M.O., May, M.J., Osborne, J.L., Perry, J.N., Roy, D.B., Woiwod, I.R. and Firbank, L.G. (2003) On the rationale and interpretation of the Farm

Scale Evaluations of genetically modified herbicide-tolerant crops. *Philosophical Transactions of the Royal Society, London B* 358, 1779–1799.

Talbot, B. (1983) Development of the National Institute of Health guidelines for recombinant DNA research. *Public Health Reports* 98, 361–368.

US EPA (United States Environmental Protection Agency) (1998) Guidelines for Ecological Risk Assessment. US Environmental Protection Agency, Risk Assessment Forum, Washington, DC, 175 pp. Available at: http://oaspub.epa.gov/eims/eimsapi.dispdetail?deid = 12460

US EPA (United States Environmental Protection Agency) (1999) Draft ECOFRAM Aquatic Report. In: Hendley, P. and Giddings, J. (eds) Available at: http://www.epa.gov/oppefed1/ecorisk/aquareport.pdf

US EPA (United States Environmental Protection Agency FIFRA) (2004) FIFRA Scientific Advisory Panel Report No. 2004–05, 8–10 June 2004. Available at: http://www.epa.gov/scipoly/sap/meetings/2004/june/final1a.pdf

US OSTP (United States Office of Science and Technology Policy) (1986) Coordinated framework for regulation of biotechnology. *Federal Register* 51, 23302–23350.

Van Leuven, C.J. (1996) Procedures of hazard and risk assessment. In: Van Leuven, C.J. and Vermeire, T.G. (eds) *Risk Assessment of Chemicals: An Introduction.* Kluwer, Dordrecht, The Netherlands, pp. 293–333.

Wilkinson, M.J., Sweet, J.B. and Poppy, G. (2003) Preventing the regulatory log jam; the tiered approach to risk assessments. *Trends in Plant Science* 8, 208–212.

WHO (World Health Organization) (2005) Modern food biotechnology, human health and development: an evidence-based study. Available at: http://www.who.int/foodsafety/publications/biotech/biotech_en.pdf

Wolt, J.D. and Peterson, R.K.D. (2000) Agricultural biotechnology and societal decision-making: the role of risk analysis. *AgBioForum* 3, 291–298.

5 Insect Resistance to Genetically Modified Crops

B.E. Tabashnik and Y. Carrière

Department of Entomology, University of Arizona, Tucson, Arizona, USA

Keywords: Evolution, resistance, *Bacillus thuringiensis*, transgenic crops, genetically engineered crops

Summary

Crops genetically modified to produce crystal (Cry) toxins from *Bacillus thuringiensis* (*Bt*) for insect control can reduce reliance on conventional insecticides. Evolution of resistance to *Bt* toxins by insect populations is the primary threat to the continued success of this approach. Resistance of lepidopteran insects to *Bt* toxins in the Cry1A family commonly entails recessive inheritance and reduced toxin binding to midgut membrane target sites. Analysis of more than a decade of data from studies monitoring resistance to *Bt* maize and *Bt* cotton shows that field-evolved resistance was detected in some US populations of *Helicoverpa zea*, but not in populations of five other major lepidopteran pests from Australia, China, Spain and the USA: *Helicoverpa armigera*, *Heliothis virescens*, *Ostrinia nubilalis*, *Pectinophora gossypiella* and *Sesamia nonagrioides*. The resistance of *H. zea* to the Cry1Ac toxin in *Bt* cotton has not caused widespread crop failures, in part because insecticide sprays and two-toxin cotton producing Cry2Ab and Cry1Ac have been used to control this pest. Field-evolved resistance also has been reported recently to *Bt* corn producing Cry1Ab in *Busseola fusca* in South Africa and to *Bt* corn producing Cry1P in *Spodoptera frugiperda* in Puerto Rico. The documented field outcomes are consistent with projections from modelling based on the population genetic principles underlying the refuge strategy. In particular, *H. zea* was expected to evolve resistance faster than other pests because it has non-recessive inheritance of resistance to Cry1Ac. In other words, the concentration of Cry1Ac is not sufficient to kill a high percentage of hybrid progeny from matings between resistant and susceptible moths. The results suggest that refuges of non-*Bt* host plants have helped to delay resistance.

Introduction

To control some key insect pests, crops have been genetically modified to produce insecticidal proteins from the common bacterium *Bacillus thuringiensis*

©CAB International 2009. *Environmental Impact of Genetically Modified Crops* (eds N. Ferry and A.M.R. Gatehouse)

(*Bt*; Schnepf *et al.*, 1998; de Maagd *et al.*, 2001). In principle, genes from microbes, plants or animals could be used in genetically modified (GM) plants to protect them from insect attack (Schuler *et al.*, 1998; Moar, 2003; Ferry *et al.*, 2004, 2006; Cohen, 2005; Gatehouse, 2008; see Chapter 6, this volume). Silencing of insect genes with ribonucleic acid (RNA) interference also has potential for defending GM plants from insects (Baum *et al.*, 2007; Mao *et al.*, 2007). However, insecticidal crystal (Cry) proteins from *Bt* are the basis of defence against insects in nearly all GM crops grown commercially to date. Transgenic *Bt* crops that kill key pests can reduce reliance on insecticide sprays, thereby providing economic, health and environmental benefits (Shelton *et al.*, 2002; Carrière *et al.*, 2003; Cattaneo *et al.*, 2006). Evolution of resistance by pests, however, diminishes the efficacy of *Bt* crops and the associated benefits (Tabashnik, 1994; Gould, 1998; van Rensburg, 2007; Matten *et al.*, 2008; Tabashnik *et al.*, 2008). Recent reviews have provided biochemical, genetic and evolutionary perspectives on insect resistance to *Bt* toxins (Griffitts and Aroian, 2005; Heckel *et al.*, 2007; Bravo and Soberón, 2008; Tabashnik and Carrière, 2008; Gassmann *et al.*, 2009). This chapter focuses on the lessons learned about insect resistance during the first 11 years of *Bt* crops.

Bt crops were first planted on a large scale in 1996, with rapid adoption leading to 42 million ha grown worldwide during 2007 and a cumulative total of >200 million ha from 1996 to 2007 (James, 2007; Fig. 5.1). Although the diversity of *Bt* toxins in GM crops has increased since 2001 (Table 5.1), the first decade of *Bt* crops consisted almost entirely of cotton producing *Bt* toxin Cry1Ac and maize producing Cry1Ab (referred to hereafter as *Bt* cotton and *Bt* maize). These toxins kill some major lepidopteran pests of cotton and maize (Table 5.1).

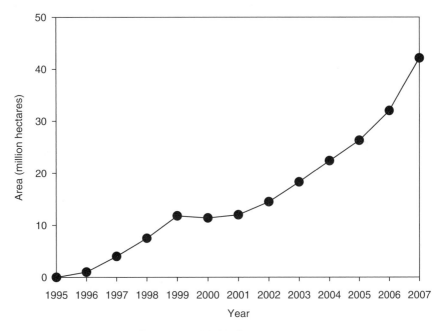

Fig. 5.1. Area planted to *Bt* crops worldwide from 1995 to 2007.

Table 5.1. Bt crops registered for commercial use in the USA. (Available at:
http://www.epa.gov/pesticides/biopesticides/pips/pip_list.htm last updated
11 January 2007, site accessed 20 March 2007.)

Bt toxin(s)	Target pests[a]	First registered
Maize		
Cry1Ab	Caterpillars	1995
Cry1F	Caterpillars	2001
Cry3Bb	Corn rootworms	2003
Cry1Ab + Cry3Bb	Caterpillars, corn rootworms	2003
Cry34Ab + Cry35Ab	Corn rootworms	2005
Cry1F + Cry34Ab + Cry35Ab	Caterpillars, corn rootworms	2005
Cry1Ab + Cry3Bb	Caterpillars, corn rootworms	2005
Cry3A	Corn rootworms	2006
Cry1Ab + Cry3A[b]	Caterpillars, corn rootworms	2007
Cotton		
Cry1Ac	Caterpillars	1995
Cry1Ac + Cry2Ab	Caterpillars	2002
Cry1Ac + Cry1F	Caterpillars	2004

[a]The specific target pests depend on the toxins produced and where the crop is grown. Major
caterpillar pests targeted by Bt maize include European corn borer (Ostrinia nubilalis) and corn
earworm (Helicoverpa zea). Major caterpillar pests targeted by Bt cotton include tobacco
budworm (Heliothis virescens), bollworm (H. zea), pink bollworm (Pectinophora gossypiella)
and cotton bollworm (Helicoverpa armigera). Corn rootworms targeted by Bt maize include the
western corn rootworm (Diabrotica virgifera virgifera), northern corn rootworm (Diabrotica
barberi) and Mexican corn rootworm (Diabrotica virgifera zea).
[b]From S. Matten, USEPA, 21 March 2007, personal communication.

Cry1Ab and Cry1Ac are so similar that evolution of resistance to one usually confers cross-resistance to the other (Tabashnik et al., 1996; Ferré and
Van Rie, 2002). Thus, from the standpoint of pest resistance, these two toxins
can be considered one type of toxin. Whereas Bt toxins in sprays degrade
quickly, commercially grown Bt crops produce toxin continuously. In effect, for
the past decade, the first generation of Bt crops exposed pest populations over
vast areas to a single type of Bt toxin throughout the growing season.

The extensive exposure to Bt toxins in GM crops represents one of the
largest, most sudden selections for resistance experienced by insects (Tabashnik
et al., 2003). Therefore, evolution of resistance by target pests is considered
the primary threat to the continued success of Bt crops (Tabashnik, 1994;
Gould, 1998; Carrière et al., 2001c; US Environmental Protection Agency
(USEPA), 2001; Ferré and Van Rie, 2002; Griffitts and Aroian, 2005;
Tabashnik et al., 2008). Resistance to a Bt toxin is a genetically based decrease
in the frequency of individuals susceptible to the toxin caused by exposure of
the population to the toxin (Tabashnik, 1994).

In the laboratory, many strains of major pests have evolved resistance to
Bt toxins (Table 5.2; Tabashnik, 1994; Ferré and Van Rie, 2002). In addition, evolution of resistance to Bt sprays outside of the laboratory is documented for two lepidopteran pests of cole crops, with evidence from
greenhouse populations of cabbage looper, Trichoplusia ni (Hübner; Janmaat

Table 5.2. Sixteen species of insect pests with *Bt* resistance. The resistant strains were produced by laboratory selection, with notable exceptions for the following five species: Resistance to Cry1Ac has evolved in field populations of *Helicoverpa zea* exposed to *Bt* cotton. Resistance to *Bt* sprays has evolved in field populations of diamondback moth and in greenhouse populations of cabbage looper. Field-evolved resistance has been reported to *Bt* cotton producing Cry1Ac in *H. zea* in the southeastern USA (Tabashnik *et al.*, 2008), to *Bt* corn producing Cry1Ab in *Busseola fusca* in South Africa (Van Rensburg, 2007), and to *Bt* corn producing Cry1F in *Spodoptera frugiperda* in Puerto Rico (Matten *et al.*, 2008). Individual studies are reviewed in Tabashnik (1994), Ferré and Van Rie (2002) and here; see Augustin *et al.* (2004) for *Chrysomela tremulae*.

Coleoptera
  Chrysomela scripta (cottonwood leaf beetle)
  Chrysomela tremulae (leaf beetle)
  Leptinotarsa decemlineata (Colorado potato beetle)

Diptera
  Culex quinquefasciatus (southern house mosquito)

Lepidoptera
  Busseola fusca (African stem borer)
  Helicoverpa armigera (cotton bollworm)
  Helicoverpa zea (bollworm)
  Heliothis virescens (tobacco budworm)
  Ostrinia nubilalis (European corn borer)
  Pectinophora gossypiella (pink bollworm)
  Plodia interpunctella (Indianmeal moth)
  Plutella xylostella (diamondback moth)
  Spodoptera exigua (beet armyworm)
  Spodoptera frugiperda (fall armyworm)
  Spodoptera littoralis (cotton leafworm)
  Trichoplusia ni (cabbage looper)

and Myers, 2003) and field populations of diamondback moth, *Plutella xylostella* [L.] (Table 5.2; Tabashnik *et al.*, 1990, 2003). Analysis of global monitoring data reveals that resistance to *Bt* toxins in GM crops has also evolved in some field populations of pests targeted by *Bt* crops (Van Rensburg, 2007; Matten *et al.*, 2008; Tabashnik *et al.*, 2008).

To better understand why some pest populations but not others have evolved resistance to *Bt* crops, we summarize information about *Bt* toxins and their mode of action, the genetic basis of *Bt* resistance, the refuge strategy for delaying pest resistance to *Bt* crops, and finally field monitoring data for pest resistance to such crops. We compare the observed field outcomes with the patterns predicted by the theory underlying the refuge strategy and consider the implications of what has been learned about insect resistance to *Bt* crops.

Bt Toxins and Their Mode of Action

Many *Bt* toxins kill certain key pests, yet unlike broad-spectrum insecticides, they have little or no toxicity to most non-target organisms including people, wildlife and most other insects (Mendelsohn *et al.*, 2003; Naranjo *et al.*, 2005; O'Callaghan *et al.*, 2005; Cattaneo *et al.*, 2006; Romeis *et al.*, 2006; Marvier *et al.*, 2007; Showalter *et al.*, 2008; see Chapters 8 and 13, this volume). Sprays of *Bt* toxins have been used safely and effectively in organic and conventional agriculture as well as forestry for decades. More than 140 *Bt* toxins are known (Crickmore *et al.*, 1998). Collectively, they can kill a wide variety of insects including the larvae of moths, beetles and flies; yet the spectrum of toxicity for each toxin is usually narrow (Schnepf *et al.*, 1998; Griffitts and Aroian, 2005).

The specificity of *Bt* toxins arises from their mode of action (Schnepf *et al.*, 1998; Bravo *et al.*, 2007). Naturally occurring *Bt* toxins are consumed and, in the case of lepidopteran larvae, dissolved in the alkaline midgut, where they are then activated by insect proteases from full-length protoxin to the active toxin. Although full-length Cry1Ac protein is produced by *Bt* cotton, truncated versions of Cry proteins are produced by at least three varieties of *Bt* maize (Mendelsohn *et al.*, 2003). Specific binding of active toxin to midgut membrane receptors is a key determinant of specificity. Much evidence implies that this specific binding causes pores in midgut membranes that ultimately lead to cell lysis and insect death (Schnepf *et al.*, 1998; Griffitts and Aroian, 2005; Soberón *et al.*, 2007). Some work with cell lines, however, suggests that specific binding initiates a magnesium-dependent cellular signalling pathway that ultimately kills the insect (Zhang *et al.*, 2006).

Genetics of Insect Resistance to *Bt* Toxins

Many genetic and biochemical mechanisms of insect resistance to *Bt* toxins are conceivable because resistance could arise from disruption of any of the steps in the mode of action (Heckel, 1994; Heckel *et al.*, 2007). Is the genetic basis of resistance to *Bt* toxins alike in many different pest species, as seen with resistance to some synthetic insecticides (ffrench-Constant *et al.*, 1998, 2000)? Or do multiple modes of resistance arise? Even though the number of examples is limited, the data show that insects can achieve resistance to *Bt* toxins by various mechanisms.

Resistance to *Bt* toxins has been studied mostly in lepidopteran pests that are primary targets of *Bt* sprays and *Bt* crops. The most common type of *Bt* resistance in Lepidoptera, called 'mode 1', entails strong resistance to one or more *Bt* toxins in the Cry1A family, limited cross-resistance to Cry1C and most other *Bt* toxins, recessive inheritance, and reduced binding of one or more Cry1A toxins to midgut membrane target sites (Tabashnik *et al.*, 1998). Mode 1 resistance is documented for some strains of six pests (cabbage looper; diamondback moth; Indianmeal moth, *Plodia interpunctella* (Hübner); tobacco budworm, *Heliothis virescens* [F.]; pink bollworm, *Pectinophora gossypiella* [Saunders]; and cotton bollworm, *Helicoverpa armigera* [Hübner]; Tabashnik *et al.*, 1998; Akhurst *et al.*, 2003; González-Cabrera *et al.*, 2003; Xu *et al.*, 2005; Wang *et al.*, 2007).

The molecular genetic basis of mode 1 resistance is known only in the latter three species. In these three cotton pests, laboratory-selected mode 1 resistance is tightly linked with a gene encoding a cadherin protein that binds Cry1Ac in susceptible insects (Gahan *et al.*, 2001; Morin *et al.*, 2003; Tabashnik *et al.*, 2005b; Xu *et al.*, 2005; Yang *et al.*, 2007). So far, at least seven different mutations of the cadherin gene that interfere with production of a full-length protein are associated with resistance to Cry1Ac in these three species. Evidence that the resistance-associated mutations block binding of *Bt* toxin is reported for *H. virescens* (Jurat-Fuentes *et al.*, 2004) and suspected in the others. Although the locations of the mutations within the cadherin gene are unique to each species, each species harbours at least one mutation that introduces a premature stop codon. Whereas only one resistance-associated cadherin mutation has been identified in tobacco budworm, pink bollworm and cotton bollworm each have at least three, including two that do not introduce a premature stop codon. GM versions of Cry1Ab and Cry1Ac designed to bypass cadherin kill resistant pink bollworm with cadherin deletion mutations (Soberón *et al.*, 2007).

Many cases of resistance to *Bt* toxins in Lepidoptera do not fit the mode 1 pattern (Tabashnik *et al.*, 1998; Ferré and Van Rie, 2002; Griffitts and Aroian, 2005). Sometimes mode 1 resistance and other modes of resistance occur in the same pest species (Tabashnik *et al.*, 1998). At least two alternatives to cadherin-based resistance to *Bt* toxins have been identified at the molecular genetic level. One of these is a different version of target site resistance involving an aminopeptidase N. A strain of beet armyworm, *Spodoptera exigua* (Hübner), with resistance to *Bt* toxin Cry1C lacks an aminopeptidase N thought to be a receptor for Cry1C (Herrero *et al.*, 2005). Unlike the aforementioned examples involving altered or absent receptors, resistance to *Bt* toxin Cry1Ab in a strain of European corn borer, *Ostrinia nubilalis* (Hübner), was due primarily to reduced trypsin-like activity of a proteinase that activates *Bt* protoxin to active toxin (Li *et al.*, 2005). The major locus conferring field-evolved mode 1 resistance in several strains of diamondback moth is not genetically linked with cadherin, several aminopeptidases N or other candidate genes tested so far (Baxter, 2005; Baxter *et al.*, 2005).

The Refuge Strategy for Delaying Pest Resistance to *Bt* Crops

Although many strategies have been proposed to delay pest resistance to *Bt* crops (Tabashnik, 1994; Roush, 1997; Gould, 1998; Bates *et al.*, 2005; Mehlo *et al.*, 2005; Zhao *et al.*, 2005; Soberón *et al.*, 2007), we focus here on the one that is most widely used, the refuge strategy. To discuss the theory underlying the refuge strategy, we begin with assumptions about the genetic basis of resistance to *Bt* crops. We assume the simplest genetic model, i.e. resistance to *Bt* crops is controlled by a single locus with two alleles, *r* for resistance and *s* for susceptibility. Although this is oversimplified, it is a reasonable starting point because mutations at single loci do confer resistance to *Bt* toxins in several well-studied cases (Tabashnik *et al.*, 1997; Gahan *et al.*, 2001; Morin *et al.*, 2003; Baxter *et al.*, 2005; Herrero *et al.*, 2005; Li *et al.*, 2005; Xu *et al.*, 2005). Also, resistance to the intense selection imposed by *Bt* crops is most likely to involve loci with major effects (Carrière and Roff, 1995; McKenzie, 1996; Groeters and Tabashnik, 2000).

The refuge strategy is mandated in most countries where *Bt* crops are grown, including the USA, which accounts for more than half of the world's *Bt* crop acreage (USEPA, 2001; Lawrence, 2005; James, 2007). The refuge strategy is based on evolutionary theory elaborated in dozens of papers (e.g. Comins, 1977; Georghiou and Taylor, 1977; Curtis *et al.*, 1978; Tabashnik and Croft, 1982; Gould, 1998; Peck *et al.*, 1999; Caprio, 2001; Carrière and Tabashnik, 2001; Onstad *et al.*, 2002; Carrière, 2003) and on small-scale experiments with diamondback moth (Liu and Tabashnik, 1997a; Shelton *et al.*, 2000; Tang *et al.*, 2001). The theory underlying the refuge strategy is to reduce heritability of resistance by: (i) providing refuges of non-*Bt* host plants that produce susceptible adults; (ii) promoting mating between resistant and susceptible adults; and (iii) decreasing the dominance of resistance.

To implement the refuge strategy, farmers grow refuges of non-*Bt* host plants near *Bt* crops to promote survival of susceptible pests. This strategy is expected to work best if resistance is conferred by rare, recessive alleles and if most of the extremely rare homozygous resistant (*rr*) adults emerging from *Bt* crops mate with homozygous susceptible (*ss*) adults from refuges. The theory predicts that such conditions will greatly delay evolution of resistance.

Refuge size and composition

Although rigorous large-scale tests of the refuge strategy are difficult, modelling results and small-scale experiments show that resistance is expected to evolve more slowly as the area of refuges relative to *Bt* crops increases. The regulations implemented in the field represent a compromise between the conflicting goals of delaying resistance (which favours large refuges) and minimizing constraints on growers (which favours minimal regulation and small refuges).

In the USA, refuge requirements for *Bt* crops have changed over time. Refuge requirements for *Bt* cotton implemented in 1996 included two options described in terms of the percentage of the total area of cotton on each farm accounted for by non-*Bt* cotton: (i) a 4% refuge of non-*Bt* cotton not sprayed with insecticides that kill the lepidopteran pests targeted by *Bt* cotton; or (ii) a 20% refuge of non-*Bt* cotton that could be sprayed with any insecticides other than *Bt*. Net production of susceptible pests in the sprayed and unsprayed refuges was expected to be similar, assuming approximately 20% survival of pests in the sprayed refuge (i.e. 0.20 refuge × 0.20 survival = 0.04; Gould and Tabashnik, 1998). Regulations adopted in 2001 varied among regions depending on the primary pests targeted by *Bt* cotton, but still included the options of unsprayed (5%) and sprayed (20%) refuges of non-*Bt* cotton near *Bt* cotton (USEPA, 2001). The percentage of maize acreage required for non-*Bt* maize refuges ranges from 20% in regions with little or no *Bt* cotton to 50% in regions with substantial amounts of *Bt* cotton (USEPA, 2001).

In June 2007, the US Environmental Protection Agency (USEPA, 2007) made a dramatic change in the refuge requirements in association with use of

GM cotton producing *Bt* toxins Cry2Ab and Cry1Ac. This two-toxin GM cotton was registered in December 2002 (USEPA, 2002) and planted on more than 1 million ha in the USA in 2006 (Monsanto, 2007). The combination or 'pyramid' of Cry1Ac and Cry2Ab is thought to be especially useful for thwarting resistance because plants producing Cry2Ab kill pests resistant to Cry1Ac (Tabashnik *et al.*, 2002b; Zhao *et al.*, 2005; Downes *et al.*, 2007). In the south eastern USA and parts of Texas where *H. virescens* and *H. zea* are the primary targets of *Bt* cotton, refuges of non-*Bt* cotton are required for cotton that produces only Cry1Ac, but not for cotton producing Cry1Ac and Cry2Ab. For these two polyphagous pests, the USEPA accepted Monsanto's proposal that host plants other than cotton – including other crops and weeds – provide sufficient refuge to delay resistance to GM cotton producing Cry1Ac and Cry2Ab. Although refuges of host plants other than cotton have been called 'natural refuges' and non-*Bt* cotton refuges have been dubbed 'structured' refuges (USEPA, 2007), we prefer the more transparent terms 'non-cotton refuges' and 'non-*Bt* cotton refuges', respectively.

Insect gene flow

The success of the refuge strategy depends on insect gene flow between refuges and *Bt* crops. The extent and impact of this gene flow is affected by many factors including pest movement and mating patterns, as well as the spatial and temporal distribution of refuges and *Bt* crops (Carrière *et al.*, 2004a,b; Sisterson *et al.*, 2004, 2005). For example, because refuges are sources of insect pests and *Bt* crop fields are sinks, movement between refuges and *Bt* crops can reduce population size in refuges and regionally (Riggin-Bucci and Gould, 1997; Onstad and Guse, 1999; Carrière *et al.*, 2003, 2004a,b; Caprio *et al.*, 2004). Moreover, movement from *Bt* crops to refuges can bring resistance alleles to refuges, which may increase heritability of resistance when the rare resistant adults emerging from *Bt* crops mate with adults from refuges bearing resistance alleles (Comins, 1977; Caprio and Tabashnik, 1992; Sisterson *et al.*, 2004).

For simplicity, many models of insecticide resistance evolution assume that insect movement is sufficient to achieve random mating between adults from refuges and areas treated with insecticide (e.g. Georghiou and Taylor, 1977). This apparently led to the claim that random mating between adults from refuges and *Bt* crops is crucial for the refuge strategy (e.g. Roush, 1997; Liu *et al.*, 1999; Glaser and Matten, 2003). Although the refuge strategy does require mating between resistant and susceptible adults, results from many modelling studies show that random mating is not essential or even optimal (e.g. Tabashnik, 1990; Caprio and Tabashnik, 1992; Gould, 1998; Peck *et al.*, 1999; Caprio, 2001; Ives and Andow, 2002; Onstad *et al.*, 2002; Carrière *et al.*, 2004b; Sisterson *et al.*, 2005). Nonetheless, models have not yielded consistent conclusions about how much movement is best for delaying resistance. In particular, modelling studies have reported that resistance evolves slowest when adult movement between refuges and *Bt* crops is either high (Peck *et al.*, 1999; Storer, 2003), low (Ives and Andow, 2002) or intermediate (Caprio, 2001).

Because models used in various studies differ in many ways, the precise cause of such contradictory conclusions among studies is not obvious. On the other hand, sensitivity analyses in which only one or a few assumptions are varied systematically in a single study enable testing of the hypothesis that interactions among factors can alter the effect of movement on the rate of resistance evolution (Tabashnik and Croft, 1982). Sensitivity analyses show that interactions among the relative abundance, spatial distribution and temporal distribution of refuges and *Bt* crop fields can alter the effects of movement on resistance evolution (Sisterson *et al.*, 2005). Furthermore, it appears that differences in conclusions among studies can be explained by differences in assumptions about the relative abundance and distribution of refuges and *Bt* crop fields (Sisterson *et al.*, 2005). One robust result is that resistance can be delayed effectively by fixing locations of refuges across years and distributing refuges uniformly to ensure that *Bt* crop fields are not isolated from refuges. However, rotating fields between refuges and *Bt* crops between years may provide better insect control and thereby reduce the need for insecticide sprays.

Dominance of resistance

In the early stages of resistance evolution, alleles conferring resistance are expected to be rare. When resistance alleles are rare, they occur mostly in heterozygotes and resistant homozygotes are extremely rare. Therefore, the response of heterozygotes – which is determined by dominance – governs the early trajectory of resistance evolution. For example, before commercialization of *Bt* cotton, the estimated frequency of alleles conferring resistance to Cry1Ac in *H. virescens* from four states of the USA was 0.0015, based on tests of progeny from single-pair matings between wild males and *rr* females from a laboratory-selected strain (Gould *et al.*, 1997). Assuming Hardy-Weinberg equilibrium in the example above, the expected frequency of *rs* (3×10^{-3}) is more than a thousand times greater than that of *rr* (2.2×10^{-6}).

Although dominance is sometimes considered an invariant genetic property, the refuge strategy exploits the principle that dominance can depend on the environment (Curtis *et al.*, 1978; Bourguet *et al.*, 1996; Tabashnik *et al.*, 2004). In particular, the dominance of resistance depends on the dose of toxin. The refuge strategy is sometimes called the 'high-dose refuge strategy' because results from many modelling studies show that refuges are most effective if the dose of toxin received by insects eating *Bt* plants is high enough to kill all or nearly all *rs* individuals (e.g. Curtis *et al.*, 1978; Tabashnik and Croft, 1982; Roush, 1997; Gould, 1998). In other words, refuges are predicted to work best if the toxin concentration in *Bt* plants is high enough to make resistance functionally recessive.

Bioassay results from some key lepidopteran pests show that the dominance of their resistance to *Bt* toxins decreases as toxin concentration increases (Tabashnik *et al.*, 1992, 2002a; Gould *et al.*, 1995; Liu and Tabashnik, 1997b; Tang *et al.*, 1997; Zhao *et al.*, 2000; Liu *et al.*, 2001b; Alves *et al.*, 2006). Two examples illustrate why this occurs (Table 5.3). At low toxin concentrations, survival is low to moderate for susceptible homozygotes (*ss*), and

Table 5.3. Dominance (*h*) of resistance to *Bacillus thuringiensis* (*Bt*) toxins as a function of toxin concentration. (Adapted from Tabashnik *et al.*, 2004.) We calculated *h* as (survival of *rs* – survival of *ss*)/(survival of *rr* – survival of *ss*) (Liu and Tabashnik, 1997b). Values of *h* vary from 0 (recessive) to 1 (dominant).

Bt toxin concentration (µg ml⁻¹ diet)	Survival of larvae exposed to *Bt* toxin (%)			Dominance (*h*)
	ss	*rs*	*rr*	
Heliothis virescens versus Cry1Ab (Gould *et al.*, 1995)				
0.32	47	100	100[a]	1.00
1.6	16	75	100	0.70
8.0	0	31	97	0.32
40.0	0	0	97	0.00
Pectinophora gossypiella versus Cry1Ac (Tabashnik *et al.*, 2002a)				
0.32	37	93	100[a]	0.89
1.0	4	52	100[a]	0.50
3.2	0	8	100[a]	0.08
10.0	0[b]	0.5	100	0.005

[a]Survival inferred to be 100% based on 100% survival at a higher toxin concentration.
[b]Survival inferred to be 0% based on 0% survival at a lower toxin concentration.

relatively high for heterozygotes (*rs*) and resistant homozygotes (*rr*). The similarity between *rs* and *rr* at low toxin concentrations yields dominant resistance. Conversely, at sufficiently high toxin concentrations, survival is low for *ss* and *rs* relative to *rr*, which yields recessive resistance. Responses of *H. virescens* to Cry1Ab incorporated in artificial diet show that inheritance of resistance varied from completely dominant to completely recessive as the concentration of *Bt* toxin increased from low to high. Although the variation is not as extreme, the trend is similar in responses of pink bollworm to Cry1Ac (Table 5.3).

To achieve high concentrations of *Bt* toxins in transgenic crop plants, *Bt* toxin genes have been modified for improved expression in plants (Mendelsohn *et al.*, 2003). For some but not all targeted pests, the toxin concentrations in *Bt* crops are high enough to render resistance functionally recessive (Tabashnik *et al.*, 2000a, 2003, 2008). Results from *H. armigera* demonstrate that the dominance of resistance to *Bt* cotton can vary as the concentration of Cry1Ac changes during the growing season (Bird and Akhurst, 2004, 2005). Whereas resistance was recessive (*h* = 0) on young cotton plants with relatively high Cry1Ac concentration, it was additive (*h* = 0.49) on older cotton plants with 75% lower Cry1Ac concentration.

Fitness costs

Fitness costs occur when fitness in refuges is lower for individuals with resistance alleles than for individuals without resistance alleles (Gassmann *et al.*,

2009). Fitness costs reflect antagonistic pleiotropy causing a trade-off across environments (Carrière et al., 1994, 1995; Carrière and Roff, 1995): resistance alleles increase fitness on Bt crops but may decrease fitness in the absence of Bt toxins. Fitness costs are often associated with resistance to Bt toxins (Groeters et al., 1994; Tabashnik, 1994; Alyokhin and Ferro, 1999; Oppert et al., 2000; Carrière et al., 2001a,b, 2005, 2006a,b; Ferré and Van Rie, 2002; Akhurst et al., 2003; Higginson et al., 2005; Raymond et al., 2006; Wenes et al., 2006; Bird and Akhurst, 2007a; but also see Tang et al., 1997; Huang et al., 2005).

Fitness costs keep resistance alleles rare before populations are exposed extensively to Bt toxins (McKenzie, 1996) and select against resistance in refuges. In principle, evolution of resistance can be delayed substantially or even reversed when fitness costs and refuges are present (Lenormand and Raymond, 1998; Carrière and Tabashnik, 2001; Carrière, 2003; Carrière et al., 2002, 2004b, 2005; Tabashnik et al., 2005a; Crowder et al., 2006; Gould et al., 2006; Gustafson et al., 2006).

Setting aside tri-trophic interactions (Gassmann et al., 2006), at least three types of pleiotropic effects might cause fitness costs. Resistance mutations affecting midgut proteins might interfere with food assimilation, increase gut permeability to toxic phytochemicals (Carrière et al., 2002, 2004b,c, 2005), or increase metabolism due to faster replacement of midgut cells in resistant versus susceptible insects (Dingha et al., 2004). Contrary to the increased metabolism hypothesis, the metabolic rate did not differ between a Bt-resistant and Bt-susceptible strain of the beet armyworm (Dingha et al., 2004).

In the initial stages of resistance evolution, the dominance of fitness costs is crucial, just as the dominance of resistance is critical. As noted above, when Bt crops are first introduced, most r alleles occur in rs individuals. If resistance is recessive, these rs individuals are killed by Bt crops and survive only in refuges. Therefore, the fitness of rs relative to ss in refuges is a key determinant of resistance evolution. Non-recessive fitness costs make fitness in refuges lower for rs than ss, favouring a decrease in resistance through selection in refuges, even though the rare rr individuals are favoured by selection in Bt crop fields (Carrière and Tabashnik, 2001; Tabashnik et al., 2005a). With large refuges, recessive costs can also significantly delay or reverse the evolution of resistance to Bt crops (Carrière and Tabashnik, 2001; Tabashnik et al., 2005a).

The magnitude and dominance of costs associated with Bt resistance are often affected by environmental factors, including variation in host plants (Bird and Akhurst, 2004, 2005, 2007; Carrière et al., 2004c, 2005, 2006a; Janmaat and Myers, 2005, 2006; Raymond et al., 2006), competition for mates (Higginson et al., 2005), crowding (Raymond et al., 2005) and natural enemies (Gassmann et al., 2006). This creates the opportunity to manipulate costs to enhance the success of the refuge strategy (Carrière et al., 2001b, 2002, 2004b,c, 2005; Pittendrigh et al., 2004; Gassmann et al., 2006; Bird and Akhurst, 2007a). Although it has been proposed that fitness costs of resistance will be higher on less suitable host plants (Bergelson and Purrington, 1996), experimental results do not show a consistent relationship between fitness costs of *Bt* resistance and host-plant suitability (Janmaat and Myers, 2005; Raymond *et al.*, 2006; Bird and Akhurst, 2007a).

While data on dominance of fitness costs are limited, available evidence suggests that host-plant quality affects dominance and that non-recessive costs may be common (Gassmann *et al.*, 2009). For *H. armigera*, fitness costs were recessive on pigeonpea, often additive on non-*Bt* cotton and frequently dominant on sorghum (Bird and Akhurst, 2007a). Plant phenology also affected dominance of costs for *H. armigera* on non-*Bt* cotton, with costs recessive on flowering cotton (Bird and Akhurst, 2005) and frequently additive on cotton in the 4–6 leaf stage as well as on cotton with bolls (Bird and Akhurst, 2004, 2007). In pink bollworm, fitness costs were usually recessive (Carrière *et al.*, 2001a,b, 2005, 2006a), but adding the cotton-defensive chemical gossypol to diet increased the magnitude and dominance of costs affecting pupal weight (Carrière *et al.*, 2004c). Costs were recessive in *T. ni* fed cucumber, tomato or pepper (Janmaat and Myers, 2005). Further study assessing the magnitude and dominance of costs on different host plants could help to provide information useful for improving resistance management by manipulating refuge quality (Showalter *et al.*, 2008; Gassmann *et al.*, 2009).

Modelling results indicate that relative to GM plants producing only one *Bt* toxin, costs may more readily delay or reverse resistance to GM plants producing two distinct toxins (Gould *et al.*, 2006). If pest survival on such two-toxin plants requires resistance-conferring mutations at two independent loci, the frequency of resistant individuals will be much lower than the frequency of individuals resistant to only one of the two toxins. If resistance at each locus is recessively inherited, then *r* alleles only confer a selective advantage in individuals homozygous for resistance at both loci. Because such individuals are much rarer than individuals with one, two or three resistance alleles (out of a possible four at the two loci), selection against resistance in refuges caused by fitness costs can easily surpass selection for resistance in *Bt* crop fields (Gould *et al.*, 2006).

Incomplete resistance

Many early models assumed that *rr* individuals are completely resistant to *Bt* crops, i.e. fitness of *rr* is equal on *Bt* and non-*Bt* host plants (Gould, 1998). Although this is true in some special cases, such as resistant diamondback moth on experimental *Bt* crucifers versus non-*Bt* crucifers (Ramachandran *et al.*, 1998; Tang *et al.*, 1999), it is not generally applicable. In particular, the fitness of *rr* is often lower on *Bt* crop plants than on their non-*Bt* counterparts (Liu *et al.*, 1999, 2001a; Bird and Akhurst, 2004; Carrière *et al.*, 2006a). The disadvantage suffered by resistant insects on transgenic plants relative to their conventional non-transgenic counterparts is called incomplete resistance (Carrière and Tabashnik, 2001). Unlike fitness costs, incomplete resistance cannot reverse resistance. However, lower fitness of *rr* in a *Bt* crop field relative to a refuge weakens selection for resistance, and thus can help to delay resistance (Carrière and Tabashnik, 2001; Carrière *et al.*, 2002; Tabashnik *et al.*, 2005a).

Monitoring Resistance to *Bt* Crops in Field Populations of Pests

Field-evolved resistance to Cry1Ac, the *Bt* toxin in first-generation GM cotton, has been documented in some field populations of *H. zea* from Arkansas and Mississippi, but not from North Carolina (Burd *et al.*, 2003; Ali *et al.*, 2006; Jackson *et al.*, 2006; Tabashnik *et al.*, 2008). In addition, field-evolved resistance has been reported to *Bt* corn producing Cry1A in *Busseola fusca* in South Africa (Van Rensburg, 2007) and to *Bt* corn producing Cry1F in *Spodoptera frugiperda* in Puerto Rico (Matten *et al.*, 2008). Published data also provide evidence that resistance to *Bt* crops has not been detected in monitored field populations of *H. armigera*, *H. virescens*, *O. nubilalis*, *P. gossypiella* and *Sesamia nonagrioides* in Australia, China, Spain and the USA (Tabashnik *et al.*, 2003, 2005a, 2006, 2008; Farinós *et al.*, 2004; Ali *et al.*, 2006; Stodola *et al.*, 2006; Bird and Akhurst, 2007b; Downes *et al.*, 2007; Gahan *et al.*, 2007; Siegfried *et al.*, 2007; Wu, 2007).

Resistance to Cry1Ac in *H. zea* from Arkansas and Mississippi was monitored during 1992–1993 (Luttrell *et al.*, 1999) and during 2002–2004 (Ali *et al.*, 2006), enabling comparison before and after commercialization of *Bt* cotton. Both studies used bioassays with toxin incorporated in diet to measure the concentration of Cry1Ac killing 50% of larvae (LC_{50}) for strains derived from field populations and susceptible laboratory strains. These data allow calculation of resistance ratios, which are LC_{50}s of field-derived strains divided by LC_{50}s of susceptible laboratory strains. Two strains of *H. zea* sampled during 2003 and 2004 in Arkansas and Mississippi had resistance ratios >500 and two others had resistance ratios >100 (Ali *et al.*, 2006), while none of the field strains from 1992 to 1993 was resistant (Luttrell *et al.*, 1999). Thus, field-evolved resistance to Cry1Ac occurred in some Arkansas and Mississippi populations 7–8 years after widespread use of *Bt* cotton in 1997. Data from field populations sampled in 2005 and 2006 also demonstrate *H. zea* resistance to Cry1Ac, yielding resistance ratios >100 for ten additional strains from Arkansas and Georgia, including two strains with resistance ratios >1000 (Luttrell and Ali, 2007).

Results for *H. armigera* from India show increases in the LC_{50} of Cry1Ac from 2000–2002 to 2004–2006 in each of the four regions studied (Gujar *et al.*, 2007). In the 2004–2006 data, eight of 12 (67%) of the field-derived strains tested had resistance ratios >10 and one strain from the Bhatinda region had a resistance ratio of 120 (Gujar *et al.*, 2007). Although Gujar *et al.* (2007) report changes in LC_{50} over time, they do not report data from a susceptible strain that was tested simultaneously with the putative resistant strains. Consequently, one cannot exclude the possibility that the observed increases in LC_{50} were caused partly or entirely by decreasing potency of the toxin over time.

The results for *H. armigera* from China summarized by Wu (2006, 2007) show no decrease in susceptibility to Cry1Ac from 1998 to 2006, based on survival and growth of >94 field-derived strains. Based on bioassays of isofemale lines, Li *et al.* (2007) estimated that the resistance allele frequency was 0.0011 in 2002 versus 0.0 in 2005 for Anci county and 0.00059 in 2002 versus 0.0023 in 2005 for Xijian county. Neither of these changes in estimated resistance allele frequency is statistically significant. Li *et al.* (2007) report significant increases from 2002 to 2005 in a relative growth index for both counties, yet similar to the limitation of the results from India noted above, they state: '[T]he possibility that the results were due to testing conditions being less stringent in each successive year cannot be ruled out.'

Evidence from Field Monitoring Versus Predictions from the Refuge Theory

As described above, the theory underlying the refuge strategy predicts that resistance will evolve slower as dominance of resistance decreases and refuge size increases. Observed variation in resistance to *Bt* crops among and within pest species is consistent with these predictions. Of the six major pests examined by Tabashnik *et al.*, (2008), *H. zea* is the only one with dominant inheritance of resistance to the *Bt* toxin in a GM crop (Burd *et al.*, 2000). As expected, *H. zea* has evolved resistance faster than pests with recessive resistance, including *H. virescens* and *P. gossypiella* (Tabashnik *et al.*, 2008). Furthermore, *H. zea* resistance to *Bt* cotton evolved faster in Arkansas and Mississippi than in North Carolina, corresponding with the lower percentage of refuges in Arkansas and Mississippi (39%) than in North Carolina (82%; Gustafson *et al.*, 2006).

As noted above, the dominance of resistance to *Bt* cotton varies in *H. armigera*, with partially dominant inheritance reported from India ($h = 0.43$) and a range from recessive to partially dominant ($h = 0–0.49$) on young to old plants in Australia (Bird and Akhurst, 2004, 2005). Accordingly, the risk of resistance is expected to be higher in *H. armigera* than in pests with completely recessive resistance to *Bt* crops ($h = 0$).

In Australia, *Bt* cotton producing Cry1Ac was limited to 30% of total cotton planted, providing a minimum 70% refuge (Downes *et al.*, 2007). In China, small fields of cotton are close to fields of several other non-*Bt* crops that are host plants for *H. armigera*, creating refuges that account for up to 95% of the available hosts for this pest in some regions (Wu *et al.*, 2002a,b). *Bt* cotton accounted for all of the cotton grown in Anci and Xijian counties of China from 2002 to 2005 (Li *et al.*, 2007). However, of the area planted with four host plants of *H. armigera* (cotton, maize, groundnut and soybean), cotton was only 9.2% in Anci and 73% in Xijian (Li *et al.*, 2007). In India, a minimum 20% refuge of non-*Bt* cotton rows around the borders of *Bt* cotton is required and several non-*Bt* crops are host plants for *H. armigera* (Ravi *et al.*, 2005; Gujar *et al.*, 2007). However, the situation in India is complicated by 'widespread illegal cultivation of spurious *Bt* cotton, growing of F2 *Bt* cotton, and poor implementation of resistance management strategies' (Gujar *et al.*, 2007).

Implications of Field-evolved Resistance to *Bt* Cotton

Analysis of monitoring data shows that some field populations of *H. zea* have evolved resistance to Cry1Ac, the toxin produced by first-generation *Bt* cotton

(Tabashnik *et al.*, 2008). Although tests of survival on *Bt* cotton plants from neonate to adult have not been reported for field-selected resistant strains of *H. zea*, the available evidence implies that survival on *Bt* cotton is higher for resistant field populations than for susceptible populations. Susceptible *H. zea* shows some survival on *Bt* cotton (USEPA, 1998; Jackson *et al.*, 2004a,b), which means that any decrease in susceptibility to Cry1Ac would be expected to increase survival. Luttrell *et al.* (2004) reported that after 4 days on *Bt* cotton leaves, survival of a field strain of *H. zea* from Mississippi with 40-fold resistance to Cry1Ac was 52% compared with 0% for a susceptible strain. Moreover, results of glasshouse experiments with *H. zea* show that survival on *Bt* cotton was significantly higher for a laboratory-selected strain with 100-fold resistance to Cry1Ac than for a feral strain (Jackson *et al.*, 2004a). Consequently, the greater than 100-fold resistance to Cry1Ac documented for 14 field populations in Arkansas, Georgia and Mississippi (Ali *et al.*, 2006; Luttrell and Ali, 2007) probably increases survival on *Bt* cotton. Indeed, increased damage to *Bt* cotton in the field caused by *H. zea* has been reported in Arkansas (James, 2006).

Nonetheless, resistance of *H. zea* to Cry1Ac has not caused widespread crop failures in the field for several reasons (Tabashnik *et al.*, 2008). First, the documented resistance is spatially limited. Resistance was not seen in North Carolina and most populations tested from Arkansas and Mississippi were not resistant to Cry1Ac (Ali *et al.*, 2006). Second, from the outset, insecticide sprays have been used to improve control of *H. zea* on *Bt* cotton because Cry1Ac alone is not effective enough to manage this pest (USEPA, 1998; Jackson *et al.*, 2004b). Such insecticide sprays would mask problems associated with reduced control of *H. zea* by *Bt* cotton. Third, even against a strain with 100-fold resistance to Cry1Ac, the Cry1Ac in *Bt* cotton killed about 60% of larvae (Jackson *et al.*, 2004a). Finally, GM cotton producing *Bt* toxins Cry2Ab and Cry1Ac was planted on more than 1 million hectares in the USA in 2006 (Monsanto, 2007). Control of *H. zea* by Cry2Ab would also limit problems associated with resistance to Cry1Ac (Jackson *et al.*, 2004a).

For pest populations with resistance to Cry1Ac, *Bt* cotton with Cry2Ab and Cry1Ac may act like single-toxin cotton, with control achieved primarily or only by Cry2Ab. If so, for these Cry1Ac-resistant populations, the potential benefits of 'pyramiding' two toxins will not be fully realized in terms of delaying resistance (Zhao *et al.*, 2005).

Conclusions

Researchers and others have attempted to gauge the risk of pest resistance to transgenic crops based on various traits of the pest–crop interaction (Tabashnik, 1994; Gould, 1998; USEPA, 1998; Tabashnik *et al.*, 2003). Consistent with the population genetic theory on which the refuge strategy is based, dominant resistance to a *Bt* crop seems to be a good indicator of high risk of rapid resistance evolution. However, it is often difficult to obtain resistant strains that survive on *Bt* crops, which is a prerequisite for measuring dominance (Tabashnik *et al.*, 2000b). Fortunately, dominance is inversely associated with a trait that can be measured readily before transgenic crops are commercialized: the efficacy of *Bt* crops against susceptible pests. In particular, *H. armigera* and *H. zea* are the only pests for which published data show laboratory-selected resistance to a commercially grown *Bt* crop that is not completely recessive ($h > 0$ for both). Also, these are the only two of the six key pests examined by Tabashnik *et al.*, (2008) for which efficacy of the *Bt* crop versus susceptible populations is <99% (USEPA, 1998; Jackson *et al.*, 2004a,b; Kranthi *et al.*, 2005; Downes *et al.*, 2007; Gujar *et al.*, 2007).

The observed inverse association between dominance of resistance and efficacy of *Bt* crops may occur because, for marginally susceptible pests (like *H. zea* versus Cry1Ac), any alleles conferring decreased susceptibility to a *Bt* toxin could increase survival on the *Bt* crop. In particular, a single resistant allele in a heterozygous individual could boost survival. Conversely, for the most susceptible pests, relatively large decreases in susceptibility to a *Bt* toxin may not increase survival on the *Bt* crop (Tabashnik *et al.*, 2003). The USEPA (2001) guidelines for managing resistance to *Bt* crops follow this line of reasoning, recommending a toxin concentration in GM plants 25 times higher than that killing 99% (LC_{99}) of susceptible pests. This standard, dubbed a 'high dose', is not met for *H. zea* by GM cotton producing only Cry1Ac (USEPA, 2001; Jackson *et al.*, 2004a,b). Moreover, field efficacy data suggest that this standard is not met for *H. zea* by GM cotton producing Cry2Ab and Cry1Ac (Jackson *et al.*, 2004b).

The results reviewed here imply that when high efficacy of *Bt* crops versus target pests is not achieved, large refuges of non-*Bt* host plants may be needed to substantially delay resistance. Therefore, refuge requirements should take into account the initial efficacy of the GM crop against susceptible pests. A notable example is the relatively low efficacy of some new varieties of *Bt* maize with toxins targeting corn rootworms (*Diabrotica* spp.; Siegfried *et al.*, 2005), which may require large refuges to thwart resistance. As we enter the second decade of GM crops, we can use knowledge gained from the first decade to minimize their risks and maximize their benefits.

Acknowledgements

We are grateful to our colleagues at the University of Arizona and elsewhere for comments and encouragement. Aaron Gassmann, David Crowder, David Onstad, Mark Sisterson, Randy Luttrell, Andrea Mathias, Fred Gould and Juan Ferré provided useful comments on previous versions of the manuscript. Portions of this chapter are reprinted with permission from: B.E. Tabashnik and Y. Carrière, 'Evolution of insect resistance to transgenic plants' in *Specialization, Speciation, and Radiation: The Evolutionary Biology of Herbivorous Insects*, edited by K. Tilmon, copyright 2008, The Regents of the University of California, published by the University of California Press. Financial support includes funds from several USDA grants (National Research Initiative grants 2003-01469, 2005-00925, 2006-35302-17365 and Biotechnology Risk Assessment research grant 2003-04371).

References

Akhurst, R.J., James, W., Bird, L.J. and Beard, C. (2003) Resistance to the Cry1Ac ∂-endotoxin of *Bacillus thuringiensis* in the cotton bollworm, *Helicoverpa armigera* (Lepidoptera: Noctuidae). *Journal of Economic Entomology* 96, 1290–1299.

Ali, M.I., Luttrell, R.G. and Young, S.Y. (2006) Susceptibilities of *Helicoverpa zea* and *Heliothis virescens* (Lepidoptera: Noctuidae) populations to Cry1Ac insecticidal protein. *Journal of Economic Entomology* 99, 164–175.

Alves, A.P., Spencer, A., Tabashnik, B.E. and Siegfried, B.D. (2006) Inheritance of resistance to the Cry1Ab *Bacillus thuringiensis* toxin in *Ostrinia nubilalis* (Lepidoptera: Crambidae). *Journal of Economic Entomology* 99, 494–501.

Alyokhin, A.V. and Ferro, D.N. (1999) Relative fitness of the Colorado potato beetle (Coleoptera, Chrysomelidae) resistant and susceptible to *Bacillus thuringiensis* Cry3A toxin. *Journal of Economic Entomology* 92, 510–515.

Augustin, S., Courtin, C., Rejasse, A., Lorme, P., Genissel, A. and Bourguet, D. (2004) Genetics of resistance to transgenic *Bacillus thuringiensis* poplars in *Chryso-mela tremulae* (Coleoptera: Chrysomelidae). *Journal of Economic Entomology* 97, 1058–1064.

Bates, S.L., Cao, J., Zhao, J.-Z., Earle, E.D., Roush, R.T. and Shelton, A.M. (2005) Evaluation of a chemically inducible promoter for developing a within-plant refuge for resistance management. *Journal of Economic Entomology* 98, 2188–2194.

Baum, J.A., Bogaert, T., Clinton, W., Heck, G.R., Feldmann, P., Ilagan, O., Johnson, S., Plaetinck, G., Munyikwa, T., Pleau, M., Vaughn, T. and Roberts, J. (2007) Control of coleopteran insect pests through RNA interference. *Nature Biotechnology* 25, 1322–1326.

Baxter, S.W. (2005) Molecular and genetic analysis of Bt and spinosad resistance in diamondback moth, *Plutella xylostella*. PhD dissertation, University of Melbourne, Melbourne, Australia.

Baxter, S.W., Zhao, J.-Z., Gahan, L.J., Shelton, A.M., Tabashnik, B.E. and Heckel, D.G. (2005) Novel genetic basis of field-evolved resistance to Bt toxins in *Plutella xylostella*. *Insect Molecular Biology* 14, 327–334.

Bergelson, J. and Purrington, C.B. (1996) Surveying patterns in the cost of resistance in plants. *American Naturalist* 148, 536–558.

Bird, L.J. and Akhurst, R.J. (2004) Relative fitness of Cry1A-resistant and -susceptible *Helicoverpa armigera* (Lepidoptera: Noctuidae) on conventional and transgenic cotton. *Journal of Economic Entomology* 97, 1699–1709.

Bird, L.J. and Akhurst, R.J. (2005) The fitness of Cry1A-resistant and -susceptible *Helicoverpa armigera* (Lepidoptera, Noctuidae) on transgenic cotton with reduced levels of Cry1Ac. *Journal of Economic Entomology* 59, 1166–1168.

Bird, L.J. and Akhurst, R.J. (2007a) Effects of host plant species on fitness costs of Bt resistance in *Helicoverpa armigera* (Lepidoptera: Noctuidae). *Biological Control* 40, 196–203.

Bird, L.J. and Akhurst, R.J. (2007b) Variation in susceptibility of *Helicoverpa armigera* (Hübner) and *Helicoverpa punctigera* (Wallengren) (Lepidoptera: Noctuidae) in Australia to two *Bacillus thuringiensis* toxins. *Journal of Invertebrate Pathology* 94, 84–94.

Bourguet, D., Prout, M. and Raymond, M. (1996) Dominance of insecticide resistance presents a plastic response. *Genetics* 143, 407–416.

Bravo, A. and Soberón, M. (2008) How to cope with insect resistance to Bt toxins? *Trends in Biotechnology* 26, 573–579.

Bravo, A., Gill, S.S. and Soberón, M. (2007) Mode of action of *Bacillus thuringiensis* Cry and Cyt toxins and their potential for insect control. *Toxicon* 49, 423–435.

Burd, A.D., Bradley, J.R., Van Duyn, J.W. and Gould, F. (2000) Resistance of bollworm, *Helicoverpa zea,* to Cry1A(c) toxin. *Proceedings of the Beltwide Cotton Conference Volume 2.* National Cotton Council, Memphis, Tennessee, pp. 923–926.

Burd, A.D., Gould, F., Bradley, J.R., Van Duyn, J.W. and Moar, W.J. (2003) Estimated frequency of nonrecessive *Bt* resistance genes in bollworm, *Helicoverpa zea* (Boddie) (Lepidoptera, Noctuidae) in eastern North Carolina. *Journal of Economic Entomology* 96, 137–142.

Caprio, M.A. (2001) Source-sink dynamics between transgenic and nontransgenic habitats and their role in the evolution of resistance. *Journal of Economic Entomology* 94, 698–705.

Caprio, M.A. and Tabashnik, B.E. (1992) Gene flow accelerates local adaptation among finite populations: simulating the evolution of insecticide resistance. *Journal of Economic Entomology* 85, 611–620.

Caprio, M.A., Fave, M.K. and Hankins, G. (2004) Evaluating the impacts of refuge width on source-sink dynamics between transgenic and non-transgenic cotton. *Journal of Insect Science* 4, 3.

Carrière, Y. (2003) Haplodiploidy, sex and the evolution of pesticide resistance. *Journal of Economic Entomology* 96, 1626–1640.

Carrière, Y. and Roff, D.A. (1995) Change in genetic architecture resulting from the evolution of insecticide resistance: a theoretical and empirical analysis. *Heredity* 75:618–629.

Carrière, Y. and Tabashnik, B.E. (2001) Reversing insect adaptation to transgenic insecticidal plants. *Proceedings of the Royal Society of London, Series B: Biological Sciences* 268, 1475–1480.

Carrière, Y., Roff, D.A. and Deland, J.P. (1995) The joint evolution of diapause and insecticide resistance: a test of an optimality model. *Ecology* 76, 1497–1505.

Carrière, Y., Deland, J.-P., Roff, D.A. and Vincent, C. (1994) Life history costs associated with the evolution of insecticide resistance. *Proceedings of the Royal Society of London, Series B: Biological Sciences* 58, 35–45.

Carrière, Y., Dennehy T.J., Ellers-Kirk, C., Holley D., Liu Y.-B., Sims M.A. and Tabashnik B.E. (2002) Fitness costs, incomplete resistance, and management of resistance to Bt crops. In: Akhurst, R.J., Beard C.E. and Hughes P.A. (eds.) *Proceedings of the 4th Pacific Rim Conference on the Biotechnology of* Bt *and its Environmental Impact.* CSIRO, Canberra. pp. 82–91.

Carrière, Y., Ellers-Kirk, C., Patin, A.L., Sims, M.A., Meyer, S., Liu, Y.-B., Dennehy, T.J. and Tabashnik, B.E. (2001a) Overwintering costs associated with resistance to transgenic cotton in the pink bollworm. *Journal of Economic Entomology* 94, 935–941.

Carrière, Y., Ellers-Kirk, C., Liu, Y.-B., Sims, M.A., Patin, A.L., Dennehy, T.J. and Tabashnik, B.E. (2001b) Fitness costs and maternal effects associated with resistance to transgenic cotton in the pink bollworm. *Journal of Economic Entomology* 94, 1571–1576.

Carrière, Y., Dennehy, T.J., Petersen, B., Haller, S., Ellers-Kirk, C., Antilla, L., Liu, Y.-B., Willot, E. and Tabashnik, B.E. (2001c) Large-scale management of insect resistance to transgenic cotton in Arizona: can transgenic insecticidal crops be sustained? *Journal of Economic Entomology* 94, 315–325.

Carrière, Y., Ellers-Kirk, C., Sisterson, M., Antilla, L., Whitlow, M., Dennehy, T.J. and Tabashnik, B.E. (2003) Long-term regional suppression of pink bollworm by *Bacillus thuringiensis* cotton. *Proceedings of the National Academy of Sciences of the USA* 100, 1519–1523.

Carrière, Y., Dutilleul, P., Ellers-Kirk, C., Pedersen, B., Haller, S., Antilla, L., Dennehy, T.J. and Tabashnik, B.E. (2004a) Sources, sinks, and zone of influence of refuges for managing insect resistance to Bt crops. *Ecological Applications* 14, 1615–1623.

Carrière, Y., Sisterson, M.S. and Tabashnik, B.E. (2004b) Resistance management for sustainable use of *Bacillus thuringiensis* crops. In: Horowitz, A.R. and Ishaaya, I. (eds) *Insect Pest Management: Field and Protected Crops*. Springer, New York, pp. 65–95.

Carrière, Y., Ellers-Kirk, C., Biggs, R., Dennehy, T.J. and Tabashnik, B.E. (2004c) Effects of gossypol on fitness costs associated with resistance to Bt cotton in pink bollworm. *Journal of Economic Entomology* 97, 1710–1718.

Carrière, Y., Ellers-Kirk, C., Biggs, R., Degain, B., Holley, D., Yafuso, C., Evans, P., Dennehy, T.J. and Tabashnik, B.E. (2005) Effects of cotton cultivar on fitness costs associated with resistance of pink bollworm (Lepidoptera, Gelechiidae) to Bt cotton. *Journal of Economic Entomology* 98, 947–954.

Carrière, Y., Ellers-Kirk, C., Biggs, R.W., Nyboer, M.E., Unnithan, G.C., Dennehy, T.J. and Tabashnik, B.E. (2006a) Cadherin-based resistance to Bt cotton in hybrid strains of pink bollworm: fitness costs and incomplete resistance. *Journal of Economic Entomology* 99, 1925–1935.

Carrière, Y., Nyboer, M., Ellers-Kirk, C., Sollome, J., Colletto, N., Antilla, L., Dennehy, T.J., Staten, R.T. and Tabashnik, B.E. (2006b) Effect of resistance to Bt cotton on pink bollworm (Lepidoptera: Gelechiidae) response to sex pheromone. *Journal of Economic Entomology* 99, 946–953.

Cattaneo, M., Yafuso, C., Schmidt, C., Huang, C., Rahman, M., Olson, C., Ellers-Kirk, C., Orr, B.J., Marsh, S.E., Antilla, L., Dutilleul, P. and Carrière, Y. (2006) Farm-scale evaluation of transgenic cotton impacts on biodiversity, pesticide use, and yield. *Proceedings of the National Academy of Sciences of the USA* 103, 7571–7576.

Cohen, J.I. (2005) Poorer nations turn to publicly developed GM crops. *Nature Biotechnology* 23, 27–33.

Comins, H.N. (1977) The development of insecticide resistance in the presence of migration. *Journal of Theoretical Biology* 64, 177–197.

Crickmore, N., Zeigler, D.R., Feitelson, J., Schnepf, E., Van Rie, J., Lereclus, D., Baum, J. and Dean, D.H. (1998)

Microbiology and Molecular Biology Reviews. 62, 807–813. http://www.lifesci.sussex.ac.uk/Home/Neil_Crickmore/Bt/

Crowder, D.W., Carrière, Y., Tabashnik, B.E., Ellsworth, P.C. and Dennehy, T.J. (2006) Modeling the evolution of resistance to pyriproxifen by the sweet potato whitefly (Hemiptera: Aleyrodidae). *Journal of Economic Entomology* 99, 1396–1406.

Curtis, C.F., Cook, L.M. and Wood, R.J. (1978) Selection for and against insecticide resistance and possible methods of inhibiting the evolution of resistance in mosquitoes. *Ecological Entomology* 3, 273–287.

de Maagd, R.A., Bravo, A. and Crickmore, N. (2001) How *Bacillus thuringiensis* has evolved specific toxins to colonize the insect world. *Trends in Genetics* 17, 193–199.

Dingha, B.N., Moar, W.J. and Appel, A.G. (2004) Effects of *Bacillus thuringiensis* Cry1C toxin on the metabolic rate of Cry1C resistant and susceptible *Spodoptera exigua* (Lepidoptera: Noctuidae). *Physiological Entomology* 29, 409–418.

Downes, S., Mahon, R. and Olsen, K. (2007) Monitoring and adaptive resistance management in Australia for Bt-cotton: current status and future challenges. *Journal of Invertebrate Pathology* 95, 208–213.

Farinós, G.P., de la Poza, M., Hernández-Crespo, P., Ortego, F. and Castañera, P. (2004) Resistance monitoring of field populations of the corn borers *Sesamia nonagrioides* and *Ostrinia nubilalis* after 5 years of Bt maize cultivation in Spain. *Entomologia Experimentata et Applicata* 110, 23–30.

Ferré, J. and Van Rie, J. (2002) Biochemistry and genetics of insect resistance to *Bacillus thuringiensis*. *Annual Review of Entomology* 47, 501–533.

Ferry, N., Edwards, M.G., Gatehouse, J.A. and Gatehouse, A.M.R. (2004) Plant–insect interactions: molecular approaches to insect resistance. *Current Opinions in Biotechnology* 15, 155–161.

Ferry, N., Edwards, M.G., Gatehouse, J., Capell, T., Christou, P. and Gatehouse, A.M.R. (2006) Transgenic plants for insect pest control: a forward looking scientific perspective. *Transgenic Research* 15, 13–19.

ffrench-Constant, R.H., Pittendrigh, B.A. and Vaughan, A. (1998) Why are there so few resistance-associated mutations in insecticide target genes? *Philosophical Transactions of the Royal Society of London, Series B: Biological Sciences* 353, 1685–1693.

ffrench-Constant, R.H., Anthony, N., Aronstein, K., Rocheleau, T. and Stilwell, G. (2000) Cyclodiene insecticide resistance: from molecular to population genetics. *Annual Review of Entomology* 48, 449–466.

Gahan, L.J., Gould, F. and Heckel, D.G. (2001) Identification of a gene associated with Bt resistance in *Heliothis virescens*. *Science* 293, 857–860.

Gahan, L.J., Gould, F., López, J.D., Micinski, S. and Heckel, D.G. (2007) A polymerase chain reaction screen of field populations of *Heliothis virescens* for a retrotransposon insertion conferring resistance to *Bacillus thuringiensis* toxin. *Journal of Economic Entomology* 100, 187–194.

Gassmann, A.J., Carrière, Y. and Tabashnik, B.E. (2009) Fitness costs of insect resistance to *Bacillus thuringiensis*. *Annual Review of Entomology* 54, 147–163.

Gassmann, A.J., Stock, S.P., Carrière, Y. and Tabashnik, B.E. (2006) Effects of entomopathogenic nematodes on the fitness cost of resistance to Bt toxin Cry1Ac in pink bollworm (Lepidoptera: Gelechiidae). *Journal of Economic Entomology* 99, 920–926.

Gatehouse, J.A. (2008) Biotechnological prospects for engineering insect-resistant plants. *Plant Physiology* 146, 881–887.

Georghiou, G.P. and Taylor, C.E. (1977) Operational influences in the evolution of insecticide resistance. *Journal of Economic Entomology* 70, 653–658.

Glaser, J.A. and Matten, S.R. (2003) Sustainability of insect resistance management strategies for transgenic Bt corn. *Biotechnology Advances* 22, 45–69.

González-Cabrera, J., Escriche, B., Tabashnik, B.E. and Ferré, J. (2003) Binding of *Bacillus thuringiensis* toxins in resistant and susceptible strains of pink bollworm (*Pectinophora gossypiella*). *Insect Biochemistry and Molecular Biology* 33, 929–935.

Gould, F. (1998) Sustainability of transgenic insecticidal cultivars: integrating pest genetics and ecology. *Annual Review of Entomology* 43, 701–726.

Gould, F. and Tabashnik, B.E. (1998) Bt-cotton resistance management. In: Mellon, M. and Rissler, J. (eds) *Now or Never: Serious New Plans to Save a Natural Pest Control*. Union of Concerned Scientists, Cambridge, Massachusetts, pp. 67–105.

Gould, F., Anderson, A., Reynolds, A., Bumgarner, L. and Moar, W. (1995) Selection and genetic analysis of a *Heliothis virescens* (Lepidoptera: Noctuidae) strain with high levels of resistance to *Bacillus thuringiensis* toxins. *Journal of Economic Entomology* 88, 545–1559.

Gould, F., Anderson, A., Jones, A., Sumerford, D., Heckel, D.G., Lopez, J., Micinski, S., Leonard, R. and Laster, M. (1997) Initial frequency of alleles for resistance to *Bacillus thuringiensis* toxins in field populations of *Heliothis virescens*. *Proceedings of the National Academy of Sciences of the USA* 94, 3519–3523.

Gould, F., Cohen, M.B., Bentur, J.S., Kennedy, G.G. and Van Duyn, J. (2006) Impact of small fitness costs on pest adaptation to crop varieties with multiple toxins: a heuristic model. *Journal of Economic Entomology* 99, 2091–2099.

Griffitts, J.S. and Aroian, R.V. (2005) Many roads to resistance: how invertebrates adapt to Bt toxins. *BioEssays* 27, 614–624.

Groeters, F.G. and Tabashnik, B.E. (2000) Roles of selection intensity, major genes, and minor genes in evolution of insecticide resistance. *Journal of Economic Entomology* 93, 1580–1587.

Groeters, F.R., Tabashnik, B.E., Finson, N. and Johnson, M.W. (1994) Fitness costs of resistance to *Bacillus thuringiensis* in the diamondback moth (*Plutella xylostella*). *Evolution* 48, 197–201.

Gujar, G.T., Kalia, V., Kumari, A., Singh, B.P., Mittal, A., Nair, R. and Mohan, M. (2007) *Helicoverpa armigera* baseline susceptibility to *Bacillus thuringiensis* Cry toxins and resistance management for Bt cotton in India. *Journal of Invertebrate Pathology* 95, 214–219.

Gustafson, D.I., Head, G.P. and Caprio, M.A. (2006) Modeling the impact of alternative hosts on *Helicoverpa zea* adaptation to Bollgard cotton. *Journal of Economic Entomology* 99, 2116–2124.

Heckel, D.G. (1994) The complex genetic basis of resistance to *Bacillus thuringiensis* in insects. *Biocontrol Science and Technology* 4, 405–417.

Heckel, D.G., Gahan, L.J., Baxter, S.W., Zhao, J.-Z., Shelton, A.M., Gould, F. and Tabashnik, B.E. (2007) The diversity of Bt resistance genes in species of Lepidoptera. *Journal of Invertebrate Pathology* 95, 192–197.

Herrero, S., Gechev, T., Bakker, P.L., Moar, W.J. and de Maagd, R.A. (2005) *Bacillus thuringiensis* Cry1Ca-resistant *Spodoptera exigua* lacks expression of one of four aminopeptidase N genes. *BMC Genomics* 6, 1–10.

Higginson, D.M., Morin, S., Nyboer, M., Biggs, R., Tabashnik, B.E. and Carrière, Y. (2005) Evolutionary trade-offs of insect resistance to Bt crops: fitness costs affecting paternity. *Evolution* 59, 915–920.

Huang, F., Buschman, L.L. and Higgins, R.A. (2005) Larval survival and development of susceptible and resistant *Ostrinia nubilalis* (Lepidoptera: Pyralidae) on diet containing *Bacillus thuringiensis*. *Agricultural and Forest Entomology* 7, 45–52.

Ives, A.R. and Andow, D.A. (2002) Evolution of resistance to Bt crops: directional selection in structured environments. *Ecology Letters* 5, 792–801.

Jackson, R.E., Bradley, J.R. Jr and Van Duyn, J.W. (2004a) Performance of feral and Cry1Ac-selected *Helicoverpa zea* (Lepidoptera: Noctuidae) strains on transgenic cottons expressing either one or two *Bacillus thuringiensis* ssp. *kurstaki* proteins under greenhouse conditions. *Journal of Entomological Science* 39, 46–55.

Jackson, R.E., Bradley, J.R. Jr, Van Duyn, J. W. and Gould, F. (2004b) Comparative production of *Helicoverpa zea* (Lepidoptera: Noctuidae) from transgenic cotton expressing either one or two *Bacillus thuringiensis* proteins with or without insecticide over-

sprays. *Journal of Economic Entomology* 97, 1719–1725.

Jackson, R.E., Gould, F., Bradley, J.R. Jr and Van Duyn, J.W. (2006) Genetic variation for resistance to *Bacillus thuringiensis* in *Helicoverpa zea* (Lepidoptera: Noctuidae) from eastern North Carolina. *Journal of Economic Entomology* 99, 1790–1797.

James, C. (2007) Global status of commercialized biotech/GM Crops: 2007. *ISAAA Briefs* No. 37. International Service for the Acquisition of Agri-biotech Applications, Ithaca, New York.

James, L. (2006) Bollworms feeding on Bt cotton in Arkansas. Delta Farm Press. July 28, 2006. http://deltafarmpress.com/news/060728-cotton-bollworms/.

Janmaat, A.F. and Myers, J.H. (2003) Rapid evolution and the cost of resistance to *Bacillus thuringiensis* in greenhouse populations of cabbage loopers, *Trichoplusia ni. Proceedings of the Royal Society of London Series B: Biological Sciences* 270, 2263–2270.

Janmaat, A.F. and Myers, J.H. (2005) The cost of resistance to *Bacillus thuringiensis* varies with the host plant of *Trichoplusia ni. Proceedings of the Royal Society of London Series B: Biological Sciences* 272, 1031–1038.

Janmaat, A.F. and Myers, J.H. (2006) The influences of host plant and genetic resistance to Bacillus thuringiensis on trade-offs between offspring number and growth rate in cabbage looper, *Tricoplusia ni. Ecological Entomology* 31, 172–178.

Jurat-Fuentes, J.L., Gahan, L.J., Gould, F.L., Heckel, D.G. and Adang, M.J. (2004) The HevCaLP protein mediates binding specificity of the Cry1A class of *Bacillus thuringiensis* toxins in *Heliothis virescens*. *Biochemistry* 43, 14299–14305.

Kranthi, K.R., Naidu, S., Dhawad, C.S., Tatwawadi, A., Mate, K., Patil, E., Bharose, A.A., Behere, G.T., Wadaskar, R.M. and Kranthi, S. (2005) Temporal and intra-plant variability of Cry1Ac expression in Bt-cotton and its influence on the survival of the cotton bollworm, *Helicoverpa armigera* (Hübner) (Lepidoptera: Noctuidae). *Current Science* 89, 291–298.

Lawrence, S. (2005) Agbio keeps growing. *Nature Biotechnology* 23, 281.

Lenormand, T. and Raymond, M. (1998) Resistance management: the stable zone strategy. *Proceedings of the Royal Society of London Series B: Biological Sciences* 265, 1985–1990.

Li, G., Wu, K., Gould, F., Wang, H., Mikao, J., Gao, X. and Guo, Y. (2007) Increasing tolerance to Cry1Ac cotton from cotton bollworm, *Helicoverpa armigera,* was confirmed in Bt cotton farming area of China. *Ecological Entomology* 32, 366–375.

Li, H., Oppert, B., Higgins, R.A., Huang, R., Buschman, L.L., Gao, J.-R. and Zhu, K.Y. (2005) Characterization of cDNAs encoding three trypsin-like proteinases and mRNA quantitative analysis in Bt-resistant and-susceptible strains of *Ostrinia nubilalis*. *Insect Biochemistry and Molecular Biology* 35, 847–860.

Liu, Y.B. and Tabashnik, B.E. (1997a) Experimental evidence that refuges delay insect adaptation to *Bacillus thuringiensis* toxin Cry1C in diamondback moth. *Applied and Environmental Microbiology* 63, 2218–2223.

Liu, Y.B. and Tabashnik, B.E. (1997b) Inheritance of resistance to *Bacillus thuringiensis* toxin Cry1C in diamondback moth. *Applied and Environmental Microbiology* 63, 2218–2223.

Liu, Y.B., Tabashnik, B.E., Dennehy, T.J., Patin, A.J. and Bartlett, A.C. (1999) Development time and resistance to *Bt* crops. *Nature* 400, 519.

Liu, Y.B., Tabashnik, B.E., Dennehy, T.J., Patin, A.J., Sims, M.A., Meyer, S.K. and Carrière, Y. (2001a) Effects of Bt cotton and Cry1Ac toxin on survival and development of pink bollworm (Lepidoptera: Gelechiidae). *Journal of Economic Entomology* 94, 1237–1242.

Liu, Y.B., Tabashnik, B.E., Meyer, S.K., Carrière, Y. and Bartlett, A.C. (2001b) Genetics of pink bollworm resistance to *Bacillus thuringiensis* toxin Cry1Ac. *Journal of Economic Entomology* 94, 248–252.

Luttrell, R.G., Wan, L. and Knighten, K. (1999) Variation in susceptibility of Noctuid (Lepidoptera) larvae attacking cotton and soybean to purified endotoxin proteins and commercial formulations of *Bacillus thuringiensis*. *Journal of Economic Entomology* 92, 21–32.

Luttrell, R.G. and Ali, M.I. (2007) Exploring selection for Bt resistance in Heliothines: results of laboratory and field studies, In Boyd, S., Huffman, M., Richter, D. and Robertson, B. (eds) *Proceedings of the 2007 Beltwide Cotton Conferences*, New Orleans, LA, January 9–12, 2007, National Cotton Council of America, Memphis, TN, pp. 1073–1086.

Luttrell, R.G., Ali, I., Allen, K.C., Young, S.Y., III, Szalanski, Williams, K., Lorenz, G. Parker, C.D., Jr. and Blanco, C. (2004) Resistance to Bt in Arkansas populations of cotton bollworm. In Richter, D.A. (ed.) *Proceedings of the 2004 Beltwide Cotton Conferences*, San Antonio, TX, January 5–9, 2004, National Cotton Council of America, Memphis, TN pp. 1373–1383.

Mao, Y.-B., Cai, W.-J., Wang, J.-W., Hong, G.-J., Tao, X.-Y., Wang, L.-J., Huang, Y.-P. and Chen, X.Y. (2007) Silencing a cotton bollworm P450 monooxygenase gene by plant-mediated RNAi impairs larval tolerance of gossypol. *Nature Biotechnology* 25, 1307–1313.

Marvier, M., McCreedy, C., Regetz, J. and Kareiva, P. (2007) A meta-analysis of effects of Bt cotton and maize on non-target invertebrates. *Science* 316, 1475–1477.

Matten, S.R., Head, G.P. and Quemada (2008) How government regulation can help or hinder the integration of Bt crops within IPM programs. In Romeis, J., Shelton, A.M. and Kennedy, G.G. (eds) *Integration of Insect-Resistant Genetically Modified Crops within IPM Programs*, Springer, New York, pp. 27–39.

McKenzie, J.A. (1996) *Ecological and Evolution-ary Aspects of Insecticide Resistance*. R. G. Landes and Academic Press, Austin, Texas.

Mehlo, L., Gahakwa, D., Nghia, P., Loc, N.T., Capell, T., Gatehouse, J.A., Gatehouse, A.M.R. and Christou, P. (2005) *Proceedings* of the National Academy of Sciences of the USA 102, 7812–7816.

Mendelsohn, M., Kough, J., Vaitizis, Z. and Matthews, K. (2003) Are Bt crops safe? *Nature Biotechnology* 21, 1003–1009.

Moar, W.J. (2003) Breathing new life into insect-resistant plants. *Nature Biotechnology* 21, 1152–1154.

Monsanto (2006) Monsanto biotechnology trait acreage: fiscal years 1996 to 2006. Updated: 11 October 2006. Available at: www.monsanto.com/monsanto/content/investor/financial/reports/2006/Q42006Acreage.pdf

Monsanto (2007) Monsanto biotechnology trait acreage: fiscal years 1996 to 2007 (Monsanto, St. Louis, updated October 10, 2007) http://www.monsanto.com/investors/presentations.asp; scroll down to Fourth-Quarter 2007 Monsanto Company Earnings Conference Call, Biotech Acres).

Morin, S., Biggs, R.W., Sisterson, M.S., Shriver, L., Ellers-Kirk, C., Higginson, D., Holley, D., Gahan, L.J., Heckel, D.G., Carrière, Y., Dennehy, T.J., Brown, J.K. and Tabashnik, B.E. (2003) Three cadherin alleles associated with resistance to *Bacillus thuringiensis* in pink bollworm. *Proceedings of the National Academy of Sciences of the USA* 100, 5004–5009.

Naranjo, S.E., Head, G. and Dively, G.P. (2005) Field studies assessing arthropod nontarget effects in Bt transgenic crops, introduction. *Environmental Entomology* 34, 1178–1180.

O'Callaghan, M., Glare, T.R., Burgess, E.P.J. and Malone, L.A. (2005) Effects of plants genetically modified for insect resistance on nontarget organisms. *Annual Review of Entomology* 50, 271–292.

Onstad, D.W. and Guse, G.A. (1999) Economic analysis of the use of transgenic crops and nontransgenic refuges for management of European corn borer (Lepidoptera: Pyralidae). *Journal of Economic Entomology* 92, 1256–1265.

Onstad, D.W., Guse, G.A., Porter, P., Buschman, L.L., Higgins, R.A., Sloderbeck, P.E., Peairs, F.B. and Gronholm, G.B. (2002) Modeling the development of resistance by stalk-boring lepidopteran insects

(Crambidae) in areas with transgenic corn and frequent insecticide use. *Journal of Economic Entomology* 95, 1033–1043.

Oppert, B., Hammel, R., Thorne, J.E. and Kramer, K.J. (2000) Fitness cost of resistance to *Bacillus thuringiensis* in the Indianmeal moth, *Plodia interpunctella*. *Entomologia Experimentata et Applicata* 96, 281–287.

Peck, S., Gould, F. and Ellner, S.P. (1999) Spread of resistance in spatially extended regions of transgenic cotton, implications for management of *Heliothis virescens* (Lepidoptera, Noctuidae). *Journal of Economic Entomology* 92, 1–16.

Pittendrigh, B.R., Gaffney, P.G., Huesing, J.E., Onstad, D.W., Roush, R.T. and Murdock, L.L. (2004) 'Active' refuges can inhibit the evolution of resistance in insects towards transgenic insect-resistant plants. *Journal of Theoretical Biology* 231, 461–474.

Ramachandran, S., Buntin, G.D., All, J.N., Tabashnik, B.E., Raymer, P.L., Adang, M.J., Pulliam, D.A. and Stewart, C.N. Jr (1998) Survival, development, and oviposition of resistant diamondback moth (Lepidoptera, Plutellidae) on transgenic canola producing a *Bacillus thuringiensis* toxin. *Journal of Economic Entomology* 91, 1239–1244.

Ravi, K. C., Mohan, K.S., Manjunath, T.M., Head, G., Patil, B.V., Angeline Greba, D.P., Premalatha, K., Peter, J. and Rao, N. G. V. (2005) Relative abundance of *Helicoverpa armigera* (Lepidoptera: Noctuidae) on different host crops in India and the role of these crops as natural refuge to *Bacillus thuringiensis* cotton. *Environmental Entomology* 34, 59–69.

Raymond, B., Sayyed, A.H. and Wright, D.J. (2005) Genes and environment interact to determine the fitness costs of resistance to *Bacillus thuringiensis. Proceedings of the Royal Society of London Series B: Biological Sciences* 272, 1519–1524.

Raymond, B., Sayyed, A.L. and Wright, D.J. (2006) Host plant and population determine the fitness costs of resistance to *Bacillus thuringiensis. Biology Letters* 3, 82–85.

Riggin-Bucci, T.M. and Gould, F. (1997) Impact of intraplot mixtures of toxic and non-toxic plants on population dynamics of diamondback moth (Lepidoptera, Plutellidae) and its natural enemies. *Journal of Economic Entomology* 90, 241–251.

Romeis, J., Meissle, M. and Bigler, F. (2006) Transgenic crops expressing *Bacillus thuringiensis* toxins and biological control. *Nature Biotechnology* 24, 63–71.

Roush, R.T. (1997) Bt-transgenic crops, just another pretty insecticide or a new start for resistance management? *Pesticide Science* 51, 328–344.

Schnepf, E., Crickmore, N., van Rie, J., Lereclus, D., Baum, J., Feitelson, J., Zeigler, D.R. and Dean, D.H. (1998) *Bacillus thuringiensis* and its pesticidal crystal proteins. *Microbiology and Molecular Biology Review* 62, 775–806.

Schuler, T.H., Poppy, G.M., Kerry, B.R. and Denholm, I. (1998) Insect-resistant transgenic plants. *Trends in Biotechnology* 16, 168–175.

Shelton, A.M., Tang, J.D., Roush, R.T., Metz, T.D. and Earle, E.D. (2000) Field tests on managing resistance to Bt-engineered plants. *Nature Biotechnology* 18, 339–342.

Shelton, A.M., Zhao, J.-Z. and Roush, R.T. (2002) Economic, ecological, food safety, and social consequences of the deployment of Bt transgenic plants. *Annual Review of Entomology* 47, 845–881.

Showalter, A.M., Heuberger, S., Tabashnik, B.E. and Carrière, Y. (2008) A primer for the use of insecticidal transgenic cotton in developing countries. *Journal of Insect Science, in press.*

Siegfried, B.D., Vaughn, T.T. and Spencer, T. (2005) Baseline susceptibility of western corn rootworm to Cry3Bb1 *Bacillus thuringiensis* toxin. *Journal of Economic Entomology* 98, 1320–1324.

Siegfried, B.D., Spencer, T., Crespo, A.L., Storer, N., Head, G.P., Owens, E.D. and Guyer, D. (2007) Ten years of Bt resistance monitoring in the European corn borer: what we know, what we don't know, and what we can do better. *American Entomologist* 53, 208–214.

Sisterson, M.S., Antilla, L., Carrière, Y., Ellers-Kirk, C. and Tabashnik, B.E. (2004) Effects of insect population size on evolution of resistance to transgenic crops. *Journal of Economic Entomology* 97, 1413–1424.

Sisterson, M.S., Carrière, Y., Dennehy, T.J. and Tabashnik, B.E. (2005) Evolution of resistance to transgenic crops: interactions between insect movement and field distribution. *Journal of Economic Entomology* 98, 1751–1762.

Soberón, M., Pardo-López, L., López, I., Gómez, I., Tabashnik, B.E. and Bravo, A. (2007) Engineering modified Bt toxins to counter insect resistance. *Science* 318:1640–1642.

Stodola, T.J., Andow, D.A., Hyden, A.R., Hinton, J.L., Roark, J.J., Buschman, L.L., Porter, P. and Cronholm, G.B. (2006) Frequency of resistance to *Bacillus thuringiensis* toxin Cry1Ab in southern United States corn belt populations of European corn borer (Lepidoptera: Crambidae). *Journal of Economic Entomology* 99, 502–507.

Storer, N.P. (2003) A spatially explicit model simulating western corn rootworm (Coleoptera: Chrysomelidae) adaptation to insect-resistant maize. *Journal of Economic Entomology* 96, 1530–1547.

Tabashnik, B.E. (1990) Modeling and evaluation of resistance management tactics. In: Roush, R.T. and Tabashnik, B.E. (eds) *Pesticide Resistance in Arthropods.* Chapman & Hall, New York, pp. 153–182.

Tabashnik, B.E. (1994) Evolution of resistance to *Bacillus thuringiensis*. *Annual Review of Entomology* 39, 47–79.

Tabashnik, B.E. and Carrière, Y. (2008) Evolution of insect resistance to transgenic plants. In: Tilmon, K. (ed.) *Specialization, Speciation, and Radiation: The Evolutionary Biology of Herbivorous Insects.* University of California Press, Berkeley, California, pp. 267–279.

Tabashnik, B.E. and Croft, B.A. (1982) Managing pesticide resistance in crop-arthropod complexes, interactions between biological and operational factors. *Environmental Entomology* 11, 1137–1144.

Tabashnik, B.E., Cushing, N.L., Finson, N. and Johnson, M.W. (1990) Field development of resistance to *Bacillus thuringiensis* in diamondback moth (Lepidoptera: Plutellidae). *Journal of Economic Entomology* 83, 1671–1676.

Tabashnik, B.E., Schwartz, J.M., Finson, N. and Johnson, M.W. (1992) Inheritance of resistance to *Bacillus thuringiensis* in diamondback moth (Lepidoptera, Plutellidae). *Journal of Economic Entomology* 85, 1046–1055.

Tabashnik, B.E., Malvar, T., Liu, Y.B., Borthakur, D., Shin, B.S., Park, S.H., Masson, L., de Maagd, R.A. and Bosch, D. (1996) Cross-resistance of diamondback moth indicates altered interactions with domain II of *Bacillus thuringiensis* toxins. *Applied and Environmental Microbiology* 62, 2839–2844.

Tabashnik, B.E., Liu, Y.B., Finson, N., Masson, L. and Heckel, D.G. (1997) One gene in diamondback moth confers resistance to four *Bacillus thuringiensis* toxins. *Proceedings of the National Academy of Sciences of the USA* 94, 1640–1644.

Tabashnik, B.E., Liu, Y.B., Malvar, T., Heckel, D.G., Masson, L. and Ferré, J. (1998) Insect resistance to *Bacillus thuringiensis*: uniform or diverse? *Philosophical Transactions of the Royal Society of London Series B* 353, 1751–1756.

Tabashnik, B.E., Patin, A.L., Dennehy, T.J., Liu, Y.B., Carrière, Y., Sims, M.A. and Antilla, L. (2000a) Frequency of resistance to *Bacillus thuringiensis* in field populations of pink bollworm. *Proceedings of the National Academy of Sciences of the USA* 97, 12980–12984.

Tabashnik, B.E., Roush, R.T., Earle, E.D. and Shelton, A.M. (2000b) Resistance to Bt toxins. *Science* 287, 42.

Tabashnik, B.E., Liu, Y.B., Dennehy, T.J., Sims, M.A., Sisterson, M.S., Biggs, R.W. and Carrière, Y. (2002a) Inheritance of resistance to Bt toxin Cry1Ac in a field-derived strain of pink bollworm (Lepidoptera, Gelechiidae). *Journal of Economic Entomology* 95, 1018–1026.

Tabashnik, B.E., Dennehy, T.J., Sims, M.A., Larkin, K., Head, G.P., Moar, W.J. and Carrière, Y. (2002b) Control of resistant pink bollworm by transgenic cotton with *Bacillus thuringiensis* toxin Cry2Ab. *Applied and Environmental Microbiology* 68, 3790–3794.

Tabashnik, B.E., Carrière, Y., Dennehy, T.J., Morin, S., Sisterson, M., Roush, R.T., Shelton, A.M. and Zhao, J.-Z. (2003) Insect resistance to transgenic Bt crops: lessons from the laboratory and field. *Journal of Economic Entomology* 96, 1031–1038.

Tabashnik, B.E., Gould, F. and Carrière, Y. (2004) Delaying evolution of insect resistance to transgenic crops by decreasing dominance and heritability. *Journal of Evolutionary Biology* 17, 904–912.

Tabashnik, B.E., Dennehy, T.J. and Carrière, Y. (2005a) Delayed resistance to transgenic cotton in pink bollworm. *Proceedings of the National Academy of Sciences of the USA* 102, 15389–15393.

Tabashnik, B.E., Biggs, R.W., Higginson, D.W., Henderson, S., Unnithan, D.C., Unnithan, G.C., Ellers-Kirk, C., Sisterson, M.S., Dennehy, T.J., Carrière, Y. and Morin, S. (2005b) Association between resistance to Bt cotton and cadherin genotype in pink bollworm. *Journal of Economic Entomology* 98, 635–644.

Tabashnik, B.E., Fabrick, J.A., Henderson, S., Biggs, R.W., Yafuso, C.A., Nyboer, M.E., Manhardt, N.M., Coughlin, L.A., Sollome, J., Carrière, Y., Dennehy, T.J. and Morin, S. (2006) DNA screening reveals pink bollworm resistance to Bt cotton remains rare after a decade of exposure. *Journal of Economic Entomology* 99, 1525–1530.

Tabashnik, B.E., Gassmann, A.J., Crowder, D.W. and Carrière, Y. (2008) Insect resistance to Bt crops: evidence versus theory. *Nature Biotechnology* 26, 199–202.

Tang, J.D., Gilboa, S., Roush, R.T. and Shelton, A.M. (1997) Inheritance, stability, and lack-of-fitness costs of field-selected resistance to *Bacillus thuringiensis* in diamondback moth (Lepidoptera: Plutellidae) from Florida. *Journal of Economic Entomology* 90, 732–741.

Tang, J.D., Collins, H.L., Roush, R.T., Metz, T.D., Earle, E.D. and Shelton, A.M. (1999) Survival, weight gain, and oviposition of resistant and susceptible *Plutella xylostella* (L.) on broccoli expressing Cry1Ac toxin of *Bacillus thuringiensis*. *Journal of Economic Entomology* 92, 47–55.

Tang, J.D., Collins, H.L., Metz, T.D., Earle, E.D., Zhao, J.-Z., Roush, R.T. and Shelton, A.M. (2001) Greenhouse tests on resistance management of Bt transgenic plants using refuge strategies. *Journal of Economic Entomology* 94, 240–247.

USEPA (US Environmental Protection Agency) (1998) *The Environmental Protection Agency's White Paper on* Bt *Plant-Pesticide Resistance Management* (EPA Publication 739-S-98-001, 1998). Available at: www.epa.gov/EPA-PEST/1998/January/Day-14/paper.pdf

USEPA (US Environmental Protection Agency) (2001) Biopesticides registration action document – *Bacillus thuringiensis* plant-incorporated protectants. Available at: http://www.epa.gov/pesticides/biopesticides/pips/bt_brad.htm

USEPA (US Environmental Protection Agency) (2002) Biopesticides registration action document – *Bacillus thuringiensis* Cry2Ab2 protein and its genetic material necessary for its production in cotton. Available at: http://www.epa.gov/pesticides/biopesticides/ingredients/factsheets/factsheet_006487.htm

USEPA (US Environmental Protection Agency) (2007) Pesticide news story: EPA approves natural refuge for insect resistance management in Bollgard II cotton. Available at: http://www.epa.gov/oppfead1/cb/csb_page/updates/2007/bollgard-cotton.htm

Van Rensburg, J.B.J. (2007) First report of field resistance by stem borer, *Busseola fusca* (Fuller) to Bt-transgenic maize. *South African Journal of Plant and Soil* 24, 147–151.

Wang, P., Zhao, J.-Z., Rodrigo-Simón, A., Kain, W., Janmaat, A.F., Shelton, A.M., Ferré, J. and Myers, J. (2007) Mechanism of resistance to *Bacillus thuringiensis* toxin Cry1Ac in a greenhouse population of the cabbage looper, *Trichoplusia ni*. *Applied and Environmental Microbiology* 73, 1199–1207.

Wenes, A.-L., Bouguet, D., Andow, D.A., Courtin, C., Carré, G., Lorme, P., Sanchez, L. and Augustin, S. (2006) Frequency and fitness cost of resistance to *Bacillus thuringiensis* in *Chrysomela tremulae* (Coleoptera: Chrysomelidae). *Heredity* 97, 127–134.

Wu, K. (2007) Monitoring and management strategy for *Helicoverpa armigera* resistance to Bt cotton in China. *Journal of Invertebrate Pathology* 95, 220–223.

Wu, K., Guo, Y. and Gao, S. (2002a) Evaluation of the natural refuge function for *Helicoverpa armigera* within *Bacillus thuringiensis* transgenic cotton growing areas in North China. *Journal of Economic Entomology* 95, 832–837.

Wu, K., Guo, Y., Lv, N., Greenplate, J.T. and Deaton, R. (2002b) Resistance monitoring of *Helicoverpa armigera* to *Bacillus thuringiensis* insecticidal protein in China. *Journal of Economic Entomology* 95, 826–831.

Wu, K., Guo, Y. and Head, G. (2006) Resistance monitoring of *Helicoverpa armigera* (Lepidoptera: Noctuidae) to Bt insecticidal protein during 2001–2004 in China. *Journal of Economic Entomology* 99, 893–898.

Xu, X., Yu, L. and Wu, Y. (2005) Disruption of a cadherin gene associated with resistance to Cry1Ac ∂-endotoxin of *Bacillus thuringiensis* in *Helicoverpa armigera*. *Applied and Environmental Microbiology* 71, 948–954.

Yang, Y., Chen, H., Wu, Y., Yang, Y. and Wu, S. (2007) Mutated cadherin alleles from a field population of *Helicoverpa armigera* confer resistance to *Bacillus thuringiensis* toxin Cry1Ac. *Applied and Environmental Microbiology* 73, 6939–6944.

Zhang, X., Candas, M., Griko, N.B., Taussig, R. and Bulla, L.A. Jr (2006) A mechanism of cell death involving an adenylyl cyclase/PKA signaling pathway is induced by the Cry1Ab toxin of *Bacillus thuringiensis*. *Proceedings of the National Academy of Sciences of the USA* 103, 9897–9902.

Zhao, J.-Z., Collins, H.L., Tang, H.Z., Cao, J., Earle, E.D., Roush, R.T., Herrero, S., Escriche, B., Ferré, J. and Shelton, A.M. (2000) Development and characterization of diamondback moth resistance to transgenic broccoli expressing high levels of Cry1C. *Applied and Environmental Microbiology* 66, 3784–3789.

Zhao, J.-Z., Cao, J., Collins, H.L., Bates, S.L., Roush, R.T., Earle, E.D. and Shelton, A.M. (2005) Concurrent use of transgenic plants expressing a single and two *Bacillus thuringiensis* genes speeds insect adaptation to pyramided plants. *Proceedings of the National Academy of Sciences of the USA* 102, 8426–8430.

6 Resistance Management of Transgenic Insect-resistant Crops: Ecological Factors

B. Raymond[1] and D.J. Wright[2]

[1]Department of Zoology, University of Oxford, Oxford, UK; [2]Department of Life Sciences, Imperial College London, London,UK

Keywords: *Bt* toxins, ecology, integrated resistance management, transgenic crops

Summary

From their introduction in 1996–2007, the genetically modified (GM) crops grown with traits for insecticide resistance (GM IR) have been predominantly maize and cotton varieties expressing *Bacillus thuringiensis* Cry1 toxins for the control of larval stages of relatively few key moth pest species. Resistance management strategies have been refined with the life history of such lepidopteran pests in mind. With the introduction from 2003 of Cry3, Cry34 and Cry35 toxins for coleopteran pests of maize and with an increasing range of GM IR crops and other *Bt* toxins being commercialized, a much wider range of insect pest species and populations will become exposed to selection. Here, we consider the influence of three broad ecological factors on the evolution and management of pest resistance to GM IR crops: population-level processes; bottom-up effects of the host plant; and top-down effects deriving from natural enemies and pathogens. The exploitation of such ecological factors in resistance management systems, developments in transgenic crops and a novel genetic method for resistance management are discussed.

Introduction

In this chapter, the term genetically modified (GM) refers to genetically modified crops with traits for herbicide tolerance and/or insecticide resistance (GM IR).

GM IR crops commercialized up to 2007 constitutively express one, two or three insecticidal crystal (Cry) delta-endotoxins from the soil bacterium *Bacillus thuringiensis* (*Bt*; Chapter 5, this volume). Individual Cry toxins show considerable selectivity, typically being active against a limited range of species within the Lepidoptera, Coleoptera or Diptera, with very low toxicity towards non-target organisms (Schnepf *et al.*, 1998; see Chapters 8 and 18, this volume).

Spray products based on *Bt* were introduced in 1938 and they have proved to be very environmentally friendly compared with most other insecticides and are recommended for use in organic agriculture.

Global Status of GM Crops

GM crops were introduced commercially in 1996 and the global area of GM crops has increased annually from <2 million ha in 1996 to 114 million ha in 2007. Of the 114 million ha grown in 2007, 42 million ha were maize and cotton crops with *Bt* traits, a threefold increase in usage over 5 years (Tabashnik *et al.*, 2003; James, 2008). The most commonly used *Bt* toxins in GM crops up to 2007 have been Cry1Ac and Cry1Ab for the control of the larval stages of the major lepidopteran pests of cotton and maize; other Cry toxins with commercial registrations are Cry1F and Cry2Ab (against Lepidoptera), Cry3A, Cry3Bb, Cry34Ab and Cry35Ab (against Coleoptera; see Chapter 5, this volume).

In addition to the global increase in GM crops cultivated, there are a number of underlying trends (James, 2008):

- In 2007, while the USA, Argentina, Brazil, Canada, India and China grew 95% of GM crops, the number of countries adopting this technology had increased to 23.
- The use of GM crops is growing faster in developing than in industrialized countries (threefold faster in 2007), and farmers in the developing world (in 12 countries) now account for 90% of users of GM crops.
- While GM varieties of soybean, maize, cotton and canola represent 64%, 24%, 43% and 20% of hectares grown for each crop, respectively, and collectively 99% of all GM crops, various other GM crops have been introduced (papaya, sweet potato, rice, lucerne, carnation, petunia and poplar).
- Stacked ('pyramided') GM crops, combining two or three traits, are the fastest-growing sector of the market; in 2007, 63% of maize, 78% of cotton and 37% of all GM crops in the USA were stacked.
- The most common trait by area for GM crops in 2007 was herbicide tolerance (63%), followed by stacked traits (19%) and insect resistance (18%).

Resistance Management in a Changing Environment

The number of GM crop countries, crop types and traits, and area cultivated are all projected to double globally between 2006 and 2015 (James, 2007). This will present new challenges for GM IR crop resistance management, particularly in regions where increased production is particularly marked. For example, in India, the largest cotton growing country in the world by area, there was a reported 124-fold increase in *Bt* cotton grown between 2002 and 2007 (James, 2008).

As the usage and diversity of *Bt* crops grown continues to increase, more insect pest species and populations, some of which may be particularly adaptable, will become exposed to selection. Mechanisms for resistance may occur that show a much greater degree of cross-resistance between *Bt* toxins compared with the most common Cry1A resistance mechanism found to date, or that are associated with lower fitness costs (below). More cases of non-recessive resistance occurring under field conditions may also arise, as found in the bollworm/corn earworm *Heliothis zea* (Boddie) in the USA. Theory and field observations suggest that non-recessive resistance to Cry1Ac in *H. zea* (Burd *et al.*, 2003) could compromise resistance management strategies for *Bt* cotton (Tabashnik *et al.*, 2008; Chapter 5, this volume; see also below).

With *H. zea*, there is evidence for selection of Cry1Ac resistance in a few field populations in Arkansas and Mississippi 7–8 years after the introduction of *Bt* cotton (Ali *et al.*, 2006; Luttrell and Ali, 2007; Chapter 5, this volume). The lack of control failures with *H. zea* has been attributed to the fact that resistant populations are of relative rarity; Cry1Ac cotton still gives up to 60% larval mortality with resistant strains; stacked *Bt* cotton, with Cry1Ac and Cry2Ab, is increasingly grown; and that insecticide sprays have continued to be used to control high-density pest populations (Tabashnik *et al.*, 2008).

Agroecosystems for existing and new *Bt* crops (including continuous cropping, lack of natural refugia) and pest biology (e.g. host-plant range, mode and rate of reproduction, dispersal behaviour) may combine in some cases to provide especially suitable conditions for the development of resistance. For example, in high-value horticultural crops, the crucifer specialist, the diamondback moth, *Plutella xylostella* (L.), has a long history of evolving high levels of resistance to insecticides in many field populations, including *Bt* spray products (Ferré and Van Rie, 2002). While the cabbage looper, *Trichoplusia ni* (Hübner), has also evolved resistance to *Bt* sprays, this has been limited to the relatively enclosed confines of commercial glasshouses in western Canada (Janmaat and Myers, 2003; Franklin and Myers, 2008). Resistance to *Bt* Cry toxins and the resistance management strategies implemented, particularly for *Bt* cotton, are discussed in detail by Tabashnik and Carrière in Chapter 5 (this volume), and will therefore only be outlined here.

In addition to the cases of field/glasshouse selection described above, resistance to *Bt* toxins has been selected under laboratory conditions in various strains of coleopteran, dipteran and lepidopteran pests (Ferré and Van Rie, 2002). Resistance to Cry toxins has been linked variously to reduced target site binding and with pre- or post-binding events during the course of poisoning (Griffitts and Aroian, 2005; Heckel *et al.*, 2007). In most of the insect strains examined, resistance to *Bt* toxins is completely or partially recessive (Ferré and Van Rie, 2002; Tabashnik *et al.*, 2008) and backcross data in most strains fit a single locus or tightly linked loci model (Ferré and Van Rie, 2002).

In the most studied lepidopteran species (*P. xylostella*, *T. ni*, *Plodia interpunctella* Hübner, *Heliothis virescens* F., *Pectinophora gossypiella* Saunders and *Helicoverpa armigera* Hübner), the commonest form of resistance found has been termed 'Mode 1'. Mode 1 resistance is characterized as being recessive, with >500-fold resistance to one or more Cry1A toxins (typically in most studies between Cry1Ac and Cry1Ab) and little or no cross-resistance to Cry1C, and is linked to reduced binding of toxin on the insect midgut membrane. In *H. virescens*, *H. armigera* and *P. gossypiella* at least, Mode 1 resistance has been associated with mutations in a toxin-binding 12-cadherin-domain protein (Heckel *et al.*, 2007). However, mapping studies have shown that Mode 1 resistance in *P. xylostella* has a different and as yet unknown molecular genetic basis (Baxter *et al.*, 2005). This type of resistance could be of greater concern than mutations in the cadherin gene which appear to incur high fitness costs (Heckel *et al.*, 2007).

Molecular genetic studies have detected an additional major Cry1A resistance gene in *H. virescens*, which does not show linkage with Cry1A resistance in *P. xylostella* (Heckel *et al.*, 2007), and Cry1A resistance linked to a protoxin-processing protease in *P. interpunctella*. Evidence for non-Mode 1 resistance has also been found in a number of *Bt*-resistant insect strains using classical genetic and biochemical methods (Ferré and Van Rie, 2002; Sayyed *et al.*, 2005). Studies on a laboratory-selected strain of *H. zea* have shown that resistance to Cry1Ac was not linked to loss of toxin binding but that proteases may be involved in resistance (Anilkumar *et al.*, 2008).

In contrast to *Bt* sprays, resistance to *Bt* crops leading to control failures has yet to evolve in a field population of insects (Bates *et al.*, 2005; Tabashnik *et al.*, 2005, 2008), although, interestingly, a number of laboratory-selected strains of insects can survive on *Bt* plants (Tabashnik *et al.*, 2003). While it is difficult to test how effective resistance management strategies are under field conditions it is certainly true that resistance management strategies for *Bt* crops have been proactive rather than reactive, as has generally been the case with conventional pesticides.

Resistance management for *Bt* crops is widely based upon a strategy of providing refugia (non-*Bt* crops) to allow some homozygous susceptible insects to survive and mate with resistant individuals, thereby reducing the dominance of resistance (Chapter 5, this volume). The refugia strategy is generally combined with the use of GM crops that produce a sufficiently high dose of toxin to kill insects that are heterogeneous for resistance, thus rendering resistance functionally recessive (Gould, 1998), a major exception being *H. zea* (Tabashnik *et al.*, 2008; see Chapter 5, this volume). The introduction from 2003 of GM maize lines expressing Cry3, Cry34 and Cry35 toxins, which do not provide the same level of control of coleopteran pests compared with Cry toxins for lepidopteran pests on cotton and maize, may require larger refuges to prevent the development of resistance (Chapter 5, this volume).

Ecology and the Evolution of Resistance

Ecological factors are extremely important for the evolution and spread of resistance genes (May, 1985; Denholm and Rowland, 1992). Population dynamics, such as the ability of populations to recover from a perturbation (e.g. a pesticide application), can alter the rate of evolution of resistance (Comins, 1977). Populations that bounce back from perturbations to a level above their equilibrium density (overcompensating populations) will tend to evolve resistance more quickly than those that recover slowly from challenges (undercompensating populations; Comins, 1977). Population structure, such as subdivision into metapopulations, can also affect rates of evolution, particularly of recessive resistance genes (Caprio and Hoy, 1994). Subdivision can lead to variation in gene frequencies between metapopulations and with local increases in the level of homozygosity. Metapopulations with high levels of homozygous resistant pests can then 'seed' neighbouring metapopulations with phenotypically resistant insects (Caprio and Hoy, 1994).

However, only a limited range of ecological factors can be manipulated in the field for the benefit of resistance management, and this is the main focus of this review. Here we consider three broad ecological factors: population levels processes, in particular migration; the ecological impact of food plant of the pest (*bottom-up effects*) and those deriving from natural enemies and pathogens (*top-down effects*).

Population-level processes and the evolution of resistance

Migration rates between *Bt* and conventional crops can, to some extent, be affected by the spatial distribution of these two crop types. The implementation of the high dose/refuge strategy for the management of resistance in GM IR *Bt* crops has been influenced by the potential impact of migration on the evolution of resistance (Chapter 5, this volume). Low migration between refugia and toxin-expressing crops will lead to non-random mating and, potentially, lead to local increases in numbers of homozygous resistant pests that can survive on *Bt* crops. It follows that refugia should be placed within the adult flight distance of the GM IR crop in order to successfully prevent the emergence of resistant homozygotes (Peck *et al.*, 1999).

The consequences of migration for the evolution of resistance are not always simple, however. While some simulation models have found that the rate of evolution of resistance declines as migration increases (Ives and Andow, 2002) and that spatial configurations of refugia that maximize mixing will be more effective at slowing the evolution of resistance (Cerda and Wright, 2004), other studies have found that intermediate levels of dispersal can lead to the most rapid rates of resistance evolution (Comins, 1977; Caprio, 2001).

Conflicting results may arise in part because dispersal disrupts non-random mating and thereby slows the evolution of resistance but dispersal can also improve the mating success of adults carrying the resistance allele and thereby increase their fitness and the spread of resistance (Ives and Andow, 2002). Sisterson *et al.* (2005) showed that the effects of dispersal on the evolution of resistance can depend upon the assumptions made about the abundance and distribution of refugia. Spatially isolated refugia are more beneficial when movement rates are high, whereas when refugia and *Bt* crops are adjacent and their locations fixed, increased dispersal increases the rate of evolution of resistance (Sisterson *et al.*, 2005). They emphasized that the optimal course for resistance management (uniform distribution of refugia and no rotation) did not provide the best pest control. In contrast with much of the *Bt* resistance management literature, Vacher *et al.* (2003) suggested that the resistance could be delayed more effectively with an aggregated distribution of refugia and low toxin expression. The validity of their models did, however, depend upon the resistance being recessive as the dose of toxins in GM crops declined, whereas the dominance of resistance is generally extremely sensitive to toxin dose (Tabashnik *et al.*, 2004). In fact, recent evolution of resistance to *Bt* crops in *H. zea* in the USA has been attributed to partial dominance of resistance in *Bt* cotton (Tabashnik *et al.*, 2008).

Bottom-up impacts on the evolution of resistance

Ecological and environmental factors are also important for the fitness costs of resistance to pesticides. These fitness costs are commonly exacerbated under stressful or extreme environmental conditions (Foster *et al.*, 1997; Kraaijeveld and Godfray, 1997; Carrière *et al.*, 2001b; Gazave *et al.*, 2001; Raymond *et al.*, 2005). Fitness costs can significantly inhibit the evolution of resistance, especially of largely recessive resistance genes (Carrière and Tabashnik, 2001). Large environmentally dependent fitness costs associated with resistance to *Bt* in the pink bollworm, *P. gossypiella,* for example, during overwintering, are very likely to be one of the factors that have prevented the increase in frequencies of *Bt* resistance in populations exposed to *Bt* cotton in recent years (Carrière *et al.*, 2001a,b, 2004; Carrière and Tabashnik, 2001). Fitness costs can also be exploited in resistance management in conventional crops by rotating alternative pesticidal products or by applying pesticides only when warranted by economic thresholds (Tabashnik, 1989; Croft, 1990; Curtis *et al.*, 1993).

One possibility for managing the size and possibly the dominance of fitness costs is through the choice of crop cultivar or species. *Bt*-resistant *T. ni* cultured on pepper, a poor-quality crop for this insect, suffers severe fitness penalties (Janmaat and Myers, 2005). Host-plant species and plant defence compounds such as gossypol, a cotton phytochemical, can increase the magnitude and dominance of fitness costs associated with *Bt* resistance in *H. armigera* and *P. gossypiella* (Carrière *et al.*, 2004; Bird and Akhurst, 2007), although increased fitness costs were not found on cotton cultivars expressing larger quantities of gossypol (Carrière *et al.*, 2005). Plant species, and potentially plant age, can alter the fitness costs in diamondback moth, *P. xylostella*, feeding on *Brassica* spp. (Raymond *et al.*, 2005, 2007b). A cautionary note for the exploitation of fitness costs in resistance management is that environmental effects on the fitness costs of resistance interact with resistance genotype (Carrière *et al.*, 2004; Raymond *et al.*, 2005, 2007b). Unless the predominant resistant genotype in the field is well characterized, laboratory results may not be particularly helpful in inferring the impact of crop cultivar on the evolution of resistance in the field.

Crop cultivars that are poorer resources for pests will, by definition, reduce pest fitness and population growth rates. In conventional crops, this may be beneficial by reducing pest population size below spray thresholds. In integrated pest management, lower-quality crop plants may also mean that natural enemies will be better able to control pest populations (Haukioja, 1991; Denno *et al.*, 2002). Mathematical modelling also indicates that reduction in overall pest fitness (not just that of resistant pests) can slow or prevent the evolution of resistance (Ives and Andow, 2002; Carrière *et al.*, 2005). Stochastic simulation models, however, indicate that relying on low pest fitness to reduce the rate of evolution of resistance is not a reliable strategy (B. Raymond, Oxford, 2008, unpublished data). If pest fitness is low enough to prevent populations replacing themselves, pests may be driven to low population sizes where the combination of intense selection pressure and variation in gene frequencies will mean that a proportion of pest populations could rapidly evolve resistance (Sisterson *et al.*, 2004; B. Raymond, Oxford, 2008, unpublished data).

Top-down impacts on the evolution of resistance

For conventional crops, one of the most obvious benefits of effective natural enemies is their ability to control populations and reduce the need to apply pesticides (Chilcutt and Tabashnik, 1999). This possibility is not open for resistance management on GM crops, although insect natural enemies such as parasitoids may be able to attack and grow within *Bt*-resistant pests on *Bt*-engineered crop plants (Schuler *et al.*, 1999, 2004). Insect pathogens, such as nucleopolyhedroviruses (NPVs) and nematodes, can also be combined with GM crops (as treatments for refugia) or with *Bt* sprays in conventional crops (Gassman *et al.*, 2006; Raymond *et al.*, 2006, 2007a).

Glasshouses or polytunnels can be particularly suitable environments for the use of pathogens. NPVs, for example, can cycle very efficiently in glasshouses, which have reduced ultraviolet (UV) light and enhanced viral survival (Huber, 1998). *T. ni*, which is already encountering resistance problems in Canadian glasshouses (Janmaat and Myers, 2005), suffers from latent NPV infections (A. Janmaat, Helsinki, 2004, personal communication). Historically, *T. ni* NPV has been shown to be quite an effective biological control agent (Jacques, 1974).

There is increasing evidence that pathogens and natural enemies can benefit resistance management programmes through improving pest control and/or by increasing the magnitude and dominance of fitness costs (Gassman *et al.*, 2006; Raymond *et al.*, 2007a). Pathogens that cycle naturally in pest populations can provide substantial benefits in managing resistance because additional rounds of infection that occur after application can provide effective population control and potentially reduce pesticide application rates (Entwistle *et al.*, 1983; Dwyer and Elkinton, 1993; Pingel and Lewis, 1997). Some combinations, however, may not be suitable. A fungal pathogen was shown to be capable of accelerating the evolution of resistance when applied to a low *Bt*-toxin expression crop, because susceptible insects had increased restlessness and acquired more infections in the presence of the toxin (Johnson *et al.*, 1997a,b). However, this type of application is generally unnecessary for the current high-expression *Bt* crops. Spraying tank mixes of *Bt* with other pathogens may also fail to provide benefits for resistance management. Maximizing the kill rate of two products in mixed sprays increases their efficacy in slowing the evolution of resistance dramatically (Curtis *et al.*, 1993; Roush, 1993). Selection experiments and modelling of a mixed-spray strategy with *Bt* and an additional pathogen confirmed that any benefits in slowing the evolution of resistance were very sensitive to relative doses (Raymond *et al.*, 2007a). The application of two very high doses of expensive bio-pesticides, with different persistence properties, may not be a workable or economically viable approach to resistance management for *Bt* sprays.

Future Developments

Increased use of 'pyramided' (stacked) *Bt* crops with two or three Cry toxins that provide activity against a wider range of target pests compared with single constructs (James 2008; above) has the additional predicted advantage of slowing the rate at which resistance evolves within pest populations, provided resistance to one or more toxin does not already exist. However, field experiments with *P. xylostella* and *Bt* broccoli plants expressing one (Cry1Ac) or two toxins (Cry1Ac and Cry1C) suggest that this advantage may be lost if plants expressing one and two toxins are grown concurrently (Zhao *et al.*, 2005).

Transgenic plants expressing different types of *Bt* toxins and non-*Bt* toxins have also been produced, although to date only one such crop (SGK cotton expressing both *Bt* and the protease inhibitor CpTI) has as yet been commercialized (see Chapter 16, this volume). However, it is anticipated that some of these others will be commercialized in the near to longer term. These will provide new opportunities for resistance management, particularly where there are no cross-resistance mechanisms between toxins (Bates *et al.*, 2005; Christou *et al.*, 2006; Kurtz *et al.*, 2007; Tabashnik *et al.*, 2008; Wu *et al.*, 2008). VipCot cotton (Syngenta) is one such product for which commercial registration is being sought (Kurtz *et al.*, 2007). This transgene expresses two different types of *Bt* toxin, Cry1Ab and a novel vegetative insecticidal protein Vip3A, toxins that are both active against *H. virescens* and *H. zea*. Studies with laboratory Cry1A-selected *H. virescens* have shown no evidence of cross-resistance to Vip3A (Jackson *et al.*, 2007).

Other developments that could reduce the selection pressure for resistance include the use of inducible promoters to control when and in which plant tissue toxin expression occurs (Bates *et al.*, 2005). Toxins may also be reengineered to overcome the resistance mechanism (Soberón *et al.*, 2007; Chapter 5, this volume).

Genetic methods

Advances in insect molecular biology are also suggesting new methods for managing resistance in field populations. For example, a biotechnological tool intended to supplant sterile insect technique, i.e. the release of insects carrying a dominant lethal gene (RIDL), genetically engineers insects so that they can only complete their development in the presence of a particular diet supplement which represses the action of a female-specific dominant lethal gene (Thomas *et al.*, 2000). Simulation modelling of the mass release of pesticide-susceptible males indicates that RIDL could prevent, or reverse, the evolution of resistance and reduce the minimum refuge size required for effective resistance management (Alphey *et al.*, 2007).

Integrated methods for resistance management

Using a diverse approach to resistant management with GM IR crops has many attractive features and has been facilitated by the reduction in the use of conventional insecticides that has accompanied their introduction, most notably with *Bt* cotton (Chapter 5, this volume).

The refuge resistance management strategy for GM crops can potentially benefit both the agroecosystem and an integrated approach to resistance management, by providing havens for natural enemies, such as predators (Paoletti and Pimentel, 2000). Biological control, together with cultural and other traditional methods of Integrated Pest Management (IPM), can play an important role in improved, integrated resistance management (IRM) strategies for transgenic crops (Fitt, 2000; Bates *et al.*, 2005). For example, in Australia, control of *H. armigera* on *Bt* cotton is complemented by the destruction of overwintering pupae by ploughing and by the adoption of narrow planting windows to reduce the number of pest generations selected as part of a broader IPM strategy (Fitt, 2000).

Bt plants could also be employed as 'dead-end' trap crops where transgenic cash crops may not be practical or desirable (Shelton *et al.*, 2008). In trials, Indian mustard plants expressing Cry1C (Cao *et al.*, 2008) were shown to be preferred over cabbage and collard crops for oviposition by *P. xylostella*, while preventing subsequent survival of larval stages (Shelton *et al.*, 2008). Such a system may be particularly useful in complex agricultural environments, for example, in South-east Asia where up to 50 types of cruciferous cash crops may be grown in relatively compact areas comprising many hundreds of small (0.5–2 ha) farms (D. Wright, Cameron, Highlands, 2000, personal observation).

The number of farmers in the developing world adopting GM crop technology will undoubtedly continue to increase (James, 2008; above) and the use of GM IR and GM herbicide-tolerant (GM HT) and other biotech crops (drought tolerance, improved varieties of local crops) may be able to contribute significantly to agricultural sustainability (Thomson, 2008). Shelton (2007) has highlighted the social, political, regulatory and biological factors that can influence the adoption of pest control measures by farmers in low- and high-income countries. If the integration of other control methods with GM crops is perceived to be unnecessarily complex or less robust than the use of biotechnology alone, the sustainability of GM crops may be compromised.

Acknowledgements

We thank the Biotechnology and Biological Sciences Research Council (UK) for funding some of the work discussed in this chapter.

References

Ali, M.I., Luttrell, R.G. and Young, S.Y. (2006) Susceptibilities of *Helicoverpa zea* and *Heliothis virescens* (Lepidoptera:Noctuidae) populations to Cry1Ac insecticidal protein. *Journal of Economic Entomology* 99, 164–175.

Alphey, N., Coleman, P.G., Donnelly, C.A. and Alphey, L. (2007) Managing insecticide resistance by mass release of engineered insects. *Journal of Economic Entomology* 100, 1642–1649.

Anilkumar, K.J., Rodrigo-Simon, A., Ferré, J., Pusztai-Carey, M., Sivasupramaniam, S. and Moar, W.J. (2008) Production and characterization of *Bacillus thuringiensis* Cry1Ac-resistant cotton bollworm *Helicoverpa zea* (Boddie). *Applied and Environmental Microbiology* 74, 462–469.

Bates, S.L., Zhao, J.-Z., Roush, R.T. and Sheton, A.M. (2005) Insect resistance management in GM crops: past, present and future. *Nature Biotechnology* 23, 57–62.

Baxter, S.W., Zhao, J.-Z., Gahan, L.J., Shelton, A.M., Tabashnik, B.E. and Heckel, D.G. (2005) Novel genetic basis of field-evolved resistance to Bt toxins in *Plutella xylostella*. *Insect Molecular Biology* 14, 327–334.

Bird, L.J. and Akhurst, R.J. (2007) Effects of host plant species on fitness costs of Bt resistance in *Helicoverpa armigera* (Lepidoptera: Noctuidae). *Biological Control* 40, 196–203.

Burd, A.D., Gould, F., Bradley, J.R., Van Duyn, J.W. and Moar, W.J. (2003) Estimated frequency of nonrecessive *Bacillus thuringiensis* resistance genes in bollworm, *Helicoverpa zea* (Boddie) (Leipdoptera; Noctuidae) in eastern North Carolina. *Journal of Economic Entomology* 96, 137–142.

Cao, J., Shelton, A.M. and Earle, E.D. (2008) Sequential transformation to pyramid two

Bt genes in vegetable Indian mustard (*Brassica juncea* L.) and its potential for control of diamondback moth larvae. *Plant Cell Reports* 27, 479–487.

Caprio, M.A. (2001) Source-sink dynamics between transgenic and non-transgenic habitats and their role in the evolution of resistance. *Journal of Economic Entomology* 94, 698–705.

Caprio, M.A. and Hoy, M.A. (1994) Metapopulation dynamics affect resistance development in the predatory mite, *Metaseiulus-occidentalis* (Acari, Phytoseiidae). *Journal of Economic Entomology* 87, 525–534.

Carrière, Y. and Tabashnik, B.E. (2001) Reversing insect adaptation to transgenic insecticidal plants. *Proceedings of the Royal Society of London Series B-Biological Sciences* 268, 1475–1480.

Carrière, Y., Ellers-Kirk, C., Liu, Y.B., Sims, M.A., Patin, A.L., Dennehy, T.J. and Tabashnik, B.E. (2001a) Fitness costs and maternal effects associated with resistance to transgenic cotton in the pink bollworm (Lepidoptera: Gelechiidae). *Journal of Economic Entomology* 94, 1571–1576.

Carrière, Y., Ellers-Kirk, C., Patin, A.L., Sims, M.A., Meyer, S., Liu, Y.B., Dennehy, T.J. and Tabashnik, B.E. (2001b) Overwintering cost associated with resistance to transgenic cotton in the pink bollworm (Lepidoptera: Gelechiidae). *Journal of Economic Entomology* 94, 935–941.

Carrière, Y., Ellers-Kirk, C., Biggs, R., Higginson, D.M., Dennehy, T.J. and Tabashnik, B.E. (2004) Effects of gossypol on fitness costs associated with resistance to *Bt* in Pink Bollworm. *Journal of Economic Entomology* 97, 1710–1718.

Carrière, Y., Ellers-Kirk, C., Biggs, R., Degain, B., Holley, D., Yafuso, C., Evans, P., Dennehy,

T.J. and Tabashnik, B.E. (2005) Effects of cotton cultivar on fitness costs associated with resistance of pink bollworm (Lepidoptera: Gelechiidae) to Bt cotton. *Journal of Economic Entomology* 98, 947–954.

Cerda, H. and Wright, D.J. (2004) Modeling the spatial and temporal location of refugia to manage resistance in Bt transgenic crops. *Agriculture Ecosystems and Environment* 102, 163–174.

Chilcutt, C.F. and Tabashnik, B.E. (1999) Simulation of integration of *Bacillus thuringiensis* and the parasitoid Cotesia plutellae (Hymenoptera: Braconidae) for control of susceptible and resistant diamondback moth (Lepidoptera: Plutellidae). *Environmental Entomology* 28, 505–512.

Christou, P., Capell, T., Kohli, A., Gatehouse, J.A. and Gatehouse, A.M.R. (2006) Recent developments and future prospects in insect pest control in transgenic crops. *Trends in Plant Science* 11, 302–308.

Comins, H.N. (1977) The development of insecticide resistance in the presence of migration. *Journal of Theoretical Biology* 64, 177–197.

Croft, B.A. (1990) Management of pesticide resistance in arthropod pests. In: Green, M.B., LeBaron, H.M. and Moberg, W.K. (eds) *Managing Resistance to Agrochemicals*. American Chemical Society, Washington, DC, pp. 149–168.

Curtis, C.F., Hill, N. and Kasim, S.H. (1993) Are there effective resistance management strategies for vectors of human disease? *Biological Journal of the Linnean Society* 48, 3–18.

Denholm, I. and Rowland, M.W. (1992) Tactics for managing pesticide resistance in arthropods: theory and practice. *Annual Review of Entomology* 37, 91–112.

Denno, R.F., Gratton, C., Peterson, M.A., Langellotto, G.A., Finke, D.L. and Huberty, A.F. (2002) Bottom-up forces mediate natural-enemy impact in a phytophagous insect community. *Ecology* 85, 1443–1458.

Dwyer, G. and Elkinton, J.S. (1993) Using simple models to predict virus epizootics in gypsy moth populations. *Journal of Animal Ecology* 62, 1–11.

Entwistle, P.F., Adams, P.H.W., Evans, H.F. and Rivers, C.F. (1983) Epizootiology of nuclear polyhedrosis virus (Baculoviridae) in European spruce sawfly (*Gilpinia hercyniae*) – spread of disease from small epicentres in comparison with spread of disease in other hosts. *Journal of Applied Ecology* 20, 473–487.

Ferré, J. and Van Rie, J. (2002) Biochemistry and genetics of insect resistance to *Bacillus thuringiensis*. *Annual Review of Entomology* 47, 501–533.

Fitt, G.P. (2000) An Australian approach to IPM in cotton: integrating new technologies to minimise insecticide dependence. *Crop Protection* 19, 793–800.

Foster, S.P., Harrington, R., Devonshire, A.L., Denholm, I., Clark, S.J. and Mugglestone, M.A. (1997) Evidence for a possible fitness trade-off between insecticide resistance and the low temperature movement that is essential for survival of UK populations of *Myzus persicae* (Hemiptera: Aphididae). *Bulletin of Entomological Research* 87, 573–579.

Franklin, M.T. and Myers, J.H. (2008) Refuges in reverse: the spread of *Bacillus thuringiensis* resistance in unselected greenhouse populations of cabbage loopers *Trichoplusia ni*. *Agricultural and Forest Entomology* 10, 119–127.

Gassman, A.J., Stock, S.P., Carrière, Y. and Tabashnik, B.E. (2006) Effect of entomopathogenic nematodes on the fitness cost of resistance to Bt toxin in pink bollworm (Lepidoptera: Gelechiidae). *Journal of Economic Entomology* 99, 920–926.

Gazave, L., Chevillon, C., Lenormand, T., Marquine, M. and Raymond, M. (2001) Dissecting the cost of insecticide resistance genes during the overwintering period of the mosquito *Culex pipiens*. *Heredity* 87, 441–448.

Gould, F. (1998) Sustainability of transgenic insecticidal cultivars: integrating pest genetics and ecology. *Annual Review of Entomology* 43, 701–726.

Griffitts, J.S. and Aroian, R.V. (2005) Many roads to resistance: how invertebrates adapt to Bt toxins. *BioEssays* 27, 614–624.

Haukioja, E. (1991) Cyclic fluctuations in density – interactions between a defoliator and its host tree. *Acta Oecologica–International Journal of Ecology* 12, 77–88.

Heckel, D.G., Gahan, L.J., Baxter, S.W., Zhao, J.-Z., Shelton, A.M., Gould, F. and Tabashnik, B.E. (2007) The diversity of Bt resistance genes in species of Lepidoptera. *Journal of Invertebrate Pathology* 95, 192–197.

Huber, J. (1998) Regional summaries: Western Europe. In: Hunter-Fujita, F.R., Entwistle, P.F., Evans, H.F. and Crook, N.E. (eds) *Insect Viruses and Pest Management.* Wiley, New York, pp. 201–215.

Ives, A.R. and Andow, D.A. (2002) Evolution of resistance to Bt crops: directional selection in structured environments. *Ecology Letters* 5, 792–801.

Jackson, R.E., Marcus, M.A., Gould, F., Bradley, J.R. and Van Duyn, J.W. (2007) Cross-resistance responses of Cry1Ac-selected *Heliothis virescens* (Lepidoptera: Noctuidae) to the *Bacillus thuringiensis* protein Vip3A. *Journal of Economic Entomology* 100, 180–186.

Jacques, R.P. (1974) Occurence and accumulation of viruses of *Trichoplusia ni* in treated field plots. *Journal of Invertebrate Pathology* 23, 140–152.

James, C. (2007) Global status of commercialized biotech/GM crops: 2006. *ISAAA Briefs* 36, 1–10.

James, C. (2008) Global status of commercialized biotech/GM crops: 2007. *ISAAA Briefs* 37, 1–11.

Janmaat, A.F. and Myers, J.H. (2003) Rapid evolution and the cost of resistance to *Bacillus thuringiensis* in greenhouse populations of cabbage loopers. *Proceedings of the Royal Society of London Series B-Biological Sciences* 270, 2263–2270.

Janmaat, A.F. and Myers, J.H. (2005) The cost of resistance to *Bacillus thuringiensis* varies with the host plant of *Trichoplusia ni.* *Proceedings of the Royal Society of London Series B-Biological Sciences* 272, 1031–1038.

Johnson, M.T., Gould, F. and Kennedy, G.G. (1997a) Effect of an entomopathogen on adaptation of *Heliothis virescens* popula-

tions to transgenic host plants. *Entomologia Experimentalis et Applicata* 83, 121–135.

Johnson, M.T., Gould, F. and Kennedy, G.G. (1997b) Effects of natural enemies on relative fitness of *Heliothis virescens* genotypes adapted and not adapted to resistant host plants. *Entomologia Experimentalis et Applicata* 82, 219–230.

Kraaijeveld, A.R. and Godfray, H.C.J. (1997) Trade-off between parasitoid resistance and larval competitive ability in *Drosophila melanogaster. Nature* 389, 278–280.

Kurtz, R.W., McCaffery, A. and O'Reilly, D.O. (2007) Insect resistance management for Syngenta's VipCot transgenic cotton. *Journal of Invertebrate Pathology* 95, 227–230.

Luttrell, R.G. and Ali, M.I. (2007) Exploring selection for Bt resistance in Heliothines: results of laboratory and field studies. *Proceedings of the 2007 Beltwide Cotton Conferences, New Orleans, Louisiana, 9–12 January 2007.* National Cotton Council of America, Memphis, Tennessee, pp. 1073–1086.

May, R.M. (1985) Evolution of pesticide resistance. *Nature* 315, 12–13.

Paoletti, M.G. and Pimentel, D. (2000) Environmental risk of pesticides versus genetic engineering for agricultural pest control. *Journal of Agricultural and Environmental Ethics* 12, 279–303.

Peck, S.L., Gould, F. and Ellner, S. (1999) Spread of resistance in spatially extended regions of transgenic cotton: implications of for management of *Heliothis virescens* (Lepidoptera: Noctuidae). *Journal of Economic Entomology* 92, 1–16.

Pingel, R.L. and Lewis, L.C. (1997) Field application of *Bacillus thuringiensis* and *Anagrapha falcifera* multiple nucleopolyhedrovirus against the corn earworm (Lepidoptera: Noctuidae). *Journal of Economic Entomology* 90, 1195–1199.

Raymond, B., Sayyed, A.H. and Wright, D.J. (2005) Genes and environment interact to determine the fitness costs of resistance to *Bacillus thuringiensis. Proceedings of the Royal Society of London Series B-Biological Sciences* 272, 1519–1524.

Raymond, B., Sayyed, A.H. and Wright, D.J. (2006) The compatibility of a nucleopolyhedrosis virus control with resistance management for *Bacillus thuringiensis*: co-infection and cross-resistance studies with the diamondback moth, *Plutella xylostella*. *Journal of Invertebrate Pathology* 93, 114–120.

Raymond, B., Sayyed, A.H., Hails, R.S. and Wright, D.J. (2007a) Exploiting pathogens and their impact on fitness costs to manage the evolution of resistance to *Bacillus thuringiensis*. *Journal of Applied Ecology* 44, 768–780.

Raymond, B., Sayyed, A.H. and Wright, D.J. (2007b) Host-plant environment and population interact to determine the fitness costs of resistance to *Bacillus thuringiensis*. *Biology Letters* 3, 82–85.

Roush, R.T. (1993) Occurrence, genetics and management of insecticide resistance. *Parasitology Today* 9, 174–179.

Sayyed, A.H., Gatsi, R., Ibiza-Palacios, M.S., Escriche, B., Wright, D.J. and Crickmore, N. (2005) Common, but complex, mode of resistance of *Plutella xylostella* to *Bacillus thuringiensis* toxins Cry1Ab and Cry1Ac. *Applied and Environmental Microbiology* 71, 6683–6689.

Schuler, T.H., Potting, R.P.J., Denholm, I. and Poppy, G.M. (1999) Parasitoid behaviour and *Bt* plants. *Nature* 400, 825–826.

Schuler, T.H., Denholm, I., Clark, S.J., Stewart, C.N. and Poppy, G.M. (2004) Effects of Bt plants on the development and survival of the parasitoid *Cotesia plutellae* (Hymenoptera: Braconidae) in susceptible and Bt-resistant larvae of the diamondback moth, *Plutella xylostella* (Lepidoptera: Plutellidae). *Journal of Insect Physiology* 50, 435–443.

Shelton, A.M. (2007) Considerations on the use of transgenic crops for insect control. *Journal of Development Studies* 43, 890–900.

Shelton, A.M., Hatch, S.L., Zhao, J.-Z., Chen, M., Earle, E.D. and Cao, J. (2008) Suppression of diamondback moth using Bt-transgenic plants as trap crops. *Crop Protection* 27, 403–409.

Schnepf, E., Crickmore, N., van Rie, J., Lereclus, D., Baum, J., Feitelson, J.,

Zeigler, D.R. and Dean, D.H. (1998) *Bacillus thuringiensis* and its pesticidal crystal proteins. *Microbiology and Molecular Biology Review* 62, 775–806.

Sisterson, M.S., Antilla, L., Carrière, Y., Ellers-Kirk, C. and Tabashnik, B.E. (2004) Effects of insect population size on evolution of resistance to transgenic crops. *Journal of Economic Entomology* 97, 1413–1424.

Sisterson, M.S., Carrière, Y., Dennehy, T.J. and Tabashnik, B.E. (2005) Evolution of resistance to transgenic crops: interactions between insect movement and field distribution. *Journal of Economic Entomology* 98, 1751–1762.

Soberón, M., Pardo-López, L., López, I., Gómez, I., Tabashnik, B.E. and Bravo, A. (2007) Engineering modified Bt toxins to counter insect resistance. *Science* 318, 1640–1642.

Tabashnik, B.E. (1989) Managing resistance with multiple pesticide tactics: theory, evidence and recommendations. *Journal of Economic Entomology* 82, 1263–1269.

Tabashnik, B.E., Carrière, Y., Dennehy, T.J., Morin, S., Sisterton, M.S., Roush, R.T., Shelton, A.M. and Zhao, J.-Z. (2003) Insect resistance to transgenic Bt crops: lessons from the laboratory and field. *Journal of Economic Entomology* 96, 1031–1038.

Tabashnik, B.E., Gould, F. and Carrière, Y. (2004) Delaying evolution of insect resistance to transgenic crops by decreasing dominance and heritability. *Journal of Evolutionary Biology* 17, 904–912.

Tabashnik, B.E., Dennehy, T.J. and Carrière, Y. (2005) Delayed resistance to transgenic cotton in pink bollworm. *Proceedings of the National Academy of Sciences of the USA* 102, 15389–15393.

Tabashnik, B.E., Gassmann, A.J., Crowder, D.W. and Carrière, Y. (2008) Insect resistance to Bt crops: evidence versus theory. *Nature Biotechnology* 26, 199–202.

Thomas, D.D., Donnelly, C.A., Wood, R.J. and Alphey, L.S. (2000) Insect population control using a dominant, repressible, lethal genetic system. *Science* 287, 2474–2476.

Thomson, J.A. (2008) The role of biotechnology for agricultural sustainability in Africa.

Philosophical Transactions of the Royal Society B-Biological Sciences 363, 905–913.

Vacher, C., Bourguet, D., Rousset, F., Chevillon, C. and Hochberg, M.E. (2003) Modelling the spatial configuration of refuges for a sustainable control of pests: a case study of Bt cotton. *Journal of Evolutionary Biology* 16, 378–387.

Wu, J.H., Luo, X., Wang, Z., Tian, Y.C., Liang, A.H. and Sun, Y. (2008) Transgenic cotton expressing synthesized scorpion insect toxin AcHIT gene confers enhanced resistance to cotton bollworm (*Helicoverpa armigera*) larvae. *Biotechnology Letters* 30, 547–554.

Zhao, J.-Z., Cao, J., Collins, H.L., Bates, S.L., Roush, R.T., Earle, E.D. and Shelton, A.M. (2005) Concurrent use of transgenic plants expressing a single and two *Bacillus thuringiensis* genes speed insect adaptation to pyramidal plants. *Proceedings of the National Academy of Sciences of the USA* 102, 8426–8430.

7 Herbicide-tolerant Genetically Modified Crops: Resistance Management

M.D.K. OWEN

Department of Agronomy, Iowa State University, Ames, Iowa, USA

Keywords: Glyphosate, herbicide, resistance, gene flow, transgenic, management

Summary

The adoption of transgenic herbicide-resistant crops has made an unfathomable change in global agriculture within the last decade. Currently, an estimated 114.3 million ha of transgenic crops are planted throughout a variety of agroecosystems in 23 developing and industrial countries. Approximately 90% of the land area with transgenic crops includes a trait for resistance to glyphosate herbicide. While there are numerous benefits to society ascribed to transgenic crops, there are also a number of risks, perceived and real, to the inclusion of transgenic crops in agriculture. Importantly, scientists and agronomists do not always address the questions posed by society and thus a disconnect between the scientific community and global consumers exists. Some of the concerns expressed by society are not upheld by the available science. However, the issue of transgene movement, either by pollen or seed, to non-transgenic crops and weeds is real and should be addressed in a manner that effectively alleviates the questions from the people. This chapter addresses some of these questions about the benefits and risks of transgenic crops, focusing specifically on glyphosate-resistant crops. The value that transgenic genetically modified crops (GMCs) provide, both economically and environmentally, will be described. Appropriately, the risks attributed to these crops will also be discussed. Other topics that will be included are the impact of transgenic crops on biological diversity, their coexistence with non-transgenic crops and the impact that transgenic crops and supporting agronomic management systems have on weed communities, specifically population shifts, evolved resistance and transgene flow to near-relative weeds.

Introduction

Commercially available genetically modified crops (GMCs) have only been available since 1996, but have dramatically changed the face of world agriculture in slightly more than a decade. The farmer adoption of GMCs has increased at double-digit values since 1996 and sustained growth is anticipated, given the new GMCs that are under development (Anonymous, 2007b). GMCs probably represent the most rapidly adopted technology in agriculture ever (Service, 2007b). This trend in GMCs is largely attributed to one herbicide, glyphosate (*N*-(phosphonomethyl)glycine) and crops that are genetically modified to have selectivity (resistance) to the glyphosate when it is applied topically to the crops. However, many new GMCs have been introduced that include multiple trans-genic traits thus providing greater value to agriculture. The reported increased economic profitability attributable to GMCs is a major factor that provides the impetus for the incredible adoption of the biotechnology worldwide in develop-ing as well as industrial nations (Meyer, 2008).

Unfortunately, GMCs are not perceived universally as beneficial to human-kind. The world community has been less receptive to GMCs and has fears that all questions of risk attributable to GMCs have not been addressed, or perhaps even identified (Krayer von Krauss *et al.*, 2004). The long-term effects of GMCs, if any, are not well understood and conducting an objective assessment of risk and benefit is difficult at best (Madsen and Sandoe, 2005). While many of these fears may be unfounded, the public debate about the potential environ-mental and public health risks is critically important (Anonymous, 2007c). Importantly, agriculture has not maintained the best record of policing issues surrounding GMCs which has not done anything positive to alleviate these public concerns (Brasher, 2008).

As suggested, the trend for adoption of GMCs is mainly due to glyphosate-resistant crops (GRCs; Duke and Powles, 2008). An estimated 90% of all GMCs worldwide are GRCs and new GRCs are being rapidly developed for commercial introduction (Duke, 2005). While many of the public fears con-cerning GMCs are about future risks, problems with GRCs have been identified and must be addressed. The focus on glyphosate and GRCs as the primary, if not the sole tactic for weed control, is not without problems. Unfortunately, the scientific reports about the implications of the widespread adoption of GRCs are mixed. The consequences of genetic engineering may be neutral and pos-sibly even beneficial to the environment (Snow and Palma, 1997). However, given the unprecedented adoption of GRCs, it is critically important to assure that the undesirable aspects of the technology are identified, understood and resolved. While some suggest that this technology will enhance the biodiversity of agriculture, others suggest that increased weediness on agricultural land and invasiveness of unmanaged areas will result from GRCs (Freckleton *et al.*, 2004; Clark, 2006). Other risks attributable to GRCs that have been identified include weed population shifts, introgression of the trait to volunteer crops and weeds and evolved weed resistance to glyphosate (Zelaya *et al.*, 2004, 2007; Owen, 2005, 2008; Gealy *et al.*, 2007). It is also suggested that growers who adopt GRCs will ignore the principles of integrated weed management (IWM) which, in itself, has significant implications on agricultural profitability and sus-tainability (Sanyal *et al.*, 2008). Some of the issues (i.e. trait introgression) attributable to GRCs can be addressed by novel strategies (Lin *et al.*, 2008). However, it is evident that issues such as evolved glyphosate-resistant weed populations have not been effectively addressed (Owen, 2008). Despite the occurrence of glyphosate-resistant weed populations in agriculture, there is still an opportunity to provide stewardship in order to enhance the sustainability of GRCs (Sammons *et al.*, 2007).

While there are a number of herbicide-tolerant GMCs that have been developed to several herbicides with different modes of phytotoxic action, the primary influence in world agriculture is GRCs (Duke, 2005; Duke and Powles, 2008). More specifically, of the 23 countries that plant GMCs, the USA, Canada, Argentina and Brazil are the countries that account for most of the hectares (Anonymous, 2007b). In these countries, the principal GRCs planted include maize (*Zea mays*), soybean (*Glycine max*), cotton (*Gossypium hirsutum*) and canola (*Brassica napus*). Thus, the focus of this chapter will be these GRCs and the implications of their adoption in agriculture in the western hemisphere. Specifically, the implications of GRCs on agronomic systems, herbicide use, evolved glyphosate resistance in weeds, gene flow and trait introgression and crop volunteers will be addressed.

Changes in Agronomic Practice

The adoption of GMCs, and specifically GRCs, has resulted in significant changes in agronomic practices. These changes include the increase in glyphosate use at the cost of other herbicides, the manner and frequency that glyphosate is used and the amount of tillage that is conducted for crop production (Young, 2006; Service, 2007a,b; Foresman and Glasgow, 2008). The reduction in tillage obviously has an important benefit of reducing the use of petroleum-based fuels as well as an implicit gain in time-use efficiency by growers. A significant reduction in pesticides use has been attributed to the adoption of GRCs (Sankula, 2006). Furthermore, the benefits ascribed to GRCs have dramatically changed the crop cultivars selected by growers and have hastened the development of new GRCs for commercial distribution worldwide (Duke, 2005; Dill *et al.*, 2008).

However, not all the changes in agronomic practice attributed to the adoption of GRCs can be considered positive. Weed management, better described as weed control, in GRCs is now generally considered simple and growers do not perceive a need to observe IWM practices nor recognize the importance of a basic understanding of weed biology as it relates to crop–weed interactions (Owen, 2000; Sanyal *et al.*, 2008). Succinctly, growers are more concerned about controlling, rather than managing, weeds; the subtle difference between these concepts has considerable economic and biologic implications on agronomic systems. The perceived simplification of weed control and the elimination of IWM have resulted in the consistent loss of crop yield due to delayed glyphosate application timing and have further affected changes in weed communities (Owen, 2000, 2006a; Cerdeira and Duke, 2006). The loss of crop yield potential has a direct negative impact on agricultural profitability, and the changes in weed communities indirectly increase the complexity of future weed management tactics which are likely to be more costly and require greater time commitment. Finally, the perceived simplicity of GRC-based systems may

contribute to changes in crop rotations such that the hectares planted to GRCs
used for biofuels increase (Searchinger *et al.*, 2008). This change in agronomic
practice is predicted to have a significant impact on greenhouse gases. A more
detailed assessment on the implications of GRCs on agronomic practices
follows.

Adoption of genetically modified crops

Since the commercial introduction of GMCs in 1996, growers have increased
their adoption of these crops at an increasing rate (Fig. 7.1; Anonymous,
2007b). In 2006, approximately 100 million ha of GMCs were planted worldwide and an estimated 80% were glyphosate GRCs (Service, 2007b). Growers
continued to increase the use of GMCs in 2007 and planted 12% more GMCs

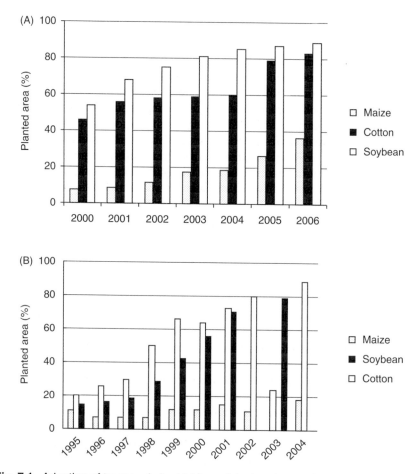

Fig. 7.1. Adoption of transgenic herbicide-resistant maize, soybean and cotton in
the USA (A) and percentage of planted area treated with glyphosate (B). (Adapted
from USDA, Anonymous, 2006b.)

Table 7.1. Countries that grow most of the genetically modified crops in the world.[a]

Country	Million hectares	Primary GMC grown
USA	57.7	Soybean, maize, cotton, canola, lucerne,[b] sugarbeet[c]
Argentina	19.1	Soybean, maize, cotton
Brazil	15.0	Soybean, cotton
Canada	7.0	Canola, maize, soybean
India	6.2	Cotton
China	3.8	Cotton, maize
Paraguay	2.6	Soybean
South Africa	1.8	Maize, soybean, cotton
Uruguay	0.5	Soybean, maize
Philippines	0.3	Cotton

[a]Data are adapted from ISSA Brief 37-2007 (Anonymous, 2007b).
[b]Returned to regulated status due to court order.
[c]Deregulated in 1999 but not commercially available until 2008.

(Anonymous, 2007b). Global adoption of GMCs in 2007 represented 114.3 million ha and included 23 countries that have approval to plant GMCs. Currently there are 29 countries that allow importation of GMCs and the number of countries, crops and traits is expected to increase dramatically in the next decade (Meyer, 2008). The primary countries that plant GMCs are the USA, Argentina, Brazil and Canada (Table 7.1). North America represents 57% of the GMCs planted globally while Central and South America contribute 33% of the total hectares. It is estimated that approximately 90% of the GMCs grown globally are GRCs (Duke and Powles, 2008). Many of the GRC cultivars contain multiple GM traits, notably *Bacillus thuringiensis* (*Bt*; see Chapters 15–17, this volume). In the USA, 90% of the soybean, 70% of the cotton and 60% of the maize currently grown are GRC cultivars (Gianessi, 2005; Service, 2007b). South America has widely adopted GRCs, primarily GR soybean, since the commercial introduction of the GR cultivars (Qaim and Traxler, 2005; Christoffoleti *et al.*, 2008). These cultivars represent in excess of 90% of soybeans grown and typically cultivation represents monocultures with glyphosate as the primary, if not sole, herbicide used. GM canola dominates the cultivars grown in Canada and the USA representing >80% of the total crop grown (Beckie and Owen, 2007). Cultivars that are resistant to glyphosate are widely grown.

Monsanto reported that their 2006 market share of GRCs included 29 million ha of the US soybean area, 13.8 million ha of maize, 4.6 million ha of cotton and 2.3 million ha of canola in the USA and Canada (Anonymous, 2006a). They anticipate that their GRC market share in US maize will approach 24.3 million ha by 2010 (Anonymous, 2006c). The dramatic increase likely reflects the interest in maize-based ethanol as well as the grower perception that GRC-based crop production systems are simple, economical and effective. What has not been considered is the enormous selection pressure that the widespread application of glyphosate on millions of hectares will have on the weed communities. This GRC revolution likely represents the largest man-caused biological experiment in history and the ramifications are only beginning to be identified and (hopefully) understood.

Several GRCs have been developed but have not been commercially introduced for a number of reasons. Notably sugarbeet (*Beta vulgaris*) cultivars that are resistant to glyphosate were deregulated in 1999, but not commercially offered until 2008 due to concerns about the acceptability of sugar refined from a GMC (Duke, 2005; Gianessi, 2005). The development of GMC rice (*Oryza sativa*) cultivars modified to be resistant to glufosinate (2-amino-4-(hydroxymethyl-phosphinyl)butanoic acid) herbicide proceeded from 1998 to 2001, but withdrawn and commercial development was terminated due to concerns about market acceptance of the GM rice (Gealy and Dilday, 1997b; Gealy et al., 2007). Similarly, wheat (*Triticum aestivum*) GR cultivars were under development but the programme was terminated in 2004 (Dill, 2005). While the GRC-based wheat production systems demonstrated excellent opportunities for improved weed management, concerns about the acceptance of the flour made from GR wheat cultivars as an export commodity in GMC-adverse countries resulted in the decision to halt further development (Stokstad, 2004; Howatt et al., 2006). Lucerne (*Medicago sativa*) cultivars resistant to glyphosate were developed and deregulated for commercial use in 2005 (Weise, 2007). The lucerne cultivars represented the fourth GRC grown in the USA and approximately 80,000 ha were planted in 2006. However, in 2007, the GR lucerne was returned to regulated status due to a federal court order (US District Court for the Northern District of California, No. C 06–01075 CRB) and seed sales was terminated immediately and planting of the GRC was prohibited after 30 March 2007 (Fisher, 2007; Harriman, 2007). The ruling indicated that USDA-APHIS had not provided sufficient evidence about the risk of growing the GRC and thus to date, GR lucerne is not commercially available for planting (Charles, 2007). Finally, GR creeping bentgrass (*Agrostis stolonifera*) cultivars were developed and field trials established in Jefferson County, Oregon, USA (Anonymous, 2002). Given that creeping bentgrass is a wind-pollinated perennial with wild relatives, there was a high potential for the introgression of the GR trait despite the mitigation plan that was established (Watrud et al., 2004). Pollen-mediated introgression of the GR trait was detected and further production of the GRC was terminated (Mallory-Smith et al., 2005; Reichman et al., 2006).

Value ascribed to genetically modified crops

Estimates suggested that GMCs (GR soybeans) created globally more than US$1.2 billion of economic surplus in 2001, of which 53% went to consumers, 34% to the technology companies and 13% to growers (Qaim and Traxler, 2005). Interestingly, in Argentina where intellectual property protection is lax at best, growers received 90% of the benefits. The adoption of GMCs lowered production costs by US$1.4 billion while increasing net returns by US$2.0 billion in 2005 (Sankula, 2006). The economic contribution of GRCs specifically was suggested to be US$1.2 billion and was attributable to savings associated with the cost of herbicides, application, tillage and alternative weed management tactics (Gianessi, 2005). The global cumulative income benefits for growers who planted GRCs from 1996 to 2004 are estimated to approach US$19,339 million with soybeans accounting for 89% of the income, maize for 3%, cotton for 4% and canola for 4% (Gianessi, 2008). With the availability of GR sugarbeet, US growers could potentially save US$93 million specifically attributable to the GRC (Gianessi, 2005). This economic windfall is primarily attributable to lower herbicide costs but also reflects improved weed control (Gianessi and Reigner, 2007).

GMCs reduced overall pesticide use by 87% and GRCs accounted for the preponderance of the reduction (Sankula, 2006). Given the dominance of the GRCs and the concomitant use of glyphosate as the primary, if not sole, herbicide used for weed management, there have been predictable outcomes within the industry (Fulton and Giannakas, 2001). Notably, research and development budgets have shifted to support the development of crop traits rather than new herbicides (Green, 2007; Service, 2007b). Furthermore, the diversity of herbicides used for weed management has lessened dramatically, initially in soybean, but other GRCs are likely experiencing the same 'simplification' of weed control tactics (Young, 2006). It is clear that the perceived simplicity of GRC-based systems is resulting in problems of growers not practising sound weed management and, in particular, not implementing appropriate weed management tactics to mitigate the evolution of herbicide-resistant weed populations. These problems, combined with the dramatic change in tillage systems that accompanied the adoption of GRC-based production systems, have contributed to dramatic changes in weed populations (Culpepper *et al.*, 2004, 2006; Sankula, 2006; Webster *et al.*, 2006; Young, 2006; Service, 2007a).

GRC cultivars are also perceived to have value not attributable to improved weed management. For example, GMCs and concomitant use of herbicides (glyphosate and glufosinate) reduced herbicide residues in water when compared to non-GMC-based production systems (Shipitalo *et al.*, 2008). Furthermore, the application of glyphosate to GRCs has demonstrated efficacy on fungi, specifically stripe rust (*Puccinia striiformus* f. sp.*tritici* (Erikss) CO Johnston) and leaf rust (*P. triticina* Erikss), which are important diseases of wheat (Feng *et al.*, 2005, 2008). In GR wheat, glyphosate demonstrated preventative activity and was also a curative of these diseases; glyphosate also demonstrated activity on Asian soybean rust (*Phakopsora pachyrhizi* Syd & P Syd). Given that glyphosate inhibits 5-enolpyruvul shikimate 3-phosphate synthase (EPSPS, EC 2.5.1.19), which is a key enzyme in the shikimate biosynthetic pathway of plants, fungi and bacteria, these observations are not unexpected and could potentially provide considerable additional value to GRCs. However, it is critical to recognize the importance of the agronomic performance of GRCs when assessing overall economic value. Profitability of GRCs is likely to be associated with yield quantity and quality and not the GM trait (Jost *et al.*, 2008; Mills *et al.*, 2008).

Perceptions of the risks associated with glyphosate-resistant crops

Weed control based on GRCs is effective and consistently excellent, and the GRC-based systems provide considerable savings of time and labour; estimates suggest that collectively, herbicides are equivalent to 70 million workers engaged in hand weeding (Gianessi and Reigner, 2007). Regardless, over the past decade the disconnect between the general public and the scientific and agricultural communities has not resolved itself despite an increasing number of GMCs becoming commercially available and the incredible adoption of GRCs world-wide (Krayer von Krauss *et al.*, 2004; Owen, 2005). Two general areas of miscommunication are suggested; risk assessment must test a specific question rather than for scientists and regulators to just present data and a wider context must be considered thus enabling 'non-scientific' questions to be included in the regulatory process (Johnson *et al.*, 2006). This is necessary to provide an opportunity for society to participate in the regulatory decisions affecting GRCs. The disconnection between the lay public and agriculture is most likely attributable to the fact that society does not directly 'experience' the benefits that GRCs provide to production agriculture. Importantly, GMCs are typically presented in a negative position by advocacy groups and society cannot, or does not, participate in the scientific debate resulting in controversial decisions about the science used to support the favourable risk assessment analyses. It is possible that a major part of the concern expressed about GRCs is that the technology is associated with herbicide use, which in itself is perceived as risky by the public sector (Madsen and Sandoe, 2005).

Often the risks perceived by society (i.e. the presumptive connection of GMCs with foot-and-mouth disease) are not supported by science, and the manner in which the risk analyses are conducted does not allow the full assessment of societal concerns (Johnson *et al.*, 2006). Furthermore, societal opinions do not often follow a logical, scientifically defendable path. For example, surveys indicated that pesticide use in gardens was deemed acceptable while similar use in agriculture was reported to be a significant concern (Crane *et al.*, 2006).

Given that GRC systems focus on the use of glyphosate, society has expressed concerns about the impact of the trait/herbicide system on human health, food quality and environmental impact. While there is some evidence supporting the potential impact of GRCs on weediness in agricultural systems, invasiveness in natural areas and other non-target impacts (Rissler and Mellon, 1996; Clark, 2006), an assessment of the effects of GMCs, including GRCs, on the environment demonstrated that the impacts of the GMCs varied by specific crop and production tactic (Champion *et al.*, 2003; Firbank *et al.*, 2003; Hawes *et al.*, 2003; Heard *et al.*, 2003a; Heard *et al.*, 2003b; Halford, 2004; Bohan *et al.*, 2005; Gibbons *et al.*, 2006). Importantly, these assessments were conducted at a large field scale so as to cover a wide range of environments and thus provide strong support about observations of GMC-based systems on biodiversity and related processes.

While the rate of adoption of GRC-based crop production systems remains high and the economic impact of the technology significant, questions about the risks associated with the technology must be assessed not to the satisfaction of just the regulatory agencies, but also the global public (Dill, 2005; Gianessi, 2005; Sankula, 2006). It is suggested that because GRC-based systems are associated specifically with the use of glyphosate and that the commercial adoption of these systems has been so great, the concerns about possible environmental risks are well founded (Rissler and Mellon, 1996; Altieri, 2000). However, the author suggests that the primary risk, environmental or ecological, associated with GRCs is the impact on weed population shifts, whether expressed as the rise in economic prominence of a new weedy species or the evolution of glyphosate-resistant weed biotypes. This position is supported by the herbicide resistance risk analysis which suggests that glyphosate-based systems are at high risk of selecting for glyphosate-resistant weeds (Vidal *et al.*, 2007). Not all research has supported the assessment that GRC-based systems are at high risk of selecting glyphosate-resistant weed populations (Werth *et al.*, 2008). However, given the unprecedented adoption of the technology, application of glyphosate to millions of hectares will do what is impossible to duplicate in scientific study (Gressel and Levy, 2006). Other agronomic risks for GRCs include the difficulty in assessing the impact of the GM trait on yield quantity and quality compared to non-GRC cultivars and the movement of the GM traits via pollen into non-GM crops and weed populations (Elmore *et al.*, 2001a,b; Raymer and Grey, 2003; Gealy *et al.*, 2007). It is important to recognize that while some GM traits (i.e. *Bt*) can directly impact pests and non-target organisms, the GR trait is likely to be functionally benign in the environment (Snow, 2002; Owen, 2008). While new GRC technologies are being developed, they do not appear to mitigate the risks associated with the widespread use of single tactic weed control (Fryer, 1981; Owen, 2005; Green *et al.*, 2008).

Effect of GRCs on biological diversity

Notably, there have been reports of marked decline in the abundance of species attributable to agriculture (Krebs *et al.*, 1999; Schutte, 2003; Benton, 2007). Regardless, there is a dearth of conclusive evidence that GRCs impact diversity or abundance of systems (Cerdeira and Duke, 2006). It is suggested that the ecosystem effects of GRCs have been minimal (Davis, 2006). However, the indirect impact of GRCs on the agroecosystem, particularly as a result of changes in tillage and weed management tactics, is important (Norris, 2005). Reports support both an increase and a reduction of species in GRC-based systems (Scursoni *et al.*, 2001; Bitzer *et al.*, 2002; Freckleton *et al.*, 2004; Cerdeira and Duke, 2006; Scursoni *et al.*, 2007). Furthermore, the effect is often specific to crop and the comparison of weed management tactics for GRCs and non-GM crops (Heard *et al.*, 2003a; Heard *et al.*, 2003b). Weed communities are affected by the efficiency of management tactics (Scursoni *et al.*, 2001); weed communities were more diverse when single glyphosate applications were used compared to multiple applications.

Interestingly, the diversity and abundance of springtails (order Collembola) were not negatively affected in GR soybean compared to conventional production systems and, in fact, abundance was often greater in the GRCs, suggesting that the GRC-based weed management systems impose no deleterious effects on this important non-insectan hexapod (Bitzer et al., 2002). Effects on butterfly population densities in GRCs reflected herbicide efficacy; where weed control was good, fewer butterflies were found likely reflecting a lack of nectar availability (Roy et al., 2003); these effects were also observed on the field margins. However, few differences were observed for bees, gastropods and other invertebrates. Generally, the importance of weed species diversity on field margins is unclear (Sosnoskie et al., 2007). It was suggested that there could be an impact of GRC-based systems on birds that use weed seeds as a food source, basing the fact that reduced weed population densities would result in less available food resources (Gibbons et al., 2006; see also Chapter 12, this volume). However, the effect was again dependent on crop and the relative effectiveness of weed control provided by the contrasting GRC versus conventional weed management tactics. Furthermore, it is difficult to assess if any cumulative differences in biodiversity attributable to GRC maize when compared to conventional systems exist, given the variability of the data (Heard et al., 2006). The effects of GRCs on the soil biota were found to be negligible (Cerdeira and Duke, 2006). Thus, there appears to be favourable and unfavourable data on the effects of GRCs on biological diversity (Duke, 2005). The critical consideration is that these effects are highly dependent on specific crop and management tactics. It is likely that any unfavourable effect on biological diversity could be ameliorated by subtle manipulation of the GRC-based system.

Coexistence of glyphosate-resistant crops with non-genetically modified crops

The importance of GRCs to agriculture is demonstrated in their adoption by growers (Fig. 7.1). However, another pervasive problem exists with the production of GRCs; their coexistence with non-GM crops (Byrne and Fromherz, 2003). The issue of coexistence includes three possibilities: (i) introgression of the trait via pollen (pollen drift); (ii) containment of plant products during the production year (grain segregation); and (iii) volunteer GRC plants in following years (Owen, 2005). While GRCs and non-GM crops can coexist, growers must go to great lengths to accomplish segregation (Anonymous, 2007a).

Grain segregation, or identity preservation (IP), while difficult to maintain, can be accomplished and will effectively minimize the impact of GRCs on non-GM crops. Given that the loss of IP can be costly to growers, establishing appropriate tactics to isolate GRCs from non-GM crops must be done (Owen, 2000). Controlling volunteer GRCs is also relatively easy, depending on the rotational crop, but does require more diligence on the part of the grower (Owen, 2005). The introgression of the GR trait via pollen movement encompasses the other possibility but management of this problem is considerably more difficult in open-pollinated crops such as maize (Luna, V. et al., 2001; Palmer et al., 2001; Westgate et al., 2002; Abud et al., 2007). A number of factors affect the success of maize pollen movement and subsequent pollination, and generally, the greater the distance between the pollen source and donor, the less likely is the introgression of the GR trait (Luna, 2001; Westgate, 2003). However, given the tolerance levels established for some GM traits in non-GM crops, the isolation distances required to mitigate the risks of gene flow are too large to be realistic (Matus-Cadiz *et al.*, 2004). Other open-pollinated crops have also been scrutinized with significant legal ramification as mentioned previously (Charles, 2007; Fisher, 2007; Harriman, 2007; Weise, 2007). It is suggested that the issues of the coexistence of GRCs with non-GM crops will continue to be a concern as long as there are economic differences between the crop cultivar types (Hurburgh, 2000, 2003; Ginder, 2001).

Sustainability of GRC-based crop production systems

The discussions about the sustainability and stewardship of GRCs typically focus on the inevitability of weed population shifts, and specifically on the evolution of glyphosate-resistant weed biotypes (Sammons *et al.*, 2007). Interestingly, the arguments stated suggest that glyphosate resistance can be effectively managed because the fold of resistance in weeds is typically low; this harkens to the original treatise that glyphosate resistance would not evolve because of the unique mechanism by which glyphosate affects susceptible plant species (Bradshaw *et al.*, 1997). Regardless, the widespread adaptation of weed communities to the selection pressure imposed by GRC-based weed management systems is testament to the fact that significant problems exist with simple, single focus pest control strategies (Christoffoleti *et al.*, 2008). This selection for adapted weed populations will only escalate, given the efforts of seed companies to incorporate glyphosate resistance in most crop cultivars and even to the extent that new mechanisms of glyphosate resistance have been developed (Green *et al.*, 2008). Regardless, crop systems based on GRCs, irrespective of the GR mechanism, will continue to change weed communities at an increasing rate. The sustainability of these GRCs will depend on the specific crop, the specific production practices available to support the crop, the economic returns on the crop and the alternative tactics available to manage the problems (Owen and Boerboom, 2004; Boerboom and Owen, 2007). For example, crop production systems in the Midwest USA offer many alternatives (i.e. tillage and herbicides) that are not available in the south-east USA.

Tillage
The biotechnological achievement of developing GRCs has been perceived by growers to provide better weed control which fed the unprecedented adoption of these crops worldwide. However, perhaps a more important impact was the perception by growers that the GRCs and glyphosate provided an effective, consistent, simple and low-risk 'system' for crop production with less tillage (Carpenter and Gianessi, 1999; Service, 2007a). Costs of energy have reinforced the desirability of fewer tillage trips in the production of crops as well as the benefits attributable to improved time management. Based on these

perceived and real benefits, crop production in no-tillage and other conservation tillage systems increased dramatically, largely due to GRCs (Cerdeira and Duke, 2006; Service, 2007a; Dill *et al.*, 2008). For example, cotton production using no-tillage or strip tillage increased almost threefold between 1997 (the year glyphosate-resistant cotton was introduced) and 2002 (Anonymous, 2004). However, more recent data suggest that conventional tillage (defined as seedbed preparation that leaves <30% residue on the soil surface) is the dominant cultural practice in GR cotton (Dill *et al.*, 2008). The reason for the dramatic shift in tillage for GR cotton is attributed in part to the evolution of glyphosate-resistant weeds (Mueller *et al.*, 2005). Dramatic increases in maize and soybean produced in no-tillage and conservation tillage systems are also noted and largely attributable to the introduction of GR cultivars (Duke, 2005; Gianessi, 2005; Young, 2006; Dill *et al.*, 2008). The reductions in tillage resulted in significant economic and time savings for growers, as well as reductions in equipment expenses (Gianessi, 2005; Gianessi and Reigner, 2007).

While the 'conservation' tillage systems represented significant economic savings for growers, a more important feature was the environmental savings from reduced soil erosion (Fawcett and Towery, 2004; Gianessi, 2005). The adoption of GR soybean in South America, for example, allowed growers to save an estimated 1t of soil per hectare in Brazil and 7.5t in Argentina (Penna and Lema, 2003; Service, 2007a). In the USA, 1997 estimates suggest that wind erosion was reduced by 31% and soil erosion due to water was reduced by 30% compared to 1982, again largely attributable to the adoption of GRCs and conservation tillage practices (Fawcett and Towery, 2004). Savings due to reduced soil sedimentation were estimated to be US$3.5 billion in 2002. Given that current adoption of GRCs has increased considerably during the recent decade, environmental savings are suggested to have increased significantly.

Weed seedbank, dormancy and germination

While GRCs are not generally credited as having a direct effect on weed seed dormancy and germination factors, there are examples of an indirect impact from the glyphosate used in the system on seed germination (Blackburn and Boutin, 2003). However, the concomitant increase in no-tillage production systems attributed to GRCs does impact the weed seedbank characteristics. Thus, specific weeds that are adapted to conservation tillage because of seedbank characteristics are likely to be favoured in GRC-based production systems and will potentially be more likely to become significant economic problems. However, while it is important to have the ability to predict weed seedbank characteristics, this is difficult to accomplish given the variability that exists within the field environment (Batlla and Benech-Arnold, 2007). Numerous scientific papers describe the changes in weed seedbanks due to tillage system. Considerable variation in weed seed germination was noted among specific locations and years across the US Corn Belt (Forcella *et al.*, 1997). The variation was suggested to be attributable to secondary dormancy which was affected by microclimate, which is in part a function of tillage. Differences in weed emergence patterns may allow species to adapt to the GRC-based system and result in a weed population shift (Hilgenfeld *et al.*, 2000).

Generally, weed seedbank depletion is higher with increasing soil disturbance (Ball, 1992; Mulugeta and Stoltenberg, 1997). The impact of tillage on common lamb's quarters (*Chenopodium album*) was suggested to be on seed distribution within the soil profile and seedbank depletion; the greater the intensity of tillage, the deeper the placement of seeds and the poorer the emergence success. Buhler observed a clear correlation between tillage and specific weed populations (Buhler, 1992). Common lamb's quarters and horseweed (*Conyza canadensis*) population densities were considerably higher in conservation tillage systems compared to conventional tillage. Horseweed seeds germinate shallow and seeds that are buried deeper than 0.5 cm do not contribute to the population (Nandula *et al.*, 2006). Horseweed also demonstrates the ability to germinate, emerge and survive in the fall (autumn) and well into the growing season (Buhler and Owen, 1997). Another subtle effect on weed seedbanks attributable to GRC-based systems is the microhabitat that is established in these systems. Changes in microhabitat influence seed predation and weed seedbanks (Orrock *et al.*, 2006). Given that GRCs support no-tillage systems which are greater in weed community diversity and microhabitat, potentially there is greater depletion of weed seedbanks in GRCs compared to non-GM crop systems.

Thus, GRCs are unlikely to have a direct effect on weed seed characteristics and seedbanks. However, GRC-based systems have an important effect on the composition of weed communities, in part due to the tillage regimes that are used and also attributable to the selection pressure imparted by glyphosate used as the predominant tactic to control weeds (Dill *et al.*, 2008). Another possible effect attributable to GRCs is the volunteer GRC seedbank (Lutman *et al.*, 2005). The relative life of the volunteer GRC seedbank is dependent upon the specific crop, the environment and the production system. Generally, seeds from GRCs do not have a long life in soil; however, GR canola can survive as long as 10 years in the seedbank (Simard *et al.*, 2002; Beckie and Owen, 2007; Black, 2008). An interesting perspective is proposed by May *et al.* who suggested that GRCs could be managed to increase weed seedbanks which would support invertebrates and birds thus representing a significant environmental benefit (May *et al.*, 2005).

Crop rotation

Crop rotation is suggested to be an effective means of managing a number of different pest complexes (Buhler *et al.*, 1997; Miller *et al.*, 2006). Crop rotation, in theory, reduces weed population densities and maintains weed species diversity (Doucet *et al.*, 1999). However, the effectiveness of crop rotation on impacting weed population density is dependent on the characteristics of the crop and the management tactics used to produce the crop. For example, the increased diversification found in complex crop rotations dilutes the selection pressure that favours specific weeds and subsequently reduces the potential for weed population shifts (Liebman and Dyck, 1993). Conversely, simple crop rotations and management systems often result in weeds that are well adapted to the agroecosystem and thus difficult to effectively manage (Culpepper *et al.*, 2004, 2006; Culpepper, 2006). Crop systems based on GRCs are generally simple and unlikely to include complexity, thus favouring adapted weeds.

However, the effects of crop rotation on weed communities are difficult to iso-late from the management tactics that are used for the production of the crops (Buhler et al., 1997). In some studies, cropping sequence is reported to be the dominant factor in affecting the weed seedbank, while in other research, crop rotation did not impact weed numbers (Ball, 1992; Felix and Owen, 1999). Factors such as herbicide use and tillage system are typically components of a crop production system and have significant impacts on the weed population dynamics that cannot be separated from the effect of the crop rotation on the weed community (Doucet et al., 1999). Regardless, crop rotations that create differences in soil disturbance and resource competition resulting in an inhospit-able and unstable agroecosystem will have a significant impact on reducing the likelihood of weed shifts (Liebman and Dyck, 1993).

Herbicide Use

Herbicide use has changed dramatically in response to the commercial avail-ability and adoption of GMCs. Glyphosate has become the most widely used herbicide globally and growth in glyphosate use continues at a significant pace (Woodburn, 2000; Duke and Powles, 2008). Market demand in China con-tinues to grow by 15–20% annually (Anonymous, 2008a). Most of this growth is the result of GRCs. In the USA and Canada, there have been dramatic changes in herbicide use attributable to the adoption of GMCs, specifically GRCs. These changes can generally be described as the grower adoption of 'simple', glyphosate-based systems at the expense of alternate herbicide use (Young, 2006). The changes in herbicide use attributable to GMCs, specifically canola, cotton, maize and soybean, in US production systems will be detailed.

Trends attributable to glyphosate-resistant crops

While GR canola cultivars historically have been more widely adopted than glu-fosinate-resistant cultivars, in 2007, 45% of the canola hectares in Canada were planted to glufosinate-resistant cultivars compared to 43% of the hectares with GR canola (Beckie et al., 2006). Herbicide-resistant GM canola accounts for 88% of the hectares in Canada and 97% of the canola grown in the USA (Beckie et al., 2006; Beckie and Owen, 2007). Thus, in canola the primary herbicides used for weed control are glyphosate or glufosinate. Historically, there have been a limited number of products available, notably clethodim 2-[(1E)-1[[[(2E)-3-chloro-2-propenyl]oxy]imino]propyl]-5-[2-(ethulthio)propyl]-3-hydroxy-2-cyclohexen-1-one) which was used for annual grass weed control. However, evolved resistance to this herbicide has limited the use. Ethametsulfuron (methyl   2-[[[[[4-ethoxy-6-(methylamino)-1m3m5-triazin-2-yl]amino]carbonyl] amino]sulfonyl]benzoate) was used for broadleaf weed control, but evolved resist-ance in weeds also limited the utility of this and other herbicides (Beckie, 2006; Beckie and Owen, 2007). Thus, GM canola and glyphosate or glufosinate is the dominant weed management 'system' in North American canola production.

However, herbicide mode of action rotation is reported to be an important management tactic in Canada canola production (Beckie, 2006).

Cotton weed management was dominated by soil-applied herbicides prior to the introduction of GR cotton cultivars. Notably, trifluralin (2,6-dinitro-*N*, *N*-dipropyl-4-(trifluoromethyl)benzenamine) and fluometuron (*N*,*N*-dimethyl-*N*-[3-(trifluoromethyl)phenyl]urea) were applied to more than 50% and 25% of the hectares from 1992 to 1999 (Young, 2006). Glyphosate replaced trifluralin as the most widely used herbicide in cotton. Interestingly, despite the wide adoption of glyphosate, the number of different herbicide active ingredients and sites of action have not declined dramatically. The use of glyphosate alone in cotton represents about 21% of the hectares planted to GR cotton cultivars (Foresman and Glasgow, 2008). Glyphosate is reportedly used in combination with other herbicides or applied sequentially after soil-applied herbicides by 21% and 52% of the growers who plant GR cotton, respectively. Current glyphosate registrations allow for a maximum in-crop single application of $0.87\,kg\ AE\ ha^{-1}$ in cotton and growers make 2.2 glyphosate applications per hectare per year (Foresman and Glasgow, 2008). The number of different active ingredients used in cotton declined from 3.1 in 1997 to 2.1 in 2001 (Young, 2006).

Herbicide use in maize has changed also, but differently than suggested for other GRCs. Atrazine (6-chloro-*N*-ethyl-*N*-(1-methylethyl)-1,3,5-triazine-2, 4-diamine) and chloracetamide (i.e. metolachlor (2-chloro-*N*-(2-ethyl-6-methyl-phenyl)-*N*-[(1S)-2-methoxy-1-methyethyl]acetamide)) were the most widely used herbicides in maize from 1992 to 2002 (Young, 2006). Importantly, there was no change in the number of herbicide active ingredients or sites of action that were used on 10% or more of the US maize hectares. However, this comprised of only four sites of herbicide action. With the introduction of GR maize cultivars in 1999, glyphosate use has increased dramatically. Glyphosate did not rank among the top five herbicides that were applied to maize in 2002 (Young, 2006). In 2006, 24% of US growers surveyed reported that they used glyphosate alone on GR maize cultivars, while 20% and 49% reported that glyphosate was used in combination with other herbicides or applied sequentially after a soil-applied herbicide (Foresman and Glasgow, 2008). The average number of glyphosate applications in US GR maize was $1.3\,ha^{-1}\ year^{-1}$ with a maximum allowable amount of $1.3\,kg\ AE\ ha^{-1}$. Glufosinate-resistant maize accounts for approximately 11% of the herbicide-resistant GMC hectares (Gianessi, 2008).

Herbicide use in soybean has also changed dramatically in response to the acceptance of GR soybean cultivars and glyphosate. In South America, GR soybeans are treated predominantly with glyphosate to the point that the use is described as an overreliance on the herbicide (Christoffoleti *et al.*, 2008). Dinitroanaline and imidazolinone herbicides were used predominantly for weed management in US soybeans until 1996 when glyphosate use began to increase substantially (Young, 2006). By 2002, glyphosate was used on 79% of soybean hectares in the USA. The use of alternative herbicides declined precipitously in response to grower decisions to rely exclusively on glyphosate for weed control in soybean. While there were 11 different herbicide active ingredients used on 10% or more of the treated hectares in 1995, only glyphosate was used by

2002 (Young, 2006). Similarly, the number of sites of herbicide action has declined from seven to one over the same time period. Weed management programmes are different for US soybean production in the north when compared to the south (Foresman and Glasgow, 2008). Survey results indicated that 56% of northern US growers used glyphosate alone for weed control while 42% of southern US growers relied on glyphosate only. Twice as many southern growers, compared with the northern growers, reported that the glyphosate was applied as a tank mixture with other herbicides, 22% versus 11%, respectively. The percentage of growers using a preemergence herbicide followed by glyphosate was similar for north versus south, 33% compared with 34%, respectively (Foresman and Glasgow, 2008). Glufosinate-resistant soybeans are anticipated to be available for commercial use in 2008 on a very limited basis, but are not expected to have a major impact on herbicide use in the USA in the short term.

Implications of these changes to 'herbicide diversity'

Given the global adoption of GRCs, and in particular, the dominance of these cultivars in North American agriculture, there has been a significant decline in the use of 'alternative' herbicides (Shaner, 2000; Young, 2006). Major reductions in the use of herbicides, which are acetyl coenzyme A carboxylase (ACCase) inhibitors, acetolactate synthase (ALS) inhibitors, and Protox (PPO) inhibitors, have impacted weed management and are the result of GRCs (Shaner, 2000). The lack of herbicide diversity has created an environment where changes in weed communities are inevitable. These changes in weed populations do not necessarily eliminate the use of glyphosate, but do provide a strong impetus for improved weed management tactics (Green, 2007). Growers in Canada have strongly embraced the need to alternate herbicide modes of action (Beckie, 2006). However, their ability to utilize this strategy is because of their more diverse crop rotations. While there are limited data from the USA, it is unlikely that growers practise herbicide rotation because, in part, the crop rotations employed are composed of GRCs. Thus, the changes in weed communities reported in the USA are a result of a lack of management (herbicide) diversity in weed management programmes.

The loss of herbicide revenues for companies because of, in part, GRCs has resulted in many corporate changes (i.e. American Cyanamid merging with BASF). Furthermore, the development of new herbicide products, and specifically research for new sites of herbicide action, has slowed significantly, again due to shrinking markets (Green, 2007; Green et al., 2008). This slowed development of new herbicides will likely be a long-term trend and not be easily changed, regardless of the weed problems that inevitably will develop.

Glyphosate, which is the major, if not sole, weed management tactic in GRCs, has increased in economic importance since commercialization in the early 1970s (Woodburn, 2000). An interesting consequence of the development of GRCs is a significant decline in the cost of the technical glyphosate acid from US$34 kg^{-1} in 1991 to US$20 kg^{-1} in 1997 for the USA, and the

1996 grower price varied considerably from US\$40 kg^{-1} in the UK to US\$9 kg^{-1} in China (Woodburn, 2000). With the widespread adoption of GRCs, the cost of glyphosate declined approximately 40% from 1999 to 2005 (Gianessi, 2008). At that same time, the cost (value) of alternative herbicides declined dramatically, attributable in part to the fact that the proprietary US patent for glyphosate expired in 2000 and a great number of generic glyphosate products were introduced. However, for the grower, there was little net economic gain because of technology fees for GR seed which offset the lower cost of glyphosate. Recently, the cost of glyphosate has increased dramatically from 30% or higher than the cost in 2007 (Gullickson, 2008). The explanation of the increase is suggested to be the result of increased global demand for glyphosate which has overstretched synthesis and formulation capabilities. It is possible that the higher glyphosate price will result in the use of alternative (older) herbicides thus potentially diluting selection pressure on the weed communities for species that are not effectively controlled by glyphosate alone.

Anticipated alternatives to augment weed management in glyphosate-resistant crops

One consistent observation with regard to grower response to issues with GRCs is that nothing will be done until the problems become of paramount importance locally (Foresman and Glasgow, 2008). Without sufficient diversity and the inclusion of alternative tactics, the sustainability of GRC-based systems is tenuous at best (Duke and Powles, 2008). Recent evidence of grower unwillingness to use alternative strategies for weed management is illustrated with the situation that developed in Georgia, USA, where, despite the evolution of glyphosate-resistant Palmer amaranth (*Amaranthus palmeri*) and significant loss of profitability, no mitigation tactics were adopted (Culpepper and York, 2007). This same perception of risk was presented by growers and commodity association at the first National Glyphosate Stewardship Forum in 2004 (Owen and Boerboom, 2004). The evidence is clear, consistent and pervasive that the only way to protect the benefits of GRCs is to adopt alternative tactics and IWM practices (Beckie, 2006; Duke and Powles, 2008). Models predict that alternative weed management practices included with GRC-based production systems can reduce the risk of weed problems to near zero and also provide considerable economic incentives. (Mueller *et al.*, 2005; Neve, 2008; Werth *et al.*, 2008).

Tactics that have been proposed to provide alternatives for weed management in GRCs are typically herbicide-based and do not reflect biological options (Liebman and Dyck, 1993). Strategies such as the development of new GM crops with resistance to alternative herbicides, novel mechanisms of glyphosate resistance and multiple-herbicide resistance seem to be the focus of new technologies (Castle *et al.*, 2004; Matringe *et al.*, 2005; Behrens *et al.*, 2007; Green, 2007; Green *et al.*, 2008). The authors suggests that these technologies (i.e. multiple-herbicide resistance in crops) will not help sustain GRCs as they are similar to the current system which is based on delaying or preventing weed population shifts and focuses on mitigation tactics after the problem appear (Sammom *et al.*, 2007). Historically, the problems of weed population shifts have been effectively addressed with supplemental alternative herbicides. However, with the decline in alternative herbicides and the general observation that weed shifts, specifically evolved herbicide resistance in response to the available alternative herbicides, have already reduced the utility of these products to augment GRC-based systems (Christoffoleti *et al.*, 2008). The question that must be addressed is the relative severity and universality of problems anticipated in GRCs. To be sure, the problems will be worse in some GRCs and regional production systems than in others. None the less, there can be no question that the sustainability of GRC-based production systems is threatened (Boerboom and Owen, 2007).

Superweeds?

The idea of a 'superweed' evolving due to acquired herbicide resistance is an interesting, albeit ecologically misguided, concept likely presented by advocates who oppose the use of GMCs and herbicides. However, the idea has gained considerable traction in the popular press and lay publications in agriculture (Gullickson, 2005). In fact, a mention of glyphosate-resistant *C. canadensis* and other 'superweeds' with evolved resistance to the 'superherbicide' glyphosate was made in *National Geographic* (Lange, 2008). Furthermore, it has been suggested that GMCs with weedy near-relatives may 'commandeer' the GM trait and spread rapidly across an ecosystem (Kaiser, 2001). This notoriety in the public media provides just enough information, better described as misinformation, to incite public concerns about the environmental stewardship and sustainability of agriculture, particularly with regard to the inclusion of GRCs. In the case of traits for herbicide resistance, the concept of 'superweeds' is inappropriate and has limited connection with transgenic glyphosate resistance trait acquisition from GRCs.

Concept of superweeds

The concept of 'superweed' is ecologically based on the presumption that herbicide resistance, specifically resistance to glyphosate, will improve the fitness of the resistant weed biotype. While genetic mutations that confer glyphosate resistance are extremely rare in weed communities, given the global adoption of GRCs and glyphosate, the probabilities suggest that it is inevitable that these mutant-resistant weeds will be detected, despite suggestions that it is unlikely to occur (Bradshaw *et al.*, 1997; Gressel and Levy, 2006; Gustafson, 2008). Interestingly, Harper predicted the evolution of herbicide resistance in weed populations over 50 years ago (Harper, 1956). He suggested that resistance to herbicides was a classic example of 'evolution in action' that occurs within a relatively short time period. However, Harper did not foresee the concerns about transgenic traits and addressed his thoughts to the evolutionary forces that drive herbicide resistance in weeds. The presumption of weeds acquiring transgenic herbicide resistance (i.e. glyphosate) is not well founded unless there are near-relatives of GRCs in weed populations. Of the current GRCs grown in North America, the only GRC with a near-relative that would support this type of trait movement is canola. Generally, the evolution of herbicide resistance in weeds is not a function of GR trait introgression from GRCs. However, as more GRCs are released (i.e. sugarbeet), the potential for transgenic glyphosate resistance in related weed populations increases considerably (Snow, 2002; Mallory-Smith *et al.*, 2005; Mallory-Smith and Zapiola, 2008).

Herbicide resistance in weeds is a function of the selection pressure brought to bear on the weed population by the use of herbicides, the frequency of the mutation in the native population and the relative impact of the herbicide resistance on plant fitness (Gressel and Segel, 1976; Friesen *et al.*, 2000; Neve, 2007). It should be assumed that, regardless of the genetic frequency of herbicide-resistant mutants in an unselected population, agriculture is such that these traits will be discovered (Gressel and Levy, 2006). The effect of fitness is a key factor in natural and selected weed populations and while difficult to assess, strongly impacts the potential evolution of herbicide-resistant weed populations (Neve, 2007; Ellegren and Sheldon, 2008). Furthermore, an assessment of the fitness of herbicide-resistant weed biotypes is critical when determining the ability of the biotype to adapt and impact agriculture, thus possibly achieving 'superweed' status.

To that end, the evolution of herbicide resistance has not typically enhanced the fitness of the herbicide-resistant weed biotypes compared to sensitive weed biotypes and in some cases has reduced the fitness of the resistant weeds (Gressel, 2002). Thus, 'Darwinian' fitness for herbicide-resistant weeds exists only in the presence of the selective herbicide. Without the herbicide, the herbicide-resistant weed biotypes behave similar to the herbicide-sensitive biotypes and thus have no significant impact on the environment. Of course, one should not ignore the remote possibility of pleiotropic effects attributable to the herbicide-resistant gene. Regardless, there is no pervasive evidence that herbicide-resistant weed populations behave differently than herbicide-sensitive weed populations in the absence of the herbicide, thus suggesting that herbicide resistance conferring 'superweed' status on these mutant populations is a considerable overstatement of the reality.

Current state of herbicide-resistant weeds

The current status of herbicide-resistant weeds from a global perspective suggests that new herbicide-resistant populations continue to evolve. Scientific publications describing some aspects of herbicide resistance in agriculture dominate a number of professional journals, and presentations about herbicide resistance at professional weed science conferences have represented the most prevalent topic for many years. There is, however, considerable difficulty in assessing the current status of herbicide-resistant weeds. The most popular and consistent information about the global status describing herbicide-resistant weeds is from the International Survey of Herbicide Resistant Weeds (www.weedscience.org; Heap, 2004). However, this tally of herbicide-resistant weeds requires that reports of new herbicide-resistant weed populations meet exact criteria for confirmation. Furthermore, weed scientists must volunteer these reports. Thus, while this survey is the best source of information on herbicide-resistant weeds, it is likely incomplete and may not include an accurate distribution description of the reported herbicide-resistant weeds. Currently, 317 herbicide-resistant biotypes are listed on the site, which includes 183 weed species and over 290,000 fields. The 183 species are divided between 110 dicotyledonous and 73 monocotyledonous weed species (Heap, 2004).

The understanding that growers have demonstrated about the factors which affect the evolution of GR weed populations is mixed and generally the level of concern expressed is not sufficient to suggest that these inevitable problems can be mitigated (Johnson and Gibson, 2006). An interesting perspective on the response of agriculture, at various levels, to GR weeds was suggested by Boerboom (2008) and based on the Kubler-Ross model; growers may progress through five phases in response to the discovery of GR weeds (Kubler-Ross, 2005; Boerboom, 2008). These phases are: (i) denial; (ii) anger; (iii) bargaining; (iv) depression; and (v) acceptance; at each phase, a specific sector or sectors of agriculture are represented. For example, industry originally denied the possibility of evolved resistance to glyphosate despite the suggestions that resistance was inevitable (Gressel, 1996; Bradshaw *et al.*, 1997; Owen, 2000; Zelaya and Owen, 2000). A typical grower response to GR weeds might be described in the third phase – bargaining. Regardless of how the situation surrounding glyphosate and weed resistance arises, there can be no question that changes in weed populations are occurring more rapidly and are widely distributed across a number of crop production systems.

Weed population shifts have often been described as different from evolved herbicide resistance. From an ecological perspective, weed shifts describe the change in relative prominence of species in a weed community in response to the total selective forces that make up the agroecosystem (Owen, 2008). Weed shifts are a general function of the relative fitness, or genetic variance, of the species and tend to evolve slowly, depending on the location environment (Ellegren and Sheldon, 2008). However, weed population shifts in response to herbicide selection can occur more rapidly and quickly become of significant economic consequence (Culpepper, 2006). What is important to recognize about weed shifts attributable to evolved glyphosate resistance is that weeds were able to demonstrate successful adaptation to a highly effective herbicide much quicker and more widely than originally reported (Padgette *et al.*, 1995). Gressel (1996) suggested that mechanisms were varied and constraints for evolved glyphosate resistance in weeds less than previously suggested (Gressel, 1996). Others reported evolved glyphosate resistance in weed populations as soon as 2 years after the strategy for glyphosate use changed to topical applications to the crop (Zelaya and Owen, 2000, 2002; VanGessel, 2001). The current state of weeds that have evolved resistance to glyphosate will be covered in the following section. Only a brief overview of these weeds that have been confirmed resistant to glyphosate will be discussed. Weeds that are described as

Table 7.2. Weeds reported with evolved resistance to glyphosate.[a]

Weed	Country (state)	Year of first report
Amaranthus palmeri (Palmer amaranth)	USA (Georgia, Arkansas, Tennessee)	2005
Amaranthus rudis (Common waterhemp)	USA (Missouria,[b] Illinois,[b] Kansas, Minnesota)	2005
Ambrosia artemisiifolia (Common ragweed)	USA (Arkansas, Missouri, Kansas)	2004
Ambrosia trifida (Giant ragweed)	USA (Ohio, Indiana, Kansas, Minnesota, Tennessee)	2004
Conyza bonariensis (Hairy fleabane)	South Africa, Spain, Brazil, Colombia, USA	2003
Conyza canadensis (Horseweed)	USA (Delaware, Kentucky, Tennessee, Indiana, Maryland, Missouri, New Jersey, Ohio,[b] Arkansas, Mississippi, North Carolina, Pennsylvania, California, Illinois, Kansas, Michigan), Brazil, China, Spain, Czech Republic	2000
Echinochloa colona (Junglerice)	Australia	2007
Eleusine indica (Goosegrass)	Malaysia[b]	1997
Euphorbia heterophylla (Wild poinsettia)	Brazil[b]	2006
Lolium multiflorum (Italian ryegrass)	Chile, Brazil, USA (Oregon, Mississippi), Spain	2001
Lolium rigidum (Rigid ryegrass)	Australia (Victoria, New South Wales, Western Australia), USA (California), South Africa,[b] France, Spain	1996
Plantago lanceolata (Buckhorn plantain)	South Africa	2003
Sorghum halepense (Johnsongrass)	Argentina, USA (Arkansas, Mississippi)	2005

[a]Adapted from www.weedscience.org, accessed 14 March 2008.
[b]Biotypes demonstrating resistance to multiple herbicide mechanisms of action.

naturally resistant or tolerant to glyphosate will not be included (Yuan *et al.*, 2002; Nandula *et al.*, 2005; Owen and Zelaya, 2005).

Currently there are 13 weeds reported and confirmed to have resistance to glyphosate (Table 7.2). In the USA, nine species have been confirmed as glyphosate-resistant. These GR weed populations are widely distributed across the USA and have evolved in a number of agricultural and non-agricultural environments. It is troubling that since 2004, six new species of GR weeds have been confirmed in the USA. Given that the first GR weed species in the USA was identified in 1998, and the first GR weed in an agroecosystem confirmed in 2000, it is clear that the evolution of GR weed species is increasing at an increasing rate. Two plant families are prominent: Compositae includes four species, and Amaranthaceae currently has two species listed with GR populations. Six of the species are dicots and three are monocots. Three other species, including *Xanthium strumarium*, *C. album* and *Kochia scoparia*, are currently suspected to have populations with evolved resistance to glyphosate (Boerboom, 2008). A brief review of prominent glyphosate-resistant weeds follows.

Conyza *species*
Populations of *C. canadensis* evolved eightfold to 13-fold resistance to glyphosate within 3 years after the adoption of GR soybeans (VanGessel, 2001). Since the first report of glyphosate resistance in Delaware, 16 states across the USA have GR *C. canadensis* populations (Table 7.2; Heap, 2004). These widely distributed occurrences of glyphosate resistance appear to be the result of independent founding events despite the fact that *Conyza* spp. produces a large number of seeds with facilitated transport (Buhler and Owen, 1997; Ozinga *et al.*, 2004). The mechanism that confers resistance to glyphosate is reported to be attributable to differential translocation (Feng *et al.*, 2004). The heritability of glyphosate resistance in *C. canadensis* is governed by a partially dominant single-locus nuclear gene (Zelaya *et al.*, 2004). Populations of *Conyza bonariensis* were reported in 2007 to have evolved resistance to glyphosate (Urbano *et al.*, 2007). These populations demonstrated tenfold resistance compared to the sensitive biotypes. Glyphosate-resistant *C. bonariensis* populations are also reported in South Africa, Brazil, Colombia and California, USA (Heap, 2004).

Euphorbia heterophylla
Glyphosate-resistant populations of *Euphorbia heterophylla* from the northern part of Rio Grande do Sul, Brazil, were reported to have threefold resistance compared to the sensitive biotypes (Vidal *et al.*, 2007). The GR *E. heterophylla* populations were selected in GR soybean where recurrent applications of low doses of glyphosate had been used (Vila-Aiub *et al.*, 2008). The selection from low recurrent doses strongly suggests the evolution of creeping (quantitative) resistance which likely is a polygenic trait. (Gressel, 1995, 2002). The specific mechanism(s) of resistance has not been described.

Eleusine indica
Populations of *Eleusine indica* with evolved resistance to glyphosate were identified in 1997 in Malaysian plantation crops where recurrent applications of glyphosate had been used for an estimated 10 years (Lee and Ngim, 2000). The GR biotype demonstrated eightfold to 12-fold resistance compared to the sensitive biotype. The mechanism of resistance was attributed to a less-sensitive EPSPS target site (Baerson *et al.*, 2002; Ng *et al.*, 2003). The heritability of the trait was described as a single incompletely dominant nuclear gene (Ng *et al.*, 2004a). Interestingly, another mechanism of glyphosate resistance in *E. indica* has been reported and suggests that glyphosate resistance in this species may be controlled by two independent mechanisms (Ng *et al.*, 2004b).

Amaranthus *species*

Resistance to glyphosate was reported in Iowa fields in 1998 for *Amaranthus rudis/Amaranthus tuberbulatus* (Zelaya and Owen, 2000, 2002). These populations were able to survive glyphosate applied at 6.72 kg ha^{-1}. While the mechanism of the resistance was not characterized, these biotypes had fivefold less accumulation of shikimate in response to glyphosate. Cursory evidence suggested that the resistance was polygenic (Zelaya and Owen, 2002). However, other reports suggest that glyphosate resistance in *A. tuberculatus* may be controlled by a single gene (Tranel, 2007). Recent evidence suggests that glyphosate resistance in *A. tuberculatus* is partially dominant and nuclear heritable (Bell *et al.*, 2007); segregation for the trait supports that more than one gene is likely involved. This report supports earlier information that glyphosate in *A. rudis/A. tuberbulatus* was a polygenic trait. Interestingly, a recent paper by (Volenberg *et al.*, 2007) suggests that in Illinois, USA, *A. tuberculatus* populations have not decreased in sensitivity to glyphosate despite significant selection pressure from recurrent glyphosate applications since 1996. However, glyphosate resistance in *A. rudis/A. tuberculatus* has been confirmed in Missouri, Illinois, Kansas and Minnesota, USA (Table 7.2). Another *Amaranthus* sp., *A. palmeri*, was confirmed to have evolved resistance to glyphosate in three US states (Table 7.2). The Georgia biotype was described to have threefold resistance in the field and up to eightfold resistance in the glasshouse (Culpepper *et al.*, 2006). The mechanism of resistance was not described, but differential absorption, translocation and the chromosome numbers were not different when the resistance and sensitive biotypes were compared.

Ambrosia *species*

Recent reports of two *Ambrosia* spp. from seven US states indicate that glyphosate-resistant biotypes are becoming increasing agronomic problems (Table 7.2; Heap, 2004). Populations of *Ambrosia trifida* and *Ambrosia artemisiifolia* are resistant to glyphosate. The *A. trifida* populations are of particular concern, given the competitiveness of this weed (Leer, 2006). Little information describing the mechanism of resistance or genetic heritability is available.

Sorghum halepense

Populations of *Sorghum halepense* have evolved resistance to glyphosate in Argentina and the USA (Arkansas and Mississippi) and are significant agronomic problems (Table 7.2; Heap, 2004). The resistant biotypes exhibited 3.5–10.5-fold resistance to glyphosate when compared to the sensitive biotype (Vila-Aiub *et al.*, 2007). These resistant biotypes resulted from the intense selection pressure of recurrent glyphosate applications in GR soybeans (Vila-Aiub *et al.*, 2008). Thus far, there is no published information about the heritability or mechanism(s) of resistance.

Lolium *species*

The evolution of glyphosate resistance in weeds was first reported in *Lolium rigidum* populations in Australia (Powles *et al.*, 1998). These populations demonstrated sevenfold to 11-fold resistance to glyphosate when compared to sensitive populations. In Chile and Brazil, evolved glyphosate resistance was described in *Lolium multiflorum* (Vila-Aiub *et al.*, 2008). Currently glyphosate resistance in *L. rigidum* is confirmed in Australia, South Africa, France, Spain and the USA while resistance in *L. multiflorum* is confirmed in Brazil, Chile, Spain and the USA (Table 7.2; Heap, 2004). The mechanism of resistance for *L. rigidum* has been described as the result of differential translocation of the glyphosate from the treated leaves to meristematic tissues (Lorraine-Colwill *et al.*, 2003; Powles and Preston, 2006; Preston and Wakelin, 2008). Differential translocation of glypho-sate is also identified in some glyphosate-resistant *L. multiflorum* biotypes (Michette *et al.*, 2005). Target site (EPSPS) resistance has also been identified in both *L. rigidum* and *L. multiflorum* populations (Powles and Preston, 2006). The change in the EPSPS that confers resistance is an amino acid shift at location 106 from a proline to either a serine or threonine. Interestingly, both of these resistance mechanisms are inherited as a single nuclear gene (Powles and Preston, 2006). It is speculated that these different mechanisms conferring resistance to glyphosate could conceivably co-occur but there are no reports to document this.

Influence of herbicide-resistant genetically modified crops on 'superweeds'

As previously suggested, the herbicide-resistant GM traits generally have not affected the relative fitness of weed populations and thus have little influence on weed populations. This presumes a direct affect of the GMC on the indigenous weed communities. It could be argued that this is a brusque assessment of the reality in that there are GMCs, more specifically GRCs and transgenic glufosinate-resistant crops, with near-weedy relatives that conceivably would be genetically compatible with the GMC. The best example of this is GM canola and the weedy Brassicaceae where there is evidence that trait movement between the GM crop and weedy relatives occurs with little effect on fitness (Hauser *et al.*, 2003; Legere, 2005). However, to date, there is little documented evidence of this interaction (Hauser *et al.*, 2003; Legere, 2005). With the introduction of new GRCs (i.e. *B. vulgaris*), the lack of direct effect on weeds from GMCs may change.

Regardless, there is clear and consistent evidence that the cultivation of GRCs has a significant impact on weed populations, whether the impact is on the evolution of GR weeds or the selection for 'naturally' GR weed populations within the weed communities (Owen, 2008). Of course, these changes in the weed community are only important when the selective pressure(s) is in effect (i.e. recurrent glyphosate applications) and in the specific agronomic environ-ment. Thus, these weeds cannot ecologically be classified as 'superweeds'; again the presumption is that the lack of response to glyphosate, irrespective of the mechanism, does not impart improved fitness in the non-selective environment *and* that pleiotropic effects from the evolved GR trait are unlikely to occur.

Having suggested that there is little/no direct influence of GMCs on weeds, there can be no question that indirectly GMCs, specifically GRCs, have a major impact on weed populations. The literature has many examples implicating the adoption of GRCs and evolved GR weed populations (Owen, 2008). However, the GR trait in itself does not impart selection pressure on the agroecosystem; if growers adopted GRCs but did not include glyphosate in the crop production programme, no change in the weed community would occur when compared to non-GRCs. Thus, GRCs facilitate a major change in the management of weed communities and farmers make the decision to impart the selection pressure from recurrent glyphosate applications. Given the scale of GRCs and ubiquitous use of glyphosate in these systems, it is not surprising that weed communities have changed relatively quickly and widely (Gressel and Levy, 2006; Owen, 2008). Some reports suggest that changes in 'naturally' resistant weeds will occur faster than changes in weeds with evolved resistance; however, current information suggests that evolved glyphosate-resistant populations are increasing at a much greater rate than the 'naturally' tolerant species (Shaner, 2000; Owen, 2008).

Culpepper (2006) surveyed weed scientists and documented numerous changes in weed communities attributable to GRCs and the concomitant use of glyphosate in these production systems (Culpepper, 2006). These shifts included weeds with evolved glyphosate resistance as well as changes in the populations of 'naturally' resistant weeds. It was noted that no changes attributable to GR maize systems were reported; however, it is probable that GR maize/GR soybean may contribute to weed community changes. The greatest impact on weed communities occurs when selective forces are consistent; continuous GR soybean or GR cotton with an emphasis on glyphosate for weed control will have the greatest affect on weed communities. In GR cotton, *Amaranthus* spp., *Commelina* spp., *Ipomoea* spp. and *Cyprus* spp. have become more difficult to manage. In GR soybean, *Ipomoea* spp., *Commelina* spp., *Amaranthus* spp., *Chenopodium* spp. and winter annual weeds (i.e. *C. canadensis*) have been more problematic (Culpepper, 2006).

The evolution of GR *C. canadensis* was a function of grower adoption of GR soybeans and GR cotton and the concomitant use of glyphosate (VanGessel, 2001; Koger *et al.*, 2004; Culpepper, 2006; Owen, 2008). Similarly, evolved resistance in *Amaranthus* spp. resulted from weed management tactics in GR soybean and GR cotton (Zelaya and Owen, 2000, 2002; Nandula *et al.*, 2005; Culpepper *et al.*, 2006; Culpepper and York, 2007; Tranel, 2007). Glyphosate-resistant *S. halepense* evolved in Brazil and Argentina in GR soybeans, and in GRC-based systems in Arkansas and Mississippi (Valverde and Gressel, 2006; Vila-Aiub *et al.*, 2007; Anonymous, 2008b; Person, 2008). South American adoption of GR soybean also facilitated the evolution of GR *E. heterophylla* (Vila-Aiub *et al.*, 2008). Thus, while GRCs do not themselves directly affect weed shifts, the resultant weed management tactic utilized to cultivate GRCs do directly enhance the opportunities for weed shifts to species adapted to glyphosate use. The increasing reliance on glyphosate for weed management in GRCs will inevitably lead to more weed populations that do not respond to glyphosate (Powles and Preston, 2006).

Genetically Modified Trait Movement

Does the movement of transgenic traits matter? Pervasive evidence suggests that GM herbicide-resistant traits moving into weed populations is not the primary problem, but rather the weed management tactics that are implemented in GRCs. Regardless, public concerns exist about the dearth of information that addresses the implications of GM trait movement in the environment in a fashion that is objective, convincing and understandable (Madsen *et al.*, 2002; Krayer von Krauss *et al.*, 2004; Madsen and Sandoe, 2005; Clark, 2006). Importantly, there have been too many examples that demonstrate the inability or unwillingness of agriculture to control GM trait movement (Hurburgh, 2000; Ginder, 2001; Hurburgh, 2003; Brasher, 2008). GM traits can move by two important routes: the movement of GMC seed and via pollen. Neither of these mechanisms of movement is easily managed and, succinctly, the containment of GM traits is difficult at best. However, if the stated assumption that GM traits conferring herbicide resistance, specifically glyphosate resistance, can generally be considered benign, why is it important to understand trait movement (Mallory-Smith and Zapiola, 2008; Owen, 2008)? The reasons include a number of considerations: (i) often GRCs include other GM traits which could potentially impact plants; (ii) it cannot be stated with 100% confidence that GRCs may not have an unintended and unanticipated consequence; and (iii) the lay public are not convinced that the science they are told is objective or correct. Finding an unregulated Starlink GM trait in your taco shells does not instil much confidence that agriculture is able to regulate itself, regardless of the actual impact of that trait (Ginder, 2001). While scientists propose novel strategies for GM trait containment, the concerns of the public may not be resolved (Lin *et al.*, 2008).

An excellent summary of the issue of GM trait movement is available in Council for Agricultural Science and Technology (CAST) Issue paper #37 (2007; Gealy *et al.*, 2007). One of the key points is that gene movement is not necessarily bad. The authors state: 'To date, all biotech crops approved for commercial use in the United States have been shown to pose minimal or negligible risk to the environment and human and animal health.' Unfortunately, global society has a sliding scale of risk assessment. The inclusion of 'to date' and 'minimal or negligible risk' in a description to address GM trait risk does not resonate well with the public who demand a zero risk for GM traits. Importantly, a review of the 25 most important food crops reports that all but four have near-relative weeds (Warwick and Stewart, 2005). Ellstand (2003a) lists 48 cultivated plants that have hybridized with wild relatives (Ellstand, 2003b). Snow (2002) provides an excellent series of reasons why the movement of GM traits is a concern and why scientists should address these issues (Snow, 2002). The identified issues included: (i) transgenes move in the environment and can be incorporated into the genomes of other plant species; (ii) the movement of genes is widespread and long-lived; (iii) risk assessment must be different for different crops/plants; (iv) there is a range of fitness affects for GM traits and this is not well understood; and (v) novel GM traits may worsen the impact of weeds on crops. The general implications of GRCs and trait movement will be addressed in crops and examples of hybridization addressed in the following sections.

Gene flow: crop to crop

Opinions and perspectives of gene flow in plants have changed considerably within the last few decades (Ellstand, 2003). Where once considered of little significance, now the movement of plant traits moves with relative ease over significant distances. However, the potential for gene flow demonstrates tremendous variability depending on the plant species (Ellstand, 2003). In the case of intraspecific gene flow in crops, transgenic traits for herbicide resistance provide an excellent and simple way to measure the frequency and distance of gene flow (Beckie *et al.*, 2006). The consequences of gene flow in crops are often difficult to determine and economic impact hard to quantify. However, when volunteer crops (i.e. maize) have acquired a transgene for herbicide resistance that was not anticipated, management costs increase and potential crop yield declines (Owen and Zelaya, 2005). If transgene flow is discovered in organic crops, the crop may be forfeited due to a zero tolerance for the biotech trait (Gealy *et al.*, 2007). Gene flow in crops can occur via pollen and via seed, the latter potentially affecting agriculture temporally and on a much larger scale than gene flow attributable to pollen (Hall *et al.*, 2003). However, in general, the risks of unintended trait movement are difficult to assess (Clark, 2006). Consider, however, that gene flow is no different in GMCs than in non-GM cultivars and that gene flow from GRCs is a reality (Mallory-Smith and Zapiola, 2008). To expect compliance of a zero GR trait tolerance is not reasonable and an acceptable tolerance must be established for the coexistence of GRC and non-GM crops.

Maize

Maize is an open-pollinated, wind-facilitated species and gene flow via pollen is well recognized (Haslberger, 2001). Thus, the movement of GM traits is a significant consideration in maize production (Luna *et al.*, 2001; Ma *et al.*, 2004). While models have been developed to predict pollen movement, they have thus far been unable to accurately predict longer-distance movement or account for environmental conditions (Ashton *et al.*, 2000). Factors that influence the distance that gene flow can occur include the longevity of pollen viability, the distance from pollen source, wind direction and speed, and synchronization of pollen shed and silk formation (Luna, 2001; Ma *et al.*, 2004). Environmental conditions that impact pollen viability include temperature and relative humidity. Generally, the introgression of GR traits in seed maize can be managed successfully ($\leq$1% outcross) by establishing isolation distances of 200 m between fields (Ma *et al.*, 2004). However, in typical maize production regions of the USA, these isolation distances are not possible and GR trait introgression into non-GR fields is prevalent. The occurrence of the GR transgene in non-GM maize can have significant economic consequences if the grower of the non-GM maize has a contract to provide a GM-free product. Furthermore, incidences of GM gene introgression in local landraces of maize in Oaxaca, Mexico have been reported (Quist and Chapella, 2001). The implications of the transgene occurrence reflect concerns for the maintenance of the genetic resource of the landrace maize. However, the initial report of transgene introgression was followed by a second report that suggested no transgenes existed in these landraces of maize (Ortiz-Garcia *et al.*, 2005). Regardless, given the adoption of GR maize, the discovery of transgene introgression into landrace maize is probably inevitable.

Soybean

Soybeans are an autogamous species with limited opportunity for pollen-directed gene flow (Palmer *et al.*, 2001). Spontaneous gene flow in cultivated soybeans ranges from 0.02% to 5% depending on distance and is facilitated by thrips (*Thrips tabaci*) and honeybees (*Apis mellifera*). While the movement of the GR transgene has been observed in soybeans, there are extremely limited opportunities for this occurrence, and pollen-mediated gene flow in GR soybeans is essentially a non-issue (Abud, 2004; Owen, 2005). However, gene flow by seed is highly probable and represents a significant economic problem (Swoboda, 2002; Owen, 2005).

Cotton

There are limited reports that cotton demonstrates introgression at low frequencies (Ellstand *et al.*, 1999). Pollen movement in cotton is dependent on insects. Cotton is predominantly self-pollinated and natural outcrossing is typically quite low (Xanthopoulos and Kechagia, 2000). Thus, there is a very low probability that the GR transgene would move into non-GR cotton cultivars via pollen flow. While gene flow via GR cotton seed can occur, the consequences of this are not thought to be important.

Canola (oilseed rape)

Canola is self-fertile but also capable of outcrossing (Rakow and Woods, 1987; Cuthbert and McVetty, 2001). Gene flow in canola has been repeatedly confirmed and thus represents a concern with regard to the movement of transgenes that confirm herbicide resistance (Legere, 2005). Furthermore, given that there are two GM traits for herbicide resistance in canola, glyphosate and glufosinate resistance, it has been documented that introgression can result in multiple GM herbicide resistance in canola. Furthermore, there does not appear to be a measureable fitness penalty from the introgression of the GM traits for herbicide resistance (Hauser *et al.*, 1998a,b; Snow *et al.*, 1999). Possibly a greater concern for gene flow in crops is the introgression of the GM traits attributable to seeds. Canola with GM traits can move great distances and persist in the environment during transport, and shattering during harvesting can result in the opportunity for trait introgression the following year (Legere, 2005).

Rice

The movement of GM transgenes in rice via pollen has been identified and it does occur, albeit over relatively short distances (Messeguer *et al.*, 2001; Rong *et al.*, 2005). The movement of traits can occur from GM rice cultivars to non-GM rice cultivars and from GM rice cultivars to weedy rice (Langevin *et al.*, 1990; Gealy and Dilday, 1997a; Shivrain *et al.*, 2007). Furthermore, the movement of GM traits via seed is highly probable. However, there are no GR rice cultivars currently available and the development of transgenic rice with herbicide resistance (GR or glufosinate) has been curtailed due to concerns in the marketplace. If GM rice resumes commercial development, the introgression of the transgene via pollen can be mitigated by conferring cleistogamy and chloroplast transformation within the GM cultivars (Kwon *et al.*, 2001). A simpler tactic that maintains pollen dispersal at very low levels is to establish spatial isolation between GM and non-GM rice cultivars (Rong *et al.*, 2007). Furthermore, the development of non-shattering GM cultivars would help resolve gene flow attributable to seed.

Sugarbeet

Sugarbeet is a biennial plant that is self-incompatible and primarily wind-pollinated (Desplanque *et al.*, 2002). Thus, the movement of the GR transgene in pollen is highly probable; gene flow over great distances via seed is also likely (Arnaud *et al.*, 2003). However, GR sugarbeet cultivars are anticipated to be commercialized in 2008 so to date, gene flow of the GR transgene to non-GM sugarbeet cultivars has not been reported.

Lucerne

Movement of the transgene conferring resistance to glyphosate has not been demonstrated; however, the potential movement of genes in lucerne via facilitated pollen transport was sufficient to cause the return to regulated status of GR lucerne (Jenczewski *et al.*, 1999; Fisher, 2007; Harriman, 2007).

Creeping bentgrass

The movement of genes from GR creeping bentgrass cultivars to wild creeping bentgrass populations was identified and the further development of the GR cultivars terminated (Watrud *et al.*, 2004; Mallory-Smith *et al.*, 2005; Reichman *et al.*, 2006; Mallory-Smith and Zapiola, 2008).

Sunflower

Sunflower (*Helianthus annuus*) is a member of the Asteraceae and is grown as a crop widely in the USA and Europe. Sunflowers are self-incompatible and pollinated by insects, primarily bees (*Apis* spp.; Berville *et al.*, 2005). Gene flow within sunflower cultivars has been reported but given that no GR sunflower cultivars exist, movement of transgenes that confer herbicide resistance is not an issue.

Wheat

No GM herbicide-resistant wheat cultivars are available and thus there is no current concern for gene flow (Beckie and Owen, 2007). While GR wheat was under development, the programme was terminated despite an apparent good technical fit (Dill, 2005; Howatt *et al.*, 2006).

Gene flow: crop to weed

Concern for gene flow from crops to weeds has increased dramatically with the inclusion of GM trait in many globally important food crops. Part of the issue focuses on the technology of the transgenes themselves and fears of the general public about increasing the prevalence of pernicious and highly invasive new weed species. Part of the concern relates to the impact that transgene introgression may have on the genetic diversity of key food crops such as maize in Mexico and soybean in China (Gepts and Papa, 2003; Lu, 2004; Raven, 2005). A large part of the concern is simply that there is now a relatively inexpensive, convenient, rapid and precise way to measure gene flow (Stewart *et al.*, 2003). While the process of gene flow from GM crops to weeds is presumed to be simple pollen movement, the actual introgression of the transgene is complex and likely requires a number of generations to complete (Stewart *et al.*, 2003). Some major food plants, including crops in the genera *Beta*, *Helianthus*, *Sorghum* and *Zea*, exist as part of a crop–weed–wild complex and can mutually influence related species by introgression of traits (Van Raamsdonk and Van der Maesen, 1996). Importantly, for gene flow between crops and weeds to occur, a near-relative wild plant must be available spatially and temporally. Introgression between subspecies appears to be more frequent than between species and thus less likely to create a more aggressive transgenic weedy species (Stewart *et al.*, 2003). The requirement of near-relatives available to receive the transgenic pollen makes some crops less risky than others (i.e. soybean and maize), while other crops should be considered higher risk (i.e. sunflower and wheat); crops that do not have transgenic cultivars represent no risk for transgene introgression into near-relative weeds. It is important to recognize, however, that crop-to-weed gene flow can still occur for these non-transgene crops, but it is difficult to identify.

Maize

While there are no near-relative weeds of maize where most of the commercial maize is produced, in Mexico and Central America, spontaneous hybridizations between maize and teosintes (*Z. mays* spp. *mexicana*, *Zea luxurians*, *Zea diploperennis*) are described (Wilkes, 1977). However, these morphological intermediates are not thought to be the result of the introgression of traits from maize into teosinte (Kato, 1997). A number of fitness penalties resulting from transgene introgression from maize into teosinte suggested that there was little risk of occurrence (Martinez-Soriano and Leal-Klevezas, 2000; Martinez Soriano *et al.*, 2002). However, new evidence strongly supported gene flow between maize and teosinte, albeit at low frequencies (Baltazar *et al.*, 2005). Following this paper, another publication refuted the results and declared that evidence did not exist to support gene flow from maize to teosinte (Ortiz-Garcia *et al.*, 2005). The basis for the concern is the loss of genetic variability in maize and concerns culturally for landrace maize in Mexico (Raven, 2005). However, the maize to teosinte hybrids, GR or *Bt* transgene(s), if they exist, are not likely to change the relative invasive characteristics of the plant. Teosinte is not considered a significant weedy competitor with crops.

Soybean

Wild plants that are thought to be morphologically similar to soybean occur spontaneously in China and Korea, the origin of soybean (Nakayama and Yamaguchi, 2002). These hybrids are the result of pollen flow between *G. max* spp. *soya* and *G. max* spp. *max* and occur at variable, but low, frequencies.

Given the global adoption of GR soybean, some concern has been expressed with regard to the preservation of soybean biodiversity which could be compromised if transgenes introgressed into wild soybean (Lu, 2004). The loss of genetic diversity and fitness as well as other unexpected ecological implications are possible outcomes of gene flow (Lu, 2005). However, given that wild-type soybeans are the only genetic near-relative of soybean and they do not represent a significant threat as weeds, nor are they widely distributed, there is little concern for crop-to-weed gene flow.

Cotton

No weedy near-relatives of cotton have been identified in the major cotton production regions of the world. While wild relatives exist and interspecific introgression between the wild and cultivated *Gossypium* spp. has been observed, no problem has been reported (Ellstand *et al.*, 1999). These near-relatives of cotton do not exist where commercial cotton production is important (Stewart *et al.*, 2003).

Canola (oilseed rape)

Gene flow from GM canola cultivars occurs to near-relatives and resultant hybrids are significant weeds (Legere, 2005). Thus, canola has been designated as a moderate risk crop with regard to the potential introgression of GM traits (Stewart *et al.*, 2003). Hybrids between *B. napus* and *Brassica rapa*, *Sinapis arvensis*, *Erucastrum gallicum* and *Raphanus raphanistrum* have been identified (Warwick *et al.*, 2000). Hybridization between cultivated and weedy Brassicaceae can result in a fitness gain in the F_1 which has significant implications in canola production (Hauser *et al.*, 2003). The implications of transgenic herbicide resistance introgression are not fully appreciated from an ecological perspective, but the transgene can persist in the population without herbicide selection suggesting that there is little if any fitness penalty attributable to the trait (Snow *et al.*, 1999). An important result of crop-to-weed gene flow in canola is the occurrence of hybrids with multiple herbicide resistance genes. These biotypes are extremely difficult to manage effectively and represent a significant problem in Canada (Beckie and Owen, 2007). Costs are attributable to contaminated seed, more complex and costly management tactics and issues with crop rotation sequence (Legere, 2005).

Gene flow between *B. napus* and *R. raphanistrum* occurs, albeit at an extremely low frequency (Rieger *et al.*, 2002; Rieger, 2002). Hybridization frequencies ranged from 10^{-7} to 3.10^{-5} (Chevre *et al.*, 2000).

Rice

Rice and near-relatives are interfertile and gene flow occurs in and near rice fields when the weedy species are present (Ellstand *et al.*, 1999). One of the most important weeds in rice is red rice (*O. sativa*) and is obviously a close relative of cultivated rice (*O. sativa*; Burgos *et al.*, 2008). The development of GM rice, including GR and glufosinate-resistant cultivars, was terminated due to concerns about the market acceptability of the GM crop by important foreign markets (Beckie and Owen, 2007). Prior to that, GM rice cultivars with resistance to glufosinate had demonstrated promise in the management of important weeds in rice (Gealy and Dilday, 1997a,b). Introgression of the transgene from GM rice cultivars to weedy red rice was demonstrated, but at a low frequency (Zhang *et al.*, 2003). Concerns about the possibility of rapid and widespread introgression of transgenic herbicide resistance from GM rice suggest that robust mitigation tactics be in place prior to any commercialization of the transgene cultivars (Olofsdotter *et al.*, 2000). Spatial separation between non-GM and GM rice cultivars may provide some resolution of introgression attributable to pollen movement and thus protect the purity in grain. (Messeguer *et al.*, 2001; Rong *et al.*, 2005). However, other novel tactics will be necessary to mitigate transgene flow from GM rice to red rice, given that the weed infestations occur spontaneously within the crop. Given the current situation with GM rice development, it appears to be a problem that does not require resolution.

Sugarbeet

Crop-to-weed gene flow has been demonstrated in sugarbeet (Andersen *et al.*, 2005). Based on the potential for introgression of genes from sugarbeet (*B. vulgaris*) to weedy beets (*B. vulgaris*) and sea beet (*B. vulgaris* L. spp. *maritime* (L.) Arcangeli), it is suggested that sugarbeets are at moderate risk for transgene flow (Stewart *et al.*, 2003), with production in Europe being at greater risk than in the USA. Weedy beets are competitive and difficult to manage in sugarbeet. Furthermore, weedy beets produce large numbers of seeds that have a long-lived seedbank (Desplanque *et al.*, 2002). Gene flow into sea beet is at a low level and of unknown importance (Andersen *et al.*, 2005). Given that sugarbeet production practices require harvesting the biennial plant after the first year of growth, only a small number of sugarbeets actually flower and thus provide a pollen base for the introgression of the trait. However, gene flow attributable to seed dispersal has also been implicated in trait introgression (Arnaud *et al.*, 2003). Monsanto intends to introduce sugarbeet GR cultivars in 2008. Thus, the likelihood of GR transgene introgression into weedy beets is high, depending on the husbandry practices demonstrated in sugarbeet production and whether or not Europe allows GR sugarbeets to be produced. The results of a GR weedy beet biotype greatly complicate weed management tactics in sugarbeets. It is suggested that novel strategies can mitigate the risk of transgene introgression into sugarbeet near-relatives by incorporating the transgene into the pollinator breeding lines (Desplanque *et al.*, 2002).

Sunflower

The genetic basis for commercial sunflower cultivars is narrow and there is considerable homology with wild sunflowers and near-relative sunflowers (Berville *et al.*, 2005). Cultivated sunflowers and wild sunflowers are the same species with only minimal differences at specific alleles. Thus, the potential for gene flow from crop to weed in sunflowers is assessed as a moderate risk (Stewart *et al.*, 2003). Evidence has been presented to support the intraspecific introgression of traits from sunflowers to wild, weedy sunflowers and these traits remain in the wild populations for a relatively long period of time (Whiton *et al.*, 1997; Linder *et al.*, 1998). Field surveys indicated that approximately

66% of cultivated sunflower fields were close to weedy sunflowers and that considerable synchrony of flowering existed (Burke *et al.*, 2002). Furthermore, morphological observations suggested a rate of natural hybridization ranging from 10% to 33% and thus support the potential for frequent gene flow from crop to weed. In fact, this directional gene flow has been implicated in the relative weediness of the wild type (Ellstand *et al.*, 1999). While no GM sunflower cultivars currently are commercially available, there can be no question that the introgression of transgenes conferring pest management traits (i.e. herbicide resistance or *Bt*) would readily and widely occur thus impacting weed management and possibly the fitness of the weedy GM biotype (Snow *et al.*, 2000).

Wheat

There is considerable evidence of spontaneous hybridization of wheat (bread and durum varieties) with near-relatives including *Aegilops* spp. (Ellstand *et al.*, 1999). Thus, wheat is ranked as a crop at moderate risk for gene flow from crop to weed (Stewart *et al.*, 2003). The weedy near-relative to wheat, *Aegilops cylindrica*, is a serious weed in wheat production and gene flow from wheat has been identified, although at a relatively low level when assessed morphologically (Rehman *et al.*, 2006). The identification of herbicide resistance in *A. cylindrica* was found to be similar to the expected frequency of natural mutation (Hanson *et al.*, 2005). However, the potential for gene flow should not be ruled out given that herbicides with a low frequency of mutation are likely to present a greater risk of escape through introgression. A field assessment of hybridization between wheat and *A. cylindrica* identified F_1 hybrid seeds at a rate of 0–8% based on specific fields (Morrison *et al.*, 2002). While the evidence for the introgression of herbicide-resistant trait from wheat to near relative weeds is less than clear, data indicate that the potential for the escape of transgenes from GR wheat cultivars, if developed, may be problematic.

Gene flow: weed to crop

Given that gene flow from GRC and GM crops to near-relative weeds is described and highly probable, depending on the specific crop, is it reasonable to suggest that reverse gene flow from weeds to crops is possible (Snow, 1997; Ellstand, 2003)? Documentation of weed-to-crop gene flow exists and, in fact, this directional movement of traits was greater than the crop-to-weed introgression in rice (Gealy *et al.*, 2003). Theoretically, the movement of GR and GM traits from weed biotypes that have acquired them from previous introgression is possible. A major assumption that must be made is that GR hybrid weeds do not demonstrate any fitness penalty compared to the wild-type population. If the GR trait is fitness neutral or positive, it would support the temporal existence of the GR biotype in the weed community; this is documented in *B. rapa* populations (Snow *et al.*, 1999). Furthermore, once established, the weed seedbank would provide a longer-term repository for the introgressed GM trait within the weed population (Legere, 2005). The questions that must be addressed are if gene flow from weeds to crop occurs and, if so, what is the impact? Volunteer sunflowers have introgressed via pollen flow traits that have modified the commercial value of the oil, thus answering both questions in a positive fashion (Berville *et al.*, 2005). Consider also that the value of an organic crop would be lost with the spontaneous introgression of a GM trait from near-relative weeds.

However, despite the fact that 12 of the 13 globally important food crops have demonstrated spontaneous hybridization with near-relative weeds, the likelihood of GR/GM transgene introgression from weeds to crops is low (Ellstand *et al.*, 1999). Only four of the 12 food crops demonstrating hybridization have GM cultivars and none have been implicated in weed evolution. However, there are crops that have supported weed evolution via hybridization and thus hypothetically represent a potential problem for weed-to-crop gene flow if GM cultivars are developed. These crops include, but are not limited to, rice, sorghum, bean and sunflower.

Gene flow: weed to weed

The potential for interspecific hybridization from the cross fertilization of two related species promotes genetic diversity and is a critical component of speciation (Abbot, 1992; Barton, 2001). While this process is well understood in respect of crop breeding, there has been considerably less attention on weed-to-weed gene flow and the implications that interspecific hybridization might have on weedy traits. Part of the problem is the difficulty in identifying hybrids conclusively. Morphological identification of interspecific hybrids is laborious and often less than conclusive. In the case of hybridization of the *Conyza* spp., a cosmopolitan genus adapted to agricultural and non-agricultural systems globally, interspecific hybrids are well documented in Europe, but not in the USA (Knobloch, 1972; McClintock and Marshall, 1988; Thebaud and Abbot, 1995). However, with evolution of heritable herbicide resistance traits, the ease and precision of identifying interspecific hybrids in weedy taxa increases. Nevertheless, there is a wide disparity of weed-to-weed transfer of herbicide resistance reported ranging in frequency from 0.15% to 85%, depending on the genera (Zelaya *et al.*, 2007).

The introgression of glyphosate resistance between *C. canadensis* (glyphosate-resistant) and *Conyza ramosissima* (glyphosate-sensitive) was investigated and confirmed with morphological comparisons as well as glyphosate assay (Zelaya *et al.*, 2007). While hybridization occurred in controlled environments, the potential for field hybridization to occur was deemed highly possible, although at a relatively low frequency. Heterosis of the glyphosate-resistant trait was observed in the hybrid; this has obvious implications with regard to the adaptations of the new taxa to agriculture environments where GRCs and glyphosate dominate.

Hybridization within the genus *Ambrosia* is also reported (Lee and Dickinson, 1980). In this study, reciprocal hybrids between *A. trifida* and *A. bidentata* were found to occur in the field at a frequency of 2% in *A. trifida* and 1.5% in *A. bidentata*. While the study did not assess the potential for the introgression of herbicide resistance traits, *A. trifida* is widely reported to have evolved resistance to ALS inhibitor herbicides and glyphosate (Heap, 2004). Thus, it is plausible to suggest that the introgression of herbicide resistance traits via interspecific hybridization of *Ambrosia* spp. is likely. Amaranthaceae is another important weedy genus that has evolved resistance to a number of herbicides including glyphosate (Heap, 2004). Hybridization of Amaranthaceae has been reported and the introgression of ALS inhibitor herbicide resistance has been described (Trucco *et al.*, 2005). Hybrids from *A. tuberculatus* and *A. hybridus* occurred at a frequency of 33% (Trucco *et al.*, 2005, 2006). Hybrids from crosses of *A. palmeri* and *A. rudis* have also been identified (Steinau *et al.*, 2003). These *Amaranthus* spp. have been reported as evolving multiple resistant populations to a number of herbicides including glyphosate, ALS inhibitors, PPO inhibitors and triazine herbicides (Heap, 2004).

The potential for weed-to-weed hybridization is dependent on a number of factors including the compatibility of the species and the ecological circumstances under which the plants developed. However, it is clear that interspecific hybridization of weeds has considerable implications with regard to the dissemination of transgenic and herbicide resistance traits, and thus programmes for the management of weeds. In the case of GRCs and the use of glyphosate, hybrid weeds with GR traits are possible and could represent a difficult new weed problem for growers to manage.

Herbicide Resistance Gene Stacking

Weed populations have demonstrated significant response to weed control programmes based on a single herbicide mode of action (Owen, 2008). Despite the obvious inevitability of weed shifts attributable to recurrent applications of glyphosate, growers rarely consider tactics to mitigate these weed population shifts proactively (Beckie, 2006). The ability to use glyphosate in-crop without concern for phytotoxicity in GRCs is a major consideration that explains, in part, the global adoption of GRCs. Grower perspectives that glyphosate-based systems are simple and consistent lull growers into a false sense of security with regard to the sustainability of GRC-based crop production systems (Carpenter and Gianessi, 1999). Thus, as weed populations shift to biotypes that are no longer controlled by glyphosate, growers look to new technology. The possibility of including more than one transgene that confers herbicide resistance in crop cultivars has been proposed as an answer to weeds shifts and evolved glyphosate resistance (Green, 2007). Furthermore, new transgenes conferring resistance to herbicides other than glyphosate and glufosinate are being developed (Matringe *et al.*, 2005; Behrens *et al.*, 2007). Given the vertical integration that has occurred in the agricultural chemical industry and acquisition of seed companies, the opportunity to develop crop cultivars with multiple herbicide transgenes is great. Monsanto has established glyphosate resistance as their platform for all seed products. Thus, any new transgene for herbicide resistance will be added to the GR cultivars. Recently, DuPont has announced their entry into the GRC marketplace. DuPont has incorporated a novel

N-acetyltransferase that confers transgenic resistance to glyphosate into maize and soybean (Green *et al.*, 2008). This enzyme was discovered in *Bacillus licheniformis* (Weigmann) Chester and, through a process termed 'gene shuffling', the glyphosate acetyltransferase activity increased 5000-fold (Castle *et al.*, 2004; Siehl *et al.*, 2005, 2007). To this transgenic event, DuPont added a robust event that confers resistance to ALS inhibitor herbicides (Green *et al.*, 2008).

Implications of multiple herbicide-resistant traits

The rationale for developing crops with multiple herbicide-resistant traits is to provide the grower with better weed control (Green *et al.*, 2008). However, given the low adoption of alternative tactics to manage weeds, the likely result of multiple herbicide-resistant traits in crops is multiple herbicide resistance in weeds. The two components of sustainability and profitability of weed management tactics (herbicides) is prevention and mitigation. It is interesting that the first multiple herbicide-resistant cultivars have resistance to glyphosate and sulfonylurea herbicides; many of the weeds that have evolved GR biotypes have previously evolved ALS inhibitor herbicide-resistant biotypes. Thus, growers will need to be at least as cautious in using the crops with multiple herbicide-resistant traits as they should have been with the single-trait crops.

Implications of *Bacillus thuringiensis* (*Bt*) Cry toxins for insect resistance

Currently, there are a number of GM crops that have stacked herbicide resistance and *Bt* events (Owen, 2006b). In some transgene events, the PAT gene (EC 2.3.1.X) is used as a marker for the *Bt*. The PAT gene also confers resistance for glufosinate herbicide. When the *Bt* trait is combined with a GR hybrid, the resultant GM cultivar is resistant to both glyphosate and glufosinate. Monsanto has announced new transgenic maize cultivars which will combine several *Bt* events plus two herbicide resistance events.

The *Bt* event functions ecologically differently compared to herbicide-resistant transgenes; the transgene for herbicide resistance can be considered benign to the weed population and has no impact until the herbicide is applied (Owen, 2008). In contrast, *Bt* exerts continuous selection pressure on the target insect whether the insect population is at economic thresholds or not. Thus, the potential impact of the *Bt* transgene may be greater than the impact of the herbicide resistance transgenes with regard to causing changes in the target pest complex. Furthermore, given the potential for the introgression of transgenes into near-relatives, *Bt* could potentially improve the fitness of compatible weed species (Snow, 1997). For example, the fitness of canola was increased by the inclusion of *Bt* (Stewart *et al.*, 1997). Given the potential for gene flow in canola and near-relatives, it is possible/likely that the trait will become incorporated into the weed population (Warwick and Stewart, 2005). This imbalance in the weed community and other affects on target and non-target insects can have significant unanticipated consequences on the agroecosystem.

Persistence of Herbicide Resistance in the Environment

The persistence of herbicide resistance reflects the longevity of the seedbank and the relative percentage of the seedbank that contains the trait for resistance. Whether the trait is attributable to the introgression from a GRC or 'traditional' selection attributable to recurrent applications of the herbicide will not matter. With few exceptions (i.e. canola) the persistence of resistance from volunteer GRCs is minimal given that crop seeds generally will not last more than 1 year in the soil. Canola, however, has demonstrated some persistence in the soil seedbank and may be a factor for several years (Legere, 2005). Other reports suggest that GM canola can persist in the soil for as long as 10 years (Black, 2008). A study in the UK demonstrated that GM canola seed losses averaged 3575 seeds m^{-2} but ranged up to 10,000 seeds m^{-2} (Lutman *et al.*, 2005). Seedbank losses were rapid immediately after harvest and accounted for 60% of the seedbank after the first year. However, thereafter, losses were considerably slower and stochastic modelling predicted a loss of 95% after 9 years.

Glyphosate resistance attributable to a weed population shift such that the GR biotype is the dominant biotype in the seedbank will require a number of years depending on the average life of the seed in the soil (Maxwell and Jasieniuk, 2000). However, once established, the GR biotype will persist for many years, depending on the environmental conditions, the weed species and effectiveness of management tactics imposed upon the weed population. If marginal weed management is imparted on the weed population, the soil seedbank increases rapidly (Bauer and Mortensen, 1992). If the weed population in question is long-lived in the soil (i.e. *Abutilon theophrasti*) and has evolved resistance to glyphosate, the problem is likely to persist indefinitely regardless of the effectiveness of subsequent management tactics. If the weed does not demonstrate a long life in the soil (i.e. *A. rudis*), remedial management tactics can reduce the population of the resistant biotype within a relatively short period of time (Steckel *et al.*, 2007). An *A. rudis* population of approximately 410 plants m^{-2} was reduced to 39%, 20%, 10% and 0.004% of the original population in 4 years of effective management. However, tillage increased the longevity of the weed seedbank. Thus, unless extraordinary weed management tactics can be consistently implemented after the establishment of an herbicide-resistant weed seedbank, it is probable that the issue will persist for a considerable time. A better practice is to keep the herbicide-resistant weed population from evolving.

Conclusions

Herbicide-resistant crops, specifically GRCs, have been globally adopted as the basis for the production of maize, soybean, cotton and canola (Dill *et al.*, 2008). The adoption of GRCs provides economic benefits to agriculture and has major positive impacts on the environment. Specifically, GRC-based systems are more likely to involve conservation tillage thus reducing soil erosion and consequently reducing herbicide losses attributable to surface water runoff

(Duke and Powles, 2008; Shipitalo *et al*., 2008). It is clear that agriculture places great value on the GRC technology and the concomitant use of glyphosate (Foresman and Glasgow, 2008). However, growers do not seem willing to proactively adopt measures that will sustain the technology (Mueller *et al*., 2005; Johnson and Gibson, 2006; Sammons *et al*., 2007; Christoffoleti *et al*., 2008). The result of this unwillingness to steward GRCs is the evolution of GR weeds (Duke and Powles, 2008; Owen, 2008). Other concerns about the current use of GRCs include the risks of transgene introgression into near-relative plants which may impact genetic diversity and fitness of these plants (Snow, 1997; Gepts and Papa, 2003). While there are tactics that are capable of mitigating some of these risks, agriculture could consider an improved process to assess the environmental risk of GRC technologies and communicating those risks to the lay public (Clark, 2006). Regardless, given the selective pressures imparted on agriculture from the global use of GRCs, weed management tactics must be more complex in order to sustain the utility of the technology (Liebman and Dyck, 1993).

References

Abbot, R.J. (1992) Plant invasions, interspecific hybridization and the evolution of new plant taxa. *Trends in Ecology and Evolution* 7, 401–405.

Abud, S., de Souza, P.I.M., Vianna, G.R., Leonardecz, E., Moreira, C.T., Faleiro, F.G., Junior, J.N., Monteiro, P.M.F.O., Rech, E.L. and Aragao, F.J.L. (2007) Gene flow from transgenic to nontransgenic soybean plants in the cerrado region of Brazil. *Genetic and Molecular Research* 6, 445–452.

Altieri, M.A. (2000) The ecological impacts of transgenic crops on agroecosystem health. *Agroecology in Action*. Available at: www.agroeco.org

Andersen, N.S., Siegismund, H.R., Meyer, V. and Jorgensen, R.B. (2005) Low level of gene flow from cultivated beets (*Beta vulgaris* L. ssp. *vulgaris*) into Danish populations of sea beet (*Beta vulgaris* L. spp. *maritima* (L.) Arcangeli). *Molecular Ecology* 14, 1391–1405.

Anonymous (2002) Oregon Administrative Rules (2002) Oregon Secretary of State, Oregon State Archives, item 603-052-1240. Oregon Department of Agriculture, Salem, Oregon.

Anonymous (2004) National Crop Residue Management Survey Data. CTIC, West Lafayette, Indiana.

Anonymous (2006a) Monsanto Biotechnology Trait Acreage: Fiscal Years 1996 to 2006. Available at: http://www.monsanto.com/pdf/pubs/2006/Q42006Acreage.pdf

Anonymous (2006b) United States Department of Agriculture National Agricultural Statistics Service Acreage Report. USDA, Washington, DC.

Anonymous (2006c) Fourth-quarter 2006 Financial Results. Monsanto, St Louis, Missouri.

Anonymous (2007a) Can biotech and organic crops coexist? Council for Biotechnology Information, Washington, DC.

Anonymous (2007b) ISAAA Brief 37-2007: Executive Summary Global Status Of Commercialized Biotech/GM crops: 2007, International Service for the Acquistion of Agri-Biotech Applications, p. 12. ISAAA.

Anonymous (2007c) Frankenfood? *The Washington Post*, p. 9.

Anonymous (2008a) China Glyphosate Industry Report 2007–2008. Available at: http://www.researchinchina.com/Report/Agriculture/5290.html

Anonymous (2008b) Glyphosate-resistant Johnsongrass Confirmed in Two Locations. AG Professional. Available at: http://www.agfax.com/news/2008/03/glyphresistjgrass0312.html

Arnaud, J.F., Viard, F., Delescluse, M. and Cuguen, J. (2003) Evidence of gene flow via seed dispersal from crop to wild relatives in *Beta vulgaris* (Chenopodiaceae): consequences for the release of genetically modified crop species with weedy lineages. *Proceedings of the Royal Society: Biological* 270, 1565–1571.

Ashton, B.A., Ireland, D.S. and Westgate, M.E. (2000) *Applicability of the Industrial Source Complex (ISC) Air Dispersion Model for Use on Corn Pollen Transport.* Ames, Iowa, p. 11.

Baerson, S.R., Rodiguez, D.J., Tran, M., Feng, Y., Biest, N.A. and Dill, G.M. (2002) Glyphosate-resistant goosegrass. Identification of a mutation in the target enzyme 5-enolpyruvylshikimate-3-phosphate synthase. *Plant Physiology* 129, 1265–1275.

Ball, D.A. (1992) Weed seedbank response to tillage, herbicides, and crop rotation sequence. *Weed Science* 40, 654–659.

Baltazar, B.M., Sanchez-Gonzalez, D.J., Cruz-Larios, L.D.L. and Schoper, J.B. (2005) Pollination between maize and teosinte: an important determinant of gene flow of Mexico. *Theoretical Applied Genetics* 110, 519–526.

Barton, N.H. (2001) The role of hybridization in evolution. *Molecular Ecology* 10, 551–568.

Batlla, D. and Benech-Arnold, R.L. (2007) Predicting changes in dormancy level in weed seed soil banks: implications for weed management. *Crop Protection* 26, 189–197.

Bauer, T.A. and Mortensen, D.A. (1992) A comparison of economic and economic optimum thresholds for two annual weeds in soybeans. *Weed Technology* 6, 228–235.

Beckie, H.J. (2006) Herbicide-resistant weeds: management tactics and practices. *Weed Technology* 20, 793–814.

Beckie, H.J. and Owen, M.D.K. (2007) Herbicide-resistant crops as weeds in North America. CAB Reviews: *Perspectives in Agriculture, Veterinary Science, Nutrition, and Natural Resources*, p. 22.

Beckie, H.J., Harker, K.N., Hall, L.M., Warwick, S.I., Legere, A., Sikkema, P.H., Clayton, A.G., Thomas, A.G., Leeson, J.Y., Seguin-Swartz, G. and Simard, M.J. (2006) A dec-

ade of herbicide-resistant crops in Canada. *Canadian Journal of Plant Science* 86, 1243–1264.

Behrens, M.R., Mutlu, N., Chakraborty, S., Bumitru, R., Jiang, W.Z., LaVallee, B.J., Herman, P.L., Clemente, T.E. and Weeks, D.P. (2007) Dicamba resistance: enlarging and preserving biotechnology-based weed management strategies. *Science* 316, 1185–1188.

Bell, M.S., Tranel, P.J., Bradely, K.W. (2007) Genetics of glyphosate resistance in a Missouri waterhemp population. In: Hartzler, R.G. (ed.) *North Central Weed Science Society*. North Central Weed Science Society (available on CD), St Louis, Missouri, p. 111.

Benton, T.G. (2007) Managing farming's footprint on biodiversity. *Science* 315, 341–342.

Berville, A., Muller, M.-H., Poinso, B. and Serieys, H. (2005) Ferality – Risks of gene flow between sunflower and other *Helianthus* species. In: Gressel, J. (ed.) *Crop Ferality and Volunteerism*. CRC Press, Boca Raton, Florida, pp. 209–230.

Bitzer, R.J., Buckelew, L.D. and Pedigo, L.P. (2002) Effects of transgenic herbicide-resistant soybean varieties and systems on surface-active springtails (Entognatha: Collembola). *Environmental Entomology* 31, 449–461.

Black, R. (2008) GM seeds can 'last for 10 years'. BBC News.

Blackburn, L.G. and Boutin, C. (2003) Subtle effects of herbicide use in the context of genetically modified crops: a case study with glyphosate (Roundup$^{(R)}$). *Ecotoxicology* 12, 271–285.

Boerboom, C. (2008) *Glyphosate Resistant Weed Update*. Wisconsin Fertilizer, Aglime and Pest Management Conference, Madison, Wisconsin, pp. 102–110.

Boerboom, C. and Owen, M. (2007) National Glyphosate Stewardship Forum II: A Call to Action. St Louis, Missouri, p. 45.

Bohan, D.A., Boffey, C.W.H., Brooks, D.R., Clark, S.J., Dewar, A.M., Firbank, L.G., Haughton, A.J., Hawes, C., Heard, M.S., May, M.J., Osborne, J.L., Perry, J.N., Rothery, P., Roy, D.B., Scott, R.J., Squire, G.R., Woiwod, I.P. and Champion, G.T. (2005) Effects on weed and invertebrate abundance and diversity of herbicide management in genetically modified herbicide-tolerant winter-sown oilseed rape. *Proceedings of the Royal Society: Biological* 272, 463–474.

Bradshaw, L.D., Padgette, S.R., Kimball, S.L. and Wells, B.H. (1997) Perspectives on glyphosate resistance. *Weed Technology* 11, 189–198.

Brasher, P. (2008) *Contaminated Biotech Corn Grown in Iowa.* The Des Moines Register, Des Moines, Iowa, p. 1D.

Buhler, D.D. (1992) Population dynamics and control of annual weeds in corn (*Zea mays*) as influenced by tillage systems. *Weed Science* 40, 241–248.

Buhler, D.D. and Owen, M.D.K. (1997) Emergence and survival of horseweed (*Conyza canadensis*). *Weed Science* 45, 98–101.

Buhler, D.D., Hartzler, R.G. and Forcella, F. (1997) Implications of weed seedbank dynamics to weed management. *Weed Science* 45, 329–336.

Burgos, N.R., Norsworthy, J.K., Scott, R.C. and Smith, K.L. (2008) Red rice (*Oryza sativa*) status after 5 years of imidazolinone-resistant rice technology in Arkansas. *Weed Technology* 22, 200–208.

Burke, J.M., Gardner, K.A. and Reieseberg, L.H. (2002) The potential for gene flow between cultivated and wild sunflower (*Helianthus annuus*) in the United States'. *American Journal of Botany* 89, 1550–1552.

Byrne, P.F. and Fromherz, S. (2003) Can GM and non-GM crops coexist? Setting a precedent in Boulder County, Colorado. *Food, Agriculture, and Environment* 1, 258–261.

Carpenter, J. and Gianessi, L. (1999) Herbicide tolerant soybeans: why growers are adopting Roundup Ready varieties. *AgBioForum* 2, 65–72.

Castle, L.A., Siehl, D.L., Gorton, R., Patten, P.A., Chen, Y.H., Bertain, S., Cho, H., Duck, N., Wong, J., Liu, D. and Lassner, M.W. (2004) Discovery and directed evolution of a glyphosate tolerance gene. *Science* 304, 1151–1154.

Cerdeira, A.L. and Duke, S.O. (2006) The current status and environmental impacts of glyphosate-resistant crops: a review. *Journal of Environmental Quality* 35, 1633–1658.

Champion, G.T., May, M.J., Bennett, S., Brooks, D.R., Clark, S.J., Daniels, R.E., Firbank, L.G., Haughton, A.J., Hawes, C., Heard, M.S., Perry, J.N., Randle, Z., Rossall, M.J., Rothery, P., Skellern, M.P., Scott, R.J., Squire, G.R. and Thomas, M.R. (2003) Crop management and agronomic context of the farm scale evaluations of genetically modified herbicide-tolerant crops. *Transactions of the Royal Society of London* 358, 1801–1818.

Charles, D. (2007) US courts say transgenic crops need tighter scrutiny. *Science* 315, 1069.

Chevre, A.M., Eber, F., Darmency, H., Fleury, A., Picault, H., Letanneur, J.C. and Renard, M. (2000) Assessment of interspecific hybridization between transgenic oilseed rape and wild radish under normal agronomic conditions. *Theoretical Applied Genetics* 100, 1233–1239.

Christoffoleti, P.J., Galli, A.J.B., Carvalho, S.J.P., Moreira, M.S., Nicolai, M., Foloni, L.L., Martins, B.A.B. and Ribeiro, D.N. (2008) Glyphosate sustainability in South American cropping systems. *Pest Management Science* 64, 422–427.

Clark, E.A. (2006) Environmental risks of genetic engineering. *Euphytica* 148, 47–60.

Crane, M., Norton, A., Leaman, J., Chalak, A., Bailey, A., Yoxon, M., Smith, J. and Fenlon, J. (2006) Acceptability of pesticide impacts on the environment: what do United Kingdom stakeholders and the public value? *Pest Management Science* 62, 5–19.

Culpepper, A.S. (2006) Glyphosate-induced weed shifts. *Weed Technology* 20, 277–281.

Culpepper, A.S. and York, A.C. (2007) *Glyphosate-Resistant Palmer Amaranth Impacts Southeast Agriculture.* Illinois Crop Protection Technology Conference. University of Illinois Urbana-Champaign, Champaign, Illinois.

Culpepper, A.S., Flanders, J.T., York, A.C. and Webster, T.M. (2004) Tropical spiderwort (*Commelina benghalensis*) control in glyphosate-resistant cotton. *Weed Technology* 18, 432–436.

Culpepper, A.S., Grey, T.L., Vencill, W.K., Kichler, J.M., Webster, T.M., Brown, S.M., York, A.C., Davis, J.W. and Hanna, W.W. (2006) Glyphosate-resistant Palmer amaranth (*Amaranthus palmeri*) confirmed in Georgia. *Weed Science* 54, 620–626.

Cuthbert, J.L. and McVetty, P.B.E. (2001) Plot-to-plot, row-to-row and plant-to-plant out-crossing studies in oilseed rape. *Canadian Journal of Plant Science* 81, 657–664.

Davis, L.C. (2006) Genetic engineering, eco-system change, and agriculture: an update. *Biotechnology and Molecular Biology Review* 1, 87–102.

Desplanque, B., Hautekeete, N. and Van Kijk, H. (2002) Transgenic weed beets: possible, probable, avoidable. *Journal of Applied Ecology* 39, 561–571.

Dill, G.M. (2005) Glyphosate-resistant crops; history, status and future. *Pest Management Science* 61, 219–224.

Dill, G.M., CaJacob, C.A. and Padgette, S.R. (2008) Glyphosate-resistant crops: adoption, use and future considerations. *Pest Management Science* 64, 326–331.

Doucet, C., Weaver, S.E., Hamill, A.L. and Zhang, J. (1999) Separating the effects of crop rotation from weed management on weed density and diversity. *Weed Science* 47, 729–735.

Duke, S.O. (2005) Taking stock of herbicide-resistant crops ten years after introduction. *Pest Management Science* 61, 211–218.

Duke, S.O. and Powles, S.B. (2008) Glyphosate: a once-in-a-century herbicide. *Pest Management Science* 64, 319–325.

Ellegren, H. and Sheldon, B.C. (2008) Genetic basis of fitness differences in natural populations. *Nature* 452, 169–175.

Ellstrand, N.C. (2003a) *Dangerous Liaisons? When Cultivated Plants Mate With Their Wild Relatives.* The Johns Hopkins University Press, Baltimore, Maryland.

Ellstand, N.C. (2003b) Current knowledge of gene flow in plants: implications for trans-gene flow. *Philosophical Transactions of the Royal Society of London B* 358, 1163–1170.

Ellstand, N.C., Prentice, H.C. and Hancock, J.F. (1999) Gene flow and introgression from domesticated plants into their wild relatives. *Annual Review of Ecological Systems* 30, 539–563.

Elmore, R.W., Roeth, F.W., Nelson, L.A., Shapiro, C.A., Klein, R.N., Knezevic, S.Z. and Martin, A. (2001a) Glyphosate-resistant soybean cultivar yields compared with sister lines. *Agronomy Journal* 93, 408–412.

Elmore, R.W., Roeth, F.W., Klein, R.N., Knezevic, S.Z., Martin, A., Nelson, L.A. and Shapiro, C.A. (2001b) Glyphosate-resistant soybean cultivar response to glyphosate. *Agronomy Journal* 93, 404–407.

Fawcett, R. and Towery, D. (2004) *Conservation Tillage and Plant Biotechnology: How New Technologies Can Improve the Environment by Reducing the Need to Plow.* Conservation Technology Information Center, West Lafayette, Indiana, p. 20.

Felix, J. and Owen, M.D.K. (1999) Weed population dynamics in land removed from the conservation reserve program. *Weed Science* 47, 511–517.

Feng, P.C.C., Tran, M., Chiu, T., Sammons, R.D., Heck, G.R. and CaJacob, C.A. (2004) Investigations into glyphosate-resistant horseweed (*Conyza canadensis*): retention, uptake, translocation, and metabolism. *Weed Science* 52, 498–505.

Feng, P.C.C., Baley, G.J., Clinton, W.P., Bunkers, G.J., Alibhai, M.F. and Paulitz, T.C. (2005) Glyphosate inhibits rust diseases in glyphosate-resistant wheat and soybean. *Proceedings of the National Academy of Science* 102, 17290–17295.

Feng, P.C.C., Clark, C., Andrade, G.C., Balbi, M.C. and Caldwell, P. (2008) The control of Asian rust by glyphosate in glyphosate-resistant soybeans. *Pest Management Science* 64, 353–359.

Firbank, L.G., Heard, M.S., Woiwod, I.P., Hawes, C., Haughton, A.J., Champion, G.T., Scott, R.J., Hill, M.O., Dewar, A.M., Squires, G.R., May, M.J., Brooks, D.R., Bohan, D.A., Daniels, R.E., Osborne, J.L., Roy, D.B., Black, H.I.J., Rothery, P. and Perry, J.N. (2003) An introduction to the Farm-Scale Evaluations of genetically modified herbicide-tolerant crops. *Journal of Applied Ecology* 40, 2–16.

Fisher, L. (2007) Growers continue to grow and use Roundup Ready alfalfa but Monsanto Company is disappointed with preliminary injunction affecting purchase and planting. Available at: http://www.monsanto.com (accessed 15/10/08)

Forcella, F., Wilson, R.G., Dekker, J., Kremer, R.S., Cardina, J., Anderson, R.L., Alm, D., Renner, K.A., Harvey, R.G., Clay, S. and Buhler, D.D. (1997) Weed seed bank emergence across the corn belt. *Weed Science* 45, 67–76.

Foresman, C. and Glasgow, L. (2008) US grower perceptions and experiences with glyphosate-resistant weeds. *Pest Management Science* 64, 388–391.

Freckleton, R.P., Stephens, P.A., Sutherland, W.J. and Watkinson, A.R. (2004) Amelioration of biodiversity impacts of genetically modified crops: predicting transient versus long-term effects. *Proceedings of the Royal Society: Biological* 271, 325–331.

Friesen, L.J.S., Ferguson, G.M. and Hall, J.C. (2000) Management strategies for attenuating herbicide resistance: untoward consequences of their promotion. *Crop Protection* 19, 891–895.

Fryer, J.D. (1981) Weed management: fact or fable? *Philosophical Transactions of the Royal Society of London B* 295, 185–197.

Fulton, M. and Giannakas, K. (2001) Agricultural biotechnology and industry structure. *AgBioForum* 4, 137–151.

Gealy, D.R. and Dilday, R.H. (1997a) *Biology of Red Rice (Oyrza sativa L.) Accessions and Their Susceptibility to Glufosinate and Other Herbicides*. Weed Science Society of America. Allen Press, Lawrence, Kansas, p. 34.

Gealy, D.R. and Dilday, R.H. (1997b) *Biology of Red Rice (Oyraz sativa L.) Accessions and Their Susceptibility to Glufosinate and Other Herbicides*. Weed Science Society of America, Allen Press, Lawrence, Kansas, p. 34.

Gealy, D.R., Mitten, D.H. and Rutger, J.N. (2003) Gene flow between red rice (*Oryza sativa*) and herbicide-resistant rice (*O. sativa*): implications for weed management. *Weed Technology* 17, 627–645.

Gealy, D.R., Bradford, K.J., Hall, L., Hellmich, R., Raybold, A., Wolt, J. and Zilberman,

D. (2007) *Implications of Gene Flow in the Scale-up and Commercial Use of Biotechnology-derived Crops: Economic and Policy Considerations*. CAST Issue Paper, Ames, Iowa, p. 24.

Gepts, P. and Papa, R. (2003) Possible effects of (trans)gene flow from crops on the genetic diversity from landraces and wild relatives. *Environmental Biosafety Research* 2, 89–103.

Gianessi, L.P. (2005) Economic and herbicide use impacts of glyphosate-resistant crops. *Pest Management Science* 61, 241–245.

Gianessi, L.P. (2008) Economic impacts of glyphosate-resistant crops. *Pest Management Science* 64, 346–352.

Gianessi, L.P. and Reigner, N.P. (2007) The value of herbicides in US crop production. *Weed Technology* 21, 559–566.

Gibbons, D.W., Bohan, D.A., Rothery, P., Stuart, R.C., Haughton, A.J., Scott, R.J., Wilson, D., Perry, J.N., Clark, S.J., Dawson, R.J.G. and Firbank, L.G. (2006) Weed seed resources for birds in fields with contrasting conventional and genetically modified herbicide-tolerant crops. *Proceedings of the Royal Society: Biological* 273, 1921–1928.

Ginder, R.G. (2001) *Channelling, Identity Preservation and the Value Chain: Lessons from the Recent Problems with Starlink Corn*. Iowa State University, Ames, Iowa.

Green, J.M. (2007) Review of glyphosate and ALS-inhibiting herbicide crop resistance and resistant weed management. *Weed Technology* 21, 547–558.

Green, J.M., Hazel, C.B., Forney, D.R. and Pugh, L.M. (2008) New multiple-herbicide crop resistance and formulation technology to augment the utility of glyphosate. *Pest Management Science* 64, 332–339.

Gressel, J. (1995) Creeping resistances: the outcome of using marginally effective or reduced rates of herbicides. In: Council, B.C.P. (ed.) *The British Crop Protection Conference Weeds*. British Crop Protection Council, Brighton, UK, pp. 587–589.

Gressel, J. (1996) Fewer constraints than proclaimed to the evolution of glyphosate-resistant weeds. *Resistant Pest Management* 8, 2–5.

Gressel, J. (2002) *Molecular Biology of Weed Control*. Taylor & Francis, London.

Gressel, J. and Levy, A.A. (2006) Agriculture: the selector of improbable mutations. *Proceedings of the National Academy of Science* 103, 12215–12216.

Gressel, J. and Segel, L.A. (1976) The paucity of genetic adaptive resistance of plants to herbicides: possible biological reasons and implications. *Journal of Theoretical Biology* 75, 349–371.

Gullickson, G. (2005) Stop superweeds – how to halt weeds resistant to multiple herbicide modes of action. *Successful Farming*, 1, p. 4.

Gullickson, G. (2008) What's behind glyphosate and roundup price hikes? AgricultureOnline. Available at: http://www.agriculture.com/ag/story.jhtml?storyid=/templatedata/ag/story/data/1204134767122.xml

Gustafson, D.I. (2008) Sustainable use of glyphosate in North American cropping systems. *Pest Management Science* 64, 409–416.

Halford, N.G. (2004) Prospects for genetically modified crops. *Annals of Applied Biology* 145, 17–24.

Hall, L.M., Good, A., Beckie, H.J. and Warwick, S.I. (2003) Gene flow in herbicide-resistant canola (*Brassica napus*): the Canadian experience. In: Lelley, T., Balazs, E. and Tepfer, M. (eds) *Ecological Impact of GMO Dissemination in Agroecosystems.* Facultas Verlagsun Buchhandels AG, Austria, pp. 57–66.

Hanson, B.D., Mallory-Smith, C.A., Price, W.J., Shafil, B., Thill, D.C. and Zemetra, R.S. (2005) Interspecific hybridization: potential for movement of herbicide resistance from wheat to jointed goatgrass (*Aegilops cylindrica*). *Weed Technology* 19, 674–682.

Harper, J.L. (1956) *The Evolution of Weeds in Relation to Resistance to Herbicides* , 3rd British Weed Control Conference. British Weed Control Council, Farnham, UK, pp. 179–188.

Haslberger, A. (2001) GMO contamination of seeds. *Nature Biotechnology* 19, 613.

Hauser, T.P., Shaw, R.B. and Ostergards, H. (1998a) Fitness of F_1 hybrids between weed *Brassica rapa* and oilseed rape (*B. napus*). *Heredity* 81, 429–435.

Hauser, T.P., Jorgensen, R.B. and Ostergards, H. (1998b) Fitness of backcross and F_2 hybrids between weedy *Brassica rapa* and oilseed rape (*B. napus*). *Heredity* 81, 436–443.

Hauser, T.P., Damgaard, C. and Jorgensen, R.B. (2003) Frequency-dependent fitness of hybrids between oilseed rape (*Brassica napus*) and weedy *B. rapa* (Brassicaceae). *American Journal of Botany* 90, 571–578.

Hawes, C., Haughton, A.J., Osborne, J.L., Roy, D.B., Clark, S.J., Perry, J.N., Rothery, P., Bohan, D.A., Brooks, D.R., Champion, G.T., Dewar, A.M., Heard, M.S., Woiwod, I.P., Daniels, R.E., Young, M.W., Parish, A.M., Scott, R.J., Firbank, L.G. and Squire, G.R. (2003) Responses of plants and invertebrate trophic groups to contrasting herbicide regimes in the Farm Scale Evaluations of genetically modified herbicide-tolerant crops. *Proceedings of the Royal Society: Biological* 358, 1988–1913.

Heap, I. (2004) The international survey of herbicide resistant weeds. Available at: www.weedscience.com.

Heard, M.S., Hawes, C, Champion, G.T., Clark, S.J., Firbank, L.G., Haughton, A.J., Parish, A.M., Perry, J.N., Rothery, P., Scott, R.J., Skellern, M.P., Squire, G.R. and Hill, M.O. (2003a) Weeds in fields with contrasting conventional and genetically modified herbicide-tolerant crops. I. Effects on abundance and diversity. *Philosophical Transactions of the Royal Society of London B* 358, 1819–1832.

Heard, M.S., Hawes, C., Champion, G.T., Clark, S.J., Firbank, L.G., Haughton, A.J., Parish, A.M., Perry, J.N., Rothery, P., Roy, D.B., Scott, R.J., Skellern, M.P., Squire, G.R. and Hill, M.O. (2003b) Weeds in fields with contrasting conventional and genetically modified herbicide-tolerant crops. II Effects on individual species. *Philosophical Transactions of the Royal Society of London B* 358, 1833–1846.

Heard, M.S., Clark, S.J., Rothery, P., Perry, J.N., Bohan, D.A., Brooks, D.R., Champion, G.T., Dewar, A.M., Hawes, C., Haughton, A.J., May, M.J., Scott, R.J., Stuart, R.S., Squire, G.R. and Firbank, L.G. (2006) Effects of successive seasons of genetically modified herbicide-tolerant maize cropping on weeds and invertebrates. *Annals of Applied Biology* 149, 249–254.

Hilgenfeld, K.L., Martin, A.R., Mortensen, D.A. and Mason, S.C. (2000) Weed species shifts in a glyphosate tolerant soybean crop. In: Hartzler, R.G. (ed.) *North Central Weed Science Society*. North Central Weed Science Society, Kansas City, Kansas, p. 59.

Howatt, K.A., Endres, G.J., Hendrickson, P.E., Aberle, E.Z., Lukach, J.R., Jenks, B.M., Riveland, N.R., Valenti, S.A. and Rystedt, C.M. (2006) Evaluation of glyphosate-resistant hard red spring wheat (*Triticum aestivum*). *Weed Technology* 20, 706–716.

Hurburgh, J.C.R. (2000) *The GMO Controversy and Grain Handling for 2000*. Iowa State University, Ames, Iowa.

Hurburgh, J.C.R. (2003) *Constraints for Isolation and Traceability of Grains*. Iowa State University, Ames, Iowa.

Jenczewski, E., Prosperi, J.-M. and Joelle, R. (1999) Evidence for gene flow between wild and cultivated *Medicago sativa* (Leguminosae) based on allozyme markers and quantitative traits. *American Journal of Botany* 86, 677–687.

Johnson, K.L., Raybould, A.F., Hudson, M.D. and Poppy, G.M. (2006) How does scientific risk assessment of GM crops fit within the wider risk analysis? *Trends in Plant Science* 12, 1–5.

Johnson, W.G. and Gibson, K.D. (2006) Glyphosate-resistant weeds and resistance management strategies: an Indiana grower perspective. *Weed Technology* 20, 768–772.

Jost, P., Shurley, D., Culpepper, S., Roberts, P., Nichols, R., Reeves, J. and Anthony, S. (2008) Economic comparison of transgenic and nontransgenic cotton production systems in Georgia. *Agronomy Journal* 100, 42–51.

Kaiser, J. (2001) Breeding a hardier weed. *Science* 293, 1425–1426.

Kato, T.A. (1997) Review of introgression between maize and teosinte. In: Serratos, J.A. Willcox, M.C. and Castillo Gonzalez, F. (eds) *Gene Flow Among Maize Landraces, Improved Maize Varieties, and Teosinte: Implications for Transgenic Maize*. CIMMYT, Mexico City, Mexico, pp. 44–53.

Knobloch, I.W. (1972) Intergeneric hybridization in flowering plants. *Taxon* 21, 97–103.

Koger, C.H., Poston, D.H., Hayes, R.M. and Montgomery, R.F. (2004) Glyphosate-resistant horseweed (*Conyza canadensis*) in Mississippi. *Weed Technology* 18, 820–825.

Krayer von Krauss, M.P., Casman, E.A. and Small, M.J. (2004) Elicitation of expert judgments of uncertainty in the risk assessment of herbicide-tolerant oilseed crops. *Risk Analysis* 24, 1515–1527.

Krebs, J.R., Wilson, J.D., Bradbury, R.B. and Siriwardena, G.M. (1999) The second silent Spring? *Nature* 400, 611–612.

Kubler-Ross, E. (2005) *On Grief and Grieving: Finding the Meaning of Grief Through the Five Stages of Loss*. Simon & Schuster, New York.

Kwon, Y.W., Kim, D.S. and Yim, K.O. (2001) Herbicide-resistant genetically modified crop: assessment and management of gene flow. *Weed Biology and Management* 1, 96–107.

Lange, K.E. (2008) Revenge of the weeds. *National Geographic*, p. 1.

Langevin, S.A., Clay, K. and Grace, J.B. (1990) The incidence and effects of hybridization between cultivated rice and its related weed rice (*Oryza sativa* L.). *Evolution* 44, 1000–1008.

Lee, Y.S. and Dickinson, D.B. (1980) Field observations on hybrids between *Ambrosia bidentata* and *A. trifida* (Compositae). *American Midland Naturalist* 103, 180–184.

Lee, L.J. and Ngim, J. (2000) A first report of glyphosate-resistant goosegrass (*Eleusine indica* (L.) Gaertn) in Malaysia. *Pest Management Science* 56, 336–339.

Leer, S. (2006) *Glyphosate-resistant Giant Ragweed Confirmed in Indiana, Ohio*. Purdue University Newsletter, West Lafayette, Indiana.

Legere, A. (2005) Risks and consequences of gene flow from herbicide-resistant crops: canola (*Brassica napus* L.) as a case study. *Pest Management Science* 61, 292–300.

Liebman, M. and Dyck, E. (1993) Crop rotation and intercropping strategies for weed management. *Ecological Applications* 3, 92–122.

Lin, C., Fang, J., Xu, X., Zhao, T., Cheng, J., Tu, J., Ye, G. and Shen, Z. (2008) A built-in strategy for containment of trangenic plants: creation of selectivity terminable transgenic rice. *Plos One* 3, 1–6.

Linder, C.R., Taha, I., Seiler, G.J., Snow, A.A. and Rieseberg, L.H. (1998) Long-term introgression of crop genes into wild sunflower populations. *Theoretical Applied Genetics* 96, 339–347.

Lorraine-Colwill, D.F., Powles, S.B., Hawkes, T.R., Hollinshead, P.H., Warner, S.A.J. and Preston, C. (2003) Investigations into the mechanism of glyphosate resistance in *Lolium rigidum*. *Pesticide Biochemistry and Physiology* 74, 62–72.

Lu, B.-R. (2004) Conserving biodiversity of soybean gene pool in the biotechnology era. *Plant Species Biology* 19, 115–125.

Lu, B.-R. (2005) Mulidirectional gene flow among wild, weedy, and cultivated soybean. In: Gressel, J. (ed.) *Crop Ferality and Volunteerism*. CRC Press, Boca Raton, Florida, pp. 137–147.

Luna, V.S., Figueroa, M.J., Baltazar, M.B., Gomez, L.R., Townsend, R. and Schoper, J.B. (2001) Maize pollen longevity and distance isolation requirements for effective pollen control. *Crop Science* 41, 1551–1557.

Lutman, P.J.W., Berry, K., Payne, R.W., Simpson, E., Sweet, J.B., Champion, G.T., May, M.J., Wightman, P., Walker, K. and Lainsbury, M. (2005) Persistence of seeds from crops of conventional and herbicide tolerant oilseed rape (*Brassica napus*). *Proceedings of the Royal Society: Biological* 272, 1909–1915.

Ma, B.L., Subedi, K.D. and Reid, L.M. (2004) Extent of cross-fertilization in maize by pollen from neighboring transgenic hybrids. *Crop Science* 44, 1273–1282.

Madsen, K.H. and Sandoe, P. (2005) Ethical reflections on herbicide-resistant crops. *Pest Management Science* 61, 318–325.

Madsen, K.H., Valverde, B.E. and Jensen, J.E. (2002) Risk assessment of herbicide resistant crops: a Latin American perspective using rice (*Oryza sativa*) as a model. *Weed Technology* 16, 215–223.

Mallory-Smith, C. and Zapiola, M. (2008) Gene flow from glyphosate-resistant crops. *Pest Management Science* 64, 428–440.

Mallory-Smith, C., Butler, M. and Campbell, C. (2005) *Gene Movement from Glyphosate-resistant Creeping Bentgrass (Agrostis stolonifera) Fields*. Weed Science Society of America, Honolulu, Hawaii, pp. 49–50.

Martinez-Soriano, J.P.R. and Leal-Klevezas, D.S. (2000) Transgenic maize in Mexico: no need for concern. *Science* 287, 1399.

Martinez-Soriano, J.P.R., Bailey, A.M. and Lara-Reyna, J. (2002) Transgenics in Mexican maize. *Nature Biotechnology*, 19.

Matringe, M., Sailland, A., Pelissier, B., Rolland, A. and Zink, O. (2005) *p*-hydroxyphenylpyruvate dioxygenase inhibitor-resistant plants. *Pest Management Science* 61, 269–276.

Matus-Cadiz, P.H., Horak, M.J. and Blomquist, L.K. (2004) Gene flow in wheat at the field scale. *Crop Science* 44, 718–727.

Maxwell, B. and Jasieniuk, M. (2000) *The Evolution of Herbicide Resistance Evolution Models*. Third International Weed Science Congress, Foz do Iguassu, Brazil, p. 172.

May, M.J., Champion, G.T., Dewar, A.M., Qi, A. and Pidgeon, J.D. (2005) Management of genetically modified herbicide-tolerant sugar beet for spring and autumn environmental benefit. *Proceedings of the Royal Society: Biological* 272, 111–119.

McClintock, D. and Marshall, J.B. (1988) On *Conyza sumatrensis* (Retz) E. Walker and certain hybrids in the genus. *Watsonia* 17, 172–173.

Messeguer, J., Fogher, C., Guiderdoni, E., Marfa, V., Catala, M.M., Baldi, G. and Mele, E. (2001) Field assessments of gene flow from transgenic to cultivated rice (*Oryza sativa* L.) using a herbicide resistance gene as a tracer marker. *Theoretical Applied Genetics* 103, 1151–1159.

Meyer, B. (2008) *Biotech Crops Gaining Favor Around the Globe*. Learfield Communications, Jefferson City, Montana. Available at: http://www.brownfieldnetwork.com/gestalt/go.cfm?objectid=8B64151A-B635–495D-08CB30A354EE3D15 (accessed 15/10/08)

Michette, P., De Prado, R., Espinosa, N. and Gauvrit, C. (2005) Glyphosate resistance in a Chilean *Lolium multiflorum*. *Communications in Agricultural Applied Biological Sciences* 70, 507–513.

Miller, D.R., Chen, S.Y., Porter, P.M., Johnson, G.A., Wyse, D.L., Stetina, S.R., Klossner, L.D. and Nelson, G.A. (2006) Rotation crop evaluations for management of the soybean cyst nematode in Minnesota. *Agronomy Journal* 98, 569–578.

Mills, C.I., Bednarz, C.W., Ritchie, G.L. and Whitaker, J.R. (2008) Yield, quality, and fruit distribution in Bollgard/Roundup Ready and Bollgard II/Roundup Ready Flex cotton. *Agronomy Journal* 100, 35–41.

Morrison, L.A., Riera-Lizarazu, O., Cremieux, L. and Mallory-Smith, C.A. (2002) Jointed goatgrass (*Aegilops cylindrica* Host) × wheat (*Triticum aestivum* L.) hybrids: hybridization dynamics in Oregon wheat fields. *Crop Science* 42, 1863–1872.

Mueller, T.C., Mitchell, P.D., Young, B.G. and Culpepper, A.S. (2005) Proactive versus reactive management of glyphosate-resistant or -tolerant weeds. *Weed Technology* 19, 924–933.

Mulugeta, D. and Stoltenberg, D.E. (1997) Seed bank characterization and emergence of a weed community in a moldboard plow system. *Weed Science* 45, 54–60.

Nakayama, Y. and Yamaguchi, H. (2002) Natural hybridization in wild soybean (*Glycine max* spp. *soja*) by pollen flow from cultivated soybean (*Glycine max* spp. *max*) in a designed population. *Weed Biology and Management* 2, 25–30.

Nandula, V.K., Reddy, K.N., Duke, S.O. and Poston, D.H. (2005) Glyphosate-resistant weeds: current status and future outlook. *Outlooks Pest Management* 12, 183–197.

Nandula, V.K., Eubank, T.W., Poston, D.H., Koger, C.H. and Reddy, K.N. (2006) Factors affecting germination of horseweed (*Conyza canadensis*). *Weed Science* 54, 898–902.

Neve, P. (2007) Challenges for herbicide resistance evolution and management: 50 years after Harper. *Weed Research* 47, 365–369.

Neve, P. (2008) Simulation modelling to understand the evolution and management of glyphosate resistance in weeds. *Pest Management Science* 64, 392–401.

Ng, C.H., Wickneswari, R., Salmijah, S., Teng, Y.T. and Ismail, B.S. (2003) Gene polymorphisms in glyphosate-resistant and -susceptible biotypes of *Eleusine indica* from Malaysia. *Weed Research* 43, 108–115.

Ng, C.H., Ratam, W., Surif, S. and Ismail, B.S. (2004a) Inheritance of glyphosate resistance in goosegrass (*Eleusine indica*). *Weed Science* 52, 564–570.

Ng, C.H., Wickneswari, R., Salmijah, S., Teng, Y.T. and Ismail, B.S. (2004b) Glyphosate resistance in *Eleusine indica* (L) Gaertn from different origins and polymerase chain reaction amplication of specific alleles. *Australian Journal of Agricultural Research* 55, 407–414.

Norris, R.F. (2005) Ecological bases of interactions between weeds and organisms in other pest categories. *Weed Science* 53, 909–913.

Olofsdotter, M., Valverde, B.E. and Madsen, K.H. (2000) Herbicide resistant rice (*Oryza sativa* L.): global implications for weedy rice and weed management. *Annals of Applied Biology* 137, 279–295.

Orrock, J.L., Levey, D.J., Danielson, B.J. and Damschen, E.I. (2006) Seed predation, not seed dispersal, explains the landscape-level abundance of an early-successional plant. *Journal of Ecology* 94, 838–845.

Ortiz-Garcia, S., Ezcurra, E., Schoel, B., Acevedo, F., Soberon, J. and Snow, A.A. (2005) Absence of detectable transgenes in local landraces of maize in Oaxaca, Mexico (2003–2004). *Proceedings of the National Academy of Science* 102, 12338–12343.

Owen, M. and Boerboom, C. (2004) *National Glyphosate Stewardship Forum*. St Louis, Missouri, p. 80.

Owen, M.D.K. (2000) Current use of transgenic herbicide-resistant soybean and corn in the USA. *Crop Protection* 19, 765–771.

Owen, M.D.K. (2005) Maize and soybeans – controllable volunteerism without ferality. In: Gressel, J. (ed.) *Crop Ferality and Volunteerism*. CRC Press, Boca Raton, Florida, pp. 149–165.

Owen, M.D.K. (2006a) *Herbicide Resistance, Weed Population Shifts, and Weed Management Stewardship: Is Anything new?* Integrated Crop Management Conference, Ames, Iowa, pp. 143–148.

Owen, M.D.K. (2006b) *Weed Management Update – Who Cares?* Integrated Crop

Management Conference, Ames, Iowa, pp. 149–153.

Owen, M.D.K (2008) Weed species shifts in glyphosate-resistant crops. *Pest Management Science* 64, 377–387.

Owen, M.D.K. and Zelaya, I.A. (2005) Herbicide-resistant crops and weed resistance to herbicides. *Pest Management Science* 61, 301–311.

Ozinga, W.A., Bekker, R.M., Schaminee, J.H.J. and Van Groenendael, J.M. (2004) Disperal potential in plant communities depends on environmental conditions. *Journal of Ecology* 92, 767–777.

Padgette, S.R., Delannay, X., Bradshaw, L., Wells, B. and Kishore, G. (1995) Development of glyphosate-tolerant crops and perspectives on the potential for weed resistance to glyphosate. In: De Prado, R., Jorrin, J. and Garcia-Torres, L. (eds) *International Symposium on Weed and Crop Resistance to Herbicides.* Cordoba, Spain, p. 92.

Palmer, R.G., Gai, J., Sun, H. and Burton, J.W. (2001) Production and evaluation of hybrid soybean. In: Janick, J. (ed.) *Plant Breeding Reviews.* Wiley, New York, pp. 263–308.

Penna, J.A. and Lema, D. (2003) Adoption of herbicide tolerant soybeans in Argentina: an economic analysis. In: Kalaitzandonakes, N. (ed.) *Economic and Environmental Impacts of Agrotechnology.* Kluwer–Plenum Publishers, New York, pp. 203–220.

Person, J. (2008) Glyphosate resistant johnsongrass confirmed in two locations, Monsanto Press release, March 12, 2008. Available at: http://www.monsanto.com (accessed 15/10/08)

Powles, S.B. and Preston, C. (2006) Evolved glyphosate resistance in plants: biochemical and genetic basis for resistance. *Weed Technology* 20, 282–289.

Powles, S.B., Lorraine-Colwill, D.F., Dellow, J.J. and Preston, C. (1998) Evolved resistance to glyphosate in rigid ryegrass (*Lolium rigidum*) in Australia. *Weed Science* 46, 604–607.

Preston, C. and Wakelin, A.M. (2008) Resistance to glyphosate from altered herbicide translocation patterns. *Pest Management Science* 64, 372–376.

Qaim, M. and Traxler, G. (2005) Roundup Ready soybean in Argentina: farm level and aggregate welfare effects. *Agricultural Economics* 32, 73–86.

Quist, D. and Chapella I.H. (2001) Transgenic DNA introgressed into traditional maize landraces in Oaxaca, Mexico. *Nature* 414, 541–543.

Rakow, G. and Woods, D.L. (1987) Outcrossing in rape and mustard under Saskatchewan prairie conditions. *Canadian Journal of Plant Science* 67, 147–151.

Raven, P.H. (2005) Transgenes in Mexican maize: desirability or inevitability? *Science* 102, 13003–13004.

Raymer, P.L. and Grey, T.L. (2003) Challenges in comparing transgenic and nontransgenic soybean cultivars. *Crop Science* 43, 1584–1589.

Rehman, M., Hansen, J.L., Brown, J., Price, W., Zemetra, R.S. and Mallory-Smith, C.A. (2006) Effect of wheat genotype on the phenotype of wheat × jointed goatgrass (*Aegilops cylindrica*) hybrids. *Weed Science* 54, 690–694.

Reichman, J.R., Watrud, L.S., Lee, E.H., Burdick, C.A., Bollman, M.A., Storm, M.J., King, G.A. and Mallory-Smith, C. (2006) Establishment of transgenic herbicide-resistant creeping bentgrass (*Agrostis stolonifera* L.) in nonagronomic habitats. *Molecular Ecology* 15, 4243–4255.

Rieger, M.A., Potter, T.D., Preston, C. and Powles, S.B. (2001) Hybridisation between *Brassica napus* L. and *Raphanus raphanistrum* L. under agronomic field conditions. *Theoretical Applied Genetics* 103, 555–560.

Rieger, M.A., Potter, T.D., Preston, C. and Powles, S.B. (2002) Gene movement between *Raphanus raphanistrum* L. and *Brassica napus* L. under Australian farming conditions. In: Wilcut, J. (ed.) *Weed Science Society of America.* Weed Science Society of America, Reno, Nevada, p. 39.

Rissler, J. and Mellon, M. (1996) *The Ecological Risks of Engineered Organisms.* MIT Press, Cambridge, Massachusetts.

Rong, J., Song, Z., Su, J., Xia, H., Lu, B.-R. and Wang, F. (2005) Low frequency of transgene flow from *Bt/CpTI* rice to its nontransgenic counterparts planted at close spacing. *New Phytologist* 168, 559–566.

Rong, J., Lu, B.-R. Song, Z., Su, J., Snow, A.A., Zhang, X., Sun, S., Chen, R. and Wang, F. (2007) Dramatic reduction of crop-to-crop gene flow within a short distance from transgenic rice fields. *New Phytologist* 173, 346–353.

Roy, D.B., Bohan, D.A., Haughton, A.J., Hill, M.O., Osborne, J.L., Clark, S.J., Perry, J.N., Rothery, P., Scott, R.J., Brooks, D.R., Champion, G.T., Hawes, C., Heard, M.S. and Firbank, L.G. (2003) Invertebrates and vegetation of field margins adjacent to crops subject to contrasting herbicide regimes in the Farm Scale Evaluations of genetically modified herbicide-tolerant crops. *Philosophical Transactions of the Royal Society of London B* 358, 1879–1898.

Sammons, R.D., Heering, D.C., Dinicola, N., Glick, H. and Elmore, G.A. (2007) Sustainability and stewardship of glyphosate and glyphosate-resistant crops. *Weed Technology* 21, 347–354.

Sankula, S. (2006) Quantification of the impacts on US agriculture of biotechnology-derived crops planted in 2005 Executive Summary, National Center for Food and Agricultural Policy. Available at: http://www. ncfap.org as a pdf file (accessed 15/10/08)

Sanyal, D., Bhowmik, P.C., Anderson, R.L. and Shrestha, A. (2008) Revisiting the perspective and progress of integrated weed management. *Weed Science* 56, 161–167.

Schutte, G. (2003) Herbicide resistance: Promises and prospects of biodiversity for European agriculture. *Agriculture and Human Values* 20, 217–230.

Scursoni, J., Peterson, D., Forcella, F., Gunsolus, J., Arnold, R.B., Owen, M., Smeda, R., Oliver, R. and Vidrine, R. (2001) Weed diversity in glyphosate-tolerant soybean from Minnesota to Lousiana. CD-ROM Computer File. In: Hartzler, R.G. (ed.) *North Central Weed Science Society*. North Central Weed Science Society, Champaign, Illinois, Milwaukee, Wisconsin.

Scursoni, J.A., Forcella, F. and Gunsolus, J. (2007) Weed escapes and delayed emergence in glyphosate-resistant soybean. *Crop Protection* 26, 212–218.

Searchinger, T., Heimlich, R., Houghton, R.A., Dong, F., Elobeid, A., Fabiosa, J., Tokgoz, S., Hayes, D. and Yu, T.-H. (2008) Use of US Croplands for biofuels increases greenhouse gases through emissions from land use change, *Sciencexpress*, p. 6.

Service, R.F. (2007a) Glyphosate – the conservationist's friend? *Science* 316, 1116–1117.

Service, R.F. (2007b) A growing threat down on the farm. *Science* 316, 1114–1117.

Shaner, D.L. (2000) The impact of glyphosate-tolerant crops on the use of other herbicides and on resistance management. *Pest Management Science* 56, 320–326.

Shipitalo, M.J., Malone, R.W. and Owens, L.B. (2008) Impact of glyphosate-tolerant sobyean and glufosinate-tolerant corn production on herbicide losses in surface runoff. *Journal of Environmental Quality* 37, 401–408.

Shivrain, V.K., Burgos, N.R., Anders, M.M., Rajguru, S.N., Moore, J. and Sales, M.A. (2007) Gene flow between Clearfield™ rice and red rice. *Crop Protection* 26, 349–356.

Siehl, D.L., Castle, L.A., Gorton, R., Chen, Y.H., Bertain, S., Cho, H., *et al.* (2005) Evolution of microbial acetyltransferase for modification of glyphosate, a novel tolerance strategy. *Pest Management Science* 61, 235–240.

Siehl, D.L., Castle, L.A., Gorton, R. and Keenan, R.J. (2007) The molecular basis of glyphosate resistance by an optimized microbial acetyltransferase. *Journal of Biological Chemistry* 282, 1146–1155.

Simard, M.-J., Legere, A., Pageau, D., Lajeunesse, J. and Warwick, S. (2002) The frequency and persistence of volunteer canola (*Brassica napus*) in Quebec cropping systems. *Weed Technology* 16, 433–439.

Snow, A.A. (2002) Transgenic crops – why gene flow matters. *Nature Biotechnology* 20, 542.

Snow, A.A. and Pedro Moran Palma (1997) Commercialization of transgenic plants: potential ecological risks. *Bioscience* 47, 86–96.

Snow, A.A., Andersen, B. and Jorgensen, R.B. (1999) Costs of transgenic herbicide resistance introgressed from *Brassica napus* into weedy *B. rapa. Molecular Ecology* 8, 605–615.

Snow, A.A., Rieseberg, L.H., Alexander, H.M., Cummings, C.H., Pilson, D. (2000) Assessment of gene flow and potential effects of genetically engineered sunflowers on wild relatives: The biosafety results of field tests of genetically modified plants and microorganisms. 5th International Symposium. Mitt. Biol. Bundesanst. Land-Forstwirtsch, Braunschweig Germany, pp. 19–25.

Sosnoskie, L.M., Luschei, E.C. and Fanning, M.A. (2007) Field margin weed-species diversity in relation to landscape attributes and adjacent land use. *Weed Science* 55, 129–136.

Steckel, L.E., Sprague, C.L., Stoller, E.W., Wax, L.M. and Simmons, F.W. (2007) Tillage, cropping system, and soil depth effects on common waterhemp (*Amaranthus rudis*) seed-bank persistence. *Weed Science* 55, 235–239.

Steinau, A.N., Skinner, D.Z. and Steinau, M. (2003) Mechanisms of extreme genetic recombination in weedy *Amaranthus* hybrids. *Weed Science* 51, 696–701.

Stewart Jr C.N., All, J.N., Raymer, P.L. and Ramachandran, S. (1997) Increased fitness of transgenic insecticidal rapeseed under insect selection pressure. *Molecular Ecology* 6, 773–779.

Stewart Jr C.N., Halfhill, M.D. and Warwick, S.I. (2003) Transgene introgression from genetically modified crops to their wild relatives. *Nature* 4, 806–817.

Stokstad, E. (2004) Monsanto pulls the plug on genetically modified wheat. *Science*, 1088–1089.

Swoboda, R. (2002) Bean quality bonus. *Wallaces Farmer* Vol 2, pp. 16–17.

Thebaud, C. and Abbot R.J. (1995) Characterization of invasive *Conyza* species (Asteraceae) in Europe: quantitative trait and isozyme analysis. *American Journal of Botany* 82, 360–368.

Tranel, P.J. (2007) *Predicting the Evolution and Spread of Glyphosate-Resistant Waterhemp.* Illinois Crop Protection Technology Conference, Champaign, Illinois, pp. 68–69.

Trucco, F., Jeschke, M.R., Rayburn, A.L. and Tranel, P.J. (2005) Promiscuity in weedy amaranths: high frequency of female tall waterhemp (*Amaranthus tuberculatus*) × smooth pigweed (*A. hybridus*) hybridization under field conditions. *Weed Science* 53, 46–54.

Trucco, F., Tatum, T., Robertson, K.R., Rayburn, A.L. and Tranel, P.J. (2006) Characterization of waterhemp (*Amaranthus tuberculatus*) × smooth pigweed (*A. hybridus*) F_1 hybrids. *Weed Technology* 20, 14–22.

Urbano, J.M., Borrego, A., Torres, V., Leon, J.M., Jimenez, C., Dinelli, G. and Barnes, J. (2007) Glyphosate-resistant hairy fleabane (*Conyza bonariensis*) in Spain. *Weed Technology* 21, 396–401.

Valverde, B.E. and Gressel, J. (2006) Dealing with the evolution and spread of *Sorghum halepense* glyphosate resistance in Argentina. A consultancy report to SENASA, p. 5.

Van Raamsdonk, L.W.D. and Van der Maesen, L.J.G. (1996) Crop-weed complexes: the complex relationship between crop plants and their wild relatives. *Acta Bot. Neerl* 45, 135–155.

VanGessel, M.J. (2001) Glyphosate-resistant horseweed from Delaware. *Weed Science* 49, 703–705.

Vidal, R.A., Trezzi, M.M., De Prado, R., Ruiz-Santaella, J.P. and Vial-Aiub, M. (2007) Glyphosate resistant biotypes of wild poinsettia (*Euphorbia heterophylla* L.) and its risk analysis on glyphosate-resistant soybeans. *Journal of Food, Agriculture & Environment* 5, 265–269.

Vila-Aiub, M.M., Balbi, M.C., Gundel, P.E., Ghersa, C.M. and Powles, S.B. (2007) Evolution of glyphosate-resistant johnsongrass (*Sorghum halepense*) in glyphosate-resistant soybean. *Weed Science* 55, 566–571.

Vila-Aiub, M.M., Vidal, R.A., Balbi, M.C., Gundel, P.E., Trucco, F. and Ghersa, C.M. (2008) Glyphosate-resistant weeds of South American cropping systems: an overview. *Pest Management Science* 64, 366–371.

Volenberg, D.S., Patzoldt, W.L., Hagar, A.G. and Tranel, P.J. (2007) Responses of contemporary and historical waterhemp (*Amaranthus tuberculatus*) accessions to glyphosate. *Weed Science* 55, 327–333.

Warwick, S.I. and Stewart Jr C.N. (2005) Crops come from wild plants – how domestification, transgenes, and linkage together shape ferality. In: Gressel, J. (ed.) *Crop Ferality and Volunteerism*. CRC Press, Boca Raton, Florida, pp. 9–30.

Warwick, S.I., Francis, A. and La Fleche, J. (2000) *Guide to Wild Germplasm of Brassica and Allied Crops (tribe Brassiceae, Brassicacea)*, 2nd Edition. Agriculture and Agri-Food Canada Research Branch Publication. ECORC, Ottawa, Canada.

Watrud, L.S., Lee, E.H., Fairbrother, A., Burdick, C., Reichman, J.R., Bollman, M., Storm, M., King, G. and Van de Water, P.K. (2004) Evidence for landscape-level, pollen-mediated gene flow from genetically modified creeping bentgrass with *CP4 EPSPS* as a marker. *Proceedings of the National Academy of Science* 101, 14533–14538.

Webster, T.M., Burton, M.G., Culpepper, A.S., Flanders, J.T., Grey, T.L. and York, A.C. (2006) Tropical spiderwort (*Commelina benghalensis* L.) control and emergence patterns in preemergence herbicide systems. *Journal of Cotton Science* 10, 68–75.

Weise, E. (2007) Effect of genetically engineered alfalfa cultivate a debate. *USA Today*. Available at: http://www.agbios.com/static/news/NEWSID8221.php

Werth, J.A., Preston, C., Taylor, I.N., Charles, G.W., Roberts, G.N. and Baker, J. (2008) Managing the risk of glyphosate resistance in Australian glyphosate-resistant cotton production systems. *Pest Management Science* 64, 417–421.

Westgate, M.E., Lizaso, J. and Batchelor, W. (2003) Quantitative relationship between pollen shed density and grain yield in maize. *Crop Science* 43, 934–942.

Whiton, J., Wolf, D.E., Arias, D.M., Snow, A.A. and Rieseberg, L.H. (1997) The persistence of cultivar alleles in wild populations of sunflowers five generations after hybridization. *Theoretical Applied Genetics* 95, 33–40.

Wilkes, H.G. (1977) Hybridization of maize and teosinte, in Mexico and Guatemala and the improvement of maize. *Economic Botany* 31, 33–40.

Woodburn, A.T. (2000) Glyphosate: production, pricing and use worldwide. *Pest Management Science* 56, 309–312.

Xanthopoulos, F.P. and Kechagia, U.E. (2000) Natural crossing in cotton (*Gossypium hirsutum* L.). *Australian Journal of Agricultural Research* 51, 979–983.

Young, B.G. (2006) Changes in herbicide use patterns and production practices resulting from glyphosate-resistant crops. *Weed Technology* 20, 301–307.

Yuan, C.I., Mou-Yen, C. and Yih-Ming, C. (2002) Triple mechanisms of glyphosate-resistance in a naturally occurring glyphosate-resistant plant *Dicliptera chinesis*. *Plant Science* 163, 543–554.

Zelaya, I. and Owen, M.D.K. (2000) *Differential Response of Common Waterhemp (Amaranthus rudis) to Glyphosate in Iowa*. Weed Science Society of America. Weed Science Society of America, Toronto, Canada, pp. 62–63.

Zelaya, I.A. and Owen, M.D.K. (2002) *Amaranthus tuberculatus* (Mq. ex DC) J.D. Sauer: potential for selection of glyphosate resistance. In: Spafford Jacob, H., Dodd, J. and Moore, J.H (eds) *13th Australian Weeds Conference*. Council of Australian Weed Science Societies, Perth, Australia, pp. 630–633.

Zelaya, I.A., Owen, M.D.K. and VanGessel, M.J. (2004) Inheritance of evolved glyphosate resistance in horseweed (*Conyza canadensis* (I.) Cronq.). *Theoretical Applied Genetics* 110, 58–70.

Zelaya, I.A., Owen, M.D.K. and VanGessel, M.J. (2007) Transfer of glyphosate resistance: Evidence of hybridization in *Conyza* (Asteraceae). *American Journal of Botany* 94, 660–673.

Zhang, N., Linscombe, S. and Oard, J. (2003) Out-crossing frequency and genetic analysis of hybrids between transgenic glufosinate herbicide-resistant rice adn the weed, red rice. *Euphytica* 130, 35–45.

8 Impact of Insect-resistant Transgenic Crops on Above-ground Non-target Arthropods

J. Romeis,[1] M. Meissle,[1] A. Raybould[2] and R.L. Hellmich[3]

[1]*Agroscope Reckenholz-Tänikon Research Station ART, Zurich, Switzerland;* [2]*Syngenta, Jealott's Hill International Research Centre, Bracknell, UK;* [3]*USDA–ARS, Corn Insects and Crop Genetics Research Unit and Department of Entomology, Iowa State University, Ames, Iowa, USA*

Keywords: Biological control, *Bt* crops, exposure assessment, Lepidoptera, predators, parasitoids, risk assessment

Summary

Genetically modified (GM) maize and cotton varieties that express insecticidal proteins derived from *Bacillus thuringiensis* (*Bt*) have become an important component in integrated pest management programmes worldwide. A number of other crops producing *Bt* toxins, or more broad-spectrum insecticidal proteins, are likely to enter commercial production in the near future. Because insecticidal GM crops target insect pests, an important part of the environmental risk assessment is their potential impact on non-target arthropods. Those include protected species and organisms providing important ecological services such as biological control of herbivores. Non-target arthropods can be exposed to the plant-produced insecticidal proteins through various routes, but mainly by feeding on GM plant material or herbivores that have consumed GM plant material. The *Bt* proteins produced in today's GM plants appear to have no direct effects on natural enemies due to their narrow spectrum of activity. Furthermore, it has become clear that in crop systems where the deployment of *Bt* varieties has led to a decline in insecticide use, biological control organisms have benefited significantly. Future GM plants that produce broader-spectrum insecticidal proteins will need to be assessed for their potential non-target effects case by case and compared to the impact of the conventional pest control methods that they replace.

Introduction

Growers use various methods to control insect pests, which generally include host-plant resistance, cultural control methods (e.g. crop rotation), biological control and chemical insecticides. Genetic modification, however, has produced a new type of control, which can be considered host-plant resistance or a type of chemical insecticide. Genetically modified (GM) plants are produced by transferring specific genes into the genome of a crop plant. For example, bacterial genes that encode insecticidal proteins have been introduced into cotton and maize. Such insect-resistant, genetically modified (IRGM) plants are commonly used today and have proven to be effective against devastating insect pests worldwide (Hellmich *et al.*, 2008; Naranjo *et al.*, 2008; Qaim *et al.*, 2008).

Crystal (Cry) proteins derived from the soil bacterium *Bacillus thuringiensis* (*Bt*) are known for their narrow spectrum of activity and long history of safe use as microbial *Bt* products (Glare and O'Callaghan, 2000). Use of IRGM maize and cotton that express *cry* genes has grown steadily worldwide since their introduction in 1996, reaching 42.1 million ha in 2007 (James, 2007). The highest commercial adoption rates of *Bt* maize varieties (percentage of maize acreage) are in Argentina (63%) and in the USA (50%; James, 2007). Commercial *Bt* maize varieties produce a Lepidoptera-specific toxin (Cry1 or Cry2) that targets stem borers such as the European corn borer (*Ostrinia nubilalis*; Lepidoptera: Crambidae), a Coleoptera-specific toxin (Cry3 or Cry34/35) for the control of corn rootworms (*Diabrotica* spp.; Coleoptera: Chrysomelidae) or both (Hellmich *et al.*, 2008). The latter are called stacked traits since the two Cry proteins target different insect pests. Potato plants that produce Cry3Aa to control the Colorado potato beetle, *Leptinotarsa decemlineata* (Coleoptera: Chrysomelidae), were introduced commercially in 1996 but withdrawn in 2001 due to marketing issues, consumer concerns and the introduction of a novel insecticide that controls beetles as well as aphids (Kaniewski and Thomas, 2004; Grafius and Douches, 2008).

Several countries have adopted *Bt* cotton varieties. The largest proportional adoption is in Australia, where in 2007 95% of the cotton acreage was planted to *Bt* varieties, followed by the USA (72%), China (69%) and India (66%; James, 2007). Current *Bt* cotton varieties produce Lepidoptera-specific Cry proteins targeting the budworm-bollworm complex, i.e. *Heliothis virescens*, *Helicoverpa* spp. (Lepidoptera: Noctuidae) and *Pectinophora gossypiella* (Lepidoptera: Gelechiidae; Naranjo *et al.*, 2008). Cotton plants that produce Cry1A and cowpea trypsin inhibitor (CpTI) are commercially grown in China (Wu and Guo, 2005; see also Chapter 16, this volume). There is published evidence that the insecticidal activity of *Bt* Cry proteins is enhanced when they are used in combination with protease inhibitors (PIs; MacIntosh *et al.*, 1990). This claim, however, has not been observed in cotton producing both CpTI and Cry1A infested by the main target pest, *Helicoverpa armigera* (Lepidoptera: Noctuidae; Wu and Guo, 2005). Besides single-gene cotton (expressing *cry1Ac*), plants that express two *Bt* genes (*cry1Ac* and *cry2Ab*, called pyramids because they target the same pest complex) have recently been released to provide an even more efficient and reliable control of the Lepidoptera pest complex and to delay the evolution of resistance in the target pest populations (Stewart *et al.*, 2001; Greenplate *et al.*, 2003; Naranjo *et al.*, 2008).

New GM plants producing other Cry proteins, vegetative insecticidal proteins (Vips) or combinations of Cry proteins with non-Cry toxins are likely to be released soon (Bates *et al.*, 2005; Malone *et al.*, 2008; see also Chapter 6, this volume). Genes expressing insecticidal proteins like PIs, α-amylase inhibitors, biotin-binding proteins or lectins have also been introduced into crop plants for controlling pests that are not susceptible to known Cry proteins (Jouanin *et al.*, 1998; Ferry *et al.*, 2004; Malone *et al.*, 2008). With the exception of the *Bt-CpTI* cotton plants grown in China, these new types of GM plants have not been commercialized. Some of these novel proteins have a broader spectrum of activity than the *Bt* Cry proteins and Vips, and consequently they have a higher potential to affect non-target organisms (O'Callaghan *et al.*, 2005; Romeis *et al.*, 2008a).

IRGM crops can be used in integrated pest management (IPM) programmes to control pest populations economically and in ways that promote sustainability (Romeis *et al.*, 2008b). Of particular interest in this respect is the impact of IRGM crops on non-target organisms. Agronomically, a non-target organism is any organism associated with the crop that does not cause economical damage. Because IRGM crops target insect pests, an important part of the environmental risk assessment is their potential impact on non-target arthropods. In particular, arthropods that provide ecosystem services like biological control, pollination or decomposition should not be harmed. Also non-target arthropods not directly associated with the IRGM crop, such as lepidopteran larvae that feed on non-crop host plants in or near the field, should be considered. Many Lepidoptera are Red List (i.e. threatened) species or species of aesthetic or cultural value, like the Monarch butterfly, *Danaus plexippus* (Lepidoptera: Danaidae), in the USA.

In this chapter, we describe how non-target organisms can be exposed to insecticidal proteins expressed by IRGM plants. We provide an overview of direct and indirect effects of IRGM plants on non-target arthropods including changes in agricultural practice. The data are discussed in the context of environmental risk assessment. We focus on above-ground arthropods, especially biological control organisms (predators and parasitoids) and non-target Lepidoptera. Pollinators and soil arthropods are covered elsewhere (see Chapters 9 and 10, this volume).

Non-target Effects Caused by the Insecticidal Proteins

Routes of exposure

Insecticidal proteins expressed by current IRGM plants target the insect midgut, and thus they need to be ingested to be effective. Non-target organisms are exposed to the insecticidal protein if they feed on the GM plant tissue or consume organisms that have eaten the toxin. Consequently, fewer non-target species are likely to be exposed to the active ingredient via GM plants than via spray insecticides. On the other hand, most IRGM crops express the active ingredient through most of the growing season, which may lengthen exposure compared with the applications of rapidly degraded chemical insecticides; potentially long exposures must be considered in environmental risk assessments for IRGM crops. The insecticidal proteins may be transferred from herbivores to predators and parasitoids; so this must be assessed when evaluating exposure of non-target arthropods. The principal routes of exposure for herbivores and natural enemies are outlined in Fig. 8.1.

Exposure through plant feeding

The most obvious way to ingest plant-expressed insecticidal protein is by feeding directly on the GM plant (Fig. 8.1, route 1). This route is particularly relevant for most non-target herbivores. However, exposure to the insecticidal protein depends to a great extent on both the herbivores' mode of feeding and on the site and time of protein expression in the plant (Dutton *et al.*, 2003). Chewing herbivores, such as caterpillars or beetles, generally ingest the insecticidal protein. Similarly, herbivores with piercing-sucking mouthparts that feed on epidermal or mesophyll cells, such as thrips, mirid bugs or spider mites, are exposed to the insecticidal protein (Dutton *et al.*, 2002, 2004; Obrist *et al.*, 2005, 2006a). This is in contrast to phloem-feeders such as aphids. Aphids have received much attention since they are common in most crop systems and provide food for many entomophagous arthropods. Different aphid species feeding on *Bt* maize, cotton and oilseed rape varieties were reported to ingest no, or at most, trace amounts of *Bt* protein. Thus, Cry proteins apparently are not transported in the plant's phloem sap (Raps *et al.*, 2001). Even though *Bt* protein has occasionally been detected in aphids (e.g. Zhang *et al.*, 2006; Burgio *et al.*, 2007), these incidences can be explained by *Bt* contamination, either with toxin-containing herbivores (e.g. spider mites), faeces of the same or tiny fractions of plant material. For example, the faeces of thrips feeding on *cry1Ab*-expressing maize plants were found to contain about tenfold higher *Bt* toxin concentrations than the fresh plant material (Obrist *et al.*, 2005). This

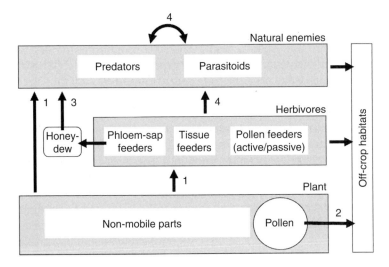

Fig. 8.1. Exposure pathways. Routes through which non-target arthropods can be exposed to insecticidal proteins expressed by IRGM plants.

demonstrates that very little contamination can easily produce false positives, especially for samples containing little or no *Bt* toxin, like aphids.

Many predators are facultative herbivores (mainly of pollen, nectar and plant juice), while parasitoids primarily feed from (extra-)floral nectaries (Dicke, 1999; Coll and Guershon, 2002; Wäckers, 2005). Insecticidal proteins have not been found in the (extra-)floral nectar of current IRGM cotton varieties. Cry protein production in pollen, however, varies with the promoter of the *cry* gene and the transformation event. Commonly grown Cry1Ab maize events *Bt*11 and MON810 produce very small amounts of toxin in the pollen (<1/100 that of leaves), which contrasts with the earlier IRGM maize variety Event *Bt*176 that was designed to produce high levels of Cry protein in pollen (similar to that of leaves; Dutton *et al.*, 2003). *Bt* protein concentrations in pollen of current corn rootworm (*Diabrotica* spp.)-resistant *Bt* maize events MON863 and MON88017 are also close to the concentration measured in leaves (Monsanto Company, 2003, 2004); on the other hand, in rootworm-resistant MIR604 maize, the concentration of *Bt* protein in pollen is at least 25 times lower than the concentration in leaves (Raybould *et al.*, 2007).

In the case of wind-pollinated plants such as maize, pollen can expose non-target organisms to the insecticidal protein both within and beyond the crop (Fig. 8.1, route 2). Organisms exposed within the crop include many biological control agents that actively seek pollen as a food source. An example is the ladybird beetle, *Coleomegilla maculata* (Coleoptera: Coccinellidae), which is abundant during maize anthesis and is known to consume maize pollen (Cottrell and Yeargan, 1998; Lundgren *et al.*, 2004, 2005). Other pollen feeders that were found to contain Cry1Ab when present in a *Bt* maize field during anthesis include *Orius* spp. (Hemiptera: Anthocoridae) and adult *Chrysoperla carnea* (Neuroptera: Chrysopidae; Obrist *et al.*, 2006b; Li *et al.*, 2008). Furthermore, organisms can passively ingest pollen deposited on their host plant. For example, larvae of butterfly species may be exposed to the *Bt* toxins both within the maize crop itself and in areas adjacent to the crop. The most prominent example is the Monarch butterfly where the larvae feed specifically on milkweed that commonly occurs in and near maize fields (Jesse and Obrycki, 2000; Oberhauser *et al.*, 2001). Passive pollen feeding has also been reported for web-building spiders that ingest pollen caught in their web when recycling the web (Ludy and Lang, 2006).

Exposure through honeydew

In contrast to current *Bt* crops, certain experimental plants expressing lectins or PIs are known to transport insecticidal proteins in the phloem. When sap-feeding Sternorrhyncha (Hemiptera), such as aphids, feed on such plants, the insecticidal proteins are likely to appear in their honeydew (Shi *et al.*, 1994; Kanrar *et al.*, 2002; Rahbé *et al.*, 2003a). This is because these insects typically possess low proteolytic activity in the gut and thus the foreign proteins are unlikely to be proteolytically degraded (Srivastava and Auclair, 1963; Rahbé *et al.*, 1995; Fig. 8.1, route 3). Similar observations have been reported for secondary plant compounds (Wink and Römer, 1986; Malcolm, 1990). Honeydew is an important food source for many arthropods including predators, parasitoids, pollinators and adult herbivores (Wäckers, 2005) and could potentially expose many non-target organisms to the insecticidal protein (Romeis *et al.*, 2003; Hogervorst *et al.*, 2009). Honeydew appears to be a minor route of exposure of non-target organisms to current *Bt* proteins, but it may become important for plants with enhanced resistance towards aphids, such as those expressing certain lectins.

Exposure through predation or parasitization

The major route of exposure to entomophagous arthropods is through their prey or hosts (Fig. 8.1, route 4). Usually the prey or host is an herbivore that feeds on the IRGM plant, but it can also be another entomophagous species. In either case, exposure through prey or host organisms is highly variable and difficult to predict for a number of reasons. As described above, the level at which different herbivores ingest the insecticidal protein depends on the site and time of toxin expression in the plant, the mode of feeding of the herbivore and the amount of plant material they ingest. Furthermore, the amount of toxin in different herbivores depends on the rate of proteolytic degradation and excretion; consequently, the amount varies considerably among species, even when feeding on the same plant. For example, the following arthropods all feed on mesophyll cells of *Bt* maize but contain different amounts of Cry protein: spider mites (toxin level similar to *Bt* maize leaf, 1X), thrips (1/6X) and leafhoppers (1/30X; Dutton *et al.*, 2004). For other examples on *Bt* maize, see Dutton *et al.* (2002) and Obrist *et al.* (2006a,b). Highly variable Cry protein concentrations among different herbivore groups also have been reported from *Bt* cotton (Head *et al.*, 2001; Torres *et al.*, 2006; Torres and Ruberson, 2008). Even within one species, toxin contents may vary considerably among life stages (Howald *et al.*, 2003; Obrist *et al.*, 2005). For example, while feeding, larval and adult stages of *Frankliniella tenuicornis* (Thysanoptera: Thripidae) contained relatively high *Bt* protein concentrations; non-feeding prepupae, pupae and newly emerged adults showed very low toxin levels (Obrist *et al.*, 2005). Immobile stages are easier to prey on by predators than moving larvae or adults, so exposure levels could be overestimated if only a mean toxin content of all stages is considered (Obrist *et al.*, 2005). The amount of insecticidal protein contained within an insect also depends on the nature of the protein. Studies on larvae of *C. carnea*, a species that is unable to excrete faeces during the larval stage, have revealed that Cry1Ab is degraded within a few days (Romeis *et al.*, 2004), while the lectin GNA remained undegraded (Hogervorst *et al.*, 2006). Another example is provided by Christeller *et al.* (2005) who studied the fate of aprotinin and avidin expressed by GM tobacco plants after ingestion by *Spodoptera litura* (Lepidoptera: Noctuidae). While both proteins were detected in the frass of *S. litura*, the biological activity of avidin was retained while around 90% of the trypsin-binding activity of aprotinin was lost. The apparent complexity in calculating the exact level at which non-target organisms are exposed to insecticidal protein produced by an IRGM plant has lead to simplifications for risk assessment by making conservative assumptions (see below).

How behaviour affects exposure

Other factors that can influence the exposure of non-target organisms to insecticidal proteins expressed by GM plants are behavioural responses to the plant itself or (in the case of natural enemies) to herbivores that have fed on the plant. There is evidence that *Bt* crops and non-*Bt* crops are similarly attractive to arthropods. For example, egg deposition of pest Lepidoptera on maize and cotton does not appear to differ between *Bt* and non-transformed varieties (Orr and Landis, 1997; Hellmich *et al.*, 1999; Pilcher and Rice, 2001; Liu *et al.*, 2002; Mellet *et al.*, 2004; Pilcher *et al.*, 2005; Torres and Ruberson, 2006; van den Berg and van Wyk, 2007). Likewise, *D. plexippus* oviposition on milkweed plants with surface-deposited pollen from *Bt* and non-*Bt* maize hybrids was similar (Tschenn *et al.*, 2001). However, insect behaviour can influence exposure in other ways. For example, there is evidence from flight chamber experiments that female *D. plexippus* oviposit less often on milkweed plants with maize pollen than those with no pollen (Tschenn *et al.*, 2001). This potentially restricts larval exposure to high (*Bt*) maize pollen densities.

It is well established that entomophagous arthropods, and parasitic wasps in particular, use volatiles that are emitted by herbivore-damaged plants for host or prey location (Vet and Dicke, 1992). Different hymenopteran parasitoids were found to respond similarly to the odours emitted by *Bt* and non-*Bt* plants that were damaged equally by mechanical wounding, or by *Bt*-resistant Lepidoptera larvae (Schuler *et al.*, 1999, 2003; Turlings *et al.*, 2005; Dean and De Moraes, 2006). A similar study with GNA revealed that the presence of the toxin in sugarcane plants does not affect the host location behaviour of the parasitoid *Cotesia flavipes* (Hymenoptera: Braconidae; Sétamou *et al.*, 2002). In contrast, sensitive Lepidoptera larvae caused less damage to *Bt* plants than to control plants, which decreased their attractiveness to parasitoids (Schuler *et al.*, 1999, 2003; Dean and De Moraes, 2006). This behavioural response has two important implications. First, parasitoids may avoid *Bt*-fed sensitive Lepidoptera larvae and thus hosts in which their offspring is unlikely to develop. Second, *Bt*-resistant larvae that cause severe feeding damage to the GM plants are highly attractive and more likely to be located and attacked with potential positive consequences for managing insect resistance. Recently, it has been established that *Bt* maize producing Cry3Bb1 for the control of *Diabrotica* spp. and non-transformed maize plants have the same ability to emit β-caryophyllene, a volatile shown to attract entomopathogenic nematodes, after chemical induction (Meissle *et al.*, 2008).

Herbivores feeding on *Bt* crops may behave, look or taste differently, which may influence predator behaviour and potentially affect prey consumption. Choice experiments with the parasitoid *Campoletis sonorensis* (Hymenoptera: Ichneumonidae) have shown no discrimination of *Bt* maize-fed *S. frugiperda* (Lepidoptera: Noctuidae) larvae despite larvae fed non-transformed maize being substantially larger (Sanders *et al.*, 2007). Choice experiments conducted with larvae of *C. carnea* and adult *Pterostichus madidus* (Coleoptera: Carabidae) revealed that the predators avoid sublethally affected *Bt*-fed Lepidoptera larvae as prey (Meier and Hilbeck, 2001; Ferry *et al.*, 2006). When *Bt*-resistant larvae were offered to *P. madidus*, however, the beetles did not differentiate between prey larvae fed *Bt* or non-*Bt* leaf tissue. However, avoidance of suboptimal *Bt*-induced prey by another carabid beetle (*Poecilus cupreus*) was not found by Meissle *et al.* (2005). Further evidence that *Bt* content in the prey does not affect predator behaviour is provided by Ferry *et al.* (2007): the predatory bee-tles *Harmonia axyridis* (Coleoptera: Coccinellidae) and *Nebria brevicollis* (Coleoptera: Carabidae) did not differentiate between *Lacanobia oleracea* (Lepidoptera: Noctuidae) larvae fed with Cry3A-expressing potato plants or non-transformed plants. These studies suggest that certain predators are able to avoid suboptimal prey when given a choice, which potentially mitigates negative indirect effects of reduced quality of prey caused by consumption of *Bt* plants.

Hazards to natural enemies

Bt proteins

There is no indication in the peer-reviewed literature or in the data submitted to regulatory agencies that the Cry proteins produced in today's *Bt* crops have a direct toxic effect on non-target organisms that are not closely related to the tar-get pests (US EPA, 2001; Romeis *et al.*, 2006; Wolfenbarger *et al.*, 2008). This high level of specificity is due to the mode of action of the Cry proteins, which need certain conditions in the insect gut to activate and bind to specific midgut receptors (Knowles, 1994; Schnepf *et al.*, 1998). Consequently, organisms that are unable to properly process the Cry proteins and do not possess the right receptors remain unaffected. Recently, Rosi-Marshall *et al.* (2007) suggested that *Bt* maize affects caddisflies (Trichoptera). Unfortunately, the authors do not state which *Bt* maize they were using for their studies. The discussion of the results, however, indicates that they have worked with a Cry1Ab-producing *Bt* maize. Phylogenetically, Trichoptera are closely related to Lepidoptera (Morse, 1997), so Cry1Ab activity would not be surprising. However, previous testing of *Bt* microbial formulations detected no adverse effects of Cry1Ab on trichopter-ans (Kreutzweiser *et al.*, 1992). There is more evidence arising that the effects reported by Rosi-Marshall *et al.* (2007) were most probably due to some other plant characteristics and not to the expression of the toxin. A new study using both Cry1Ab- and Cry1Ab-and-Cry3Bb1-stacked maize plants could not repeat the reported finding (G. Dively, San Diego, 2007, personal communication).

In the near future, it is expected that some IRGM crops will express novel *cry* genes from *B. thuringiensis* either individually, pyramided/stacked with other *cry* genes or stacked with other traits such as herbicide tolerance. In addition, IRGM plants that express Vips also derived from *B. thuringiensis* are close to commer-cialization. Currently, the US Environmental Protection Agency (EPA) is evaluat-ing maize and cotton plants expressing Vip3A, which has been reported to be specific to the order of Lepidoptera (Estruch *et al.*, 1996; Warren, 1997). In con-trast to Cry proteins, Vips do not need to be solubilized in the insect gut before they can act, and they bind to different receptors from those of Cry proteins (Lee *et al.*, 2003, 2006). Given their specificity, Vip-producing GM plants are likely to cause non-target effects similar to current Cry1-producing crops. This hypothesis has recently been corroborated in field studies with Vip3A maize (Dively, 2005; Fernandes *et al.*, 2007) and Vip3A cotton (Whitehouse *et al.*, 2007).

Experimental IRGM plants have been developed that produce broader-spectrum insecticidal proteins with a higher potential for direct toxicity to non-target organisms (Jouanin *et al.*, 1998; Carlini and Grossi-de-Sá, 2002; Ferry *et al.*, 2004; O'Callaghan *et al.*, 2005; Malone *et al.*, 2008). The best-studied examples include plants that express lectins or PIs.

Lectins

Plant lectins are carbohydrate-binding proteins known to be components of the plant's defence system (Peumans and Van Damme, 1995). Relatively little is known about the mode of action of plant lectins (Czapla, 1997). The proteins are reported to bind to the insect's midgut epithelial cells causing morphological changes that are thought to affect nutrient absorption (Powell *et al.*, 1998; Bandyopadhyay *et al.*, 2001). While lectin binding is considered to be a prerequisite for toxicity, binding does not necessarily result in toxicity (Harper *et al.*, 1995). *In vitro* and *in planta*, several lectins have been demonstrated to affect important life-table parameters in insect species in many different orders, including Lepidoptera, Coleoptera, Diptera and Hemiptera (Czapla, 1997; Jouanin *et al.*, 1998; Carlini and Grossi-de-Sá, 2002; Ferry *et al.*, 2004; Malone *et al.*, 2008). Activity against aphids (Aphididae), planthoppers (Delphacidae) and leafhoppers (Cicadellidae) makes lectins particularly interesting, since no *Bt* proteins are known to control these pests.

Snowdrop lectin (*Galanthus nivalis* agglutinin (GNA)) has received much attention because it was the first lectin found to deleteriously affect aphids (Down *et al.*, 1996), and planthoppers and leafhoppers (Rao *et al.*, 1998; Foissac *et al.*, 2000) when expressed in GM plants. Due to concerns regarding the food safety of this compound, albeit somewhat unfounded, research has partly moved to closely related lectins derived from edible plant species, such as garlic (*Allium sativum*). The garlic leaf lectin (ASAL) in particular was shown to affect lepidopteran pests (Sadeghi *et al.*, 2008), planthoppers, leafhoppers (Saha *et al.*, 2006) and aphids (Dutta *et al.*, 2005a,b; Sadeghi *et al.*, 2007) when expressed in different GM plants.

Given their mode of action, it is not surprising that lectins are not as specific as Cry or Vip proteins. A number of studies have shown that certain lectins have direct effects on parasitoids and predators when provided in artificial diets. Romeis *et al.* (2003), Bell *et al.* (2004) and Hogervorst *et al.* (2009) reported direct effects on the longevity and fecundity of four different species of parasitic wasp fed with sugar solution containing purified GNA. Similarly, growth of the predatory bug *Podisus maculiventris* (Hemiptera: Pteromalidae) was significantly reduced when fed with prey larvae injected with GNA (Bell *et al.*, 2003); and larvae of two ladybird species and the lacewing *C. carnea* had reduced longevity when exclusively fed with GNA dissolved in a sugar solution (Hogervorst *et al.*, 2006). When *C. carnea* larvae were fed alternately with sucrose solution containing GNA and insect eggs, their developmental time was significantly prolonged (Lawo and Romeis, 2008). Compared to these direct feeding studies, less pronounced effects have been reported from predators and parasitoids exposed to GNA via their prey or hosts (Malone *et al.*, 2008), which is probably due to reduced exposure levels.

Protease inhibitors

Plant PIs play a potent defensive role against insect herbivores and pathogens (Ryan, 1990). Serine PIs and cysteine PIs have received the most attention since they affect Lepidoptera and Coleoptera (Jouanin *et al.*, 1998; Carlini and Grossi-de-Sá, 2002; Malone *et al.*, 2008). More recently, aphids also have been reported to be affected by PIs of serine and cysteine proteases (Ceci *et al.*, 2003; Rahbé *et al.*, 2003a,b; Azzouz *et al.*, 2005a,b; Ribeiro *et al.*, 2006).

Potential direct effects of PIs deployed in IRGM plants on natural enemies have rarely been investigated. Studies with the larval and adult stages of the aphid parasitoids, *Aphidius ervi* (Hymenoptera: Braconidae) and *Aphelinus abdominalis* (Hymenoptera: Aphelinidae), revealed that the digestive proteolytic activity predominantly relies on serine proteases, especially those with chymotrypsin-like activities (Azzouz *et al.*, 2005a,b). This could partly explain the detrimental effects of the Soybean Bowman-Birk inhibitor (SBBI), a dual serine PI (inhibiting both chymotrypsin and trypsin), on these parasitoids (Azzouz *et al.*, 2005a,b). Also, the cysteine PI, oryzacystatin-1 (OC-1), was shown to be detrimental to these parasitoids when host aphids were dosed with the inhibitor via artificial diet. Another case is *Eulophus pennicornis* (Hymenoptera: Braconidae), an ectoparasitoid of *L. oleracea* larvae. *In vitro* studies with larvae of the parasitoid have revealed a strong activity of trypsin and chymotrypsin-like serine proteases (Down *et al.*, 1999). Nevertheless, the serine-type cowpea trypsin inhibitor (CpTI) did not markedly inhibit the larval proteolytic enzymes (Down *et al.*, 1999) and also did not cause an effect on adult longevity in direct feeding studies (Bell *et al.*, 2004).

Studies on adults and larvae of the two ladybird beetles *Adalia bipunctata* (Walker *et al.*, 1998) and *H. axyridis* (Ferry *et al.*, 2003; both Coleoptera: Coccinellidae) and the stinkbug *Perillus bioculatus* (Hemiptera: Pentatomidae) revealed that their major digestive proteolytic activity is cysteine-based (Ashouri *et al.*, 1998; Overney *et al.*, 1998). In contrast, carabid beetles have been found to rely upon serine proteases (both trypsin-like and chymotrypsin-like) for protein digestion (Gooding and Huang, 1969; Terra and Cristofoletti, 1996; Ferry *et al.*, 2005) and *N. brevicollis* adults exhibited both serine and cysteine digestive protease activity (Burgess *et al.*, 2002).

Similar to herbivorous insects (Jongsma and Bolter, 1997), several predators appear to adapt their proteolytic digestive metabolism to counteract the presence of PIs in their food. Examples include: the ladybird *H. axyridis* (Ferry *et al.*, 2003), different carabid beetles (Burgess *et al.*, 2002; Ferry *et al.*, 2005; Mulligan, 2006) and the predatory stinkbug *P. bioculatus* (Bouchard *et al.*, 2003a,b). Indirect evidence for such an adaptation is also reported for larvae of *C. carnea*. Despite serine proteases dominating the digestive tract (Mulligan, 2006), feeding a high dose of soybean trypsin inhibitor (SBTI) did not affect larval development and survival (Lawo and Romeis, 2008). This potential for adaptation should be taken into account when assessing the potential risks that PI-expressing GM plants pose to natural enemies.

Hazards to Lepidoptera larvae

Lepidoptera may passively ingest insecticidal protein from GM maize via pollen deposited on their larval host plants within or close to the crop (Fig. 8.1, exposure route 2). Consequently, several studies have focused on pollen from *Bt* maize varieties that produce Lepidoptera-active Cry proteins. Early instar Monarch (*D. plexippus*) butterflies, consuming diet with either Cry1Ab or Cry1Ac toxins showed delayed development and increased mortality; yet they were not affected by Cry1F toxins, even at high doses relative to worst-case exposure in the field (Hellmich *et al.*, 2001). Results from laboratory and semi-field studies varied with type of *Bt* maize when *D. plexippus* larvae were fed milkweed leaves with surface-deposited pollen. Short-term (4–5 days) studies conducted with Event *Bt*176 maize (high concentrations of Cry1Ab in pollen) showed negative effects (delayed development and increased mortality) with as few as ten pollen grains per square centimetre, but Event *Bt*11 and MON810 plants, which produce much smaller amounts of Cry1Ab toxin in pollen, showed little if any effects even with pollen densities of 1000 grains cm^{-2} (Wraight *et al.*, 2000; Hellmich *et al.*, 2001; Stanley-Horn *et al.*, 2001). Effects on *D. plexippus*, however, were found when they were continuously exposed throughout larval development to natural deposits of *Bt* pollen from both *Bt*11 and MON810 hybrids (Dively *et al.*, 2004).

Monarch larvae have also been tested with two other types of maize pollen: one from an experimental hybrid that produces pyramided proteins Cry1Ab and Cry2Ab2 and another from a hybrid that produces the Coleoptera-active Cry3Bb1 protein (Mattila *et al.*, 2005). Delayed development and increased mortality effects were found with the pyramided proteins but no effects were found with the Coleoptera-active protein. Also, Lee *et al.* (2003) detected no adverse effects of Vip3A on Monarch larvae. Studies that tested larvae from other Lepidoptera species that were fed host plants with surface-applied *Bt* pollen also found negative effects with pollen from Event *Bt*176 hybrids (Zangerl *et al.*, 2001; Felke *et al.*, 2002; Shirai and Takahashi, 2005; Lang and Vojtech, 2006) and, as with *D. plexippus*, little if any effects with pollen from MON810 hybrids (Wraight *et al.*, 2000; Li *et al.*, 2005). Interestingly, larvae of the milkweed tiger moth, *Euchaetes egle* (Lepidoptera: Arctiidae), were not affected when they consumed milkweed leaves with deposits of pollen from Event *Bt*176 hybrids (Jesse and Obrycki, 2002). One general conclusion that can be made from the data available is that susceptibility to Cry proteins differs among Lepidoptera.

Indirect effects

Entomophagous arthropods can suffer indirectly as a consequence of toxin effects on the target herbivores. This includes the absence of herbivores as prey or hosts and the presence of sublethally compromised (sick) prey or host herbivores. A significant reduction of the target pest population in the crop is the aim of all pest control programmes, including those that use IRGM plants. High adoption rates of *Bt* varieties can lead to region-wide pest suppression as has been reported for *P. gossypiella* and *O. nubilalis* (Carrière *et al.*, 2003; Chu *et al.*, 2006; Storer *et al.*, 2008). Similar area-wide suppression has been reached by conventional pest control methods including insecticides. A prime example is the boll weevil eradication programme in the USA (http://www.cotton.org/tech/pest/bollweevil/index.cfm; accessed 15 April 2008). This reduction in prey or host abundance will have a large impact on the population dynamics of natural enemies that attack these herbivores. The decline in abundance or activity of specialist natural enemies that depend on the target pests as prey or hosts has been observed in a number of field studies with *Bt* crops (Romeis *et al.*, 2006, 2008a; Wolfenbarger *et al.*, 2008). Most predators, however, are generalists consuming a broader range of prey, which allows them to shift between prey species (Symondson *et al.*, 2002). Consequently, generalist predators are less affected by suppression of a particular prey species. Only one study has reported a consistent and significant reduction of generalist predators when lepidopteran prey was reduced in *Bt* cotton (Naranjo, 2005a). Interestingly, this reduction in abundance by about 20% had no effect on the biological control function in the crop (Naranjo, 2005b).

Under laboratory or glasshouse conditions, prey- or host-quality-mediated effects are commonly observed. When herbivores are exposed to insecticidal proteins, mortality and sublethal effects (e.g. extended development or decreased body size) increase. Sublethal effects are likely to be accompanied by changes in herbivore physiology, which also can affect higher trophic levels; while in the field, predators may compensate for reduced prey quality by feeding on more prey or by shifting to alternative prey species. Parasitoids, however, usually complete their development in a single host and thus are very sensitive to changes in host quality. Such indirect effects on natural enemies are expected if susceptible herbivores ingest a toxin at sublethal levels (Romeis *et al.*, 2006). A prominent example is discussed in detail in Box 8.1 and Fig. 8.2. Many studies conducted under contained conditions (i.e. in the laboratory, climate chamber or glasshouse) that have investigated the impact of *Bt* plants on natural enemies through a herbivore have used susceptible species as prey or host. Negative effects on the natural enemies, as expected, became evident in some of the studies (Romeis *et al.*, 2006). Recently, Chen *et al.* (2008) were able to separate direct from indirect, host-quality effects to a parasitoid. Using populations of the diamondback moth, *Plutella xylostella* (Lepidoptera: Plutellidae), that were resistant to Cry1C, the authors showed that GM broccoli plants and purified Cry1C had no direct toxicity to the larval parasitoid *Diadegma insulare* (Hymenoptera: Ichneumonidae) when it fed inside its host after the host had consumed either *Bt* plants or leaves dipped in Cry1C solution.

Prey- or host-quality-mediated effects also have been reported for IRGM plants other than *Bt* plants. Studies on the interactions of GNA-fed aphids and aphid parasitoids suggested that indirect host-size-mediated effects caused subsequent deleterious effects on the parasitoids, rather than the GNA itself (Couty *et al.*, 2001a,b). Similarly, studies assessing the impact of GNA-fed aphids on the predator *A. bipunctata* suggested that negative effects on the predator were due to feeding on sublethally affected aphids or due to unintended effects of transformation in those particular experimental plants rather than the GNA itself (Birch *et al.*, 1999; Down *et al.*, 2000). Indirect effects have also been reported for honeydew-feeding arthropods, such as parasitic wasps (Hogervorst *et al.*, 2009). Aphids that were sublethally impaired by GNA ingestion produced inferior honeydew with consequences on honeydew-feeding aphid parasitoids.

Tritrophic studies including *L. oleracea* larvae fed with CpTI-expressing potatoes revealed a detrimental effect on different parasitoid life-table parameters (Bell *et al.*, 2001a). The study indicated that the observed effects on the

---

Box 8.1. The case of the green lacewing

Tritrophic effects of Cry1Ab-expressing *Bt* maize on larvae of the green lacewing, *Chrysoperla carnea* (Neuroptera: Chrysopidae), were assessed by Dutton *et al.* (2002). They reared different prey organisms on maize leaves and fed them to newly emerged lacewing larvae. Lacewings feeding on aphids (*Rhopalosiphum padi*; Hemiptera: Aphididae) and on spider mites (*Tetranychus urticae*; Acari: Tetranychidae) had a low mortality and did not show differences between prey reared on *Bt* maize and on control plants (near-isolines; Fig. 8.2). In contrast, when *Spodoptera littoralis* (Lepidoptera: Noctuidae) caterpillars were given as prey, the mortality in the control treatment was relatively high and significantly increased in the *Bt* treatment (Fig. 8.2). Also the development time for lacewings fed caterpillars from *Bt* maize was longer than in the other treatments. In order to explain the mechanisms behind those effects, a series of additional experiments was conducted.

First, Dutton *et al.* (2002) asked whether the different prey items ingested the Cry1Ab protein (Fig. 8.2). Measurements revealed that aphids contained little if any Cry1Ab, consequently no protein-related effects on aphids and lacewings were observed. In contrast, caterpillars contained 0.72 µg Cry1Ab g^{-1} (fresh weight) and spider mites an even higher concentration of 2.5 µg g^{-1} (fresh weight).

Next, the researchers asked whether those herbivores that ingested Cry protein when feeding on *Bt* maize were susceptible to the toxin (Fig. 8.2). Feeding experiments showed that the

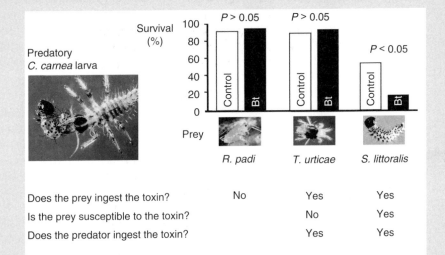

Does the prey ingest the toxin?	No	Yes	Yes
Is the prey susceptible to the toxin?		No	Yes
Does the predator ingest the toxin?		Yes	Yes

Conclusions:
The predator is not susceptible to the toxin, observed effects are prey-quality mediated

Fig. 8.2. Survival of *Chrysoperla carnea* larvae fed with different prey species (adapted from Dutton *et al.*, 2002). To interpret the results, a set of questions was answered for each of the prey species (according to Romeis *et al.*, 2006).

Continued

Box 8.1. Continued.

growth rate was similar for *T. urticae* that consumed *Bt* maize or control maize. In contrast, when *S. littoralis* larvae consumed *Bt* maize, mortality was significantly higher and development was delayed compared to those consuming control maize (Dutton *et al.*, 2002).

In a follow-up study, Obrist *et al.* (2006a) tested whether the protein is transferred from the prey to the predator (Fig. 8.2). Measurements showed that lacewing larvae feeding on caterpillars or spider mites contained about 1/2 to 1/3 of the prey Cry protein concentration, thus lacewings ingest Cry1Ab when feeding on those prey species. Furthermore, they confirmed biological activity of the protein contained in the herbivorous prey with a sensitive insect bioassay. Consequently, lacewing larvae are exposed to active Cry1Ab when feeding on *T. urticae* and when feeding on *S. littoralis* reared on *Bt* maize. However, detrimental effects on survival and development time were only seen in the caterpillar treatment, even though the concentration of Cry1Ab was much higher in the spider mite treatment.

These experiments led to the conclusion that lacewing larvae were not affected by the Cry1Ab protein itself and effects observed in the caterpillar treatment were mediated by prey quality. The high mortality of lacewing larvae, fed *S. littoralis* larvae from control maize, indicates these caterpillars were an inferior food source compared to aphids or spider mites. Thus, caterpillars compromised from ingesting *Bt* maize ('sick prey') were likely an even lower-quality food source for predators than healthy larvae from control maize. This example shows the importance of carefully formulating the study objectives and, accordingly, properly designing the study. Earlier work by Hilbeck *et al.* (1998) assessed the impact of *Bt* (Cry1Ab) maize on *C. carnea* larvae by only using caterpillars as the prey. Consequently, this study had caused some confusion since it appeared to prove that *Bt* maize causes a risk to this non-target organism.

Testing the hazard potential of a transgenic protein on a non-target species can be done with high-dose toxicity tests (e.g. 10× the dose expressed in a transgenic plant), by mixing purified protein into an artificial diet. Such tests have been performed with *C. carnea* by Romeis *et al.* (2004), Rodrigo-Simón *et al.* (2006) and Lawo and Romeis (2008), showing no indication of negative Cry1A impact on *C. carnea* larvae. Furthermore, Rodrigo-Simón *et al.* (2006) did not find binding of the Cry protein to lacewing gut membranes, a prerequisite of toxicity, in both histopathological and *in vitro* binding studies. A lack of effects has recently been confirmed also for the pollen-feeding adult stage of *C. carnea* (Li *et al.*, 2008).

parasitoid were likely to be indirect since *L. oleracea* larvae are susceptible to CpTI (Bell *et al.*, 2001a), while the parasitoid was found not to be sensitive to this particular PI (Down *et al.*, 1999; Bell *et al.*, 2004). In contrast, studies that have investigated the performance of aphid parasitoids on aphids that were sublethally affected when feeding on artificial diet containing OC-1 or Soybean Bowman-Birk inhibitor (SBBI) are difficult to interpret, since the parasitoids also were sensitive to the PIs used (Azzouz *et al.*, 2005a,b).

Sublethally affected herbivores may also lead to positive effects for natural enemies: host defence behaviour against attacking natural enemies may be altered, leading to a higher parasitization or predation frequency; longer development may result in longer availability of hosts or prey (i.e. a larger 'window of opportunity' for parasitoids and predators to attack); and a weaker immune system of a host may result in a lower encapsulation rate of endoparasitoid eggs and consequently a higher parasitization rate. Such positive indirect effects were reported by Johnson and Gould (1992) and Johnson (1997) who observed a significantly higher parasitization rate of *H. virescens* larvae by *C. sonorensis* on *Bt* tobacco plants compared with non-*Bt* controls. A similar effect was shown by Bell *et al.*

(1999, 2001b), who reported that the parasitoid *E. pennicornis* benefited when its host *L. oleracea* was sublethally affected by a GNA-containing diet.

Non-target Effects Caused by Changes in Agricultural Practice

The deployment of IRGM varieties often has an impact on the overall pest control strategy with consequences for non-target organisms. A replacement of chemical spray insecticides in particular has a strong effect on the arthropod community (Romeis *et al.*, 2006; Marvier *et al.*, 2007; Wolfenbarger *et al.*, 2008). This is an important consideration when the agricultural impacts of an IRGM variety are assessed. Whether or not the introduction of an IRGM variety influences the use of chemical insecticides varies with crop, region, target pest complex and commonly used conventional pest control (Head *et al.*, 2005; Fitt, 2008).

The introduction of maize varieties that produce Cry proteins (e.g. Cry1Ab, Cry1F) for stem borer control has resulted in lower use of insecticides on many farms in the US Western Corn Belt (Nebraska, Kansas, Colorado; Hunt *et al.*, 2007), but only modest reductions on farms in the rest of the Corn Belt (Carpenter and Gianessi, 2001; Phipps and Park, 2002) as many farmers did not use insecticides to control stem borers. In some areas of Spain, significant reductions in insecticide use have been reported. On average, conventional-maize farmers applied 0.86 sprays per year compared with 0.32 for *Bt* maize farmers; the overall percentage of farmers who applied no insecticides was 70% for *Bt* maize farmers and 42% for conventional farmers (Gómez-Babero *et al.*, 2008). In contrast to field maize, insecticide applications for Lepidoptera control are common in sweetcorn and the use of *Bt* varieties can potentially reduce the number of insecticide sprays by 70–90% (Musser and Shelton, 2003; Rose and Dively, 2007). *Bt* varieties have been found to be a very effective control option with fewer side effects on natural enemies compared with commonly used broad-spectrum spray insecticides (Musser and Shelton, 2003; Hoheisel and Fleischer, 2007; Leslie *et al.*, 2007; Rose and Dively, 2007). However, the advantages were less obvious when compared to insecticides with a more specific mode of action (e.g. indoxacarb, spinosad) or application (e.g. imidacloprid, a systemic insecticide; Musser and Shelton, 2003).

Maize varieties that produce Cry3 toxins to control larvae of *Diabrotica* spp. have the potential to drastically reduce the use of chemical insecticides (Rice, 2004; Ward *et al.*, 2005). In the USA, corn rootworms are controlled largely by application of broad-spectrum soil insecticides including organophosphates, carbamates, pyrethroids and phenyl pyrazoles. Field studies have confirmed that non-target arthropods are more abundant in fields or plots planted with *Bt* varieties when compared to those treated with insecticides for corn rootworm control (Bhatti *et al.*, 2005a,b).

Similarly, *Bt* potato cultivars supported greater and more diverse natural enemy communities (Hoy *et al.*, 1998; Reed *et al.*, 2001; Duan *et al.*, 2004; Kalushkov and Nedvěd, 2005). As with maize, the advantages of the IRGM potato varieties were most pronounced when compared with sprays of broad-spectrum insecticides than with soil applications or systemic insecticides (Reed et al., 2001; Duan et al., 2004).

The most prominent example of reduced environmental impacts due to IRGM crops is cotton, where chemical insecticides are heavily used mainly to control a complex of pest Lepidoptera. The introduction of *Bt* cotton varieties has resulted in substantial reductions in insecticide use in many countries, including the USA, China, Australia, India, Argentina and South Africa (Fitt, 2008; Naranjo et al., 2008). While insecticide reduction varied between 25% and 80% for single-gene (Cry1Ac) varieties, results from four cropping seasons in Australia indicated that due to increased efficacy, pyramid (Cry1Ac and Cry2Ab) cotton varieties can reduce the use of insecticidal active ingredients by 65–75% (with a 80–90% reduction in number of sprays; Fitt, 2008). A number of studies have revealed a significant increase in the abundance of non-target organisms in unsprayed *Bt* cotton fields when compared to insecticide-treated conventional cotton (e.g. Hagerty et al., 2005; Head et al., 2005; Naranjo, 2005a,b; Torres and Ruberson, 2005; Whitehouse et al., 2005; Cattaneo et al., 2006; Sisterson et al., 2007).

Recent studies from China indicate that significant reductions in insecticide use can also be expected when *Bt* rice for the control of stem borers is introduced (Huang et al., 2005, 2008).

A reduction in insecticide use in *Bt* crops has important consequences for the management of insect populations beyond the target pests. The data available indicate that this reduction can contribute to natural enemy conservation and increase the biological control function they provide (Romeis et al., 2006, 2008a). These natural enemies can help to suppress secondary pest populations, as has been suggested for aphids in *Bt* cotton, maize and potato fields (Hoy et al., 1998; Reed et al., 2001; Wu and Guo, 2003; Bhatti et al., 2005b) and for *Spodoptera* spp. in *Bt* cotton (Head et al., 2005). In addition, the deployment of IRGM plants can reduce the insecticide-induced resurgence of secondary pests, as reported from aphids in *Bt* cotton (Wu and Guo, 2003) and planthoppers in *Bt* rice (Chen et al., 2006).

However, secondary pest outbreaks also have been reported from *Bt* crops. This includes in particular regional outbreaks of plant and stinkbugs (Miridae and Pentatomidae) in *Bt* cotton fields. Those pests had earlier been controlled by insecticides used to control the Lepidoptera pest complex (Greene et al., 2001; Wu et al., 2002; Men et al., 2005). Thus, replacing a broad-spectrum control method (i.e. insecticides) by a more specific method (i.e. a *Bt* variety) may lead to secondary pest outbreaks, which again may require the application of control measures. While these outbreaks are not unexpected, there is little evidence that secondary pests have emerged as major problems requiring significant increases in insecticide to the extent that the reduction in insecticide requirement from the use of *Bt* cotton has been nullified (Fitt, 2008). The occurrence of secondary pests shows that IRGM plants should not be seen as a stand-alone control measure but as one component of an IPM programme for sustainable pest control (Kennedy, 2008).

While the reduction in broad-spectrum insecticide use is likely the most important factor explaining the increase in secondary pests, other factors also appear to play a role. Secondary pest species may benefit from reduced competition with the target pests. A possible example of this is the case of the increase in the frequency and severity of attacks by the western bean cutworm, *Striacosta albicosta* (Lepidoptera: Noctuidae), which is little affected by Cry1Ab and appears to benefit from the control of other Lepidoptera (Catangui and Berg, 2006; Storer *et al.*, 2008).

Another factor to consider is the overall improved health of the plant as a consequence of being protected from insect damage. An example has been reported from Vip cotton where a higher abundance of mirids could be explained by higher numbers of bolls and flowers when compared to the unprotected control (Whitehouse *et al.*, 2007). Another interesting observation is that some species benefit from herbivore (e.g. *O. nubilalis*) damage in non-*Bt* maize fields when compared to *Bt* maize. Examples are certain species of Nitidulidae (Coleoptera; Daly and Buntin, 2005; Bruck *et al.*, 2006) that are fungivores and frequently found in tunnels made by *O. nubilalis* and saprophagous beetles and flies (Candolfi *et al.*, 2004; Dively, 2005).

Broad-spectrum spray insecticides not only affect pests in the crop, but potentially also herbivores (e.g. Lepidoptera) on non-crop plants in and around the treated fields. For example, Stanley-Horn *et al.* (2001) found that all Monarch larvae on milkweed plants in a sweet-corn field sprayed with a pyrethroid insecticide were killed and larvae on plants 3 m from the edge of the field had low survival due to spray drift. Gathmann *et al.* (2006) found reduced abundances of lepidopteran larvae that occurred in strips near maize fields when the maize was treated with a pyrethroid insecticide. On the other hand, in the same study they found no negative impacts on lepidopteran larvae due to MON810 maize hybrids. Thus, reducing insecticide spraying may also benefit arthropods not directly related to the crop plant.

These examples show how important it is to select the appropriate end point and comparator when assessing the non-target effects of IRGM crops. Various risk assessment frameworks, including Annex III of the Cartagena Protocol on Biosafety (CBD Secretariat, 2000), refer to the importance of assessing risks of GM crops in the context of the risks posed by the conventional agricultural practice (Conner *et al.*, 2003; Nap *et al.*, 2003; Sanvido *et al.*, 2007). Insecticide treatments, the dominant current pest control strategy, should be considered as one baseline for risk assessment. Alternative control methods (e.g. biological control by released natural enemies) or no pest control should be included in comparison, but only if they are of practical relevance.

Implications for Regulatory Risk Assessment

The cultivation of GM crops is strictly regulated worldwide (see Chapter 4, this volume), and before their seeds can be sold and cultivated without restriction, a licence must be obtained from a regulatory authority. The decision to license a GM crop for commercial cultivation uses risk analysis, a general method for regulatory decision making. Risk analysis comprises two activities: risk assessment – the determination of the probability of specified harmful effects following a proposed action; and decision making – the evaluation of whether the risk, and the uncertainty associated with its estimation, is acceptable. Acceptability depends on the objectives of public policy, which can include the beneficial effects of the action, along with the ability to manage and communicate the risk (Wolt and Peterson, 2000; Johnson *et al.*, 2007; Raybould, 2007a).

Risk assessment should begin with the identification of a problem, not with the acquisition of data (Raybould, 2006). As with other regulated agricultural practices, for cultivation of IRGM crops the problem for environmental risk assessment is the protection of attributes of the environment that are valued and require protection. Once the valuable environmental attributes have been identified, the next step is the creation of risk hypotheses about how the IRGM crops may harm those valued attributes; and risk is characterized by testing these hypotheses. A single cycle of risk characterization may be sufficient for decision making, in which case testing can stop; or risk characterization may identify further problems and so be the beginning of another round of hypothesis formulation, testing and characterization (Romeis *et al.*, 2008c). Thus, an environmental risk assessment for proposed cultivation of an IRGM crop is simple in concept: decide what needs protection; assess how cultivation may cause harm to the entities requiring protection; and collect data to predict the probability and magnitude of harm following cultivation of the IRGM crop.

The protection objectives are not deducible by science alone; they are derived from the objectives of public policy, which will be based on political, economic, social and ethical, as well as scientific, criteria (Wolt and Peterson, 2000; Johnson *et al.*, 2007; Raybould, 2007a). Policy related to regulation of GM crops is usually enacted in laws that have the objective of protecting the environment. In the USA, pesticidal proteins produced in IRGM plants are regarded as pesticides and therefore are regulated under the Federal Insecticide, Fungicide and Rodenticide Act (FIFRA), which seeks to 'protect the public health and environment from the misuse of pesticides by regulating the labelling and registration of pesticides and by considering the costs and benefits of their use'. In the European Union, GM crops for commercial cultivation are regulated under Directive 2001/18/EC, which requires that risk assessments 'identify and evaluate potential adverse effects of the GMO, either direct [or] indirect, immediate or delayed, on human health and the environment which the deliberate release or placing on the market of GMOs may have'.

Both laws seek to protect the environment; however, 'environmental protection' is too vague a concept to be analysed scientifically. Specific targets for protection, called 'assessment end points', must be identified; for these targets to be analysable scientifically, they must comprise an entity and some property of that entity (Newman, 1998). For example in the UK, the objective of protecting biodiversity is represented by an assessment end point of an index of the population sizes of bird species common on farmland (Gregory *et al.*, 2004). In regulatory risk assessments of IRGM plants, the abundance and diversity of non-target arthropods that provide biological control is a common assessment end point (Raybould, 2007a; Romeis *et al.*, 2008c); however, this end point is often implicit rather than explicit.

A simple and effective conceptual model of how IRGM crops may cause harm is that the abundance of non-target organisms is reduced by exposure to toxic substances in the IRGM crop. The model makes two important assumptions. First, reductions in the abundance of predators and parasitoids that result from control of the target pest are not considered harmful. Such effects are common for all pest control methods, including insecticides, biological control and conventional host-plant resistance, and are generally not regarded as 'adverse' (Romeis *et al.*, 2006). As a consequence, the Organization for Economic Cooperation and Development (OECD) has stated that 'such secondary effects as a consequence of pest control achieved with a trait should be of minor concern where the trait serves primarily to bring the pest population down to an ecologically more natural level' (OECD, 1993). Similarly, the European Food Safety Authority (EFSA) has stated that 'it is important that food chain effects due to reductions in target prey species (e.g. declines in parasitoid populations) are differentiated from, for example, population declines due to the effects of GM toxin accumulation in food chains' (EFSA, 2006). We argue that the same applies for non-target effects on the natural enemy due to reduced nutritional quality of the susceptible hosts or prey. Romeis *et al.* (2006) have proposed that direct and indirect effects should be addressed separately. Probably the best-studied model species is the green lacewing, *C. carnea*, for which a series of experiments have allowed us to clearly distinguish between effects conferred by the *Bt* toxin itself (direct) and those caused as a result of reduced prey quality (indirect; Box 8.1). Similar to *Bt* crops, a range of studies have investigated the impact of PI- or lectin-expressing IRGM plants on natural enemies attacking sublethally affected herbivores. In most cases, effects have been detected (O'Callaghan *et al.*, 2005). However, a clear separation of direct and indirect (prey- or host-quality-mediated) effects is difficult, if not impossible, since both effects are likely to co-occur. It is thus important to investigate whether observed effects are more pronounced on the target pest species or on the associated natural enemies.

A second assumption of the conceptual model of how IRGM crops may cause harm to non-target organisms is the fact that any direct effects of non-GM counterparts of IRGM crops (near-iso lines) are acceptable. These assumptions greatly simplify the risk assessment as only differences in the composition of the GM and non-GM crop need to be assessed for their effects on non-target organisms.

The next stage of the risk assessment is to formulate risk hypotheses. The purpose of these hypotheses is to assist decision making, and therefore risk hypotheses should be formulated such that corroboration under rigorous testing indicates low risk with high confidence (Raybould, 2006). Tests of three risk hypotheses provide an effective method for evaluating risks to non-target organisms from IRGM crops (Raybould, 2007b):

1. There are no ecotoxicologically relevant differences between the IRGM crop and non-GM counterparts, apart from expression of the intended insecticidal protein.

2. There is no exposure to the insecticidal protein.

3. If there is exposure, the expected environmental concentration (EEC) of the insecticidal protein is below the no observable adverse effect concentration (NOAEC) for each exposed non-target organism.[1]

While the hypotheses are not stated explicitly, in effect, they form the basis for most regulatory risk assessments of IRGM plants.

Hypothesis 1 is tested in two main ways. First, a detailed molecular characterization of the inserted DNA and the genomic regions flanking that DNA is undertaken to ensure that no new open-reading frames have been created that could produce unintended new proteins. Second, the composition of the IRGM crop and an appropriate non-GM counterpart is compared. If there are no significant differences in the concentrations of certain nutrients, anti-nutrients and toxins (e.g. Kuiper *et al.*, 2002), the GM plant may be regarded as 'substantially equivalent' to the non-GM plant, and the risk assessment can consider the effect of the insecticidal protein only (Raybould, 2006; Romeis *et al.*, 2008c).

Hypothesis 2 is tested by measurements of the concentration of the insecticidal protein in various tissues (i.e. spatial expression) and considerations of exposure pathways given above. Expression of the insecticidal protein in leaves of the IRGM plant may mean that non-target arthropods are exposed to the protein through consumption of herbivores that eat leaves; on the other hand, if the protein is not detected in nectar or pollen, one may conclude that pollinators will not be exposed to the protein. For regulatory risk assessments, it is not necessary to have precise estimates of the concentrations of the insecticidal protein in the diet of the non-target organisms because conservative assumptions can be made. For this purpose, the highest mean protein expression level in any plant tissue is often taken as the worst-case EEC in regulatory risk assessments (Raybould *et al.*, 2007).

The highest mean concentration of the protein in the IRGM crop may be suitable for protecting individuals, such as those of endangered species, which could consume a diet comprising only plant material. In many cases, the objective of the risk assessment is protection of populations of non-target organisms that are omnivorous or predators; for these species, an EEC based on the highest expression in the IRGM plant may be adequately conservative. Many studies have revealed a dilution of Cry proteins along the food chain (Head *et al.*, 2001; Harwood *et al.*, 2005; Vojtech *et al.*, 2005; Obrist *et al.*, 2006a,b; Torres *et al.*, 2006; Torres and Ruberson, 2008). This dilution also seems to apply to other insecticidal proteins (e.g. Bell *et al.*, 2003; Christeller *et al.*, 2005). An exception to this finding is spider mites, which appear to contain Cry protein

---

[1] Hypothesis 2 is a special case of hypothesis 3. If the organism is not exposed, the EEC is not greater than zero and hence must be lower than the NOAEC.

levels comparable to, or greater than, those in green plant tissue (Dutton *et al.*, 2002; Obrist *et al.*, 2006b,c; Torres and Ruberson, 2008). Because of the dilution of insecticidal proteins and the mixed diet of many predators, Raybould *et al.* (2007) suggested a realistic EEC could be 0.2× the highest mean concentration in plant tissue for populations of non-target arthropods.

More precise refinements of exposure are needed only if a potential problem is indicated by conservative assumptions about exposure. Long-term exposure to *Bt* maize pollen, for example, has been shown to affect Lepidoptera larvae such as those of the Monarch butterfly (Dively *et al.*, 2004). Consequently, exposure assessment became critical to address a potential risk for Monarch populations. Considering the entire range of the US Corn Belt, the impact on Monarch populations was estimated as being negligible. This primarily is due to the limited overlap between maize anthesis and the presence of Monarch larvae (Dively *et al.*, 2004). Other studies that have focused on exposure (temporal and spatial overlap, pollen dispersal) in general have found that the potential impact of *Bt* maize pollen on lepidopteran larvae is small to negligible; and where impacts are possible, they are limited to specific geographies (Oberhauser *et al.*, 2001; Sears *et al.*, 2001; Wolt *et al.*, 2003; Gathmann *et al.*, 2006; Li *et al.*, 2005; Shirai and Takahashi, 2005; Peterson *et al.*, 2006).

Hypothesis 3 is tested using laboratory methods in the first instance; species that are representative indicators of potentially exposed non-target arthropods are exposed to concentrations of insecticidal protein in excess of conservative estimates of exposure in the field (often 10× the EEC, Rose, 2007). If no harmful effect is observed, hypothesis 3 is corroborated and the risk to non-target arthropods can be considered minimal. If effects are seen at high concentrations, the risk can be characterized further in experiments that use more realistic exposures (Garcia-Alonso *et al.*, 2006; Romeis *et al.*, 2008c).

Overall, the studies conducted to assess the direct impact of different insecticidal proteins on non-target organisms have revealed a high level of specificity of the deployed Cry proteins (US EPA, 2001; Romeis *et al.*, 2006). In contrast, other more broad-spectrum insecticidal proteins, such as lectins or PIs, are more likely to have direct effects on a wider range of non-target organisms (Malone *et al.*, 2008). Thus, it is necessary that each protein is assessed separately, on a case-by-case basis.

The IRGM maize and cotton varieties currently grown express insecticidal Cry proteins that are well understood in respect to their mode of action and spectrum of activity. Due to their specificity, the deployment of *Bt* (Cry) varieties appears to be safe to biological control organisms (Romeis *et al.*, 2006; Wolfenbarger *et al.*, 2008) and other non-target species (O'Callaghan *et al.*, 2005; Sanvido *et al.*, 2007). The studies conducted in the public sector overall confirm the negligible non-target risk conclusion from regulatory risk assessment. In many cases, the deployment of IRGM varieties has resulted in a substantial decrease in the use of chemical insecticides with clear benefits for non-target arthropods and biological control. Secondary pest outbreaks that have for example been reported from *Bt* cotton appear to be largely because Lepidoptera-resistant *Bt* varieties have replaced broad-spectrum insecticide sprays that had kept secondary pests under control (Naranjo *et al.*, 2008).

There is evidence that a healthy biocontrol community in *Bt* crops can help to control potential secondary pests such as aphids (Romeis *et al.*, 2008a). Overall, the currently deployed IRGM plants appear to be well compatible with biological control and thus can form an important part of a sustainable IPM system.

References

Ashouri, A., Overney, S., Michaud, D. and Cloutier, C. (1998) Fitness and feeding are affected in the two-spotted stinkbug, *Perillus bioculatus*, by the cysteine proteinase inhibitor, oryzacystatin I. *Archives of Insect Biochemistry and Physiology* 38, 74–83.

Azzouz, H., Cherqui, A., Campan, E.D.M., Rahbé, Y., Duport, G., Jouanin, L., Kaiser, L. and Giordanengo, P. (2005a) Effects of plant protease inhibitors, oryzacystatin I and soybean Bowman-Birk inhibitor, on the aphid *Macrosiphum euphorbiae* (Homoptera, Aphididae) and its parasitoid *Aphelinus abdominalis* (Hymenoptera, Aphelinidae). *Journal of Insect Physiology* 51, 75–86.

Azzouz, H., Campan, E.D.M., Cherqui, A., Saguez, J., Couty, A., Jouanin, L., Giordanengo, P. and Kaiser, L. (2005b) Potential effects of plant protease inhibitors, oryzacystatin I and soybean Bowman-Birk inhibitor, on the aphid parasitoid *Aphidius ervi* Haliday (Hymenoptera, Braconidae). *Journal of Insect Physiology* 51, 941–951.

Bandyopadhyay, S., Roy, A. and Das, S. (2001) Binding of garlic (*Allium sativum*) leaf lectin to the gut receptors of homopteran pests is correlated to its insecticidal activity. *Plant Science* 161, 1025–1033.

Bates, S.L., Zhao, J.-Z., Roush, R.T. and Shelton, A.M. (2005) Insect resistance management in GM crops: past, present and future. *Nature Biotechnology* 23, 57–62.

Bell, H.A., Fitches, E.C., Down, R.E., Marris, G.C., Edwards, J.P., Gatehouse, J.A. and Gatehouse, A.M.R. (1999) The effect of snowdrop lectin (GNA) delivered via artificial diet and transgenic plants on *Eulophus pennicornis* (Hymenoptera: Eulophidae), a parasitoid of the tomato

moth *Lacanobia oleracea* (Lepidoptera: Noctuidae). *Journal of Insect Physiology* 45, 983–991.

Bell, H.A., Fitches, E.C., Down, R.E., Ford, L., Marris, G.C., Edwards, J.P., Gatehouse, J.A. and Gatehouse, A.M.R. (2001a) Effect of dietary cowpea trypsin inhibitor (CpTI) on the growth and development of the tomato moth *Lacanobia oleracea* (Lepidoptera: Noctuidae) and on the success of the gregarious ectoparasitoid *Eulophus pennicornis* (Hymenoptera: Eulophidae). *Pest Management Science* 57, 57–65.

Bell, H.A., Fitches, E.C., Marris, G.C., Bell, J., Edwards, J.P., Gatehouse, J.A. and Gatehouse, A.M.R. (2001b) Transgenic GNA expressing potato plants augment the beneficial biocontrol of *Lacanobia oleracea* (Lepidoptera; Noctuidae) by the parasitoid *Eulophus pennicornis* (Hymenoptera; Eulophidae). *Transgenic Research* 10, 35–42.

Bell, H.A., Down, R.E., Fitches, E.C., Edwards, J.P. and Gatehouse, A.M.R. (2003) Impact of genetically modified potato expressing plant-derived insect resistance genes on the predatory bug *Podisus maculiventris* (Heteroptera: Pentatomidae). *Biocontrol Science and Technology* 13, 729–741.

Bell, H.A., Kirkbride-Smith, A.E., Marris, G.C., Edwards, J.P. and Gatehouse, A.M.R. (2004) Oral toxicity and impact on fecundity of three insecticidal proteins on the gregarious ectoparasitoid *Eulophus pennicornis* (Hymenoptera: Eulophidae). *Agricultural and Forest Entomology* 6, 215–222.

Bhatti, M.A., Duan, J., Head, G., Jiang, C.J., McKee, M.J., Nickson, T.E., Pilcher, C.L. and Pilcher, C.D. (2005a) Field evaluation of the impact of corn rootworm (Coleoptera: Chrysomelidae)-protected

Bt corn on ground-dwelling invertebrates. *Environmental Entomology* 34, 1325–1335.

Bhatti, M.A., Duan, J., Head, G.P., Jiang, C.J., McKee, M.J., Nickson, T.E., Pilcher, C.L. and Pilcher, C.D. (2005b) Field evaluation of the impact of corn rootworm (Coleoptera: Chrysomelidae)-protected *Bt* corn on foliage-dwelling arthropods. *Environmental Entomology* 34, 1336–1345.

Birch, A.N.E., Geoghegan, I.E., Majerus, M.E.N., McNicol, J.W., Hackett, C.A., Gatehouse, A.M.R. and Gatehouse, J.A. (1999) Tri-trophic interactions involving pest aphids, predatory 2-spot ladybirds and transgenic potatoes expressing snowdrop lectin for aphid resistance. *Molecular Breeding* 5, 75–83.

Bouchard, E., Michaud, D. and Cloutier, C. (2003a) Molecular interactions between an insect predator and its herbivore prey on transgenic potato expressing a cysteine proteinase inhibitor from rice. *Molecular Ecology* 12, 2429–2437.

Bouchard, E., Cloutier, C. and Michaud, D. (2003b) Oryzacystatin I expressed in transgenic potato induces digestive compensation in an insect natural predator via its herbivorous prey feeding on the plant. *Molecular Ecology* 12, 2439–2446.

Bruck, D.J., Lopez, M.D., Lewis, L.C., Prasifka, J.R. and Gunnarson, R.D. (2006) Effects of transgenic *Bacillus thuringiensis* and permethrin on nontarget arthropods. *Journal of Agricultural and Urban Entomology* 23, 111–124.

Burgess, E.P.J., Lovei, G.L., Malone, L.A., Nielsen, I.W., Gatehouse, H.S. and Christeller, J.T. (2002) Prey-mediated effects of the protease inhibitor aprotinin on the predatory carabid beetle *Nebria brevicollis*. *Journal of Insect Physiology* 48, 1093–1101.

Burgio, G., Lanzoni, A., Accinelli, G., Dinelli, G., Bonetti, A., Marotti, I. and Ramilli, F. (2007) Evaluation of *Bt*-toxin uptake by the non-target herbivore, *Myzus persicae* (Hemiptera: Aphididae), feeding on transgenic oilseed rape. *Bulletin of Entomological Research* 97, 211–215.

Candolfi, M.P., Brown, K., Grimm, C., Reber, B. and Schmidli, H. (2004) A faunistic approach to assess potential side-effects of genetically modified *Bt*-corn on non-target arthropods under field conditions. *Biocontrol Science and Technology* 14, 129–170.

Carlini, C.R. and Grossi-de-Sá, M.F. (2002) Plant toxic proteins with insecticidal properties. A review on their potentialities as biopesticides. *Toxicon* 40, 1515–1539.

Carpenter, J. and Gianessi, L.P. (2001) *Agricultural Biotechnology: Updated Benefit Estimates*. National Center for Food and Agricultural Policy, Washington, DC.

Carrière, Y., Ellers-Kirk, C., Sisterson, M., Antilla, L., Whitlow, M., Dennehy, T.J. and Tabashnik, B.E. (2003) Long-term regional suppression of pink bollworm by *Bacillus thuringiensis* cotton. *Proceedings of the National Academy of Sciences of the USA* 100, 1519–1523.

Catangui, M.A. and Berg, R.K. (2006) Western bean cutworm, *Striacosta albicosta* (Smith) (Lepidoptera: Noctuidae), as a potential pest of transgenic Cry1Ab *Bacillus thuringiensis* corn hybrids in South Dakota. *Environmental Entomology* 35, 1439–1452.

Cattaneo, M.G., Yafuso, C., Schmidt, C., Huang, C.Y., Rahman, M., Olson, C., Ellers-Kirk, C., Orr, B.J., Marsh, S.E., Antilla, L., Dutilleu, P. and Carrière, Y. (2006) Farm-scale evaluation of the impacts of transgenic cotton on biodiversity, pesticide use, and yield. *Proceedings of the National Academy of Sciences of the USA* 103, 7571–7576.

CBD Secretariat (2000) *Cartagena Protocol on Biosafety to the Convention on Biological Diversity: Text and Annexes*. Secretariat of the Convention on Biological Diversity, Montreal, Canada.

Ceci, L.R., Volpicella, M., Rahbé, Y., Gallerani, R., Beekwilder, J. and Jongsma, M.A. (2003) Selection by phage display of a variant mustard trypsin inhibitor toxic against aphids. *The Plant Journal* 33, 557–566.

Chen, M., Zhao, J.-Z., Ye, G.-Y., Fu, Q. and Shelton, A.M. (2006) Impact of insect-resistant transgenic rice on target insect pests and non-target arthropods in China. *Insect Science* 13, 409–420.

Chen, M., Zhao, J.-Z., Colins, H.L., Earle, E.D., Cao, J. and Shelton, A.M. (2008)

A critical assessment of the effects of Bt transgenic plants on parasitoids. *PLoS ONE*, 3(5), e 2284.

Christeller, J.T., Malone, L.A., Todd, J.H., Marshall, R.M., Burgess, E.P.J. and Philip, B.A. (2005) Distribution and residual activity of two insecticidal proteins, avidin and aprotinin, expressed in transgenic tobacco plants, in the bodies and frass of *Spodoptera litura* larvae following feeding. *Journal of Insect Physiology* 51, 1117–1126.

Chu, C.-C., Natwick, E.T., Leon-Lopez, R., Dessert, J.R. and Henneberry, T.J. (2006) Pink bollworm moth (Lepidoptera: Gelechiidae) catches in the Imperial Valley, California from 1989 to 2003. *Insect Science* 13, 469–475.

Coll, M. and Guershon, M. (2002) Omnivory in terrestrial arthropods: mixing plant and prey diets. *Annual Review of Entomology* 47, 267–297.

Conner, A.J., Glare, T.R. and Nap, J.P. (2003) The release of genetically modified crops into the environment – Part II. Overview of ecological risk assessment. *The Plant Journal* 33, 19–46.

Cottrell, T.E. and Yeargan, K.V. (1998) Effect of pollen on *Coleomegilla maculata* (Coleoptera: Coccinellidae) population density, predation, and cannibalism in sweet corn. *Environmental Entomology* 27, 1402–1410.

Couty, A., de la Viña, G., Clark, S.J., Kaiser, L., Pham-Delègue, M.-H. and Poppy, G.M. (2001a) Direct and indirect sublethal effects of *Galanthus nivalis* agglutinin (GNA) on the development of a potato-aphid parasitoid, *Aphelinus abdominalis* (Hymenoptera: Aphelinidae). *Journal of Insect Physiology* 47, 553–561

Couty, A., Clark, S.J. and Poppy, G.M. (2001b) Are fecundity and longevity of female *Aphelinus abdominalis* affected by development in GNA-dosed *Macrosiphum euphorbiae*? *Physiological Entomology* 26, 287–293

Czapla, T.H. (1997) Plant lectins as insect control proteins in transgenic plants. In: Carozzi, N. and Koziel, M. (eds) *Advances in Insect Control: The Role of Transgenic Plants*. Taylor & Francis, London, pp. 123–138.

Daly, T. and Buntin, D. (2005) Effect of *Bacillus thuringiensis* transgenic corn for Lepidoptera control on nontarget arthropods. *Environmental Entomology* 34, 1292–1301.

Dean, J.M. and De Moraes, C.M. (2006) Effects of genetic modification on herbivore-induced volatiles from maize. *Journal of Chemical Ecology* 32, 713–724.

Dicke, M. (1999) Direct and indirect effects of plants on performance of beneficial organisms. In: Ruberson, J.R. (ed.) *Handbook of Pest Management*. Marcel Dekker, New York, pp. 105–153.

Dively, G.P. (2005) Impact of transgenic VIP3A × Cry1Ab lepidopteran-resistant field corn on the non-target arthropod community. *Environmental Entomology* 34, 1267–1291.

Dively, G.P., Rose, R., Sears, M.K., Hellmich, R.L., Stanley-Horn, D.E., Russo, J.M., Calvin, D.D. and Anderson, P.L. (2004) Effects on Monarch butterfly larvae (Lepidoptera: Danaidae) after continuous exposure to Cry1Ab-expressing corn during anthesis. *Environmental Entomology* 33, 1116–1125.

Down, R.E., Gatehouse, A.M.R., Hamilton, W.D.O. and Gatehouse, J.A. (1996) Snowdrop lectin inhibits development and fecundity of the glasshouse potato aphid (*Aulacorthum solani*) when administered *in vitro* and via transgenic plants both in laboratory and glasshouse trials. *Journal of Insect Physiology* 42, 1035–1045.

Down, R.E., Ford, L., Mosson, H.J., Fitches, E., Gatehouse, J.A. and Gatehouse, A.M.R. (1999) Protease activity in the larval stage of the parasitoid wasp, *Eulophus pennicornis* (Nees) (Hymenoptera: Eulophidae); effects of protease inhibitors. *Parasitology* 119, 157–166.

Down, R.E., Ford, L., Woodhouse, S.D., Raemaekers, R.J.M., Leitch, B., Gatehouse, J.A. and Gatehouse, A.M.R. (2000) Snowdrop lectin (GNA) has no acute toxic effect on a beneficial insect predator, the 2-spot ladybird (*Adalia bipunctata* L.). *Journal of Insect Physiology* 46, 379–391.

Duan, J.J., Head, G., Jensen, A. and Reed, G. (2004) Effects of transgenic *Bacillus thur-*

ingiensis potato and conventional insecticides for Colorado potato beetle (Coleoptera: Chrysomelidae) management on the abundance of ground-dwelling arthropods in Oregon potato ecosystems. *Environmental Entomology* 33, 275–281.

Dutta, I., Majumder, P., Saha, P., Ray, K. and Das, S. (2005a) Constitutive and phloem specific expression of *Allium sativum* leaf agglutinin (ASAL) to engineer aphid (*Lipaphis erysimi*) resistance in transgenic Indian mustard (*Brassica juncea*). *Plant Science* 169, 996–1007.

Dutta, I., Saha, P., Majumder, P., Sarkar, A., Chakraborti, D., Banerjee, S. and Das, S. (2005b) The efficacy of a novel insecticidal protein, *Allium sativum* leaf lectin (ASAL), against homopteran insects monitored in transgenic tobacco. *Plant Biotechnology Journal* 3, 601–611.

Dutton, A., Klein, H., Romeis, J. and Bigler, F. (2002) Uptake of Bt-toxin by herbivores feeding on transgenic maize and consequences for the predator *Chrysoperla carnea*. *Ecological Entomology* 27, 441–447.

Dutton, A., Romeis, J. and Bigler, F. (2003) Assessing the risks of insect resistant transgenic plants on entomophagous arthropods: Bt-maize expressing Cry1Ab as a case study. *BioControl* 48, 611–636.

Dutton, A., Obrist, L., D'Alessandro, M., Diener, L., Müller, M., Romeis, J. and Bigler, F. (2004) Tracking Bt-toxin in transgenic maize for risk assessment on non-target arthropods. *IOBC/WPRS Bulletin* 27(3), 57–63.

Estruch, J.J., Warren, G.W., Mullins, M.A., Nye, G.J., Craig, J.A. and Koziel, A.G. (1996) Vip3A, a novel *Bacillus thuringiensis* vegetative insecticidal protein with a wide spectrum of activity against lepidopteran insects. *Proceedings of the National Academy of Sciences of the USA* 93, 5389–5394.

European Food Safety Authority (EFSA) (2006) Guidance document of the Scientific Panel on Genetically Modified Organisms for the risk assessment of genetically modified plants and derived food and feed. *EFSA Journal* 99, 1–100. Available at: http://www.efsa.europa.eu/EFSA/Scientific_Document/gmo_guidance_gm_plants_en,0.pdf

Felke, M., Lorenz, N. and Langenbruch, G.A. (2002) Laboratory studies on the effects of pollen from Bt-maize on larvae of some butterfly species. *Journal of Applied Entomology* 126, 320–325.

Fernandes, O.A., Faria, M., Martinelli, S., Schmidt, F., Carvalho, V.F. and Moro, G. (2007) Short-term assessment of Bt maize on non-target arthropods in Brazil. *Scientia Agricola* 64, 249–255.

Ferry, N., Raemaekers, R.J.M., Majerus, M.E.N., Jouanin, L., Port, G., Gatehouse, J.A. and Gatehouse, A.M.R. (2003) Impact of oilseed rape expressing the insecticidal cysteine protease inhibitor oryzacystatin on the beneficial predator *Harmonia axyridis* (multicoloured Asian ladybeetle). *Molecular Ecology* 12, 493–504.

Ferry, N., Edwards, M.G., Mulligan, E.A., Emami, K., Petrova, A.S., Frantescu, M., Davison, G.M. and Gatehouse, A.M.R. (2004) Engineering resistance to insect pests. In: Christou, P. and Klee, H. (eds) *Handbook of Plant Biotechnology*, Vol. 1. Wiley, Chichester, UK, pp. 373–394.

Ferry, N., Jouanin, L., Ceci, L.R., Mulligan, A., Emami, K., Gatehouse, J.A. and Gatehouse, A.M.R. (2005) Impact of oilseed rape expressing the insecticidal serine protease inhibitor, mustard trypsin inhibitor-2 on the beneficial predator *Pterostichus madidus*. *Molecular Ecology* 14, 337–349.

Ferry, N., Mulligan, E.A., Stewart, C.N., Tabashnik, B.E., Port, G.R. and Gatehouse, A.M.R. (2006) Prey-mediated effects of transgenic canola on a beneficial, non-target, carabid beetle. *Transgenic Research* 15, 501–514.

Ferry, N., Mulligan, E.A., Majerus, M.E.N. and Gatehouse, A.M.R. (2007) Bitrophic and tritrophic effects of Bt Cry3A transgenic potato on beneficial, non-target, beetles. *Transgenic Research* 16, 795–812.

Fitt, G. (2008) Have *Bt* crops led to changes in insecticide use patterns and impacted IPM? In: Romeis, J., Shelton, A.M. and Kennedy G.G. (eds) *Integration of Insect-Resistant Genetically Modified Crops within IPM Programs*. Springer, Dordrecht, The Netherlands, pp. 303–328.

Foissac, X., Loc, N.T., Christou, P., Gatehouse, A.M.R. and Gatehouse, J.A. (2000) Resistance to Green Leafhopper (*Nephotetix virescens*) and Brown planthopper (*Nilaparvata lugens*) in transgenic rice expressing snowdrop lectin (*Galanthus nivalis* agglutinin; GNA). *Journal Insect Physiology* 46, 573–583.

Garcia-Alonso, M., Jacobs, E., Raybould, A., Nickson, T.E., Sowig, P., Willekens, H., van der Kouwe, P., Layton, R., Amijee, F., Fuentes, A.M. and Tencalla, F. (2006) A tiered system for assessing the risk of genetically modified plants to non-target organisms. *Environmental Biosafety Research* 5, 57–65.

Gathmann, A., Wirooks, L., Hothorn, L.A., Bartsch, D. and Schuphan, I. (2006) Impact of Bt maize pollen (MON810) on lepidopteran larvae living on accompanying weeds. *Molecular Ecology* 15, 2677–2685.

Glare, T.R. and O'Callaghan, M. (2000) *Bacillus Thuringiensis: Biology, Ecology and Safety*. Wiley, Chichester, UK.

Gómez-Babero, M., Berbel, J. and Rodríguez-Cerezo, E. (2008) *Bt* corn in Spain – the performance of the EU's first GM crop. *Nature Biotechnology* 26, 384–386.

Gooding, R.H. and Huang, C.T. (1969) Trypsin and hymotrypsin from the beetle *Pterostichus melanarius*. *Journal of Insect Physiology* 15, 325–339.

Grafius, E.J. and Douches, D.S. (2008) The present and future role of insect-resistant genetically modified potato cultivars in potato IPM. In: Romeis, J., Shelton, A.M. and Kennedy, G.G. (eds) *Integration of Insect-Resistant Genetically Modified Crops within IPM Programs*. Springer, Dordrecht, The Netherlands, pp. 195–221.

Greene, J.K., Turnipseed, S.G., Sullivan, M.J. and May, O.L. (2001) Treatment thresholds for stink bugs (Hemiptera: Pentatomidae) in cotton. *Journal of Economic Entomology* 94, 403–409.

Greenplate, J.T., Mullins, J.W., Penn, S.R., Dahm, A., Reich, B.J., Osborn, J.A., Rahn, P.R., Ruschke, L. and Shappley, Z.W. (2003) Partial characterization of cotton plants expressing two toxin proteins from *Bacillus*

thuringiensis: relative toxin contribution, toxin interaction, and resistance management. *Journal of Applied Entomology* 127, 340–347.

Gregory, R.D., Noble, D.G. and Custance, J. (2004) The state of play of farmeland birds: population trends and conservation status of lowland farmland birds in the United Kingdom. *Ibis* 146 (Supplement 2), 1–13.

Hagerty, A.M., Kilpatrick, A.L., Turnipseed, S.G., Sullivan, M.J. and Bridges, W.C. (2005) Predaceous arthropods and lepidopteran pests on conventional, Bollgard, and Bollgard II cotton under untreated and disrupted conditions. *Environmental Entomology* 34, 105–114.

Harper, S.M., Crenshaw, R.W., Mullins, M.A. and Privalle, L.S. (1995) Lectin binding to insect bush border membranes. *Journal of Economic Entomology* 88, 1197–1202.

Harwood, J.D., Wallin, W.G. and Obrycki, J.J. (2005) Uptake of Bt endotoxins by non-target herbivores and higher order arthropod predators: molecular evidence from a transgenic corn agroecosystem. *Molecular Ecology* 14, 2815–2823.

Head, G., Brown, C.R., Groth, M.E. and Duan, J.J. (2001) Cry1Ab protein levels in phytophagous insects feeding on transgenic corn: implications for secondary exposure risk assessment. *Entomologia Experimentalis et Applicata* 99, 37–45.

Head, G., Moar, M., Eubanks, M., Freeman, B., Ruberson, J., Hagerty, A. and Turnipseed, S. (2005) A multi-year, large-scale comparison of arthropod populations on commercially managed *Bt* and non-*Bt* cotton fields. *Environmental Entomology* 34, 1257–1266.

Hellmich, R.L., Higgins, L.S., Witkowski, J.F., Campbell, J.E. and Lewis, L.C. (1999) Oviposition by European corn borer (Lepidoptera. Crambidae) in response to various transgenic corn events. *Journal of Economic Entomology* 92, 1014–1020.

Hellmich, R.L., Siegfried, B.D., Sears, M.K., Stanley-Horn, D.E., Mattila, H.R., Spencer, T., Bidne, K.G., Daniels, M.J. and Lewis, L.C. (2001) Monarch larvae sensitivity to *Bacillus thuringiensis*-purified proteins and pollen. *Proceedings of the National*

Academy of Sciences of the USA 98, 11925–11930.

Hellmich, R.L., Albajes, R., Bergvinson, D., Prasifka, J.R., Wang, Z.-Y. and Weiss, M.J. (2008) The present and future role of insect-resistant genetically modified maize in IPM. In: Romeis, J., Shelton, A.M. and Kennedy, G.G. (eds) *Integration of Insect-Resistant Genetically Modified Crops within IPM Programs.* Springer, Dordrecht, The Netherlands, pp. 119–158.

Hilbeck, A., Baumgartner, M., Fried, P.M. and Bigler, F. (1998) Effects of transgenic *Bacillus thuringiensis* corn-fed prey on mortality and development time of immature *Chrysoperla carnea* (Neuroptera: Chrysopidae). *Environmental Entomology* 27, 480–487.

Hogervorst, P.A.M., Ferry, N., Gatehouse, A.M.R., Wäckers, F.L. and Romeis, J. (2006) Direct effects of snowdrop lectin (GNA) on larvae of three aphid predators and fate of GNA after ingestion. *Journal of Insect Physiology* 52, 614–624.

Hogervorst, P.A.M., Wäckers, F.L., Woodring, J. and Romeis, J. (2009) Snowdrop lectin (*Galanthus nivalis* agglutinin) in aphid honeydew negatively affects survival of a honeydew-consuming parasitoid. *Agricultural and Forest Entomology* 11 (in press).

Hoheisel, G.A. and Fleischer, S.J. (2007) Coccinellids, aphids, and pollen in diversified vegetable fields with transgenic and isoline cultivars. *Journal of Insect Science* 7, 64. Available at: http://www. insectscience.org/7.61/

Howald, R., Zwahlen, C. and Nentwig, W. (2003) Evaluation of Bt oilseed rape on the non-target herbivore *Athalia rosae. Entomologia Experimentalis et Applicata* 106, 87–93.

Hoy, C.W., Feldman, J., Gould, F., Kennedy, G.G., Reed, G. and Wyman, J.A. (1998) Naturally occurring biological controls in genetically engineered crops. In: Barbosa, P. (ed.) *Conservation Biological Control.* Academic Press, San Diego, California, pp. 185–205.

Huang, J.K., Hu, R.F., Rozelle, S. and Pray, C. (2005) Insect-resistant GM rice in farmers'

fields: assessing productivity and health effects in China. *Science* 308, 688–690.

Huang, J., Hu, R., Rozelle, S. and Pray, C. (2008) Genetically modified rice, yields, and pesticides: assessing farm-level productivity effects in China. *Economic Development and Cultural Change* 56, 241–264.

Hunt, T.E., Buschman, L.L. and Sloderbeck, P.E. (2007) Insecticide use in Bt and non-Bt field corn in the western Corn Belt: as reported by crop consultants in a mail survey. *American Entomologist* 53, 86–93.

James, C. (2007) *Global Status of Commercialized Biotech/GM Crops: 2007.* ISAAA Briefs No. 37. International Service for the Acquisition of Agri-Biotech Applications, Ithaca, New York.

Jesse, L.C.H. and Obrycki, J.J. (2000) Field deposition of Bt transgenic corn pollen: lethal effects on the monarch butterfly. *Oecologia* 125, 241–248.

Jesse, L.C.H. and Obrycki, J.J. (2002) Assessment of the non-target effects of transgenic Bt corn pollen and anthers on the milkweed tiger moth *Euchatias egle* Drury (Lepidoptera: Arctiidae). *Journal of the Kansas Entomological Society* 75, 55–58.

Johnson, M.T. (1997) Interaction of resistant plants and wasp parasitoids of tobacco budworm (Lepidoptera: Noctuidae). *Environmental Entomology* 26, 207–214.

Johnson, M.T. and Gould, F. (1992) Interaction of genetically engineered host plant-resistance and natural enemies of *Heliothis virescens* (Lepidoptera, Noctuidae) in tobacco. *Environmental Entomology* 21, 586–597.

Johnson, K.L., Raybould, A.F., Hudson, M.D. and Poppy, G.M. (2007) How does scientific risk assessment of GM crops fit within the wider risk analysis? *Trends in Plant Sciences* 12, 1–5.

Jongsma, M.A. and Bolter, C. (1997) The adaptation of insects to plant protease inhibitors. *Journal of Insect Physiology* 43, 885–895.

Jouanin, L., Bonadé-Bottino, M., Girard, C., Morrot, G. and Giband, M. (1998) Transgenic plants for insect resistance. *Plant Science* 131, 1–11.

Kalushkov, P. and Nedvěd, O. (2005) Genetically modified potatoes expressing

Cry 3A protein do not affect aphidophagous coccinellids. *Journal of Applied Entomology* 129, 401–406.

Kaniewski, W. and Thomas, P. (2004) The potato story. *AgBioForum* 7(1&2), 41–46.

Kanrar, S., Venkateswari, J., Kirti, P.B. and Chopra, V.L. (2002) Transgenic Indian mustard (*Brassica juncea*) with resistance to the mustard aphid (*Lipaphis erysimi* Kalt.). *Plant Cell Reports* 20, 976–981.

Kennedy, G.G. (2008) Integration of insect-resistant genetically modified crops within IPM programs. In: Romeis, J., Shelton, A.M. and Kennedy, G.G. (eds) *Integration of Insect-Resistant Genetically Modified Crops within IPM Programs*. Springer, Dordrecht, The Netherlands, pp. 1–26.

Knowles, B.H. (1994) Mechanisms of action of *Bacillus thuringiensis* insecticidal δ-endotoxins. *Advances in Insect Physiology* 24, 275–308.

Kreutzweiser, D.P., Holmes, S.B., Capell, S.S. and Eichenberg, D.C. (1992) Lethal and sublethal effects of *Bacillus thuringiensis* var. *kurstaki* on aquatic insects in laboratory bioassays and outdoor stream channels. *Bulletin of Environmental Contamination and Toxicology* 49, 252–258.

Kuiper, H.A., Kleter, G.A., Noteborn, H.P.J.M. and Kok, E.J. (2002) Substantial equivalence – an appropriate paradigm for the safety assessment of genetically modified foods? *Toxicology* 181, 427–431.

Lang, A. and Vojtech, E. (2006) The effects of pollen consumption of transgenic Bt maize on the common swallowtail, *Papilio machaon* L. (Lepidoptera, Papilionidae). *Basic and Applied Ecology* 7, 296–306.

Lawo, N.C. and Romeis, J. (2008) Assessing the utilization of a carbohydrate food source and the impact of insecticidal proteins on larvae of the green lacewing, *Chrysoperla carnea*. *Biological Control* 44, 389–398.

Lee, M.K., Walters, F.S., Hart, H., Palekar, N. and Chen, J.S. (2003) Mode of action of the *Bacillus thuringiensis* vegetative insecticidal protein Vip3A differs from that of Cry1Ab delta-endotoxin. *Applied and Environmental Microbiology* 69, 4648–4657.

Lee, M.K., Miles, P. and Chen, J.S. (2006) Brush border membrane binding properties of *Bacillus thuringiensis* Vip3A toxin to *Heliothis virescens* and *Helicoverpa zea* midguts. *Biochemical and Biophysical Research Communications* 339, 1043–1047.

Leslie, T.W., Hoheisel, G.A., Biddinger, D.J., Rohr, J.R. and Fleisher, S.J. (2007) Transgenes sustain epigeal insect biodiversity in diversified vegetable farm systems. *Environmental Entomology* 36, 234–244.

Li, Y., Meissle, M. and Romeis, J. (2008) Consumption of Bt maize pollen expressing Cry1Ab or Cry3Bb1 does not harm adult green lacewings, *Chrysoperla carnea* (Neuroptera: Chrysopidae). *PLoS ONE* 3(8), e2909.

Li, W., Wu, K., Wang, X., Wang, G. and Guo, Y. (2005) Impact of pollen grains from Bt transgenic corn on the growth and development of Chinese tussah silkworm, *Antheraea pernyi* (Lepidoptera: Saturniidae). *Environmental Entomology* 34, 922–928.

Liu, Y.-B., Tabashnik, B.E., Dennehy, T.J., Carrière, Y., Sims, M.A. and Meyer, S.K. (2002) Oviposition and mining in bolls of Bt- and non-Bt cotton by resistant and susceptible pink bollworm (Lepidoptera: Gelechiidae). *Journal of Economic Entomology* 95, 143–148.

Ludy, C. and Lang, A. (2006) Bt maize pollen exposure and impact on the garden spider, *Araneus diadematus*. *Entomologia Experimentalis et Applicata* 118, 145–156.

Lundgren, J.G., Razzak, A.A. and Wiedenmann, R.N. (2004) Population responses and food consumption by predators *Coleomegilla maculata* and *Harmonia axyridis* (Coleoptera: Coccinellidae) during anthesis in an Illinois cornfield. *Environmental Entomology* 33, 958–963.

Lundgren, J.G., Huber, A. and Wiedenmann, R.N. (2005) Quantification of consumption of corn pollen by the predator *Coleomegilla maculata* (Coleoptera: Coccinellidae) during anthesis in an Illinois cornfield. *Agricultural and Forest Entomology* 7, 53–60.

MacIntosh, S.C., Kishore, G.M., Perlak, F.J., Marrone, P.G., Stone, T.B., Sims, S.R. and Fuchs, R.L. (1990) Potentiation of *Bacillus thuringiensis* insecticidal activity by serine protease inhibitors. *Journal of Agricultural and Food Chemistry* 38, 1145–1152.

Malcolm, S.B. (1990) Chemical defence in chewing and sucking insect herbivores: plant-derived cardenolides in the monarch butterfly and orleander aphid. *Chemoecology* 1, 12–21.

Malone, L.A., Gatehouse, A.M.R. and Barratt, B.I.P. (2008) Beyond *Bt*: alternative strategies for insect-resistant genetically modified crops. In: Romeis, J., Shelton, A.M. and Kennedy, G.G. (eds) *Integration of Insect-Resistant Genetically Modified Crops within IPM Programs*. Springer, Dordrecht, The Netherlands, pp. 357–417.

Marvier, M., McCreedy, C., Regetz, J. and Kareiva, P. (2007) A meta-analysis of effects of Bt cotton and maize on nontarget invertebrates. *Science* 316, 1475–1477.

Mattila, H.R., Sears, M.K. and Duan, J.J. (2005) Response of *Danaus plexippus* to pollen of two new Bt corn events via laboratory bioassay. *Entomologia Experimentalis et Applicata* 116, 31–41.

Meier, M.S. and Hilbeck, A. (2001) Influence of transgenic *Bacillus thuringiensis* corn-fed prey on prey preference of immature *Chrysoperla carnea* (Neuroptera: Chrysopidae). *Basic and Applied Ecology* 2, 35–44.

Meissle, M., Vojtech, E. and Poppy, G.M. (2005) Effects of Bt maize-fed prey on the generalist predator *Poecilus cupreus* L. (Coleoptera: Carabidae). *Transgenic Research* 14, 123–132.

Meissle, M., Hiltpold, I., Turlings, T.C.J. and Romeis, J. (2008) Belowground volatile emission of *Bt* maize after induction of plant defence. *IOBC/WPRS Bulletin* 33, 85–92.

Mellet, M.A., Schoeman, A.S. and Broodryk, S.W. (2004) Bollworm (*Helicoverpa armigera* (Hübner), Lepidoptera: Noctuidae) occurrences in Bt- and non-Bt-cotton fields, Marble Hall, Mpumalanga, South Africa. *African Entomologist* 12, 107–115.

Men, X., Ge, F., Edwards, C.A. and Yardim, E.N. (2005) The influence of pesticide applications on *Helicoverpa armigera* Hubner and sucking pests in transgenic Bt cotton and non-transgenic cotton in China. *Crop Protection* 24, 319–324.

Monsanto Company (2003) Safety Assessment of YieldGard Rootworm Corn. Available at: http://www.monsanto.com/pdf/products/yieldgard_rw_pss.pdf

Monsanto Company (2004) Petition for the Determination of Nonregulated Status for MON 88017 Corn. Monsanto Petition # 04-CR-108U; OECD Unique Identifier: MON-88017-3. Available at: http://www.aphis.usda.gov/brs/aphisdocs/04_12501p.pdf

Morse, J.C. (1997) Phylogeny of Trichoptera. *Annual Review of Entomology* 42, 427–450.

Mulligan, E.A. (2006) A system approach to comparing the impacts of genetically modified and conventional pest control on beneficial insects. Dissertation, University of Newcastle, UK.

Mulligan, E.A., Ferry, N., Jouanin, L., Walters, K.F.A., Port, G.R. and Gatehouse, A.M.R. (2006) Comparing the impact of conventional pesticide and use of a transgenic pest-resistant crop on the beneficial carabid beetle *Pterostichus melanarius*. *Pest Management Science* 62, 999–1012.

Musser, F.R. and Shelton, A.M. (2003) *Bt* sweet corn and selective insecticides: impacts on pests and predators. *Journal of Economic Entomology* 96, 71–80.

Nap, J.P., Metz, P.L.J., Escaler, M. and Conner, A.J. (2003) The release of genetically modified crops into the environment – Part I. Overview of current status and regulations. *The Plant Journal* 33, 1–18.

Naranjo, S.E. (2005a) Long-term assessment of the effects of transgenic *Bt* cotton on the abundance of non-target arthropod natural enemies. *Environmental Entomology* 34, 1193–1210.

Naranjo, S.E. (2005b) Long-term assessment of the effects of transgenic *Bt* cotton on the function of the natural enemy community. *Environmental Entomology* 34, 1211–1223.

Naranjo, S.E., Ruberson, J.R., Sharma, H.C., Wilson, L. and Wu, K. (2008) The present and future role of insect-resistant genetically modified cotton in IPM. In: Romeis, J., Shelton, A.M. and Kennedy, G.G. (eds) *Integration of Insect-Resistant Genetically Modified Crops within IPM Programs*.

Springer, Dordrecht, The Netherlands, pp. 159–194.

Newman, N.C. (1998) *Fundamentals of Ecotoxicology*. Ann Arbor Press, Chelsea, Michigan.

Oberhauser, K.S., Prysby, M., Mattila, H.R., Stanley-Horn, D.E., Sears, M.K., Dively, G.P., Olson, E., Pleasants, J.M., Lam, W.-K.F. and Hellmich, R.L. (2001) Temporal and spatial overlap between monarch larvae and corn pollen. *Proceedings of the National Academy of Sciences of the USA* 98, 11913–11918.

Obrist, L.B., Klein, H., Dutton, A. and Bigler, F. (2005) Effects of Bt maize on *Frankliniella tenuicornis* and exposure of thrips predators to prey-mediated Bt toxin. *Entomologia Experimentalis et Applicata* 115, 409–416.

Obrist, L., Dutton, A., Romeis, J. and Bigler, F. (2006a) Fate of Cry1Ab toxin expressed by Bt maize upon ingestion by herbivorous arthropods and consequences for *Chrysoperla carnea*. *BioControl* 51, 31–48.

Obrist, L.B., Dutton, A., Albajes, R. and Bigler, F. (2006b) Exposure of arthropod predators to Cry1Ab toxin in Bt maize fields. *Ecological Entomology* 31, 143–154.

Obrist, L.B., Klein, H., Dutton, A. and Bigler, F. (2006c) Assessing the effects of Bt maize on the predatory mite *Neoseiulus cucumeris*. *Experimental and Applied Acarology* 38, 125–139.

O'Callaghan, M., Glare, T.R., Burgess, E.P.J. and Malone, L.A. (2005) Effects of plants genetically modified for insect resistance on nontarget organisms. *Annual Review of Entomology* 50, 271–292.

Organisation for Economic Cooperation and Development (OECD) (1993) *Safety Considerations for Biotechnology: Scale-Up of Crop Plants*. Organisation for Economic Cooperation and Development, Paris.

Orr, D.B. and Landis, D.A. (1997) Oviposition of European corn borer (Lepidoptera: Pyralidae) and impact of natural enemy populations in transgenic versus isogenic corn. *Journal of Economic Entomology* 90, 905–909.

Overney, S., Yelle, S. and Cloutier, C. (1998) Occurrence of digestive cysteine proteases in *Perillus bioculatus*, a natural predator of the Colorado potato beetle.

Comparative Biochemistry and Physiology B 120, 191–196.

Peterson, R.K.D., Meyer, S.J., Wolf, A.T., Wolt, J.D. and Davis, P.M. (2006) Genetically engineered plants, endangered species, and risk: a temporal and spatial exposure assessment for karner blue butterfly larvae and Bt maize pollen. *Risk Analysis* 26, 845–858.

Peumans, W.J. and Van Damme, E.J.M. (1995) Lectins as plant defense proteins. *Plant Physiology* 109, 347–352.

Phipps, R.H. and Park, J.R. (2002) Environmental benefits of genetically modified crops: global and European perspectives on their ability to reduce pesticide use. *Journal of Animal and Feed Sciences* 11, 1–18.

Pilcher, C.D. and Rice, M.E. (2001) Effect of planting dates and *Bacillus thuringiensis* corn on the population dynamics of European corn borer (Lepidoptera: Crambidae). *Journal of Economic Entomology* 94, 730–742.

Pilcher, C.D., Rice, M.E. and Obrycki, J.J. (2005) Impact of transgenic *Bacillus thuringiensis* corn and crop phenology on five non-target arthropods. *Environmental Entomology* 34, 1302–1316.

Powell, K.S., Spence, J., Bharathi, M., Gatehouse, J.A. and Gatehouse, A.M.R. (1998) Immunohistochemical and developmental studies to elucidate the mechanism of action of the snowdrop lectin on the rice brown planthopper, *Nilaparvata lugens* (Stal). *Journal of Insect Physiology* 44, 529–539.

Qaim, M., Pray, C.E. and Zilberman, D. (2008) Economic and social considerations in the adoption of Bt crops. In: Romeis, J., Shelton, A.M. and Kennedy, G.G. (eds) *Integration of Insect-Resistant Genetically Modified Crops within IPM Programs*. Springer, Dordrecht, The Netherlands, pp. 329–356.

Rahbé, Y., Sauvion, N., Febvay, G., Peumans, W.J. and Gatehouse, A.M.R. (1995) Toxicity of lectins and processing of ingested proteins in the pea aphid *Acyrthosiphon pisum*. *Entomologia Experimentalis et Applicata* 76, 143–155.

Rahbé, Y., Deraison, C., Bonadé-Bottino, M., Girard, C., Nardon, C. and Jouanin, L.

(2003a) Effects of the cysteine protease inhibitor oryzacystatin (OC-I) on different aphids and reduced performance of *Myzus persicae* on OC-I expressing transgenic oilseed rape. *Plant Science* 164, 441–450.

Rahbé, Y., Ferrasson, E., Rabesona, H. and Quillien, L. (2003b) Toxicity to the pea aphid *Acyrthosiphon pisum* of anti-chymotrypsin isoforms and fragments of Bowman-Birk protease inhibitors from pea seeds. *Insect Biochemistry and Molecular Biology* 33, 299–306.

Rao, K.V., Rathore, K.S., Hodges, T.K., Fu, X.D., Stoger, E., Sudhakar, D., Williams, S., Christou, P., Bown, D.P., Powell, K.S., Spence, J., Bharathi, M., Gatehouse, A.M.R. and Gatehouse, J.A. (1998) Expression of snowdrop lectin (GNA) in the phloem of transgenic rice plants confers resistance to rice brown planthopper. *The Plant Journal* 14, 469–477.

Raps, A., Kehr, J., Gugerli, P., Moar, W.J., Bigler, F. and Hilbeck, A. (2001) Immunological analysis of phloem sap of *Bacillus thuringiensis* corn and of the non-target herbivore *Rhopalosiphum padi* (Homoptera: Aphididae) for the presence of Cry1Ab. *Molecular Ecology* 10, 525–533.

Raybould, A. (2006) Problem formulation and hypothesis testing for environmental risk assessments of genetically modified crops. *Environmental Biosafety Research* 5, 119–125.

Raybould, A. (2007a) Ecological versus ecotoxicological methods for assessing the environmental risks of transgenic crops. *Plant Science* 173, 589–602.

Raybould (2007b) Environmental risk assessment of genetically modified crops: general principles and risks to non-target organisms. *BioAssay* 2:8. Available at: http://www.bioassay.org.br/articles/2.8/BA2.8.pdf

Raybould, A., Stacey, D., Vlachos, D., Graser, G., Li, X. and Joseph, R. (2007) Non-target organisms risk assessment of MIR604 maize expressing mCry3A for control of corn rootworms. *Journal of Applied Entomology* 131, 391–399.

Reed, G.L., Jensen, A.S., Riebe, J., Head, G. and Duan, J.J. (2001) Transgenic *Bt* potato and conventional insecticides for Colorado potato beetle management: comparative efficacy and non-target impacts. *Entomologia Experimentalis et Applicata* 100, 89–100.

Ribeiro, A.P.O., Pereira, E.J.G., Galvan, T.L., Picanço, M.C., Picoli, E.A.T., da Silva, D.J.H., Fári, M.G. and Otoni, W.C. (2006) Effect of eggplant transformed with oryzacystatin gene on *Myzus persicae* and *Macrosiphum euphorbiae*. *Journal of Applied Entomology* 130, 84–90.

Rice, M.E. (2004) Transgenic rootworm corn: assessing potential agronomic, economic, and environmental benefits. *Online. Plant Health Progress.* doi:10.1094/PHP-2004-0301-01-RV. Available at: http://www.plantmanagementnetwork.org/pub/php/review/ 2004/rootworm/

Rodrigo-Simón, A., de Maagd, R.A., Avilla, C., Bakker, P.L., Molthoff, J., Gonzalez-Zamora, J.E. and Ferré, J. (2006) Lack of detrimental effects of *Bacillus thuringiensis* Cry toxins on the insect predator *Chrysoperla carnea*: a toxicological, histopathological, and biochemical approach. *Applied and Environmental Microbiology* 72, 1595–1603.

Romeis, J., Babendreier, D. and Wäckers, F.L. (2003) Consumption of snowdrop lectin (*Galanthus nivalis* agglutinin) causes direct effects on adult parasitic wasps. *Oecologia* 134, 528–536.

Romeis, J., Dutton, A. and Bigler, F. (2004) *Bacillus thuringiensis* toxin (Cry1Ab) has no direct effect on larvae of the green lacewing *Chrysoperla carnea* (Stephens) (Neuroptera: Chrysopidae). *Journal of Insect Physiology* 50, 175–183.

Romeis, J., Meissle, M. and Bigler, F. (2006) Transgenic crops expressing *Bacillus thuringiensis* toxins and biological control. *Nature Biotechnology* 24, 63–71.

Romeis, J., Van Driesche, R.G., Barratt, B.I.P. and Bigler, F. (2008a) Insect-resistant transgenic crops and biological control. In: Romeis, J., Shelton, A.M. and Kennedy, G.G. (eds) *Integration of Insect-Resistant Genetically Modified Crops within IPM Programs*. Springer, Dordrecht, The Netherlands, pp. 87–117.

Romeis, J., Shelton, A.M. and Kennedy, G.G. (eds) (2008b) *Integration of Insect-Resistant Genetically Modified Crops within IPM Programs*. Springer, Dordrecht, The Netherlands.

Romeis, J., Bartsch, D., Bigler, F., Candolfi, M.P., Gielkens, M.M.C., Hartley, S.E., Hellmich, R.L., Huesing, J.E., Jepson, P.C., Layton, R., Quemada, H., Raybould, A., Rose, R.I., Schiemann, J., Sears, M.K., Shelton, A.M., Sweet, J., Vaituzis, Z. and Wolt, J.D. (2008c) Assessment of risk of insect-resistant transgenic crops to nontarget arthropods. *Nature Biotechnology* 26, 203–208.

Rose, R.I. (ed.) (2007) White paper on tier-based testing for the effects of proteinaceous insecticidal plant-incorporated protectants on non-target invertebrates for regulatory risk assessment. USDA-APHIS and US Environmental Protection Agency, Washington, DC. Available at: http://www. epa.gov/pesticides/biopesticides/pips/ non-target-arthropods.pdf

Rose, R. and Dively, G.P. (2007) Effects of insecticide-treated and lepidopteran-active Bt transgenic sweet corn on the abundance and diversity of arthropods. *Environmental Entomology* 36, 1254–1268.

Rosi-Marshall, E.J., Tank, J.L., Royer, T.V., Whiles, M.R., Evans-White, M., Chambers, C., Griffiths, N.A., Pokelsek, J. and Stephen, M.L. (2007) Toxins in transgenic crop byproducts may affect headwater stream ecosystems. *Proceedings of the National Academy of Sciences of the USA* 104, 16204–16208.

Ryan, C.A. (1990) Protease inhibitors in plants: genes for improving defenses against insects and pathogens. *Annual Review of Phytopathology* 28, 425–449.

Sadeghi, A., Broeders, S., De Greve, H., Hernalsteens, J.-P., Peumans, W.J., Van Damme, E.J.M. and Smagghe, G. (2007) Expression of garlic leaf lectin under the control of the phloem-specific promoter *Asus*1 from *Arabidopsis thaliana* protects tobacco plants against the tobacco aphid (*Myzus nicotianae*). *Pest Management Science* 63, 1215–1223.

Sadeghi, A., Smagghe, G., Broeders, S., Hernalsteens, J.-P., De Greve, H., Peumans, W.J. and Van Damme, E.J.M. (2008) Ectopically expressed leaf and bulb lectins from garlic (*Allium sativum* L.) protect transgenic tobacco plants against cotton leafworm (*Spodoptera littoralis*). *Transgenic Research* 17, 9–18.

Saha, P., Majumder, P., Dutta, I., Ray, T., Roy, S.C. and Das, S. (2006) Transgenic rice expressing *Allium sativum* leaf lectin with enhanced resistance against sap-sucking insect pests. *Planta* 223, 1329–1343.

Sanders, C.J., Pell, J.K., Poppy, G.M., Raybould, A., Garcia-Alonso, M. and Schuler, T.H. (2007) Host-plant mediated effects of transgenic maize on the insect parasitoid *Campoletis sonorensis* (Hymenoptera: Ichneumonidae). *Biological Control* 40, 362–369.

Sanvido, O., Romeis, J. and Bigler, F. (2007) Ecological impacts of genetically modified crops: ten years of field research and commercial cultivation. *Advances in Biochemical Engineering and Biotechnology* 107, 235–278.

Schnepf, E., Crickmore, N., van Rie, J., Lereclus, D., Baum, J., Feitelson, J., Zeigler, D.R. and Dean, D.H. (1998) *Bacillus thuringiensis* and its pesticidal crystal proteins. *Microbiology and Molecular Biology Reviews* 62, 775–806.

Schuler, T.H., Potting, R.P.J., Denholm, I. and Poppy, G.M. (1999) Parasitoid behaviour and Bt plants. *Nature* 400, 825–826.

Schuler, T.H., Potting, R.P.J., Denholm, I., Clark, S.J., Clark, A.J., Stewart, C.N. and Poppy, G.M. (2003) Tritrophic choice experiments with Bt plants, the diamondback moth (*Plutella xylostella*) and the parasitoid *Cotesia plutellae*. *Transgenic Research* 12, 351–361.

Sears, M.K., Hellmich, R.L., Stanley-Horn, D.E., Oberhauser, K.S., Pleasants, J.M., Mattila, H.R., Siegfried, B.D. and Dively, G.P. (2001) Impact of Bt corn pollen on monarch butterfly populations: a risk assessment. *Procceedings of the National Academy of Science of the USA* 98, 11937–11942.

Sétamou, M., Bernal, J.S., Legaspi, J.C. and Mirkov, T.E. (2002) Parasitism and location of sugarcane borer (Lepidoptera: Pyralidae) by *Cotesia flavipes* (Hymenoptera: Braconidae) on transgenic and conventional sugarcane. *Environmental Entomology* 31, 1219–1225.

Shi, Y., Wang, M.B., Powell, K.S., Van Damme, E., Hilder, V.A., Gatehouse, A.M.R., Boulter, D. and Gatehouse, J.A. (1994)

Use of the rice sucrose synthase-1 promotor to direct phloem-specific expression of β-glucuronidase and snowdrop lectin genes in transgenic tobacco plants. *Journal of Experimental Botany* 45, 623–631.

Shirai, Y. and Takahashi, M. (2005) Effects of transgenic Bt corn pollen on a non-target lycaenid butterfly, *Pseudozizeeria maha*. *Applied Entomology and Zoology* 40, 151–159.

Sisterson, M.S., Biggs, R.W., Manhardt, N.M., Carrière, Y., Dennehy, T.J. and Tabashnik, B.E. (2007) Effects of transgenic Bt cottonon insecticide use and abundance of two generalist predators. *Entomologia Experimentalis et Applicata* 124, 305–311.

Srivastava, P.N. and Auclair, J.L. (1963) Characteristics and nature of proteases from the alimentary canal of the pea aphid, *Acyrthosiphon pisum* (Harr) (Homoptera, Aphididae). *Journal of Insect Physiology* 9, 469–474.

Stanley-Horn, D.E., Dively, G.P., Hellmich, R.L., Sears, M.K., Rose, R., Jesse, L.C.H., Losey, J.E., Obrycki, J.J. and Lewis, L.C. (2001) Assessing the impact of Cry1Ab-expressing corn pollen on monarch butterfly larvae in field studies. *Proceedings of the National Academy of Sciences of the USA* 98, 11931–11936.

Stewart, S.D., Adamczyk, J.R., Knighten, K.S. and Davis, F.M. (2001) Impact of Bt cottons expressing one or two insecticidal proteins of *Bacillus thuringiensis* Berliner on growth and survival of Noctuid (Lepidoptera) larvae. *Journal of Economic Entomology* 94, 752–760.

Storer, N.P., Dively, G.P. and Herman, R.A. (2008) Landscape effects of insect-resistant genetically modified crops. In: Romeis, J., Shelton, A.M. and Kennedy, G.G. (eds) *Integration of Insect-Resistant Genetically Modified Crops within IPM Programs*. Springer, Dordrecht, The Netherlands, pp. 273–302.

Symondson, W.O.C., Sunderland, K.D. and Greenstone, M.H. (2002) Can generalist predators be effective biocontrol agents? *Annual Review of Entomology* 47, 561–594.

Terra, W.R. and Cristofoletti, P.T. (1996) Midgut proteinases in three divergent species of Coleoptera. *Comparative Biochemistry and Physiology B* 113, 725–730.

Torres, J.B. and Ruberson, J.R. (2005) Canopy- and ground-dwelling predatory arthropods in commercial Bt and non-Bt cotton fields: patterns and mechanisms. *Environmental Entomology* 34, 1242–1256.

Torres, J.B. and Ruberson, J.R. (2006) Spatial and temporal dynamics of oviposition behavior of bollworm and three of its predators in Bt and non-Bt cotton fields. *Entomologia Experimentalis et Applicata* 120, 11–22.

Torres, J.B. and Ruberson, J.R. (2008) Interactions of *Bacillus thuringiensis* Cry1Ac toxin in genetically engineered cotton with predatory heteropterans. *Transgenic Research* 17, 345–354.

Torres, J.B., Ruberson, J.R. and Adang, M.J. (2006) Expression of *Bacillus thuringiensis* Cry1Ac protein in cotton plants, acquisition by pests and predators: a tritrophic analysis. *Agricultural and Forest Entomology* 8, 191–202.

Tschenn, J., Losey, J.E., Jesse, L.H., Obrycki, J.J. and Hufbauer, R. (2001) Effects of corn plants and corn pollen on monarch butterfly (Lepidoptera: Danaidae) oviposition behavior. *Environmental Entomology* 30, 495–500.

Turlings, T.C.J., Jeanbourquin, P.M., Held, M. and Degen, T. (2005) Evaluating the induced-odour emission of a Bt maize and its attractiveness to parasitic wasps. *Transgenic Research* 14, 807–816.

United States Environmental Protection Agency (US EPA) (2001) Biopesticide registration action document. *Bacillus thuringiensis* (*Bt*) plant-incorporated protectants. Available at: http://www.epa.gov/oppbppd1/biopesticides/pips/bt_brad.htm

Van den Berg, J. and Van Wyk, A. (2007) The effect of Bt maize on *Sesamia calamistis* in South Africa. *Entomologia Experimentalis et Applicata* 122, 45–51.

Vet, L.E.M. and Dicke, M. (1992) Ecology of infochemical use by natural enemies in a tritrophic context. *Annual Review of Entomology* 37, 141–172.

Vojtech, E., Meissle, M. and Poppy, G.M. (2005) Effects of Bt maize on the herbivore

Spodoptera littoralis (Lepidoptera: Noctuidae) and the parasitoid *Cotesia marginiventris* (Hymenoptera: Bracondiae). *Transgenic Research* 14, 133–144.

Wäckers, F.L. (2005) Suitability of (extra-) floral nectar, pollen, and honeydew as insect food sources. In: Wäckers, F.L., van Rijn, P.C.J. and Bruin, J. (eds) *Plant-Provided Food for Carnivorous Insects: a Protective Mutualism and its Applications.* Cambridge University Press, Cambridge, pp. 17–74.

Walker, A.J., Ford, L., Majerus, M.E.N., Geoghegan, I.E., Birch, N., Gatehouse, J.A. and Gatehouse, A.M.R. (1998) Characterisation of the mid-gut digestive proteinase activity of the two-spot ladybird (*Adalia bipunctata* L.) and its sensitivity to proteinase inhibitors. *Insect Biochemistry and Molecular Biology* 28, 173–180.

Ward, D.P., DeGooyer, T.A., Vaughn, T.T., Head, G.P., McKee, M.J., Astwood, J.D. and Pershing J.C. (2005) Genetically enhanced maize as a potential management option for corn rootworm: YieldGard rootworm maize case study. In: Vidal, S., Kuhlmann, U. and Edwards, C.R. (eds) *Western Corn Rootworm: Ecology and Management.* CAB International, Wallingford, UK, pp. 239–262.

Warren, G.W. (1997) Vegetative insecticidal proteins: novel proteins for control of corn pests. In: Carozzi, N. and Koziel, M. (eds) *Advances in Insect Control: The Role of Transgenic Plants.* Taylor & Francis, London, pp. 109–121.

Whitehouse, M.E.A., Wilson, L.J. and Fitt, G.P. (2005) A comparison of arthropod communities in transgenic *Bt* and conventional cotton in Australia. *Environmental Entomology* 34, 1224–1241.

Whitehouse, M.E.A., Wilson, L.J. and Constable, G.A. (2007) Target and non-target effects on the invertebrate community of Vip cotton, a new insecticidal transgenic. *Australian Journal of Agricultural Research* 58, 273–285.

Wink, M. and Römer, P. (1986) Acquired toxicity – the advantages of specializing on alkaloid-rich lupins to *Macrosiphon albifrons* (Aphidae). *Naturwissenschaften* 73, 210–212.

Wolfenbarger, L.L., Naranjo, S.E., Lundgren, J.G., Bitzer, R.J. and Watrud, L.S. (2008) Bt crop effects on functional guilds of non-target arthropods: a meta-analysis. *PLoS ONE,* 3(5), 22118.

Wolt, J.D. and Peterson, K.D. (2000) Agricultural biotechnology and societal decision-making: the role of risk analysis. *AgBioForum* 3(1), 39–46.

Wolt, J.D., Peterson, R.K.D., Bystrak, P. and Meade, T. (2003) A screening level approach for nontarget insect risk assessment: transgenic Bt corn pollen and the monarch butterfly (Lepidoptera: Danaidae). *Environmental Entomology* 32, 237–246.

Wraight, C.L., Zangerl, A.R., Carroll, M.J. and Berenbaum, M.R. (2000) Absence of toxicity of *Bacillus thuringiensis* pollen to black swallowtails under field conditions. *Proceedings of the National Academy of Sciences of the USA* 97, 7700–7703.

Wu, K.M. and Guo, Y.Y. (2003) Influences of *Bacillus thuringiensis* Berliner cotton planting on population dynamics of the cotton aphid, *Aphis gossypii* Glover, in northern China. *Environmental Entomology* 32, 312–318.

Wu, K.M. and Guo, Y.Y. (2005) The evolution of cotton pest management practices in China. *Annual Review of Entomology* 50, 31–52.

Wu, K., Li, W., Feng, H. and Guo, Y. (2002) Seasonal abundance of the mirids, *Lygus lucorum* and *Adelphocoris* spp. (Hemiptera: Miridae) on Bt cotton in northern China. *Crop Protection* 21, 997–1002.

Zangerl, A.R., McKenna, D., Wraight, C.L., Carroll, M., Ficarello, P., Warner, R. and Berenbaum, M.R. (2001) Effects of exposure to event 176 *Bacillus thuringiensis* corn pollen on monarch and black swallowtail caterpillars under field conditions. *Proceedings of the National Academy of Sciences of the USA* 98, 11908–11912.

Zhang, G.-F., Wan, F.H., Lövei, G.L., Liu, W.-X. and Guo, J.-Y. (2006) Transmission of *Bt* toxin to the predator *Propylaea japonica* (Coleoptera: Coccinellidae) through its aphid prey feeding on transgenic *Bt* cotton. *Environmental Entomology* 35, 143–150.

9 Impact of Genetically Modified Crops on Pollinators

L.A. Malone and E.P.J. Burgess

Horticulture and Food Research Institute of New Zealand Ltd, Mt Albert Research Centre, Auckland, New Zealand

Keywords: Honeybees, *Apis mellifera,* bumblebees, *Bombus* spp., osmia bees, insect-resistant transgenic plants

Summary

Evaluating the potential for genetically modified (GM) crops to have a negative impact on pollinating insects has long been recognized as an important part of risk assessment of such plants. Extensive field experience with commercial GM crops, bred for herbicide tolerance or insect resistance using *Bacillus thuringiensis* (*Bt*) crystal (Cry) toxin genes, has shown no deleterious effects on pollinators. Many other insecticidal GM plants, not yet commercialized, have been studied to see if they could be hazardous to bees. Of these, only some of the protease inhibitors and lectins could present dose-dependent hazards to bees if there were realistic routes for sufficiently high exposure. A good understanding of crop pollination biology is essential for adequately assessing risks of GM plants to pollinators. In addition, information on crop pollination will aid the study of transgene flow since pollinators could play a role in the transfer of transgenes from GM plants to non-GM crops or wild relatives.

Introduction

Many of the world's crops depend on insects for pollination and it is critically important that agricultural biotechnology does not disrupt this essential 'eco-system service'. Pollinating insects may be 'managed' species, such as honeybees kept by beekeepers and other bee species for which mass rearing techniques and nest boxes are available, or they may be wild species of bees, flies, beetles or other insects living in the crop or in adjacent habitats.

At present, pollination as a service to agriculture faces two serious threats: the global spread of honeybee diseases and a loss of wild pollinators due to habitat destruction (Cane and Tepedino, 2001; Kevan and Phillips, 2001; Staffan-Dewenter *et al.*, 2005; Biesmeijer *et al.*, 2006). The seriousness of this worldwide decline in pollinators has been acknowledged by the establishment of the International Initiative for the Conservation and Sustainable Use of Pollinators under the Cartagena Convention on Biological Diversity (CBD, 2005).

Honeybees (*Apis mellifera* L.) are the best-known domesticated pollinators of crops, and for many years this species has been a standard test organism for assessing the non-target impacts of chemical pesticides for agricultural use. In most countries, environmental protection regulations governing the use of genetically modified (GM) plants (particularly those intended to be pesticidal) require laboratory toxicity tests with honeybees prior to field release (e.g. the US Federal Insecticide Fungicide and Rodenticide Act (FIFRA; USEPA, undated)). The toxicity and potential sub-lethal impacts of several types of GM plants and the proteins they express on honeybees and other bees have therefore been investigated and reported in the scientific literature. These are summarized in the sections below. Apart from honeybees, there are a range of pollinator species that are kept commercially for crop pollination including bumblebees, osmia bees, alkali bees, leafcutter bees, carpenter bees, stingless bees and blue bottle flies (e.g. Green, 2004). Of these 'alternative pollinators', only bumblebees and osmia bees have been studied so far in relation to responses to GM plants (Tables 9.1 and 9.2). Studies with non-managed pollinator species have not yet been performed.

Currently, four high-acreage arable plant species predominate among commercial GM crops: soybean (58.6 million ha), cotton (15 million ha), oilseed rape (canola; 5.5 million ha) and maize (maize; 35.2 million ha) (James, 2007). These have been modified to be either insect-resistant (expressing *Bacillus thuringiensis* (*Bt*) insecticidal proteins or, in the case of cotton, *Bt*-CpTI; see Chapter 16, this volume) and/or herbicide-tolerant (expressing PAT or BAR proteins; Herouet *et al.*, 2005). Honeybees may visit each of these plant species, and of these, oilseed rape, cotton and soybean are important nectar sources for honey production in some countries (Crane and Walker, 1986; Free, 1993). However, many more plant species have been genetically modified with a much wider variety of traits for pre-commercial trials. Many of these plant species require insect pollination and are more attractive to domesticated and wild bees than the four currently planted commercially, e.g. sunflower, apple, clover. Use of transgenes encoding other proteins, some of which will have broader insecticidal activities, such as protease inhibitors (PIs), lectins, bacterial toxins from nematode-infesting bacteria, spider venom and biotin-binding proteins, are also under development and may feature in future GM crops. Given their mechanisms of action against insects, these have a greater potential than *Bt* for effects on pollinators, if expressed in a way that allows sufficient exposure. The potential impacts of some of these GM plants on honeybees and other pollinators have been investigated and these results are presented below. In addition, the potential roles of pollinators as agents of transgene flow are discussed.

More recently, the experimental production of 'metabolically' engineered plants with altered levels and types of secondary compounds such as lignins, lipids and anthocyanins potentially poses more complex challenges to pollinating insects. Non-target biosafety testing of such plants may require a different

Table 9.1. Effects of *Bt* proteins and GM plants on bees.[a]

Purified protein or GM plant	Type of experiment	Results
Cry1Ab, Cry1Ac, Cry9C	Larval survival	Not toxic (Sims, 1995; USEPA, 2001)
Cry1Ab	Development of hypopharyngeal glands	No effect (Babendreier *et al.*, 2005)
Cry1Ab	Intestinal bacterial communities in adults	No effect (Babendreier *et al.*, 2007)
Cry1Ba, Cry1Ac	Adult survival (in laboratory and in colony)	Not toxic (Sims, 1995; Malone *et al.*, 1999)
Cry1Ba	Adult food consumption, development of hypopharyngeal glands	No effects (Malone *et al.*, 1999, 2004)
Cry1Ba	Adult flight activity (protein fed to colony)	No effect (Malone *et al.*, 2001)
Cry1F	Larval behaviour and survival	No effects (USEPA, 2001)
Cry2Ab2	Larval and adult survival	No effects (USEPA, 2002)
Cry3B	Larval survival, pupal weight (protein fed to colony)	Not toxic (Arpaia, 1996)
Cry3A	Larval survival	No effects (USEPA, 2001)
VIP3A	Sensitivity (life stage and measurements not stated)	Not sensitive (OGTR, 2005)
VIP maize pollen (VIP3A)	Larval survival	No effect (USEPA, 2005a)
VIP maize (VIP3A)	Brood production, food stores, new bee recruitment (semi-field study with colonies)	No effects (USEPA, 2005a)
Bt maize (Cry1Ab)	Larval development, adult survival, foraging frequency (in field)	No effects (Schur *et al.*, 2000; USEPA, 2001)
Bt sweetcorn (Cry1Ab)	Adult weight, foraging activity, pollen cake consumption, amount of capped brood, pollen and honey stores, number of bees per hive (in field)	No effects (Rose *et al.*, 2007)
Bt oilseed rape (Cry1Ac)	Foraging activity on potted plants in cage: frequency and duration of flower visits, frequency of movements among flowers	No effects, even though transgenic flowers produced less nectar with lower sugar content than isogenic flowers (Tesoriero *et al.*, 2004)
Bt maize pollen (Cry1Ab, Cry1F)	Larval and pupal survival, pupal weight, haemolymph protein concentration in new adults	No effects (Hanley *et al.*, 2003)
Bt maize pollen (Cry1Ab)	Development of hypopharyngeal glands	No effect (Babendreier *et al.*, 2005)
Bt maize pollen (Cry1Ab)	Intestinal bacterial communities in adults	No effect (Babendreier *et al.*, 2007)

Continued

Table 9.1. Continued.

Purified protein or GM plant	Type of experiment	Results
Bt sweetcorn pollen (Cry1Ab)	Adult weight and survival (in laboratory)	No effects (Rose *et al.*, 2007)
Bt cotton pollen (Cry1Ac)	Adult survival, superoxide dismutases activity	No effects (Liu *et al.*, 2005)
Cry1Ac	Bumblebee (*Bombus occidentalis* and *Bombus impatiens*) colonies: pollen consumption, worker weights, colony size, amount of brood, numbers of adult offspring; foraging ability (*B. impatiens* only)	No effects (Morandin and Winston, 2003)
Cry1Ab	Bumblebees (*Bombus terrestris*): time spent foraging on artificial flowers	No effect (Babendreier *et al.*, 2008)
Cry 1Ab	Bumblebee (*B. terrestris*) microcolonies: survival of workers and drone offspring	No effects (Babendreier *et al.*, 2008)
Bt maize pollen (Cry1Ab)	Bumblebee (*B. terrestris*) microcolonies: worker survival, pollen and sugar syrup consumption, numbers of offspring, development of offspring	No effects (Malone *et al.*, 2007)
Bt oilseed rape (Cry1Ac, Green Fluorescent Protein (GFP) marker)	Bumblebee (*B. terrestris*) colonies in glasshouse: foraging behaviour	Variable results: adults visited more flowers on control plants one year, but no difference next year (Arpaia *et al.*, 2004)
Bt oilseed rape (Cry1Ac, GFP marker)	Field abundance and foraging behaviour of various bees	Variable results: no effects on flower-visiting behaviours; marginally more visits to *Bt* plants; more of the least common bee species on the control plants, otherwise no differences (Arpaia *et al.*, 2004)
Cry1Ab	Mason bee (*Osmia rufa*) larval development	No effect (Konrad and Babendreier, 2006)

[a]All bees tested are *Apis mellifera* unless otherwise stated.

methodology from the toxicity-testing approach that has been used with GM plants and bees thus far. For example, behavioural responses to the plants may be altered, and there is also the potential for positive effects on pollinators, e.g. better nutrition, with some of these plants. As yet, there are no published studies of interactions between metabolically engineered plants and pollinating insects.

Table 9.2. Effects of non-*Bt* insecticidal proteins and GM plants on bees.[a]

Purified protein or GM plant	Type of experiment	Results
Serine protease inhibitors		
Aprotinin[b], BBI[b], CpTI[b], POT-1[b], POT-2[b], SBTI[b]	Adult survival (in laboratory and in colony)	High concentrations reduce survival by a few days; low concentrations have no effect (Belzunces *et al.*, 1994; Malone *et al.*, 1995, 1998; Burgess *et al.*, 1996; Sandoz, 1996; Picard-Nizou *et al.*, 1997; Girard *et al.*, 1998; Pham-Delegue *et al.*, 2000; Sagili *et al.*, 2005)
Aprotinin, POT-1, POT-2, SBTI	Adult digestive proteases	Inhibition of some proteases (Malone *et al.*, 1995, 1998; Burgess *et al.*, 1996; Sagili *et al.*, 2005)
Aprotinin	Development of hypopharyngeal glands	No effect (Malone *et al.*, 2004)
Aprotinin	Adult flight activity (protein fed to colony)	Flight activity begins a few days earlier (when fed a high concentration; Malone *et al.*, 2001)
BBI	Artificial flower visits by adults from colonies (choice tests)	No effect of 100 µg ml^{-1} in sucrose (Dechaume-Moncharmont *et al.*, 2005)
BBI, CpTI, SBTI	Olfactory learning response	One inhibitor offered in sugar reward reduced ability to learn; others did not (Picard-Nizou *et al.*, 1997; Girard *et al.*, 1998; Jouanin *et al.*, 1998; Pham-Delegue *et al.*, 2000)
SBTI	Development of hypopharyngeal glands	Glands smaller than controls after 10 days (Babendreier *et al.*, 2005) High concentrations reduce gland protein content, lower concentrations do not (Sagili *et al.*, 2005)
SBTI	Larval survival	High concentrations reduce survival (Brodsgaard *et al.*, 2003)
SBTI	Intestinal bacterial communities in adults	High concentration (sufficient to cause mortality) altered bacterial communities (Babendreier *et al.*, 2007)
Aprotinin, POT-1, POT-2, SBTI	Bumblebee (*Bombus terrestris*) worker survival, digestive proteases, pollen and sugar syrup consumption	Aprotinin: no effects Other PIs: high concentrations reduce survival; low concentrations have no effect; inhibition of some proteases (Malone *et al.*, 2000)
SBTI	Bumblebee (*B. terrestris*) microcolonies: worker survival, pollen and sugar syrup consumption, numbers of offspring, development of offspring	High concentration reduces survival of workers and numbers of adult offspring; low concentration has no effect (Babendreier *et al.*, 2004, 2008)

Continued

Table 9.2. Continued.

Purified protein or GM plant	Type of experiment	Results
SBTI	*B. terrestris* time spent foraging on artificial flowers	No effect (Babendreier *et al.*, 2008)
Cysteine protease inhibitors		
OCI[c]	Olfactory learning response	No effect (Girard *et al.*, 1998; Jouanin *et al.*, 1998; Pham-Delegue *et al.*, 2000)
OCI, chicken egg white cystatin	Adult survival	No effect (Sandoz, 1996; Girard *et al.*, 1998)
OCI-expressing oilseed rape	Foraging behaviour	No effect (Grallien *et al.*, 1995)
OCI, OCI-expressing oilseed rape	Mason bee (*Osmia rufa*) larval development	No effect (Konrad and Babendreier, 2006)
Other novel proteins		
Chitinase	Adult survival	No effect (Picard-Nizou *et al.*, 1997)
Chitinase	Olfactory learning response	No effect (Picard-Nizou *et al.*, 1997)
Chitinase	Foraging behaviour (sugar feeder with chitinase added)	No effect (Picard *et al.*, 1991)
Chitinase	Bumblebee (*Bombus occidentalis* and *Bombus impatiens*) colonies: pollen consumption, worker weights, colony size, amount of brood, numbers of adult offspring Foraging ability (*B. impatiens* only)	No effect (Morandin and Winston, 2003)
Chitinase-expressing oilseed rape	Foraging behaviour	No effect (Picard-Nizou *et al.*, 1995)
β-1,3 glucanase	Adult survival	No effect (Picard-Nizou *et al.*, 1997)
β-1,3 glucanase	Olfactory learning response	No effect (Picard-Nizou *et al.*, 1997)
β-1,3 glucanase	Foraging behaviour (sugar feeder with β-1,3 glucanase added)	No effect (Picard *et al.*, 1991)
Biotin-binding protein (avidin)	Adult survival; adult food consumption; larval development and survival	No effect (Malone et al., 2002b)
Biotin-binding protein (avidin)	Development of hypopharyngeal glands	No effect (Malone *et al.*, 2004)
GNA[d]	Bumblebee (*B. terrestris*) microcolonies: worker survival, pollen and sugar syrup consumption, numbers of offspring, development of offspring	High concentration reduces food consumption, survival of workers and their adult offspring; low concentration has no effect (Babendreier *et al.*, 2004, 2008)

Continued

Table 9.2. Continued.

Purified protein or GM plant	Type of experiment	Results
GNA	*B. terrestris* time spent foraging on artificial flowers	No effect (Babendreier *et al.*, 2008)
GNA	Mason bee (*O. rufa*) larval development	Lower conversion of food mass into larval mass; otherwise no effects (Konrad and Babendreier, 2006)

[a]All bees tested are *Apis mellifera* unless otherwise stated.
[b]Aprotinin, also known as BPTI (bovine pancreatic trypsin inhibitor) or BSTI (bovine spleen trypsin inhibitor); BBI, Bowman-Birk trypsin inhibitor; CpTI, cowpea trypsin inhibitor; POT-1, potato proteinase inhibitor 1; POT-2, potato proteinase inhibitor 2; SBTI, Kunitz soybean trypsin inhibitor.
[c]OCI, oryzacystatin.
[d]GNA, *Galanthus nivalis* agglutinin or snowdrop lectin.

In addition to the impacts of GM plants on pollinator species, the ability of these species to collect and transport pollen between GM and non-GM plants needs to be examined. Recombinant proteins are usually expressed at low levels in the pollen of transgenic plants due to the use of specific 'constitutive' promoters in the gene constructs (see below). However, the transgenes themselves are present in DNA contained in the pollen, and therefore may be transferred to other, related plants via the activities of the pollinator species. The potential role of pollinators as agents of transgene flow is discussed below.

Toxicity Risk

GM plants can impact pollinators in two ways: 'directly', via the GM plant itself posing a hazard to the pollinator; or 'indirectly', via the use of a GM crop affecting other ecological requirements of the pollinator such as availability of weeds among a crop for forage, or the inherent 'attractiveness' of flowers from particular transgenic plants.

Two conditions must be fulfilled for GM plants to pose a direct toxicity risk to pollinating insects. First, the plant must present some kind of hazard to the insect. For example, the GM plant may produce a substance which could kill or alter the physiology, development or behaviour of the individual insects in such a way and at sufficient scale to affect a pollinator detrimentally. Second, there must be a realistic route (or routes) by which the insect could become exposed to this hazard.

Potential Hazards for Pollinators from GM Plants

Herbicide-tolerant GM plants

Weed management practices, altered by the use of GM herbicide-tolerant (HT) plants, may have an indirect affect on pollinators since many utilize weeds as forage. However, there is no evidence that the GM plants themselves represent a direct, toxicological hazard to bees or other insects. GM HT plant varieties, such as soybean and oilseed rape, contain the *pat* or *bar* transgenes originally isolated from *Streptomyces* bacteria. Direct effects of the encoded proteins on pollinators have not been reported. For registration of HT oilseed rape, Canada's Plant Biosafety Office required an assurance that honeybees would not be harmed and tests showed that the PAT protein was not detected in pollen grains or in honey produced by bees foraging on a test plot of these plants and that bee foraging and brood development were unaffected by the GM plants (CFIA, 1995). Similar results have been reported from small-scale field trials with honeybee colonies (Chaline *et al.*, 1999; Huang *et al.*, 2004). There have been no directly toxic effects of commercial HT GM crops on pollinating insects reported in the 10 years since such plants were first commercialized.

GM plants expressing *Bt*

Compared with HT GM crops, *Bt*-expressing GM plants have received greater attention in relation to possible impacts on pollinators, since the purpose of *Bt* GM plants is to control insect pests. Many different crystalline (Cry) insecticidal proteins (also known as delta endotoxins) have been isolated from various strains of *B. thuringiensis* and found to be specifically insecticidal to particular groups of insects (Glare and O'Callaghan, 2000). Even within an insect order, different species may have different responses to the same Cry protein. To date, most *Bt* GM plants have been developed to control lepidopteran pests and they express proteins from the Cry1, Cry2 or Cry9 groups, most of which are known to be quite specific to Lepidoptera. More recently, GM plants expressing Cry3 proteins have been produced specifically to control coleopteran pests, as has a hybrid gene incorporating domains from Cry1Ia and Cry1Ba (Naimov *et al.*, 2003).

A selection of purified Cry1, Cry2, Cry9 and Cry3 *Bt* proteins and pollen from *Bt*-expressing plants have been tested for effects on honeybees and bumblebees (Table 9.1). These experiments have measured not only direct toxicity to adult and larval bees but also some sub-lethal effects and behavioural characteristics of the bees. Measurements include food consumption rates (to check for repellency), pupal weights, adult weights, development of hypopharyngeal glands in adult bees (these secrete food for larval bees), haemolymph protein concentrations in new adults, intestinal bacterial communities in adults, brood numbers, pollen and honey stores in colonies, flight and foraging activity. There have been no adverse effects on bees noted in any of these studies (see Table 9.1 for references), or in a meta-analysis of 25 separate studies of the effects of Cry toxins (lepidopteran- and coleopteran-active) on honeybees (Duan *et al.*, 2008). This is not surprising, as it is very likely that bees and perhaps other Hymenoptera lack the appropriate gut receptors for *Bt* proteins currently being used, or under development for use, in transgenic crop plants. The mechanism of action of Cry proteins is well understood and involves the proteins binding to specific receptors in the insect's gut after ingestion (Schnepf *et al.*, 1998) causing pore formation, and at high toxin concentrations, lysis of the midgut epithelium and rapid death (Glare and O'Callaghan, 2000). Without this binding there is no insecticidal action.

In a 2-year field study, variable results were obtained when flower-visiting insects were studied on *Bt* oilseed rape expressing Cry1Ac and a green fluorescent protein (GFP) marker versus control plants (Arpaia *et al.*, 2004). The honeybee was the most abundant species on both transgenic and control flowers. Analysis of the composition of pollinator guilds showed a higher presence of less common species on the control plants. However, there were no significant differences in pollinator behaviour on the flowers of *Bt* and non-*Bt* plants. Honeybee colonies placed in *Bt* sweetcorn fields, and also supplied with pollen cakes made from *Bt* sweetcorn pollen, did not differ from control colonies in non-transgenic maize fields, in terms of their brood development or foraging performance (Rose *et al.*, 2007).

Recently, a new class of insecticidal proteins (vegetative insecticidal proteins or VIPs) from *B. thuringiensis* has been identified and their encoding genes are being exploited in GM plants. Those being commercialized at present target lepidopteran pests, but have a different mode of action from the Cry proteins, thus providing a compatible alternative or adjunct which should delay the onset of pest resistance to the Cry proteins (OGTR, 2004). Honeybee toxicity tests have been conducted with at least one VIP (Table 9.1) and no negative effects were noted (USEPA, 2005a).

New insecticidal gene constructs which use *Bt* genes in combination with each other and with other genetic material are also being developed in order to delay resistance or to broaden efficacy against a variety of pest insects. For example, *cry1Ab* and *cry1Ac* genes have been fused to produce 'hybrid' *Bt* cotton (Yao *et al.*, 2006). Since each *Bt* toxin in this case is known to be safe for honeybees, combining them in this way is unlikely to present a new hazard to these and other non-lepidopteran pollinators. However, this may not be the case with other fusions. For example, maize plants transformed with a *cry1Ac* gene fused to the galactose-binding domain of the non-toxic ricin B-chain gene produced significant mortality in homopteran leafhoppers, extending the range of toxicity of this *Bt* toxin beyond the Lepidoptera (Mehlo *et al.*, 2005). Potential effects of such plants on hymenopteran and other pollinators should therefore be investigated.

Transgenic plants expressing other insecticidal proteins

Bt has proven to be a rich source of insecticidal genes for incorporation into GM plants, but there are concerns over the evolution of resistance to *Bt* toxins and there is still a need to control a wide array of herbivorous pest species, including secondary pests which have arisen with the use of *Bt* crops (Moar and Schwartz, 2003). For example, mirids and stinkbugs may become significant cotton pests when *Bt* cotton is used instead of broad-spectrum sprays to control lepidopteran pests (Wu *et al.*, 2002). As yet, no *Bt*-derived proteins have been found to control these sucking insects. For these reasons, considerable effort is being expended on finding alternatives to *Bt*.

PIs from a variety of plant and animal sources have been shown to protect GM plants from insect attack (Lawrence and Koundal, 2002). Their mode of action involves the inhibition of digestive proteases in the target insect's gut. Because bees must secrete proteases in order to digest pollen, which is rich in protein, the effects of PIs on these insects have been studied extensively (Table 9.2). Results with purified proteins and with PI-expressing plants indicate that serine PIs may affect bees in a dose-dependent fashion, whereas cysteine PIs apparently have no effect. Serine PIs may alter protease activities in the midguts of honeybees and bumblebees, may reduce the size of hypopharyngeal glands (which secrete a jelly for feeding to larvae) in honeybees and may reduce adult bee longevity by a few days, if administered at high-enough concentrations. The exception is aprotinin, a serine PI which affects honeybees but not bumblebees (Burgess *et al.*, 1996; Malone *et al.*, 2000). Thus, the safety of pollinators visiting GM plants expressing PIs will depend on the degree to which they are exposed to the PI itself (see below).

Since chitin is a compound found only in insects and fungi, chitinases are being investigated for their potential as a vertebrate-safe means to control pest insects and fungal diseases of plants via GM plant varieties (e.g. Hao *et al.*, 2005; McCafferty *et al.*, 2006). Purified chitinases fed to honeybees and bumblebees have had no effect on a range of measured characteristics, and chitinase-expressing GM plants have not affected honeybee foraging behaviour (Table 9.2). A glucanase, which is a candidate for controlling fungal diseases of plants, has also been shown to have no effect on honeybees (Picard *et al.*, 1991; Picard-Nizou *et al.*, 1997).

The biotin-binding proteins, avidin from chicken egg white and streptavidin from the bacterium *S. avidinii*, have been shown to be effective controllers of pest insects when expressed in GM plants or fed in purified form to a wide range of insect species (Morgan *et al.*, 1993; Kramer *et al.*, 2000; Burgess *et al.*, 2002; Malone *et al.*, 2002a; Markwick *et al.*, 2003; Yoza *et al.*, 2005). They are thought to operate by binding with dietary biotin and depriving the insect of this essential vitamin. Laboratory studies with purified avidin fed to honeybee adults and larvae at levels similar to, and higher than, those expected in GM avidin plants have shown no effects on these pollinators (Table 9.2). This is likely to be because bees receive high levels of biotin from their diets of pollen and hypopharyngeal gland secretions, whereas many pest insects have their biotin needs met only marginally by the foliage and plant parts that they eat.

Plant-derived lectins are another class of proteins that have been shown to have insecticidal impacts on a wide range of insects, including pests such as aphids that feed only on sap and are normally difficult to control (Legaspi *et al.*, 2004). The snowdrop lectin, GNA (*Galanthus nivalis* agglutinin), has been fed in purified form to bumblebees and mason bees where it has had dose-dependent negative impacts on food consumption, survival and development (Babendreier *et al.*, 2004, 2008; Konrad and Babendreier, 2006). As with the PIs, the actual risks to pollinators of lectin-expressing GM plants will depend very much on potential exposure levels.

A number of other novel insecticidal GM plants are under development, although there are no reports of their potential impacts on pollinators as yet. Examples include plants expressing insecticidal toxins from *Photorhabdus luminescens*, a bacterium which lives in the guts of entomophagous nematodes (Bowen *et al.*, 1998; Liu *et al.*, 2003), plants expressing spider venoms (Khan *et al.*, 2006) and those expressing plant alpha-amylase inhibitors (Franco *et al.*, 2002).

Metabolically engineered plants

'Metabolic engineering', or the ability to alter biosynthetic pathways in plants, offers many possibilities in terms of new crop traits for the future. For example, altering the type of lignin produced by trees could allow for more efficient and less-polluting processing of timber (Baucher *et al.*, 2003), raising the levels of anthocyanins in food crops could make them more appealing (red colours) and healthful (high antioxidant levels; Schijlen *et al.*, 2004), as could altering the types and levels of lipids or starches expressed in crop plants (e.g. Cahoon *et al.*, 2006; Kohno-Murase *et al.*, 2006; Murphy, 2006). Obviously, some of these alterations have the potential to have an impact on pollinators as well. Investigations with such plants and bees have not yet been published, but one may speculate, for example, that plants with high levels of anthocyanins may have altered flower colours and pollen with altered nutritional properties, and pollen from GM plants with different levels of lipids or starch may have altered nutritional value for bees. For such novel future crops, biosafety-testing regimes for pollinators may need to extend beyond the current oral toxicity tests presently required by regulators for *Bt* toxins.

Potential Exposure Routes

For a pollinator to be at risk from a GM plant there must be not only an identifiable potential hazard but also a realistic exposure route. There must be temporal and spatial overlap between the expression of the hazard by the plant and the occurrence of populations of the pollinating insect, and the individual insects must interact with the plant in such a way as to be exposed to the hazard.

By definition, pollinators will visit flowers of the crop and will make sufficient contact for the effective transfer of pollen. Most pollinators do not use crop plants for shelter, as oviposition sites or in other ways that may result in indirect exposure to a hazard. They simply visit the flowers and ingest pollen and/or nectar from them. Thus, pollinators will be exposed only to traits that are expressed in the flowers; obviously, a GM plant expressing an insecticidal toxin only in the leaves or roots will not affect a pollinating insect. In addition, if the plant has been modified in such a way that the flowers are altered (e.g. different petal colour, flower structure, nectar volume, nectar concentration, pollen production or pollen nutritive value), then there may also be effects on pollinators.

Transgene expression in pollen

Where a GM plant poses a risk to a pollinator via the ingestion of a potentially hazardous substance, information on the level of expression of the novel product in pollen is vital for assessing that risk. Pollen represents the most likely vehicle for bees to be exposed to such proteins, since it is a relatively rich source of plant proteins and novel proteins have been detected in the pollen of some GM plants (Table 9.3). Thus far, expression levels of various novel proteins intended for use in GM plants have been very low or even undetectable in pollen. Interestingly, 'constitutive' promoters such as cauliflower mosaic virus 35S, maize ubiquitin and actin from various plant species appear to be relatively poor promoters of expression in pollen tissue, compared with leaf tissue (Table 9.3).

Nectar is an unlikely site for the production of novel proteins, since it is a plant secretion, not a tissue, and has no cellular content. Most nectars contain no protein, being composed principally of sugars and sometimes free amino acids. The few exceptions have very low concentrations of protein (0.024% of total nectar in tobacco (Carter *et al.*, 1999) and 0.022% in leeks (Peumans *et al.*, 1997)). Examination of GM plant nectars have confirmed the low likelihood of novel proteins occurring in this secretion. Bowman-Birk soybean trypsin inhibitor (BBI) could not be detected in the nectar of GM canola plants containing the *BBI* gene (Jouanin *et al.*, 1998) and VIP3A could not be detected in GM cotton nectar (OGTR, 2005).

Adult honeybees consume significant quantities of pollen during their first week after emergence and so might be exposed to novel proteins from some GM plants. Bee larvae also ingest pollen, especially during the later instars, but their food is composed largely of glandular secretions from nurse adult bees. Recent studies (Malone *et al.*, 2002b; Babendreier *et al.*, 2004) have estimated the amounts of pollen that honeybee larvae consume in order to better assess the potential risks that GM plants may pose to this life stage. In contrast to honeybees, bumblebees consume relatively uniform amounts of pollen throughout adult life (Malone *et al.*, 2000), and the larvae are supplied directly with pollen by the adults. Solitary bees may also provision their larval-rearing cells with pollen, thus exposing both larvae and adults to any novel proteins expressed in pollen.

To date, all but one published study measuring the responses of pollinators to pollen from GM plants or whole flowering GM plants (*Bt*, OCI, chitinase and HT) have shown no significant effects on honeybees or bumblebees (Tables 9.1 and 9.2). The exception is a study with *Bt* oilseed rape plants, which gave variable results showing some differences between GM and non-GM plants in some years (Arpaia *et al.*, 2004). These GM plants expressed a GFP marker in addition to the *Bt* toxin, which may have affected the results.

There is no evidence that current commercial GM plants, expressing *Bt* toxins and/or PAT protein, have had negative impacts on pollinators in the field. Tests with purified proteins have shown that the proteins present no hazard to honeybees (Table 9.1) and published measurements of gene expression in the pollen of such plants (Table 9.3) suggest that pollinators will be exposed to only extremely low or negligible levels of these non-hazardous proteins.

Table 9.3. Expression of novel proteins in pollen of GM plants.

Plant	Novel protein encoded by transgene	Promoter	Concentration of novel protein in pollen (as stated in original studies)	Standardized concentration of novel protein in pollen (estimated percentage of total soluble protein)[a]	Reference
Maize	*Bt* toxin	Maize pollen-specific and PEP[b] (leaf-specific) promoters	260–418 ng mg⁻¹ (of total soluble protein)	0.026–0.0418	Koziel *et al.* (1993)
Maize	*Bt* toxin Cry1Ab	Maize pollen-specific and PEP promoters	1100–2400 ng g⁻¹ fresh weight	0.00044–0.0096	Fearing *et al.* (1997)
Maize (Event 176)	*Bt* toxin Cry1Ab	Maize pollen-specific and PEP promoters	<7.1 µg g⁻¹ of pollen	<0.00284	Stanley-Horn *et al.* (2001)
Maize	*Bt* toxin	CaMV 35S[c]	Nil	0	Koziel *et al.* (1993)
Maize (*Bt* 11)	*Bt* toxin Cry1Ab	CaMV 35S	<90 ng g⁻¹ dry weight	N/A	USEPA (2001)
Maize (MON 810)	*Bt* toxin Cry1Ab	CaMV 35S	<90 ng g⁻¹ dry weight	N/A	USEPA (2001)
Maize (Starlink)	*Bt* toxin Cry9C	CaMV 35S	0.24 µg g⁻¹ fresh weight	0.000096	USEPA (2000)
Maize (Event TC1507)	*Bt* toxin Cry1F	Maize polyubiquitin promoter	31–33 ng mg⁻¹ dry weight	N/A	USEPA (2005b)
Maize (Event MON 863)	*Bt* toxin Cry3Bb1	CaMV 35S	89.2 µg g⁻¹ fresh weight	0.036	Mattila *et al.* (2005)
Maize (Event MON810 × MON 84006)	*Bt* toxins Cry1Ab and Cry2Ab2	CaMV 35S	Cry1Ac: <0.08 µg g⁻¹ fresh weight; Cry12Ab2: 0.06–0.12 µg g⁻¹ fresh weight	<0.000032 0.000024–0.000048	Mattila *et al.* (2005)
Cotton	*Bt* toxin Cry1Ac	CaMV 35S	0.6 µg g⁻¹ fresh weight	0.00024	Greenplate (1997)
Cotton	*Bt* toxin Cry1Ac	CaMV 35S	11 ng g⁻¹ fresh weight	0.0000044	USEPA (2001)
Cotton	VIP3A(a)	*Arabidopsis thaliana* actin-2 promoter	1.1 µg g⁻¹ dry weight	N/A	OGTR (2005)
Rice	Fused *Bt* toxin Cry1Ac/Cry1b	Rice actin I promoter	7.24 µg g⁻¹	0.013	Yao *et al.* (2006)
Oilseed rape	Oryzacystatin I	CaMV 35S	Nil	0	Bonade Bottino *et al.* (1998)
Oilseed rape	Bowman–Birk trypsin inhibitor	CaMV 35S	Nil	0	Jouanin *et al.* (1998)

[a] Values expressed as a proportion of fresh pollen weight in the original reference have been converted using the assumption that fresh pollen is 25% protein.
[b] Phosphoenolpyruvate.
[c] Cauliflower mosaic virus 35S promoter.

Tests with bees and experimental GM oilseed rape plants transformed with the cysteine PI, OCI or a chitinase have similarly revealed no adverse impacts or hazard/exposure evidence that such impacts might occur (Tables 9.2 and 9.3). However, other traits such as serine PIs or lectins could pose a hazard to bees and the actual risks to pollinators from such plants will depend very much on the expression levels in pollen and the amounts of pollen ingested by various life stages of the relevant pollinating insects. Metabolic engineering, especially that intended to alter nutritional qualities of plants, could also affect pollen quality and this could have flow-on effects, both positive and negative, on pollen-dependent pollinating insects.

Floral phenotypic changes

Floral phenotypic changes in GM plants may pose a potential risk to pollinators, especially if the flowers are less attractive to them. This could occur with changes in petal colour, flower structure, flower volatiles, nectar volume, nectar concentration, pollen production or pollen nutritive value. If these changes are not intentional, but simply the result of insertional mutagenesis, they can be removed from the plants before release by line selection, as is presently carried out with conventionally bred plants. If the flower change is a pleiotropic consequence of the genetic modification, or is in fact intentional, for example a GM plant with anthocyanin-rich (and therefore redder) flowers, then further tests will be required to ascertain the potential impacts of such a change.

Role of Pollinators in Transgene Flow

Pollinating insects may play a role in the transfer of transgenes from GM plants to non-GM crops of the same species or to wild relatives of the crop (reviewed in Williams, 2001; Poppy and Wilkinson, 2005). Pollen movement and gene flow in GM plants is of particular interest because GM food-labelling requirements have necessitated the development of coexistence strategies for growing GM and non-GM crops. Furthermore, there has been concern that some HT GM crops, such as oilseed rape, may cross-pollinate with wild weedy relatives, thereby creating HT weeds which would be difficult to control.

Provided that the genetic modification has not altered the pollen's physical properties, as has been shown for the unchanged weight of HT GM oilseed rape pollen grains when compared with weights from unmodified plants (Pierre *et al.*, 2003), or its attractiveness or nutritive value to bees, then data gathered from studies of pollen movement among conventionally bred crop plants will be relevant when assessing the risk of transgene flow from GM crops. Approaches include plot-to-plot field studies of pollen movement at various scales (Ramsay *et al.*, 2003; Cresswell and Osborne, 2004; Hayter and Cresswell, 2006; Klein *et al.*, 2006) and mathematical modelling to simulate pollen flow under different conditions (e.g. wind direction and speed, plot size and shape, landscape features; Cresswell *et al.*, 2002; Meagher *et al.*, 2003; Walklate *et al.*, 2004; Cresswell and Hoyle, 2006).

Pollen dispersal patterns

Pollen dispersal curves are typically exponential, with the amount of pollen deposited declining asymptotically with increasing distance up to about 100 m from the source plot or plant, and this appears to hold true for both insect- and wind-pollinated crops (e.g. Zhang *et al.*, 2005; Cresswell and Hoyle, 2006; Funk *et al.*, 2006; Pla *et al.*, 2006). Very-long-distance dispersal patterns (over several kilometres) may be more random however, as recorded in a study of HT GM oilseed rape in Australia (Rieger *et al.*, 2002), and the involvement of insects has been invoked to explain this.

With insect-pollinated plants, pollination patterns are influenced by more than just plot size and shape, and wind direction and speed. Temporal and spatial overlap of the flowering plants with pollinator populations will affect cross-pollination. For example, nest site availability can be an important factor in determining numbers of flower visitors when wild pollinators predominate (e.g. Svensson *et al.*, 2000), although this can be manipulated artificially in the case of domesticated bee species. Seasonal variation in pollinator abundance has been shown to influence pollen dispersal in winter- and spring-sown oilseed rape (Hayter and Cresswell, 2006). The behaviour of individual pollinating insects is also important in determining pollination efficiency, and this has been shown to be influenced by the size, density and location of patches of flowering plants (e.g. Cresswell and Osborne, 2004). Insect-mediated pollination is additionally complicated by the ability of insects to transfer pollen to each other before depositing it on a flower (e.g. Hoyle and Cresswell, 2006).

Each of these dispersal mechanisms must be considered when determining the likelihood of transgene flow for each GM plant species. Models of pollen dispersal patterns may aid the prediction of likely gene transfer routes.

Insect foraging distances

Honeybees have a strong tendency to forage at the nearest source of flowering plants in an area (e.g. Osborne *et al.*, 2001). Thus, most honeybees in agricultural areas forage within a few hundred metres of their hives, although significant populations have been found 3.7 km away (Winston, 1987). 'Distant flight' behaviour is also observed in agricultural areas where attractive crops are planted in widely dispersed fields. In such circumstances, significant bee populations may be found at least 6.5 km from an apiary. In forested regions, they forage at a median radius of 1.7 km from the hive and most can be found within 6 km (Winston, 1987). Moyes and Dale (1999) recorded mean foraging distances of 1.66 km and 557 m for bees foraging on flowering carrots and onions, respectively, and maximum distances for these crops of 6.17 and 4.25 km, respectively. Ramsay *et al.* (2003) noted bees flying 5 km to reach an oilseed rape field and also evidence of oilseed rape cross-pollination over 26 km, although the agent responsible was not identified. Williams (2001) found that honeybees can be recruited to feeding stations up to 10 km from a hive if there are no competing food sources, and Gary (1992) recruited honeybees in a desert 13.7 km from their hive when there were no other food sources available. These studies reveal the ability of honeybees to act as agents for transgene flow, although the likely transfer distances are relatively low.

Bumblebee foraging distances have been investigated using marked bees (Walther-Hellwig and Frankl, 2000), homing experiments (Goulson and Stout, 2001), modelling (Cresswell et al., 2002), harmonic radar (Osborne et al., 1999) and microsatellite markers (Knight et al., 2005). Results obtained varied, but indicate that different bumblebee species have different foraging ranges. Knight et al. (2005) reported 'minimum estimated maximum foraging ranges' of 758 m for B. terrestris, 674 m for Bombus pratorum, 450 m for Bombus lapidarius and 449 m for Bombus pascuorum. Darvill et al. (2004) noted B. terrestris making foraging trips of up to 625 m and B. pascuorum up to 312 m, while Chapman et al. (2003) estimated 0.62–2.8 km for the former species and 0.51–2.3 km for the latter.

Like the social bees, solitary bees also tend to favour foraging near their nesting sites. Maximum foraging distances between 150 m and 1.2 km have been estimated for 17 different species of European solitary bees, with the larger species tending to undertake longer flights than the smaller species (Gathmann and Tscharntke, 2002). Sick et al. (2004) detected GM HT oilseed rape pollen in Osmia rufa nests up to 100 m from the source crop.

The potential for nectar-feeding moths to effect long-distance transgene flow has been investigated by Richards et al. (2005), who found that cotton pollen lost viability faster when carried on the probosci of Helicoverpa armigera adults than on control surfaces. They also found that there were no differences in this respect between pollen from non-GM and Bt-expressing cotton plants. Very few pollen grains were retained by these moths, which can travel hundreds of kilometres, and it was concluded that this posed little risk for movement of Bt-cotton transgenes, since a cotton stigma may need 100–600 pollen grains to set seed.

Insect pollination of GM crops

Of the four major commercialized GM crops, maize, soybeans, cotton and oilseed rape, oilseed rape has been by far the most-studied in relation to pollen dispersal by insects.

Maize is a wind-pollinated crop and although various bees may use maize pollen as a food source, they are not known to transfer pollen to the silks. Honeybees have been reported to forage on maize in Switzerland (Wille et al., 1985; Charriere et al., 2006), as have bumblebees in the USA (Gross and Carpenter, 1991), and various species of wild bees on maize and other grasses in Cameroon (Fohouo et al., 2002, 2004). Maize pollen is large, prone to desiccation and loses viability rapidly (within 1–4 h of dehiscence; Luna et al., 2001; Aylor, 2004). Insect-mediated cross-pollination of maize has not been reported.

Soybean flowers are attractive to honeybees and this plant can be a useful source of nectar for honey production. Soybean is autogamic, with most flowers being pollinated and fertilized before the flowers open. It is generally considered that bees are not needed for soybean production, although they are capable of cross-pollinating this plant and may play a role in the production of hybrids. About 5% of plants have been estimated to be cross-pollinated (McGregor, 1976). A recent study in Brazil has shown that honeybees are responsible for most insect pollination in this crop and that significantly fewer soybean flowers abort when access to insects is provided (Chiari *et al.*, 2005), suggesting that bee pollination may be more important for this crop than previously thought. Thus, the possibility of bee-mediated cross-pollination between GM and non-GM soybean exists and may warrant further investigation.

Cotton has flowers and extra-floral nectaries that are bee-attractive, and cotton honey is a useful hive product in some areas. There are four species and many different varieties of cotton (*Gossypium* spp.) cultivated for fibre and seed; all are mainly self-pollinating with some cross-pollination (McGregor, 1976). Pre-zygotic barriers prevent cotton from crossing with wild relatives (OGTR, 2002), and so the main concern with GM cotton is its coexistence with non-GM cotton. While bees are not essential for cotton production, studies have shown that fibre yield can be increased by honeybee visits (Rhodes, 2002) and insects are thought to be important in longer-range dispersal of cotton pollen because the pollen grains are too large for wind dispersal. Field studies with *Bt* cotton have shown that GM/non-GM cross-pollination occurs at very low frequencies only a few metres beyond the edge of the crop, e.g. <1% at 7–25 m (Umbeck *et al.*, 1991), 0.08% at 20 m and 0% at 50 m (Zhang *et al.*, 2005), suggesting that long-range dispersal by insects will not pose significant hurdles to the coexistence of GM and non-GM cotton crops.

It has now been established that insects (honeybees and bumblebees in particular) are very important vectors of oilseed rape pollen, even though wind-pollination is also possible (e.g. Cresswell *et al.*, 2004). Because oilseed rape can hybridize with wild weedy relatives, and the commercial GM varieties are HT, there has been a considerable amount of research conducted on gene flow in this crop (see Poppy and Wilkinson, 2005 for a recent review). As with many crop species, most cross-pollination occurs within a few metres of the source plant (e.g. Ramsay *et al.*, 2003; Funk *et al.*, 2006). Various models have been proposed to predict transgene flow over longer distances (up to 1 km; e.g. Cresswell, 2005; Cresswell and Hoyle, 2006; Hayter and Cresswell, 2006; Klein *et al.*, 2006), but a standard method applicable to the wide range of circumstances under which oilseed rape is grown has not yet been agreed upon. Some longer-range, very-low-frequency cross-pollination events have been recorded from field studies, with insects the presumed vectors. Ramsay *et al.* (2003) recorded cross-pollination events 5 and 26 km from source at two Scottish sites, while Rieger *et al.* (2002) reported cross-pollination events at up to 3 km from source at three different Australian sites. These studies demonstrate the importance of a good understanding of crop pollination biology for making robust assessments of transgene flow risk from GM plants.

Conclusions

All reports thus far suggest that current commercial GM crops, bred for herbicide tolerance or insect resistance, do not have negative impacts on insect pollinators. However, some future transgenic traits could pose particular risks for these insects. Studies with PIs and lectins have shown that these proteins could present dose-dependent hazards to bees and care may need to be taken in restricting bee exposure to them. Other modifications, involving traits conferring wide-spectrum insecticidal properties or altering the metabolic profiles of plants, will need to be researched further to quantify the risks to pollinators from these plants.

Such research should focus on understanding the modes of action of the novel traits, and how these may or may not have an impact on pollinators. For many new GM traits, standard toxicity-testing regimes may not be appropriate or particularly helpful. For example, a better understanding of bees' physiological responses to various biochemicals may be required. Information gained will aid in the formulation of realistic risk hypotheses as the basis for further testing.

Thorough investigation of potential exposure routes and expression levels will also be vital for assessing the risks of new GM plants to pollinators. Toxicity and other such tests reveal only half the picture. In some cases, much of the effort expended on researching impacts of extremely high levels of transgene products may have been better spent determining pollen expression levels at an early stage of the investigation.

In order to determine potential exposure routes, and to estimate levels of insect-mediated pollen transfer for transgene flow assessment, more research is needed on basic pollination biology and ecology. Even the interactions between flower-visiting insects and widely grown crops such as soybeans are not yet well understood.

Acknowledgements

We thank Anne Gunson, Ngaire Markwick, Philippa Stevens and Jacqui Todd, all of the Horticulture and Food Research Institute of New Zealand Ltd, for their helpful comments on this manuscript.

References

Arpaia, S. (1996) Ecological impact of Bt-transgenic plants: 1. assessing possible effects of CryIIIB toxin on honey bee (*Apis mellifera* L.) colonies. *Journal of Genetics and Breeding* 50, 315–319.

Arpaia, S., Clemente, A., Leo, G.M.D. and Fiore, M.C. (2004) Pollinator abundance and foraging behaviour on Bt-expressing transgenic canola plants. In: Bernardinelli, I. and Milani, N. (eds) *EurBee*. 1–23 September 2004, agf, Udine, Italy, p. 126.

Aylor, D.E. (2004) Survival of maize (*Zea mays*) pollen exposed in the atmosphere. *Agricultural and Forest Meteorology* 123, 125.

Babendreier, D., Kalberer, N., Romeis, J., Fluri, P. and Bigler, F. (2004) Pollen consumption in honey bee larvae: a step forward in the risk assessment of transgenic plants. *Apidologie* 35, 293–300.

Babendreier, D., Kalberer, N.M., Romeis, J., Fluri, P., Mulligan, E. and Bigler, F. (2005) Influence of Bt-transgenic pollen, Bt-toxin and protease inhibitor (SBTI) ingestion on development of the hypopharyngeal glands in honeybees. *Apidologie* 36, 585–594.

Babendreier, D., Joller, D., Romeis, J., Bigler, F. and Widmer, F. (2007) Bacterial community structures in honeybee intestines and their response to two insecticidal proteins. *FEMS Microbiology and Ecology* 59, 600–610.

Babendreier, D., Reichhart, B., Romeis, J. and Bigler, F. (2008) Impact of insecticidal proteins expressed in transgenic plants on bumblebee microcolonies. *Entomologia Experimentalis et Applicata* 126, 148–157.

Baucher, M., Halpin, C., Petit-Conil, M. and Boerjan, W. (2003) Lignin: genetic engineering and impact on pulping. *Critical Reviews in Biochemistry and Molecular Biology* 38, 305.

Belzunces, L., Lenfant, C., Pasquale, S. and Di Colin, M.E. (1994) *In vivo* and *in vitro* effects of wheat germ agglutinin and Bowman-Birk soybean trypsin inhibitor, two potential transgene products, on midgut esterase and protease activities from *Apis mellifera*. *Comparative Biochemistry and Physiology* 109, 63–69.

Biesmeijer, J., Roberts, S., Reemer, M., Ohlemuller, R., Edwards, M., Peeters, T., Schaffers, A., Potts, S., Kleukers, R., Thomas, C., Settele, J. and Kunin, W. (2006) Parallel declines in pollinators and insect-pollinated plants in Britain and The Netherlands. *Science* 313, 351–354.

Bonade Bottino, M., Girard, C., Jouanin, L., Le Metayer, M., Picard-Nizou, A.L., Sandoz, G., Pham-Delegue, M.H. and Lerin, J. (1998) Effects of transgenic oilseed rape expressing proteinase inhibitors on pest and beneficial insects. *Acta Horticulturae* 459, 235–239.

Bowen, D., Rocheleau, T., Blackburn, M., Andreeve, O., Golubeva, E., Bhartia, R. and French-Constant, R. (1998) Insecticidal

toxins from the bacterium *Photorhabdus luminescens*. *Science* 280, 2129–2132.

Brodsgaard, H.F., Brodsgaard, C.J., Hansen, H. and Lovei, G.L. (2003) Environmental risk assessment of transgene products using honey bee (*Apis mellifera*) larvae. *Apidologie* 34, 139–145.

Burgess, E.P.J., Malone, L.A. and Christeller, J.T. (1996) Effects of two proteinase inhibitors on the digestive enzymes and survival of honey bees (*Apis mellifera*). *Journal of Insect Physiology* 42, 823–828.

Burgess, E.P.J., Malone, L.A., Christeller, J.T., Lester, M.T., Murray, C., Philip, B.A., Phung, M.M. and Tregidga, E.L. (2002) Avidin expressed in transgenic tobacco leaves confers resistance to two noctuid pests, *Helicoverpa armigera* and *Spodoptera litura*. *Transgenic Research* 11, 185–198.

Cahoon, E.B., Dietrich, C.R., Meyer, K., Damude, H.G., Dyer, J.M. and Kinney, A.J. (2006) Conjugated fatty acids accumulate to high levels in phospholipids of metabolically engineered soybean and *Arabidopsis* seeds. *Phytochemistry* 67, 1166–1176.

Cane, J.H. and Tepedino, V.J. (2001) Causes and extent of declines among native North American invertebrate pollinators: detection, evidence, and consequences. *Conservation Ecology* 5, 2.

Carter, C., Graham, R.A. and Thornburg, R.W. (1999) Nectarin I is a novel, soluble germin-like protein expressed in the nectar of *Nicotiana* sp. *Plant Molecular Biology* 41, 207–216.

CBD (2005) Agricultural Biodiversity. International Initiative for the Conservation and Sustainable Use of Pollinators. Available at: http://www.biodiv.org/programmes/areas/agro/pollinators.aspx#

CFIA (1995) Decision document DD95-01: Determination of Environmental Safety of AgrEvo Canada Inc.'s Glufosinate Ammonium-Tolerant Canola. Available at: http://www.inspection.gc.ca/english/plaveg/bio/dd/dd9501e.shtml

Chaline, N., Decourtye, A., Marsault, D., Lechner, M., Champolivier, J., Van Waetermeulen, X., Viollet, D. and Pham-Delègue, M.H. (1999) Impact of a novel herbicide resistant transgenic oilseed rape

on honey bee colonies in semi-field conditions. In: *Conference Internationale sur les
Ravageurs en Agriculture* 7–9 December
1999, Montpellier, France, pp. 905–912.

Chapman, R.E., Wang, J. and Bourke, A.F.G.
(2003) Genetic analysis of spatial foraging
patterns and resource sharing in bumble
bee pollinators. *Molecular Ecology* 12,
2801–2808.

Charriere, J., Imdorf, A., Koenig, C., Gallmann, S.
and Kuhn, R. (2006) Which influence has sunflower on the development of bee colonies
(*Apis mellifera*)? *Agrarforschung* 13, 380–385.

Chiari, W.C., Toledo, V.d.A.A.d., Kotaka,
M.C.C., Sakaguti, E.S. and Magalhães,
H.R. (2005) Floral biology and behaviour
of Africanized honeybees *Apis mellifera* in
soybean (*Glycine max* L. Merril). *Brazilian
Archives of Biology and Technology* 48,
367–378.

Crane, E. and Walker, P. (1986) Pollination
Directory for World Crops. International Bee
Research Association, London, UK. *New
Zealand Journal of Botany* 24, 355–356.

Cresswell, J. (2005) Accurate theoretical prediction of pollinator-mediated gene dispersal. *Ecology* 86, 574–578.

Cresswell, J. and Hoyle, M. (2006) A mathematical method for estimating patterns of
flower-to-flower gene dispersal from a simple field experiment. *Functional Ecology*
20, 245–251.

Cresswell, J., Davies, T., Patrick, M., Russells,
F., Pennel, C., Vicot, M. and Lahoubi, M.
(2004) Aerodynamics of wind pollination in
a zoophilous flower, *Brassica napus*.
Functional Ecology 18, 861–866.

Cresswell, J.E. and Osborne, J.L. (2004) The
effect of patch size and separation on
bumblebee foraging in oilseed rape: implications for gene flow. *Journal of Applied
Ecology* 41, 539–546.

Cresswell, J.E., Osborne, J.L. and Bell, S.A.
(2002) A model of pollinator-mediated
gene flow between plant populations with
numerical solutions for bumblebees pollinating oilseed rape. *Oikos* 98, 375–384.

Darvill, B., Knight, M.E. and Goulson, D.
(2004) Use of genetic markers to quantify
bumblebee foraging range and nest density. *Oikos* 107, 471–478.

Dechaume-Moncharmont, F.-X., Azzouz, H.,
Pons, O. and Pham-Delègue, M.-H. (2005)
Soybean proteinase inhibitor and the for-
aging strategy of free-flying honeybees.
Apidologie 36, 421–430.

Duan, J.J., Marvier, M., Huesing, J., Dively, G.
and Huang, Z.Y. (2008) A meta-analysis of
effects of Bt crops on honey bees
(Hymenoptera: Apidae). *PLoS ONE* 3:
e1415. doi:10.1371/journal.pone.0001415

Fearing, P.L., Brown, D., Vlachos, D., Meghji,
M. and Privalle, L. (1997) Quantitative
analysis of CryIA(b) expression in Bt maize
plants, tissues, and silage and stability of
expression over successive generations.
Molecular Breeding 3, 169–176.

Fohouo, F.-N.T., Messi, J. and Pauly, A. (2002)
L'activité de butinage des Apoïdes sau-
vages (Hymenoptera Apoidea) sur les
fleurs de maïs à Yaoundé (Cameroun) et
réflexions sur la pollinisation des gram-
inées tropicales. *Biotechnology Agronomy
Society and Environment* 6, 87–98.

Fohouo, F., Pauly, A., Messi, J., Bruckner, D.,
Tinkeu, L. and Basga, E. (2004) An afro-
tropical bee specialized in the collection of
grass pollen (Poaceae): *Lipotriches nota-
bilis* (Schletterer 1891) (Hymenoptera
Apoidea Halictidae). *Annales de la Societe
Entomologique de France* 40, 131–143.

Franco, O.L., Rigden, D.J., Melo, F.R. and
Grossi-de-Sa, M.F. (2002) Plant alpha-
amylase inhibitors and their interaction
with insect alpha-amylases: structure,
function and potential for crop protection.
European Journal of Biochemistry 269,
397–412.

Free, J.B. (1993) *Insect Pollination of Crops.*
Academic Press, London.

Funk, T., Wenzel, G. and Schwarz, G. (2006)
Outcrossing frequencies and distribution
of transgenic oilseed rape (*Brassica napus*
L.) in the nearest neighbourhood. *European
Journal of Agronomy* 24, 26–34.

Gary, N.E. (1992) Activities and behavior of
honey bees. In: Graham, J.M. (ed.) *The
Hive and the Honey Bee.* Dadant and
Sons, Hamilton, Illinois, pp. 269–373.

Gathmann, A. and Tscharntke, T. (2002)
Foraging ranges of solitary bees. *Journal
of Animal Ecology* 71, 757–764.

Girard, C., Picard-Nizou, A.L., Grallien, E., Zaccomer, B., Jouanin, L. and Pham-Delegue, M.H. (1998) Effects of proteinase inhibitor ingestion on survival, learning abilities and digestive proteinases of the honeybee. *Transgenic Research* 7, 239–246.

Glare, T.R. and O'Callaghan, M. (2000) *Bacillus thuringiensis: Biology, Ecology and Safety.* Wiley, UK.

Goulson, D. and Stout, J.C. (2001) Homing ability of the bumblebee *Bombus terrestris* (Hymenoptera: Apidae). *Apidologie* 32, 105–111.

Grallien, E., Marilleau, R., Pham-Delegue, M.H., Picard Nizou, A.L., Jouanin, L. and Marion-Poll, F. (1995) Impact of pest insect resistant oilseed rape on honeybees. In: *9th International Rapeseed Congress, 'Rapeseed Today and Tomorrow',* 4–7 July 1995, Cambridge, pp. 784–786.

Green, D.L. (2004) Non-*Apis* bee and other pollinator vendors. Available at: http://pollinator.com/alt_polvendors.htm

Greenplate, J. (1997) Response to reports of early damage in 1996 commercial Bt transgenic cotton (Bollgard) plantings. *Society for Invertebrate Pathology* 29, 15–18.

Gross, H.R. and Carpenter, J.E. (1991) Role of the fall armyworm (Lepidoptera: Noctuidae) pheromone and other factors in the capture of bumblebees (Hymenoptera: Aphidae) by universal moth traps. *Environmental Entomology* 20, 377–381.

Hanley, A.V., Huang, Z.Y. and Pett, W.L. (2003) Effects of dietary transgenic Bt corn pollen on larvae of *Apis mellifera* and *Galleria mellonella. Journal of Apicultural Research* 42, 77–81.

Hao, C., Chai, B., Wang, W., Sun, Y. and Liang, A. (2005) Polyclonal antibody against *Manduca sexta* chitinase and detection of chitinase expressed in transgenic cotton. *Biotechnology Letters* 27, 97–102.

Hayter, K.E. and Cresswell, J.E. (2006) The influence of pollinator abundance on the dynamics and efficiency of pollination in agricultural *Brassica napus*: implications for landscape-scale gene dispersal. *Journal of Applied Ecology* 43, 1196.

Herouet, C., Esdaile, D.J., Mallyon, B.A., Debruyne, E., Schulz, A., Currier, T.,

Hendrickx, D., van der Klis, R.J. and Rouan, D. (2005) Safety evaluation of the phosphinothricin acetyltransferase proteins encoded by the *pat* and *bar* sequences that confer tolerance to glufosinate-ammonium herbicide in transgenic plants. *Regulatory Toxicology and Pharmacology* 41, 134.

Hoyle, M. and Cresswell, J. (2006) Remobilization of initially deposited pollen grains has negligible impact on gene dispersal in bumble bee-pollinated *Brassica napus. Functional Ecology* 20, 958–965.

Huang, Z.Y., Hanley, A.V., Pett, W.L., Langenberger, M. and Duan, J.J. (2004) Field and semifield evaluation of impacts of transgenic canola pollen on survival and development of worker honey bees. *Journal of Economic Entomology* 97, 1517–1523.

James, C. (2007) Global Status of Commercialized Biotech/GM Crops: 2007. *The International Service for the Acquisition of Agri-biotech Applications* (ISAAA) Ithaca, New York.

Jouanin, L., Girard, C., Bonade-Bottino, M., Le Metayer, M., Picard Nizou, A., Lerin, J. and Pham-Delegue, M. (1998) Impact of oilseed rape expressing proteinase inhibitors on coleopteran pests and honey bees. *Cahiers Agriculture* 7, 531–536.

Kevan, P.G. and Phillips, T.P. (2001) The economic impacts of pollinator declines: an approach to assessing the consequences. *Conservation Ecology* 5(1), 8.

Khan, S., Zafar, Y., Briddon, B., Malik, K. and Mukhtar, Z. (2006) Spider venom toxin protects plants from insect attack. *Transgenic Research* 15, 349–357.

Klein, E.K., Lavigne, C., Picault, H., Renard, M. and Gouyon, P.H. (2006) Pollen dispersal of oilseed rape: estimation of the dispersal function and effects of field dimension. *Journal of Applied Ecology* 43, 141–151.

Knight, M.E., Martin, A.P., Bishop, S., Osborne, J.L., Hale, R.J., Sanderson, R.A. and Goulson, D. (2005) An interspecific comparison of foraging range and nest density of four bumblebee (*Bombus*) species. *Molecular Ecology* 14, 1811–1820.

Kohno-Murase, J., Iwabuchi, M., Endo-Kasahara, S., Sugita, K., Ebinuma, H. and Imamura, J. (2006) Production of *trans*-10, *cis*-12 conjugated linoleic acid in rice. *Transgenic Research* 15, 95–100.

Konrad, R. and Babendreier, D. (2006) Are solitary bees affected when feeding on transgenic insect-resistant crop plants? In: Vesely, V., Vorechovska, M. and Titera, D. (eds) *Proceedings of the Second European Conference on Apidology*, Eurbee 2006, 10–16 September 2006. Bee Research Institute, Dol, Prague, Czech Republic, pp. 74–75.

Koziel, M.G., Beland, G.L., Bowman, C., Carozzi, N.B., Crenshaw, R., Crossland, L., Dawson, J., Desai, N., Hill, M., Kadwell, S., Launis, K., Lewis, K., Maddox, D., McPherson, K., Meghji, M.R., Merlin, E., Rhodes, R., Warren, G.W., Wright, M. and Evola, S.V. (1993) Field performance of elite transgenic maize plants expressing an insecticidal protein derived from *Bacillus thuringiensis*. *Bio/Technology* 11, 194–200.

Kramer, K.J., Morgan, T.D., Throne, J.E., Dowell, F.E., Bailey, M. and Howard, J.A. (2000) Transgenic avidin maize is resistant to storage insect pests. *Nature Biotechnology* 18, 670–674.

Lawrence, P.K. and Koundal, K.P. (2002) Plant protease inhibitors in control of phytophagous insects. *Electronic Journal of Biotechnology* 5, 1–9.

Legaspi, J.C., Legaspi, Jr B.C. and Setamou, M. (2004) Insect-resistant transgenic crops expressing plant lectins. *Transgenic Crop Protection: Concepts and Strategies*. Science Publishers, Enfield, New Hampshire, pp. 85–116.

Liu, B., Xu, C.G., Yan, F.M. and Gong, R.Z. (2005) The impacts of the pollen of insect-resistant transgenic cotton on honeybees. *Biodiversity and Conservation* 14, 3487.

Liu, D., Burton, S., Glancy, T., Li, Z.S., Hampton, R., Meade, T. and Merlo, D.J. (2003) Insect resistance conferred by 283-kDa *Photorhabdus luminescens* protein TcdA in *Arabidopsis thaliana*. *Nature Biotechnology* 21, 1222–1228.

Luna, S., Figueroa, J., Baltazar, B., Gomez, R., Townsend, R. and Schoper, J.B. (2001) Maize pollen longevity and distance isola-tion requirements for effective pollen control. *Crop Science* 41, 1551.

Malone, L.A., Giacon, H.A., Burgess, E.P.J., Maxwell, J.Z., Christeller, J.T. and Laing, W.A. (1995) Toxicity of trypsin endopeptidase inhibitors to honey bees (Hymenoptera: Apidae). *Journal of Economic Entomology* 88, 46–50.

Malone, L.A., Burgess, E.P.J., Christeller, J.T. and Gatehouse, H.S. (1998) *In vivo* responses of honey bee midgut proteases to two protease inhibitors from potato. *Journal of Insect Physiology* 44, 141–147.

Malone, L.A., Burgess, E.P.J. and Stefanovic, D. (1999) Effects of a *Bacillus thuringiensis* toxin, two *Bacillus thuringiensis* biopesticide formulations, and a soybean trypsin inhibitor on honey bee (*Apis mellifera* L.) survival and food consumption. *Apidologie* 30, 465–473.

Malone, L.A., Burgess, E.P.J., Stefanovic, D. and Gatehouse, H.S. (2000) Effects of four protease inhibitors on the survival of worker bumblebees, *Bombus terrestris* L. *Apidologie* 31, 25–38.

Malone, L.A., Burgess, E.P.J., Gatehouse, H.S., Voisey, C.R., Tregidga, E.L. and Philip, B.A. (2001) Effects of ingestion of a *Bacillus thuringiensis* toxin and a trypsin inhibitor on honey bee flight activity and longevity. *Apidologie* 32, 57–68.

Malone, L.A., Burgess, E.P.J., Mercer, C.F., Christeller, J.T., Lester, M.T., Murray, C., Phung, M.M., Philip, B.A., Tregidga, E.L. and Todd, J.H. (2002a) Effects of biotin-binding proteins on eight species of pasture invertebrates, *New Zealand Plant Protection* Volume 55, 2002. Proceedings of a conference, Centra Hotel, Rotorua, New Zealand, 13–15 August 2002. New Zealand Plant Protection Society, Rotorua New Zealand, pp. 411–415.

Malone, L.A., Tregidga, E.L., Todd, J.H., Burgess, E.P.J., Philip, B.A., Markwick, N.P., Poulton, J., Christeller, J.T., Lester, M.T. and Gatehouse, H.S. (2002b) Effects of ingestion of a biotin-binding protein on adult and larval honeybees. *Apidologie* 33, 447–458.

Malone, L.A., Todd, J.H., Burgess, E.P.J. and Christeller, J.T. (2004) Development of hypopharyngeal glands in adult honey bees fed with a Bt toxin, a biotin-binding protein and a protease inhibitor. *Apidologie* 35, 655–664.

Malone, L.A., Scott-Dupree, C.D., Todd, J.H. and Ramankutty, P. (2007) No sub-lethal toxicity to bumblebees, *Bombus terrestris*, exposed to Bt-corn pollen, captan and novaluron. *New Zealand Journal of Crop and Horticultural Science* 35, 435–439.

Markwick, N.P., Docherty, L.C., Phung, M.M., Lester, M.T., Murray, C., Yao, J.L., Mitra, D.S., Cohen, D., Beuning, L.L., Kutty-Amma, S. and Christeller, J.T. (2003) Transgenic tobacco and apple plants expressing biotin-binding proteins are resistant to two cosmopolitan insect pests, potato tuber moth and lightbrown apple moth, respectively. *Transgenic Research* 12, 671–681.

Mattila, H., Sears, M. and Duan, J. (2005) Response of *Danaus plexippus* to pollen of two new Bt corn events via laboratory bioassay. *Entomologia Experimentalis et Applicata* 116, 31–41.

McCafferty, H.R.K., Moore, P.H. and Zhu, Y.J. (2006) Improved *Carica papaya* tolerance to carmine spider mite by the expression of *Manduca sexta* chitinase transgene. *Transgenic Research* 15, 337–347.

McGregor, S. (1976) Insect pollination of cultivated crop plants. First and only virtual beekeeping book updated continuously. Available at: http://gears.tucson.ars.ag.gov/book/index.html

Meagher, T.R., Belanger, F.C. and Day, P.R. (2003) Using empirical data to model transgene dispersal. *Philosophical Transactions of the Royal Society of London. Series B, Biological Sciences* 358, 1157–1162.

Mehlo, L., Gahakwa, D., Nghia, P.T., Loc, N.T., Capell, T., Gatehouse, J.A., Gatehouse, A.M. and Christou, P. (2005) An alternative strategy for sustainable pest resistance in genetically enhanced crops. *Proceedings of the National Academy of Sciences of the USA* 102, 7812–7816.

Moar, W.J. and Schwartz, J.-L. (2003) Workshop on the ethics, legal, and regulatory concerns of transgenic plants. *Journal of Invertebrate Pathology* 83, 91–117.

Morandin, L.A. and Winston, M.L. (2003) Effects of novel pesticides on bumble bee (Hymenoptera: Apidae) colony health and foraging ability. *Environmental Entomology* 32, 555–563.

Morgan, T.D., Oppert, B., Czapla, T.H. and Kramer, K.J. (1993) Avidin and streptavidin as insecticidal and growth-inhibiting dietary proteins. *Entomologia Experimentalis et Applicata* 69, 97–108.

Moyes, C.L. and Dale, P.J. (1999) Organic farming and gene transfer from genetically modified crops. MAFF research project OF0157. Available at: www.gmissues.org/orgreport.htm

Murphy, D.J. (2006) Molecular breeding strategies for the modification of lipid composition. *In Vitro Cellular and Developmental Biology-Plant* 42, 89–99.

Naimov, S., Dukiandjiev, S. and Maagd, R.A.d. (2003) A hybrid *Bacillus thuringiensis* delta-endotoxin gives resistance against a coleopteran and a lepidopteran pest in transgenic potato. *Plant Biotechnology Journal* 1, 51–57.

OGTR (2002) The biology and ecology of cotton (*Gossypium hirsutum*) in Australia. August 2002. Office of the Gene Technology Regulator, Australian Government. Available at: http://www.ogtr.gov.au/pdf/ir/biologycotton.pdf

OGTR (2004) Gene Technology Technical Advisory Committee. Communique No. 12. Office of the Gene Technology Regulator, Australian Government. Available at: http://www.ogtr.gov.au/rtf/committee/12th commgttac.rtf

OGTR (2005) Risk Assessment and Risk Management Plan for DIR 058/2005. Limited and controlled release of insect resistant (VIP) GM cotton, pp. 99. Office of the Gene Technology Regulator, Australian Government. Available at: http://www.ogtr.gov.au/pdf/ir/dir058finalrarmp1.pdf

Osborne, J.L., Clark, S.J., Morris, R.J., Williams, I.H., Riley, J.R., Smith, A.D., Reynolds, D.R. and Edwards, A.S. (1999) A landscape-scale study of bumble bee foraging range and constancy, using harmonic radar. *Journal of Applied Ecology* 36, 519–533.

Osborne, J.L., Williams, I.H., Marshall, A.H. and Michaelson-Yeates, T.P.T. (2001) Pollination and gene flow in white clover, growing in a patchy habitat. *Acta Horticulturae* 561, 35–40.

Peumans, W.J., Smeets, K., Nerum, K.V., Leuven, F.V. and van Damme, E.J.M. (1997) Lectin and alliinase are the predominant proteins in nectar from leek (*Allium porrum* L.) flowers. *Planta* 201, 298–302.

Pham-Delegue, M.H., Girard, C., Le Metayer, M., Picard-Nizou, A.L., Hennequet, C., Pons, O. and Jouanin, L. (2000) Long-term effects of soybean protease inhibitors on digestive enzymes, survival and learning abilities of honeybees. *Entomologia Experimentalis et Applicata* 95, 21–29.

Picard-Nizou, A.L., Pham-Delegue, M.H., Kerguelen, V., Douault, P., Marilleau, R., Olsen, L., Grison, R., Toppan, A. and Masson, C. (1995) Foraging behaviour of honey bees (*Apis mellifera* L.) on transgenic oilseed rape (*Brassica napus* L. var. *oleifera*). *Transgenic Research* 4, 270–276.

Picard-Nizou, A.L., Grison, R., Olsen, L., Pioche, C., Arnold, G. and Pham-Delegue, M.H. (1997) Impact of proteins used in plant genetic engineering: toxicity and behavioral study in the honeybee. *Journal of Economic Entomology* 90, 1710–1716.

Picard, A.L., Pham-Delegue, M.H., Douault, P. and Masson, C. (1991) Transgenic rapeseed (*Brassica napus* L. var. *oleifera* Metzger): effect on the foraging behavior of honeybees. *Acta Horticulturae* 288, pp. 435–439.

Pierre, J., Marsault, D., Genecque, E., Renard, M., Champolivier, J. and Pham-Delegue, M.H. (2003) Effects of herbicide-tolerant transgenic oilseed rape genotypes on honey bees and other pollinating insects under field conditions. *Entomologia Experimentalis et Applicata* 108, 159–168.

Pla, M., Paz, J.L.I., Penas, G., Garcia, N., Palaudelmas, M., Esteve, T., Messeguer, J. and Mele, E. (2006) Assessment of real-time PCR based methods for quantifica-tion of pollen-mediated gene flow from GM to conventional maize in a field study. *Transgenic Research* 15, 219–228.

Poppy, G.M. and Wilkinson, M.J. (2005) *Gene Flow from GM Plants.* Blackwell Publishing, Oxford.

Ramsay, G., Thompson, C. and Squire, G. (2003) Quantifying landscape-scale gene flow in oilseed rape. *Department for Environment, Food and Rural Affairs.* Dundee, UK.

Rhodes, J. (2002) Cotton pollination by honey bees. *Australian Journal of Experimental Agriculture* 42, 513–218.

Richards, J.S., Stanley, J.N. and Gregg, P.C. (2005) Viability of cotton and canola pollen on the proboscis of *Helicoverpa armigera*: implications for spread of transgenes and pollination ecology. *Ecological Entomology* 30, 327–333.

Rieger, M.A., Lamond, M., Preston, C., Powles, S.B. and Roush, R.T. (2002) Pollen-mediated movement of herbicide resistance between commercial canola fields. *Science* 296, 2386–2388.

Rose, R., Dively, G.P. and Pettis, J. (2007) Effects of Bt corn pollen on honey bees: emphasis on protocol development. *Apidologie* 38, 368–377.

Sagili, R., Pankiw, T. and Zhu-Salzman, K. (2005) Effects of soybean trypsin inhibitor on hypopharyngeal gland protein content, total midgut protease activity and survival of the honey bee (*Apis mellifera* L.). *Journal Insect Physiology* 51, 953–957.

Sandoz, G. (1996) Etude des effects d'inhibiteurs de proteases sur un insecte pollinisateur, l'abeille domestique *Apis mellifera* L. Diplome d'Agronomie Approfondie Thesis. Institut National Agronomique Paris-Grignon, Paris.

Schijlen, E.G.W.M., Vos, C.H.R.d., Tunen, A. J.v. and Bovy, A.G. (2004) Modification of flavonoid biosynthesis in crop plants. *Phytochemistry* 65, 2631–2648.

Schnepf, E., Crickmore, N., Van Rie, J., Lereclus, D., Baum, J., Feitelson, J., Zeigler, D.R. and Dean, D.H. (1998) *Bacillus thuringiensis* and its pesticidal crystal proteins. *Microbiology and Molecular Biology Reviews* 62, 775–806.

Schur, A., Tornier, I. and Neumann, C. (2000) Bt-Mais und non Bt-Mais: vergleichende Untersuchungen an Honigbienen (Tunnel-zeltversuch). In 47th Annual Meeting of the Institutes for Bee Research, 3–5 April 2000, Blaubeuren bei Ulm, Germany.

Sick, M., Kuhne, S. and Hommel, B. (2004) Transgenic rape pollen in larval food of bees – component of a model study on the probability of horizontal plant-to-bacteria gene transfer. *Mitteilungen der Deutschen Gesellschaft fur allgemeine und angewandte Entomologie* 14, 423–426.

Sims, S. (1995) *Bacillus thuringiensis* var. *kurstaki* CryIA(c) protein expressed in transgenic cotton: effects on beneficial and other non-target insects. *Southwestern Entomologist* 20, 493–500.

Staffan-Dewenter, I., Potts, S. and Packer, L. (2005) Pollinator diversity and crop pollination services are at risk. *Trends in Ecology and Evolution* 20, 651–652.

Stanley-Horn, D.E., Dively, G.P., Hellmich, R.L., Mattila, H.R., Sears, M.K., Rose, R., Jesse, L.C.H., Losey, J.E., Obrycki, J.J. and Lewis, L. (2001) Assessing the impact of Cry1Ab-expressing corn pollen on monarch butterfly larvae in field studies. *Proceedings of the National Academy of Sciences of the USA* 98, 11931–11936.

Svensson, B., Lagerlof, J. and Svensson, B. (2000) Habitat preferences of nest-seeking bumble bees (Hymenoptera: Apidae) in an agricultural landscape. *Agriculture Ecosystems and Environment* 77, 247–255.

Tesoriero, D., Sgolastra, F., Dall'Asta, S., Venier, F., Sabatini, A., Burgio, G. and Porrini, C. (2004) Effects of Bt-oilseed rape on the foraging activity of honey bees in confined environment. *Redia* 87, 195–198.

Umbeck, P., Barton, K., Nordheim, E., McCarty, J., Parrott, W. and Jenkins, J. (1991) Degree of pollen dispersal by insects from a field-test of genetically engineered cotton. *Journal of Economic Entomology* 84, 1943–1950.

USEPA (2000) Bt plant-pesticides registration action document. Available at: http://www. epa.gov/scipoly/sap/2000/october/brad_2_scienceassessment.pdf

USEPA (2001) Bt Plant-Incorporated Protectants 15 October 2001 Biopesticides Registration Action Document. Available at: http://www. epa.gov/oppbppd1/biopesticides/pips/bt_brad2/3-ecological.pdf

USEPA (2002) *Bacillus thuringiensis* Cry2Ab2 protein and the genetic material necessary for its production in cotton (006487). Fact sheet. Available at: http://www.epa.gov/oppbppd1/biopesticides/ingredients/factsheets/factsheet_006487.htm

USEPA (2005a) Attachment III: Environmental Risk Assessment of Plant Incorporated Protectant (PIP) Inert Ingredients. Available at: http://www.epa.gov/oscpmont/sap/meetings/2005/december/pipinertenvironmentalriskassessment11-18-05.pdf

USEPA (2005b) Biopesticide Registration Action Document, *Bacillus thuringiensis* Cry1F Corn. Updated August 2005. Available at: http://www.epa.gov/opp00001/biopesticides/ingredients/tech_docs/brad_006481.pdf

USEPA (undated) United States Federal Insecticide Fungicide and Rodenticide Act (FIFRA). Available at: http://www.access.gpo.gov/uscode/title7/chapter6_.html

Walklate, P.J.J., Hunt, C.R., Higson, H.L. and Sweet, J.B. (2004) A model of pollen-mediated gene flow for oilseed rape. *Proceedings of the Royal Society of London, Series B – Biological Sciences* 271, 441–449.

Walther-Hellwig, K. and Frankl, R. (2000) Foraging habitats and foraging distances of bumblebees, *Bombus* spp. (Hym., Apidae), in an agricultural landscape. *Journal of Applied Entomology* 124, 299–306.

Wille, H., Wille, M., Kilchenmann, V., Imdorf, A. and Buhlmann, G. (1985) Pollenernte und Massenwechsel von drei *Apis mellifera* Volkern auf demselben Bienenstand in zwei aufeinanderfolgenden Jahren. *Revue Suisse Zoologie* 92, 897–914.

Williams, I.H. (2001) Bee-mediated pollen and gene flow from GM plants. *Acta Horticulturae*, 561, 25–33.

Winston, M.L. (1987) *The Biology of the Honey Bee*. Harvard University Press, Cambridge, Massachusetts.

Wu, K., Li, W., Feng, H. and Guo, Y. (2002) Seasonal abundance of the mirids, *Lygus lucorum* and *Adelphocoris* spp. (Hemiptera: Miridae) on Bt cotton in northern China. *Crop Protection* 21, 997–1002.

Yao, H.W., Ye, G.Y., Jiang, C.Y., Fan, L.J., Datta, K., Hu, C. and Datta, S.K. (2006) Effect of the pollen of transgenic rice line, TT9-3 with a fused cry1Ab/cry1Ac gene from *Bacillus thuringiensis* Berliner on non-target domestic silkworm, *Bombyx mori* Linnaeus (Lepidoptera: Bombyxidae). *Applied Entomology and Zoology* 41, 339–348.

Yoza, K., Imamura, T., Kramer, K.J., Morgan, T.D., Nakamura, S., Akiyama, K., Kawasaki, S., Takaiwa, F. and Ohtsubo, K. (2005) Avidin expressed in transgenic rice confers resistance to the stored-product insect pests *Tribolium confusum* and *Sitotroga cerealella*. *Bioscience Biotechnology and Biochemistry* 69, 966–971.

Zhang, B.H., Pan, X.P., Guo, T.L., Wang, Q.L. and Anderson, T.A. (2005) Measuring gene flow in the cultivation of transgenic cotton (*Gossypium hirsutum* L.). *Molecular Biotechnology* 31, 11.

10 Impact of Genetically Modified Crops on Soil and Water Ecology

R. WHEATLEY

Scottish Crop Research Institute, Dundee, UK

Keywords: Ecosystem functioning, primary production, system drivers, nutrient cycling, biodiversity, DNA dissemination

Summary

Soil ecosystem functioning is vital to the sustained functioning of the biosphere. Many processes occur in soils. These include the recycling of nutrients from previous crop residues to maintain primary production, and the consequent introduction of the sun's energy into the biosphere. Other processes that enhance environmental quality such as the suppression of plant pathogens, carbon sequestration and bioremediation also occur. Soils support one of the most diverse ecosystems in the biosphere, with a population consisting of many thousands of species of bacteria, protozoa, fungi and micro-, meso- and macro-fauna, which are responsible for a large number of functional processes. The soil population varies greatly in types, numbers and functional expression both temporally and spatially, and is responsive to a variety of inputs and physical conditions. These are dynamic systems that are constantly evolving and changing, in which many components are intrinsically interconnected, and conversely others are completely autonomous. Plant inputs are the major drivers of soil ecosystem functioning, and responses to these inputs vary according to both the quantity of these inputs and the constituent chemicals. These inputs come from roots, root exudates and debris, and residues from the aerial parts during active plant growth and from senescing material after the plant dies. They provide energy from the breakdown of the carbon fixed during primary production in the plant, and frequently, although consequentially, nitrogen, in many forms. Inputs differ in both amounts and types of compounds according to plant species. Hence, different plants can have differing effects on soil process dynamics. Soil organisms are also involved in other nutrient acquisitions such as phosphorous mobilization and metal chelation to enable plant uptake, as well as mycorrhizal associations. Their activities also impact on soil health and quality, with subsequent impacts on water quality and supply. Without these soil-ecosystem-driven phenomena primary production would cease. Here the possible impacts of genetically modified (GM) crops on agricultural and natural systems are considered. However, it should be borne in mind that all the topics considered will apply equally to the introduction of new, or indeed different, cultivars or crops, since it is the consequences of changed inputs from these that are relevant. Similarly, many changes in agronomic practice, such as cultivation technique and timing, planting and sowing rates, pesticide usage, etc., must also be considered. The environmental impacts of these should receive equal attention as these large perturbations that are routinely applied to soil ecosystems during normal agricultural management practices have significant impacts on soil communities and functions (Buckley and Schmidt, 2001). These and other factors such as plant type being cultivated, season, climate and geography may well be much greater drivers of community structure in the rhizosphere than any changes in inputs resulting from GM of the crop plant (Donegan et al., 1999; Griffiths et al., 2000; Lukow et al., 2000; Dunfield and Germida, 2001; Hopkins et al., 2001). When the environmental impacts of GM plants are being assessed, comparisons must be drawn against the background of such 'normal' baseline variations, in both structure and function.

Soil Ecosystems

Many processes that maintain both primary productivity and environmental quality and are therefore vital to the continued functioning and sustainability of the biosphere occur in soil ecosystems. Soil ecosystems support one of the most diverse ecosystems in the biosphere. Within soil there are many thousands of species of bacteria, protozoa, fungi and micro- and macro-fauna. The composition of the soil population, in types, numbers and functional expression, varies greatly both temporally and spatially. Complex interactions occur between these organisms, within the restrictions of the soil matrix. Soils also support plant growth and development by providing a physical matrix for root development that also contains and provides water and nutrients. Many other vital ecosystem processes also take place in soils. These include the transformation and improved availability of many macronutrients and micronutrients from plant residues, soil organic matter and inorganic complexes. Processes in the soil ecosystem are also involved in the suppression of plant pathogens, via microbial interactions, carbon sequestration and the bioremediation of toxic waste products. Functional dynamics in all of these processes are determined by complex interactions between some of the many physical, chemical and biological factors in the soil.

Physical conditions in this soil matrix, such as water and oxygen concentrations, can vary greatly over very short distances, dependent on such factors as soil texture and porosity, as can nutrient availability. These can also change according to the type and degree of microbial function occurring, combined with plant uptake rates, rainfall inputs, etc. These all work together to produce scenarios in which selection pressures are constantly changing, sometimes to a greater, and sometimes to a lesser, extent. This heterogeneity and variability combine to produce the most diverse of habitats that consequently favour the maintenance of a large number of species.

Bacterial and fungal communities form the numerically dominant groups in the system, and these perform many functions and transformations. The soil micro-, meso- and macro-fauna, which includes organisms such as nematodes, arthropods and earthworms, feed on living and dead plant tissue and play a vital part in soil nutrient cycling. Their activities break down the plant material into smaller pieces and redistribute it within the soil, so making it more available for microbial activity and also enhancing its incorporation into the soil organic matter.

The diversity of the soil biota can result in extremely complex food webs, over several trophic levels that are subject to a wide range of interactive influences. The first trophic level is composed of the primary producers, the plants and autotrophic bacteria, and their remnants (detritus). The second is composed of the decomposers and detritivores; and includes mutualists, pathogens, parasites and root feeders, while the third includes the shredders, distributors, grazers and first predators. Predators and higher predators comprise the further levels (Fig. 10.1). These biologically mediated processes are essential to the continued functioning of the biosphere, as they provide the resources for sustained plant growth and crop production. In a mutually beneficial relationship, plants utilize the nutrients released during these transactions for growth, and in turn provide energy to the other members of the system through the

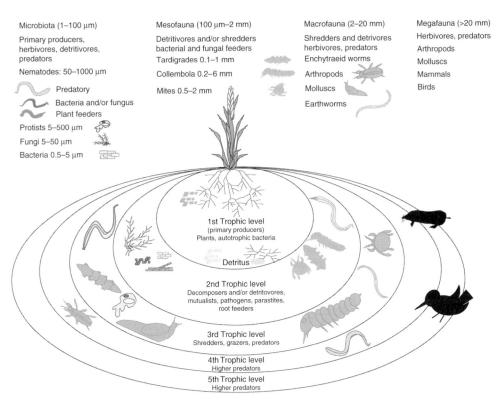

Fig.10.1. Soil food web showing an outline classification of the soil biota based on type, body size and trophic level. These components have a variety of interactions ranging from the competitive or predatorial to the cooperative and symbiotic. Size class is based on the width of the organism according to Swift. (After Lilley *et al.*, 2006.)

products of primary production. This is a major route by which the energy from the sun enters the biosphere. As well as providing this energy source, the organic matter produced by the plants, together with the activities of the soil biota, profoundly affect soil structure and health, and consequently the soil's water-holding properties and atmospheric parameters.

Plant inputs are a major driver of soil ecosystem functioning, both quantitatively and qualitatively, by providing energy, from the breakdown of carbon fixed during primary production in the plant, and, frequently consequentially, nitrogen, in many forms. These inputs come from roots and root exudates, root debris and plant residues from the aerial parts, during active plant growth, and from root and above-ground residues after the plant dies. The amounts of, and constituents in, these inputs vary according to species, cultivar and growth stage of the plant. Thus, it can be expected that a GM crop will have a specific effect on soil ecosystem dynamics, but how this varies from the effects of the non-transformed plant requires investigation. The greatest microbial activity occurs at the root surface and in the region immediately around it, the rhizosphere. Generally, the highest concentrations of bacteria and fungi are to be found in the rhizosphere, and as the plant senesces microbial population size reverts to that in the bulk soil. The activity, numbers and types of microorganisms found in the rhizosphere are influenced by plant species, and therefore by root exudate composition, plant age and other environmental factors such as soil type, moisture, temperature and pH. There is an interdependent association in that the plant exudates influence rhizosphere population development and in turn, the activities of these organisms affect the host plant's growth and development. So soil-plant ecosystems are dynamic, constantly evolving and changing, and have many components that are intrinsically interconnected to varying degrees.

Nutrient Cycling

There are a great many different types of soil, each of which is essentially defined by texture, which in turn is determined by the relative proportions of mineral particles, sand, silt and clay that are present. Soils also contain atmospheric spaces, water and organic matter. In these highly heterogeneous environments, physical conditions and chemical gradients change rapidly both spatially and temporally. This allows a wide variety of biogeochemical processes to occur. Thus, species and functional process diversity in plant–soil systems is immense. A vast range of compounds is produced, many of which are then further transformed in other processes. Such spatial and temporal variability, over small to large scales, means that system predictions and modelling are extremely difficult.

The bacterial and fungal communities form the numerically dominant groups in the soil ecosystem, and these perform many functions and transformations. Other groups such as the meso- and macro-fauna, molluscs, earthworms, ants, etc. increase the dynamics of these former groups' functions by breaking up the organic matter into smaller pieces, increasing the available surface areas for activity, and distributing it through the system, and so bringing more organisms into direct contact with it. The functions include the release and transformation of mineral nitrogen and phosphorus mobilization for plant growth, nitrogen fixation, plant growth promotion and pathogen inhibition (Van Elsas *et al.*, 1997).

Most functions in the nitrogen cycle are directly dependent on carbon inputs from plants for their energy requirements. The organisms responsible for nitrogen fixation, which is the second fundamental reaction after carbon fixation required for the introduction of essential elements from the troposphere into the biosphere, acquire their energy from the plants, particularly when in a nodulated association. Some nitrogen-fixing organisms are 'free-living' in the soil, but these too are dependent on fixed carbon from plants. Other processes reliant on plant outputs include mycorrhizal associations and the chelation of insoluble trace elements to allow plant uptake, and similarly organisms require carbon inputs for denitrification to occur. The one autotrophic process in the cycle, nitrification, although not dependent on fixed carbon for energy, is still responsive to fixed carbon inputs (Wheatley *et al.*, 2001). These relationships suggest that selection from these functions for environmental impact assessments will be useful.

Organic Matter Inputs

The global reservoir of organic carbon in soils is larger than the combined total of that present in both the biological and atmospheric components (Schimel, 1995). However, activity in soil ecosystems is frequently limited by the availability of energy, which is obtained from this organic carbon, especially in the bulk soil. Soil organic matter is formed from the residues of plants, microbes and animals that have been modified by both biological and abiotic phenomena. On a global scale, the maintenance of this pool of soil carbon, with the potential to reduce carbon fluxes to the atmosphere, is of great importance. At the smaller, more local scale, interactions between the soil organisms and organic matter have profound effects on functional dynamics involved in sustainability of soil fertility, maintenance of soil structure and waste disposal. Plant inputs are the major drivers of activity in soil ecosystems as providers of the fixed-carbon compounds, i.e. products of primary production, required for energy. These inputs come from the plant roots, as root exudates, cellular remains and root debris, and plant residues that fall on to the soil surface, such as leaves, stems, flowers and fruit. Total energy-flow estimates for terrestrial ecosystems suggest that between 60% and 90% of net primary production is dissipated by the respiration of decomposer organisms (Brady and Weil, 1999). Although some microbial activity can occur in the bulk soil, it is much greater close to, and on, the root, in the rhizosphere and rhizoplane. Inputs to the soil from the plant are greatest in these regions, and as these inputs change, with both plant type and growth stage, so do microbial functional dynamics.

In addition to being providers of energy, these carbon inputs can also affect microbial function in more subtle ways, as they are involved in microbial interactions and signalling. Temporally variable interactions that are mediated by these plant inputs occur between the bulk microbial population and specific functional groups of microorganisms, such as the autotrophic nitrifiers, in arable soils. Microbial processes have been shown to be particularly responsive to introduced protein substrates (Wheatley et al., 1997, 2001) over a range of concentrations, from that comparable to root exudate levels to those after residue incorporation.

Microbial Diversity in Soils

Earliest investigations of soil ecosystems established that the microbial populations within them, both in type and function, were immensely diverse. More recent studies, frequently using an ever-expanding suite of nucleic acid-based methodologies not only confirm this but have further emphasized the vast extent of the potential range of species and functional clones of microorganisms in such populations, and the tremendous variation in them over time and space. It has been estimated that less than 5% of the microbial species in soils have been described of an estimated 10^4 species in 1 g of soil (Torsvik et al., 1990). The reasons for, and the roles of, such biodiversity in ecosystem functioning and population dynamics, and their spatial–temporal variations have been, and are frequently, investigated.

The many different soil types combined with atmospheric, moisture and organic matter variations combine to form a vast array of different highly heterogeneous environments. These environments show great variations in chemical gradients and physical conditions, both spatially, from micro to large scale, and temporally, diurnally to annually. The vast range of spatially separated habitats resulting from this possibly explains why completion does not exclude more species. The outcome of a combination of this phenomenon and the great number of selective evolutionary pressures enforced by this vast number of ecological niches, over the thousands of years that soils have developed, is a microbial population with immense species richness and diversity. This wide range of variable physical and chemical conditions also facilitates the occurrence of a wide variety of biogeochemical processes. As a consequence there is immense species and functional diversity in plant–soil systems and many compounds are formed, many of which can be further transformed in other processes. The complexity of soil systems makes system predictions and modelling extremely difficult, because such spatial and temporal variations occur over a very wide range of scales, small to large.

This large biodiversity of species, of which the greater majority are unknown, suggests that attempts to describe the system, or changes to it, using species-specific parameters will probably be both impractical and unreliable. However, there are means to assess changes in population structure that, although still somewhat related to species parameters, can give useful indications of changes in constituent members. The effects of different crop plants, including GM crops, on microbial community structure can be studied using recently developed methods that assess the structural diversity of that community. Depending on the choice of primers the whole or constituent parts, such as specific genera or functional groups, of the microbial population can be described. Diribonucleic acid (DNA) extracts from the relevant soil are amplified by polymerase chain reaction (PCR), using specific primers. Then the products can be separated into bands on a gel by differential gradient gel electrophoresis (DGGE) or temperature gradient gel electrophoresis (TGGE; Muyzer and Smalla, 1998). Such techniques overcome the limitations of culture-based techniques, where only a small fraction, which may not be representative of the whole, may be successfully grown. These methods have revealed soil microbial population changes both during the growth of the plant and in response to changes in crop and cultivar (Fig. 10.2; Smalla *et al.*, 2001; Pennanen *et al.*, 2003). Possibly a more relevant approach would be to examine effects on functional dynamics. But there is a requirement to select key functional properties that rank high in importance in the sustainability of soil ecosystems. The process of decomposition is an obvious choice as it is one of the two major life-generating processes, the other is photosynthesis. Other processes that are

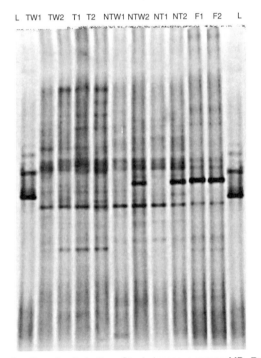

L = marker (from top to bottom *Staphylococcus aureus* MB, *Bacillus subtillis* IS 75, *Escherichia coli* HB101); TW = tillage with winter cover crop; T = tillage without winter cover crop; NTW = no tillage with winter cover crop; NT = no tillage without winter cover crop; F = native forest; 1 = first sampling; 2 = second sampling.

Fig. 10.2. DGGE banding pattern of *rpo*B PCR amplification of soil samples (0–5 cm depth). (Image courtesy of Raquel Peixoto, MSc dissertation, Federal University of Rio de Janeiro; after Mendonça-Hagler *et al.*, 2006.)

known to be responsive to plant inputs and critical to the sustained functioning of the biosphere, such as transformation processes within the nitrogen cycle, are also good candidates.

Soil Processes and the Impact of GM Crops

A combination of plant species and genotype, together with soil physical conditions, determines the amounts and types of compounds entering the soil, and so the choice of prevalent microbially driven functions and rates. Large differences in the relative proportions of individual bacterial genera have been reported on different plant species and cultivars in different soil types (Grayston *et al.*, 1998; Siciliano *et al.*, 1998; Dunfield and Germida, 2001). Soil ecosystem functional dynamics are also affected by other factors such as weather and management, but soil fertility depends primarily on microbial activity, which is responsive to plant inputs, particularly in the plant-rooting zone.

In this chapter, the possible impacts of GM crops on agricultural and natural systems are being considered. However, it should be borne in mind that all the topics considered will apply equally to the introduction of new, or indeed different, cultivars or crops, and any changes in agronomic practice, such as cultivation techniques and timing, including planting and sowing rates and time, pesticide usage, etc. The introduction or addition of anything to any biological system will, of consequence, change the dynamics in some dimension either trivially or greatly, even catastrophically, within that system and between the elements that constitute that system. So there is a serious requirement for improving both techniques and methods used to assess the impact of all agro-management regimes, whether GM crops are included or not, on soil.

Soil ecosystem populations and functional dynamics will change when GM crops are cultivated, but there will also be a similar response to the growth of any crop. The question maybe is whether such impacts are beneficial or damaging, and how such changes compare to those seen when a non-transformed crop is cultivated. Also of importance is whether these changes are transient or reversible? Further, should the risk assessment design be responsive to the consequences of the purpose of the GM transformation? For example, what ecosystem parameters should be measured when plants modified for insecticidal properties, e.g. *Bacillus thuringiensis* (Bt) transformations, are being assessed? And will these be different from the parameters assessed when the GM crop events show greater nutrient uptake rates, or drought and disease resistance? These evaluations are necessary and required to ensure that competent protocols are used to identify any and all possible effects, both beneficial and harmful, to soil ecosystems.

Possible greater attention will be paid to revealing potentially harmful effects, as the need for caution is probably greater than that for innovation. Our limited understanding of the drivers of soil ecosystem functioning should not result in such caution leading to requirements for a zero risk as the only acceptable outcome.

There is a need to define parameters that should be examined. In such complex systems are both broad and specific characterizations required, and what should these cover?

With the methods presently available, covering functional, molecular and physical parameters, a reasonably detailed description of population constituents, functional expression and soil health and fertility can be achieved. Present molecular methods allow definition of populations in great detail, but what does this mean, and is a functional assessment more useful? Such questions require decisions.

Assessment of the Impact of GM Crops

For the assessment of the possibility of GM crops impacting on soil ecosystems, approaches that utilize both broad and specific assessments of soil systems, both in type and function, together with investigations of effects on soil health and quality have been recommended by several investigators (Birch *et al.*, 2004; Lilley *et al.*, 2006; Mendonça-Hagler *et al.*, 2006).

The vast complexity of soil ecosystems makes it impractical and also unreliable, due to the difficulties inherent in species definitions, to assess impacts and compare systems using species-based rationales. Although there are numerically far fewer functional properties, it would still be impractical to attempt to measure them all. So a choice of keystone indicators is required. Considering the likely impacts of plants, whether GM or not, on soil ecosystems, investigating parameters, both biological and structural, that relate to carbon inputs and flows should provide the best indication and definition of consequential effects. Any effects of different crops on microbial community structure could be investigated using molecular techniques such as DGGE, or TGGE. Other similar techniques such as terminal restriction fragment length polymorphism (T-RFLP; Liu *et al.*, 1997) could also be used.

Other key parameters that can be determined by functional assays include the decomposition of organic matter and concomitant nitrogen transformation processes, such as ammonification and nitrification. Other microbial processes that can be of equal importance, dependent on the crop and cultivation practices, are nitrogen fixation and mycorrhizal associations. A range of other physiologically based approaches can also be used, including community-level physiological profiling (CLPP) using Biolog plates (Griffiths *et al.*, 2000; Buyer *et al.*, 2001). As the Biolog method is based on growth in different substrates, it will give some information on species diversity but is more informative of changes in functional properties. Changes in the phenotypic composition of communities can also be examined using phospholipid fatty acid profiles (PFLA; Blackwood and Buyer, 2004). Different subsets of the microbial community show different PFLA profiles, so changes in the relative ratios of these can be determined, for example, in response to changes in the inputs from a growing crop.

Hopkins *et al.* (2001) reported that residues of tobacco plants in which lignin biosynthesis had been modified were decomposed significantly quicker than residues of the wild type. In a review of studies of the responses of soil microorganisms to GM plants, Kowalchuk *et al.* (2003) concluded that a relative lack of significant non-target effects had been found after many studies of a wide range of microbial traits. Although some minor non-target effects had been reported in some studies, in many others no significant effects were reported. Similarly, Lilley et al. (2006) state that although most transgenic plants have detectable effects on the soil system, these are relatively minor compared to differences between cultivars, or those associated with weather or season, and are transient. Studies of a range of different crops have shown differences in the microbial community structure, both fungal and bacterial, in the rhizosphere of the GM crop compared to the non-transgenic control (Dunfield and Germida, 2004), but these were also transient, there being no detectable differences in the field soil during the next growing season. Consideration must be given to the consequences, possibly adverse, of the continual cultivation of a GM crop or crops expressing a common transgene product, such as Bt maize and Bt cotton, when the effects on the microbial population of the continual input may not be transient. Also as the transgene products become more radical and ambitious, such as for the synthesis of pharmaceuticals, the potential for more marked changes may well be increased.

Fate of free DNA in the environment

In a review, De Vries and Wackernagel (2004) stated that intraspecific and interspecific horizontal gene transfers, conjugation and transduction, are part of the lifestyle of prokaryotes and have shaped microbial genomes throughout evolution. However, because of the great complexity and variability in type, functional expression and interactions in soils, they conclude that although transfers between plants and bacteria are possible in principle, each of the many steps involved in the release of intact DNA from the plant to integration into the prokaryotic genome has such a low probability of completion that successful transfer events will be extremely rare. Free DNA and competent bacteria will have to be in close proximity for natural transformations to occur in soils (Smalla et al., 2001), which is an obvious possibility in the rhizosphere. Marker genes from some transformed plants have been detected on occasions in soil, for example, such from tobacco and potato plants were still detectable 77 and 137 days after the crop (Widmer et al., 2001). Similarly, DNA from transgenic sugarbeet plants could be detected for several months in the field (Nielsen et al., 2000).

Recombination with transgenic plant DNA fragments has been detected in sterile soil that had been inoculated with a naturally transformed bacterium of *Acinetobacter* sp. when cultured under optimal conditions in the laboratory (Gebhard and Smalla, 1999; Nielsen et al., 2000). Also in other investigations, non-competent *Acinetobacter* sp. cells have been made competent by the application of a variety of inorganic salts and simple carbon sources similar to those found in root exudates (Neilsen and van Elsas, 2001). Despite this theoretical possibility that genes might be transferred from GM-transformed plants to indigenous soil bacteria, there are no reports of such occurring in field soils. Indeed, Schlüter et al. (1995) did not detect horizontal gene transfer under conditions that mimicked a 'natural infection'. However, transfer did occur under optimized laboratory conditions, at a frequency of 6.3×10^{-2}, which reduced as conditions were altered to idealized natural conditions to 2.0×10^{-17}. They argued that this demonstrated that horizontal gene transfer is so rare in natural soil conditions that it is essentially irrelevant to any realistic risk assessment relating to transgenic plants. A report by Kim *et al.* (2004) supported this view and strongly rejected the possibility of gene transfer between plants and bacteria. They found that none of the transgenes introduced into genetically modified organism (GMO) soybeans could be detected in the soil bacteria *Rhizobium leguminosarum*.

There are other reports that horizontal gene transfer between GM plants and the indigenous microbial population may occur at very low frequencies (Nielsen *et al.*, 1998; Gebhard and Smalla, 1999). But the relevancy of such occurrences to the integrity of the soil ecosystem will be entirely dependent on the viability of any organisms that have received the material. However, it can be argued that the vast species resource in soil is a reflection of the low selective pressures operating in soils. There are many possible reasons for this, including spatial separation and changing biophysical gradients. So the possibility that such transforms may persist in soils must be considered.

Role of the soil micro-, meso- and macro-fauna as detritivores and disseminators

Soil micro- (e.g. protozoa and nematodes with a body width <0.1 mm), meso- (e.g. micro-arthropods and enchytraeids, 0.1–2 mm body width) and macro- (e.g. earthworms, termites and millipedes, >2 mm body width) faunal communities are extremely complex, and not particularly well described in soils. As well as being grouped by size, they can also be put into functional groups based on food choice and lifestyle. These groups also assemble into complex food webs. Similar to the microbial population, there are a vast number of species in all these groupings, many of which have not been identified or logged.

These communities have members that are beneficial to biosphere functioning and crop production, and others that are detrimental. Beneficial members include the detrivores and disseminators. The dynamics of organic matter decomposition are greatly increased by the activities of these organisms. Some, such as beetles and molluscs, break down the organic material into small pieces, so increasing the surface area for microbial activity. Others such as the earthworms and ants move pieces of material around the system, again improving the opportunity for further microbial activity. Others in these groups, considered pests, can cause damage to the crop, with subsequent loss of yield. Indeed some GM-transformed crops have been developed with the intention of controlling such pests. However, transgenic plant material, products and metabolites from such GM crops may affect non-target species and ecosystem functional dynamics (Saxena *et al.*, 1999; Saxena and Stotzky, 2000; Hilbeck, 2002; Zwahlen *et al.*, 2003a,b). As well as any direct effect, the transgenic plant material may affect non-target species indirectly through another organism, such as an herbivore (Birch et al., 1999; Hilbeck et al., 1999) or honeydew from homopteran species such as aphids, scales or white-flies (Raps et al., 2001; Bernal et al., 2002) in a food web. The number of possible pathways is immense, as it has been estimated that there are more than 250 different pathways by which a transgenic product could affect a secondary consumer, and only a few are direct (Andow and Hilbeck, 2004). Protocols for potential exposure have to accommodate this multitude of potential exposure pathways.

Agricultural practice

Global requirements for agricultural production, for food, fuel and manufacturing materials have never been greater and will continue to increase. The major environmental challenge is to balance optimal production with sustainability in functional and species biodiversity. We know that all agricultural activities, drainage, fertilizer use, the application of pesticides, etc. will have very significant effects on the dynamics of soil ecosystems. One of the greatest effects results from monoculture of crop plants, which has obvious limitations for biodiversity of plants within that system, and also sequentially for all the other contributors, be they insects, mammals or microbes.

Concerns over the possible effects of the commercial introduction of GM crops on the environment have caused great debate. The basic questions in any risk assessment related to the introduction of any plant are the same regardless of the method of plant breeding used. There are two important considerations: The first, the direct effect of the crop on the environment, and second, the consequences, be they in changed cultivation methods or in changed primary inputs, to the food web. GM crops raise particular concerns in this last area, especially when designed to impact on the dynamics of the food chain, e.g. Bt-expressing crops for insect control. There are concerns about 'non-target' organisms, where insects that are not pests are affected, with consequences for other vital processes such as decomposition and the recycling of plant nutrients for sustained system functioning. Other areas of concern include the possibility of outcrossing, where the recombinant DNA may transfer to some other organism, microbial or plant, and produce uncontrollable weeds; or other adverse effects on wild life.

All agricultural activity impacts on the environment. Presently environmental impacts of GM crops are compared against a baseline of the non-GM-isoline. Perhaps this comparison should be broadened to compare the impacts of GM crops to those of traditional, integrated or organic farming practices. Presently there is no convincing body of evidence that GM crops impact on the environment any differently to non-GM crops. But as agricultural practices profoundly affect the environment any way, perhaps this is not surprising. However, without doubt the potential of the products of GM crops will present new challenges, and probably lead to the introduction of innovative and creative new agro-managerial practices.

References

Andow, D. and Hilbeck, A. (2004) Science-based risk assessment for non-target effects of transgenic crops. *Bioscience* 54, 637–649.

Bernal, C.C., Aguda, R.M. and Cohen, M. (2002) Effect of rice lines transformed with *Bacillus thuringensis* toxin genes on the brown planthopper and its predator *Cyrtortinus lividipennis. Entomologia Experimentalis et Applicata* 102, 21–28.

Birch, A.N.E., Geoghegan, I.E., Majerus, M.E.N., McNicol, J.W., Hackett, C.A., Gatehouse, A.M.R. and Gatehouse, J. (1999) Tritrophic interactions involving pest aphids, predatory 2-spot ladybirds and transgenic potatoes expressing snowdrop lectin for aphid resistance. *Molecular Breeding* 5, 75–83.

Birch, A.N.E., Wheatley, R.E., *et al.* (2004) Biodiversity and non-target impacts: a case study of Bt maize in Kenya. In: Hilbeck, A. and Andow, D.A. (eds) *Environmental Risk Assessment of Genetically Modified Organisms Vol. 1: A Case Study of Bt Maize in Kenya.* CAB International, Wallingford, UK.

Blackwood, C.B. and Buyer, J.S. (2004) Soil microbial communities associated with Bt and non-Bt corn in three soils. *Journal of Environmental Quality* 33, 832–836.

Brady, N.C. and Weil, R.R. (1999) *Nature and Properties of Soils.* Prentice-Hall, London.

Buckley, D.H. and Schmidt, T.M. (2001) The structure of microbial communities in soils and the lasting impact of cultivation. *Microbial Ecology* 42, 11–21.

Buyer, J.S., Roberts, D.P., Millner, P. and Russek-Cohen, E. (2001) Analysis of fungal communities by single carbon utilization profiles. *Journal of Microbiological Methods* 45, 53–60.

De Vries, J. and Wackernagel, W. (2004) Microbial horizontal gene transfer and the DNA release from transgenic crop plants. *Plant and Soil* 266, 91–104.

Donegan, K.K., Seidler, R.J., Doyle, J.D., Porteous, L.A., DiGiovanni, G., Widmer, F. and Watrud, L.S. (1999) A field study with genetically engineered alfalfa inoculated with recombinant Sinorhizobium meliloti; effects on the soil ecosystem. *Journal of Applied Ecology* 36, 920–936.

Dunfield, K.E. and Germida, J.J. (2001) Diversity of bacterial communities in the rhizosphere and root-interior of field-grown genetically modified Brassica napus. *FEMS Microbiology Ecology* 82, 1–9.

Dunfield, K.E. and Germida, J.J. (2004) Impact of genetically modified crops on soil- and plan-associated microbial communities. *Journal of Environmental Quality* 33, 806–815.

Gebhard, F. and Smalla, K. (1999) Monitoring field releases of genetically modified sugar beets for persistence of transgenic plant DNA and horizontal gene transfer. *FEMS Microbiology Ecology* 28, 261–272.

Grayston, S.J., Wang, S., Campbell, C.D. and Edwards, A.C. (1998) Selective influence of plant species on microbial diversity in the rhizosphere. *Soil Biology and Biochemistry* 30, 369–378.

Griffiths, B.S., Geoghegan, I.E. and Robertson, W.M. (2000) Testing genetically engineered potato, producing the lectins GNA and Con A, on non-target soil organisms and processes. *Journal of Applied Ecology* 37, 159–170.

Hilbeck, A. (2002) Transgenic host plant resistance and non-target effects. In: *Genetically Engineered Organisms: Assessing Environmental and Human Health Effects.* Letourneau, D.K. and Burrows, B.E. (eds) CRC Press, New York, pp. 167–185.

Hilbeck, A., Moar, W.J., Pusztai-Carrey, M., Filippini, A. and Bigler, F. (1999) Prey mediated effects of Cry1Ab toxin and protoxin and Cry2A protoxin on the predator *Chrysoperia carnea. Entomologia Experimentalis et Applicata* 91, 305–316.

Hopkins, D.W., Webster, E.A., Chudek, J.A. and Halpin, C. (2001) Decomposition in soil of tobacco plants with genetic modifications to lignin biosynthesis. *Soil Biology and Biochemistry* 33, 1455–1462.

Kim, Y.T., Park, B.K., Hwang, E.I. (2004) Investigation of possible gene transfer to soil microorganisms for environmental risk assessment of genetically modified organisms. *Journal of Microbiology and Biotechnology* 14, 498–502.

Kowalchuk, G.A., Bruinsma, M. and van Veen, J.A. (2003) Assessing responses of soil microorganisms to GM plants. *Trends in Ecology and Evolution* 18, 403–410.

Lilley, A.K., Bailey, M.J., Cartwright, C., Turner, S.L. and Hirsch, P.R. (2006) Life in earth: the impact of GM plants on soil ecology? *Trends in Biotechnology* 24, 9–14.

Liu, W.T., Marsh, T.L., Cheng, H. and Forney, L.J. (1997) Characterization of microbial diversity by determining terminal restriction fragment length polymorphisms of genes encoding 16s rRNA. *Applied Environmental Microbiology* 63, 4516–4522.

Lukow, T., Dunfield, P.F. and Liesack, W. (2000) Use of T-RLFP technique to assess spatial and temporal changes in the bacterial community structure within an agricultural soil planted with transgenic and non-transgenic potato plants. *FEMS Microbiology Ecology* 32, 241–247.

Mendonça-Hagler, L.C., de Melo, I.S., Valadares-Inglis, M.C., Anyango, B.M., Van Toan, P. and Wheatley, R.E. (2006) Non-target and biodiversity in soil. In: Hilbeck, A., Andow, D.A. and Fontes, E.M.G. (eds) *Environmental Risk Assessment of Genetically Modified Organisms Vol. 2: Methodologies for Assessing Bt Cotton in Brazil.* CAB International, Wallingford, UK.

Muyzer, G. and Smalla, K. (1998) Application of differential gradient gel electrophoresis (DGGE) in microbial ecology. *Antonie van Leeuwenhoek* 73, 127–141.

Nielsen, K. and van Elsas, M.J.D. (2001) Stimulatory effects of compounds present in the rhizosphere on natural transformation of *Acinetobacter* sp BD413 in soil. *Soil Biology and Biochemistry* 33, 345–357.

Nielsen, K.M., Bones, A.M., Smalla, K. and van Elsas, J.D. (1998) Horizontal gene transfer from transgenic plants to terrestrial bacteria – a rare event? *FEMS Microbiological Reviews* 22, 79–103.

Nielsen, K.M., van Elsas, J.D. and Smalla, K. (2000) Transformation of *Acinetobacter* sp

strain BD413(pFG4 Delta nptII) with transgenic plant DNA in soil microcosms and effects of kanamycin on selection of transformants. *Applied and Environmental Microbiology* 66, 1237–1242.

Pennanen, T., Caul, S., Daniell, T.J., Griffiths, B.S., Ritz, K. and Wheatley, R.E. (2003) Community-level responses of metabolically active soil microbes to variations in quantity and quality of substrate inputs. *Soil Biology and Biochemistry* 36, 841–848.

Raps, A., Kehr, J., Gugerli, P., Moar, W.J., Bigler, F. and Hilbeck, A. (2001) Immunilogical analysis of phloem sap of *Bacillus thuringiensis* corn and of the non-target herbivore *Rhopalosiphum padi* (Homoptera: Aphidae) for presence of Cry1Ab. *Molecular Ecology* 10, 525–533.

Saxena, D. and Stotzky, G. (2000) Insecticidal toxin from Bacillus thuringiensis is released from roots of transgenic Bt corn *in vitro* and *in situ*. *FEMS Microbiology Ecology* 33, 35–39.

Saxena, D., Flores, S. and Stotzky, G. (1999) Transgenic plants – Insecticidal toxin in root exudates from Bt corn. *Nature* 402, 480–480.

Schimel, D.S. (1995) Terrestrial ecosystems and the carbon cycle. *Global Change Biology* 1, 77–91.

Schlüter, K., Futterer, J. and Potrykus, I. (1995) Horizontal gene transfer from a transgenic potato line to a bacterial pathogen (*Erwinia chrysanthemi*) occurs, if at all, at an extremely low frequency. *Biotechnology* 13, 1094–1098.

Siciliano, S.D., Theoret, C.M., Freitas, J.R. de, Hucl, P.J. and Germida, J.J. (1998) Differences in the microbial communities associated with the roots of different cultivars of canola and wheat. *Canadian Journal of Microbiology* 44, 844–851.

Smalla, K., Wieland, G., Buchner, A., Zock, A., Parzy, J., Kaiser, S., Roskot, N., Heuer, H. and Berg, G. (2001) Bulk and rhizosphere soil bacterial communities studied by denaturing gradient gel electrophoresis: Plant-dependent enrichment and seasonal shifts revealed. *Applied and Environmental Microbiology* 67, 4742–4751.

Torsvik, V., Goksoyr, J. and Daae, F. (1990) High diversity in DNA in soil bacteria. *Applied and Environmental Microbiology* 56, 782–787.

Van Elsas, J.D., Trevors, J.T. and Wellington, E.M.H. (1997) *Modern Soil Microbiology.* Marcel Dekker, New York.

Wheatley, R.E., Hackett, C., Bruce, A. and Kundzewicz, A. (1997) Effect of substrate composition on production and inhibitory activity against wood decay fungi of volatile organic compounds from *Trichoderma* spp. *International Biodeterioration and Degradation* 39, 199–205.

Wheatley, R.E., Ritz, K., Crabb, D. and Caul, S. (2001) Temporal variations in potential nitrification dynamics related to differences in rates and types of carbon inputs. *Soil Biology and Biochemistry* 33, 2135–2144.

Widmer, F., Fliessbach, A., Laczko, E., Schulze-Aurich, J. and Zeyer, J. (2001) Assessing soil biological characteristics: a comparison of bulk soil community DNA-, PLFA-, and Biolog-analyses. *Soil Biology and Biochemistry* 33, 1029–1036.

Zwahlen, C., Hilbeck, A., Gugerli, P. and Nentwig, W. (2003a) Degradation of the Cry1Ab protein within transgenic *Bacillus thuringiensis* corn tissue in the field. *Molecular Ecology* 12, 765–775.

Zwahlen, C., Hilbeck, A., Howald, R. and Nentwig, W. (2003b) Effects of transgenic Bt corn litter on the earthworm *Lumbricus terrestris. Molecular Ecology* 12, 1077–1086.

11 Biodiversity and Genetically Modified Crops

K. AMMANN

Delft University of Technology, The Netherlands

Keywords: Biodiversity, biodiversity management in agriculture, sustainable development, ecosystems, genetic introgression, exotic/alien species, transgenics

Summary

Biological diversity (*biodiversity*) has emerged in the past decade as a key area of concern for sustainable development. At its highest level of organization, biodiversity is characterized as ecosystem diversity, which can be classified in the following three categories: *natural ecosystems* (ecosystems free of human activities); *semi-natural ecosystems* (in which human activity is limited); *managed ecosystems* (systems that are managed by humans to varying degrees of intensity from the most intensive, conventional agriculture and urbanized areas, to less intensive systems including some forms of agriculture in emerging economies). Yet, despite its importance, biodiversity in agriculture, i.e. crop biodiversity, which represents a variety of food supply choice for balanced human nutrition and a critical source of genetic material allowing the development of new and improved crop varieties, is rarely considered. Species and genetic diversity within any agricultural field will inevitably be more limited than in a natural or semi-natural ecosystem. Biodiversity in agricultural settings is particularly important in areas where the proportion of land allocated to agriculture is high, as seen in continental Europe. Under these circumstances, changes in agrobiological management will have a major influence on biodiversity. Innovative thinking about how to enhance biodiversity on the level of regional landscapes, coupled with bold action, is thus critical in dealing with the loss of biodiversity. Biodiversity should act as a biological insurance for ecosystem processes, except when mean trophic interaction strength increases strongly with diversity. The conclusion, yet to be validated from field studies, is that in tropical environments with a natural high biodiversity the interactions between potentially invasive hybrids of transgenic crops and their wild relatives should be buffered through the complexity of the surrounding ecosystems. Taken together, theory and data suggest that compared to intertrophic interaction and habitat loss, competition from introduced species is not likely to be a common cause of extinctions in long-term resident species at global, metacommunity and even most community levels.

The Needs for Biodiversity: the General Case

Biological diversity (often contracted to *biodiversity*) has emerged in the past decade as a key area of concern for sustainable development, but crop biodiversity, the subject of this chapter, is rarely considered. The author's important contribution to the discussion of crop biodiversity in this volume should be considered as part of the general case for biodiversity. Biodiversity provides a source of significant economic, aesthetic, health and cultural benefits. It is assumed that the well-being and prosperity of earth's ecological balance as well as human society directly depend on the extent and status of biological diversity (Table 11.1). Biodiversity plays a crucial role in all the major biogeochemical cycles of the planet. Plant and animal diversity ensures a constant and varied source of food, medicine and raw material of all sorts for human populations. Biodiversity in agriculture represents a variety of food supply choice for balanced human nutrition and a critical source of genetic material allowing the development of new and improved crop varieties. In addition to these direct-use benefits, there are enormous other less tangible benefits to be derived from natural ecosystems and their components. These include the values attached to the persistence, locally or globally, of natural landscapes and wildlife, values which increase as such landscapes and wildlife become more scarce. The relationships between biodiversity and ecological parameters, linking the value of biodiversity to human activities, are partially summarized in Table 11.1.

Biological diversity may refer to diversity in a gene, species, community of species, or ecosystem, or even more broadly to encompass the earth as a whole. Biodiversity comprises all living beings, from the most primitive forms of viruses to the most sophisticated and highly evolved animals and plants. According to the 1992 International Convention on Biological Diversity, biodiversity means 'the variability among living organisms from all sources including, terrestrial, marine, and other aquatic ecosystems and the ecological complexes of which they are part' (CBD, 1992). It is important not to overlook the various scale-dependent perspectives of biodiversity, as this can lead to many misunderstandings in the debate about biosafety. It is not a simple task to evaluate the needs for biodiversity, especially to quantify the agroecosystem biodiversity versus total biodiversity (Purvis and Hector, 2000; Tilman, 2000).

One example may be sufficient to illustrate the difficulties: Biodiversity is indispensable to sustainable structures of ecosystems. But sustainability has many facets, among others also the need to feed, and to organize proper health care for, the poor. This last task is of utmost importance and has to be balanced against biodiversity per se, such as in the now classic case of the misled total ban on DDT, which caused hundreds of thousands of malaria deaths in Africa in recent years; the case is summarized in many publications, and here is a small selection: Taverne (1999), Attaran and Maharaj (2000), Attaran *et al.* (2000), Curtis and Lines (2000), Horton (2000), Roberts *et al.* (2000), Smith (2000), Tren and Bate (2001), Curtis (2002), WHO (2005).

Table 11.1. Primary goods and services provided by ecosystems.

Ecosystem	Goods	Services
Agroecosystems	Food crops Fibre crops Crop genetic resources	Maintain limited watershed functions (infiltration, flow control, partial soil protection) Provide habitat for birds, pollinators, soil organisms important to agriculture Build soil organic matter Sequester atmospheric carbon Provide employment
Forest ecosystems	Timber Fuelwood Drinking and irrigation water Fodder Non-timber products (vines, bamboos, leaves, etc.) Food (honey, mushrooms, fruit, and other edible plants; game) Genetic resources	Remove air pollutants, emit oxygen Cycle nutrients Maintain array of watershed functions (infiltration, purification, flow control, soil stabilization) Maintain biodiversity Sequester atmospheric carbon Generate soil Provide employment Provide human and wildlife habitat Contribute aesthetic beauty and provide recreation
Freshwater ecosystems	Drinking and irrigation water Fish Hydroelectricity Genetic resources	Buffer water flow (control timing and volume) Dilute and carry away wastes Cycle nutrients Maintain biodiversity Sequester atmospheric carbon Provide aquatic habitat Provide transportation corridor Provide employment Contribute aesthetic beauty and provide recreation
Grassland ecosystems	Livestock (food, game, hides, fibre) Drinking and irrigation water Genetic resources	Maintain array of watershed functions (infiltration, purification, flow control, soil stabilization) Cycle nutrients Remove air pollutants, emit oxygen Maintain biodiversity Generate soil Sequester atmospheric carbon Provide human and wildlife habitat Provide employment Contribute aesthetic beauty and provide recreation

Continued

Table 11.1. Continued.

Ecosystem	Goods	Services
Coastal and marine ecosystems	Fish and shellfish	Moderate storm impacts (mangroves; barrier islands)
	Fishmeal (animal feed)	Provide wildlife (marine and terrestrial) habitat
	Seaweeds (for food and industrial use)	Maintain biodiversity
	Salt	Dilute and treat wastes
	Genetic resources	Sequester atmospheric carbon
		Provide harbours and transportation routes
	Petroleum, minerals	Provide human and wildlife habitat
		Provide employment
		Contribute aesthetic beauty and provide recreation
Desert ecosystems	Limited grazing, hunting	Sequester atmospheric carbon
	Limited fuelwood	Maintain biodiversity
	Genetic resources	Provide human and wildlife habitat
	Petroleum, minerals	Provide employment
		Contribute aesthetic beauty and provide recreation
Urban ecosystems	Space	Provide housing and employment
		Provide transportation routes
		Contribute aesthetic beauty and provide recreation
		Maintain biodiversity
		Contribute aesthetic beauty and provide recreation

Types, Distribution and Loss of Biodiversity

Genetic diversity

In many instances genetic sequences, the basic building blocks of life, encoding functions and proteins are almost identical (highly conserved) across all species. The small unconserved differences are important, as they often encode the ability to adapt to specific environments. Still, the greatest importance of genetic diversity is probably in the combination of genes within an organism (the genome), the variability in phenotype produced, conferring resilience and survival under selection. Thus, it is widely accepted that natural ecosystems should be managed in a manner that protects the untapped resources of genes within the organisms needed to preserve the resilience of the ecosystem. Much work remains to be done to both characterize genetic diversity and understand how best to protect, preserve and make wise use of genetic biodiversity (Raikhel and Minorsky, 2001; Mattick, 2004; Mallory and Vaucheret, 2006; Baum *et al.*, 2007; Batista *et al.*, 2008; Cattivelli *et al.*, 2008; Witcombe *et al.*, 2008).

The number of metabolites found in one species exceeds the number of genes involved in their biosynthesis. The concept of one gene–one transcript–one protein–one product needs modification. There are many more proteins than genes in cells because of post-transcriptional modification. This can partially explain the multitude of living organisms that differ in only a small portion of their genes. It also explains why the number of genes found in the few organisms sequenced is considerably lower than anticipated.

Species diversity

For most practical purposes, measuring species biodiversity is the most useful indicator of biodiversity, even though there is no single definition of what a species is. Nevertheless, a species is broadly understood to be a collection of populations that may differ genetically from one another to some extent, but whose members are usually able to mate and produce fertile offspring. These genetic differences manifest themselves as differences in morphology, physiology, behaviour and life histories; in other words, genetic characteristics affect expressed characteristics (phenotype). Today, about 1.75 million species have been described and named but the majority remain unknown. The global total might be ten times greater, most being undescribed microorganisms and insects (May, 1990).

Ecosystem diversity

At its highest level of organization, biodiversity is characterized as ecosystem diversity, which can be classified into the following three categories.

Natural ecosystems are ecosystems free of human activities. These are composed of what has been broadly defined as 'Native Biodiversity'. It is a matter of debate whether any truly natural ecosystem exists today, as human activity has influenced most regions on earth. It is unclear why so many ecologists seem to classify humans as being 'unnatural'.

Semi-natural ecosystems are ecosystems in which human activity is limited. These are important ecosystems that are subject to some level of low-intensity human disturbance. These areas are typically adjacent to managed ecosystems.

Managed ecosystems are the third broad classification of ecosystems. Such systems can be managed by humans to varying degrees of intensity from the most intensive, conventional agriculture and urbanized areas, to less intensive systems including some forms of agriculture in emerging economies or sustainably harvested forests.

Beyond simple models of how ecosystems appear to operate, we remain largely ignorant of how ecosystems function, how they might interact with each other and which ecosystems are critical to the services most vital to life on earth. For example, the forests have a role in water management that is crucial to urban drinking water supply, flood management and even shipping.

Because we know so little about the ecosystems that provide our life support, we should be cautious and work to preserve the broadest possible range of ecosystems, with the broadest range of species having the greatest spectrum of genetic diversity within the ecosystems. Nevertheless, we know enough about the threat to, and the value of, the main ecosystems to set priorities in conservation and better management. We have not yet learnt enough about the threat to crop biodiversity, other than to construct gene banks, which can only serve as an ultimate ratio – we should not indulge into the illusion that large seedbanks could really help to preserve crop biodiversity. The only sustainable way to preserve a high crop diversity, i.e. also as many landraces as possible, is to actively cultivate and breed them further on. This has been clearly demonstrated by the studies of Berthaud and Bellon (Berthaud, 2001; Bellon *et al.*, 2003; Bellon and Berthaud, 2004, 2006). Even here we have much to learn, as the vast majority of the deposits in gene banks are varieties and landraces of the four major crops. The theory behind patterns of general biodiversity related to ecological factors such as productivity is rapidly evolving, but many phenomena are still enigmatic and far from understood (Schlapfer *et al.*, 2005; Tilman *et al.*, 2005), as for example why habitats with a high biodiversity are more robust towards invasive alien species.

The global distribution of biodiversity

Biodiversity is not distributed evenly over the planet. Species richness is highest in warmer, wetter, topographically varied, less seasonal and lower elevation areas. There are far more species in total per unit area in temperate regions than in polar ones, and far more again in the tropics than in temperate regions. Latin America, the Caribbean, the tropical parts of Asia and the Pacific together host 80% of the ecological mega-diversity of the world. An analysis of global biodiversity on a strictly metric basis demonstrates that besides the important rainforest areas there are other hotspots of biodiversity related to tropical dry forests for example (Kuper *et al.*, 2004; Kier *et al.*, 2005; Lughadha *et al.*, 2005).

Within each region, every specific type of ecosystem will support its own unique suite of species, with their diverse genotypes and phenotypes. In numerical terms, global species diversity is concentrated in tropical rainforests and tropical dry forests. Amazon basin rainforests can contain up to nearly 300 different tree species per hectare and support the richest (often frugivorous) fish fauna known, with more than 2500 species in the waterways. The submontane tropical forests in tropical Asia and South America are considered to be the richest per unit area in animal species in the world (Vareschi, 1980).

The case of agro-biodiversity

Species and genetic diversity within any agricultural field will inevitably be more limited than in a natural or semi-natural ecosystem. Many of the crops growing in farming systems all over the world have surprisingly enough ancestral parent traits which lived originally in natural monocultures (Wood and Lenne, 2001). This is after all most probably the reason why our ancestral farmers have chosen those major crops. There are many examples of natural monocultures, such as the classic stands of Kelp, *Macrocystis pyrifera*, already analysed by Darwin (1845), and more relevant to agriculture: It has now been recognized by ecologists that simple, monodominant vegetation exists throughout nature in a wide variety of circumstances. Indeed, Fedoroff and Cohen (1999) reporting Janzen (1998, 1999) use the term 'natural monocultures' in analogy with crops. Monodominant stands may be extensive. As one example of many, Harlan recorded that for the blue grama grass (*Bouteloua gracilis*): 'stands are often continuous and cover many thousands of square kilometers' of the high plains of central USA. It is of the utmost importance for the sustainability of agriculture to determine how these extensive, monodominant and natural grassland communities persist when we might expect their collapse. Although numerous examples are given in the literature (Wood and Lenne, 1999), here only a few cases will be cited: wild species – *Picea abies, Spartina townsendii*, various species of bamboos, *Arundinaria* ssp. (Gagnon and Platt, 2008), *Sorghum verticilliflorum, Phragmites communis* and *Pteridium aquilinum*. Ancestral cultivars are cited extensively by Wood and Lenne (2001): wild rice, *Oryza coarctata*, reported in Bengal as simple, oligodiverse pioneer stands of temporarily flooded riverbanks (Prain, 1903); Harlan described *Oryza* (Harlan, 1989) and illustrated harvests from dense stands of wild rice in Africa (*Oryza barthii*, the progenitor of the African cultivated rice, *Oryza glaberrima*). *O. barthii* was harvested wild on a massive scale and was a local staple across Africa from the southern Sudan to the Atlantic. Evans (1998) reported that the grain yields of wild rice stands in Africa and Asia could exceed $0.6\,t\,ha^{-1}$ – an indication of the stand density of wild rice.

Botanists and plant collectors have, according to Wood and Lenne (2001), repeatedly and emphatically noted the existence of dense stands of wild relatives of wheat. For example, in the Near East, Harlan (1992) noted that 'massive stands of wild wheats cover many square kilometers'. Hillmann (1996) reported that wild einkorn (*Triticum monococcum* ssp. *boeoticum*) in particular tends to form dense stands, and when harvested its yields per square metre often match those of cultivated wheats under traditional management. Harlan and Zohary (1966) noted that wild Einkorn 'occurs in massive stands as high as 2000 meters [altitude] in south-eastern Turkey and Iran'. Wild emmer (*Triticum turgidum* ssp. *dicoccoides*) 'grows in massive stands in the northeast' of Israel, as an annual component of the steppe-like herbaceous vegetation and in the deciduous oak park forest belt of the Near East (Nevo, 1998). According to Wood and Lenne (2001) they are the strongest examples embracing wild progenitors of wheat: Anderson (1998) recorded wild wheat growing in Turkey and Syria in natural, rather pure stands with a density of $300\,m^{-2}$.

Nevertheless, agricultural ecosystems can be dynamic in terms of species diversity over time due to management practices. This is often not understood by ecologists who involve themselves in biosafety issues related to transgenics. They still think in ecosystems close (or seemingly close) to nature. Biodiversity in agricultural settings can be considered to be important at country level in areas where the proportion of land allocated to agriculture is high (Ammann in Wolfenbarger *et al.*, 2004). This is the case in continental Europe for example, where 45% of the land is dedicated to arable and permanent crops or permanent pasture. In the UK, this figure is even higher, at 70%. Consequently, biodiversity has been heavily influenced by humans for centuries, and changes in agrobiological management will influence biodiversity in such countries overall. Innovative thinking about how to enhance biodiversity in general coupled with bold action is critical in dealing with the loss of biodiversity. High potential to enhance biodiversity considerably can be seen on the level of regional landscapes, as is proposed by Dollaker *et al.* (Dollaker, 2006; Dollaker and Rhodes, 2007); with the help of remote sensing methods it should be possible to plan for a much better biodiversity management in agriculture (Mucher *et al.*, 2000).

Centres of biodiversity are a controversial matter, and even the definition of centres of crop biodiversity is still debated. Harlan (1971) proposed a theory that agriculture originated independently in three different areas and that, in each case, there was a system composed of a centre of origin and a non-centre, in which activities of domestication were dispersed over a span of 5000–10,000 km. One system was in the Near East (the Fertile Crescent) with a non-centre in Africa; another centre includes a north Chinese centre and a non-centre in South-east Asia and the south Pacific, with the third system including a Central American centre and a South American non-centre. He suggests that the centres and the non-centres interacted with each other.

There is a widespread view that centres of crop origin should not be touched by modern breeding because these biodiversity treasures are so fragile that these centres should stay free of modern breeding. This is an erroneous opinion, based on the fact that regions of high biodiversity are particularly susceptible to invasive processes, which is wrong. On the contrary, there are studies showing that a high biodiversity means more stability against invasive species, as well as against genetic introgression (Morris *et al.*, 1994; Whitham *et al.*, 1999; Tilman *et al.*, 2005). The introduction of new predators and pathogens has caused well-documented extinctions of long-term resident species, particularly in spatially restricted environments such as islands and lakes. One of the (in)famous cases of an extinction of an endemic rare moth is documented from Hawaii; it has been caused by a failed attempt of biological control (Howarth, 1991; Henneman and Memmott, 2001). However, there are surprisingly few instances of extinctions of resident species that can be attributed to competition from new species. This suggests either that competition-driven extinctions take longer to occur than those caused by predation or that biological invasions are much more likely to threaten species through intertrophic than through intratrophic interactions (Davis, 2003). This also fits well with agricultural experience, which builds on much faster ecological processes.

Loss of Biodiversity

Biodiversity is being lost in many parts of the globe, often at a rapid pace. It can be measured by loss of individual species, groups of species or decreases in numbers of individual organisms. In a given location, the loss will often reflect the degradation or destruction of a whole ecosystem. The unchecked rapid growth of any species can have dramatic effects on biodiversity. This is true of weeds, elephants but especially humans, who being at the top of the chain can control the rate of proliferation of other species, as well as their own, when they put their mind to it.

Habitat loss due to the expansion of human urbanization and the increase in cultivated land surfaces is identified as a main threat to 85% of all species described as being close to extinction. The shift from natural habitats towards agricultural land paralleled population growth, often thoroughly and irreversibly changing habitats and landscapes, especially in the developed world. Many from the developed world are trying to prevent such changes from happening in developing nations, to the consternation of many of the inhabitants of the developing world who consider this to be eco-imperialism, promulgated by those unable to correct their own mistakes. A clear decline of biodiversity due to agricultural intensification is documented by Robinson and Sutherland (2002) for the post-war period in Great Britain.

Today, more than half of the human population lives in urban areas, a figure predicted to increase to 60% by 2020 when Europe and the Americas will have more than 80% of their population living in urban zones. Five thousand years ago, the amount of agricultural land in the world was believed to have been negligible. Now, arable and permanent cropland covers approximately 1.5 billion ha of land, with some 3.5 billion ha of additional land classed as permanent pasture. The sum represents approximately 38% of total available land surface of 13 billion ha according to FAO statistics.

Habitat loss is of particular importance in tropical regions of high biological diversity where at the same time food security and poverty alleviation are key priorities. The advance of the agricultural frontier has led to an overall decline in the world's forests. While the area of forest in industrialized regions remained fairly unchanged, natural forest cover declined by 8% in developing regions. It is ironical that the most biodiverse regions are also those of greatest poverty, highest population growth and greatest dependence upon local natural resources (Lee and Jetz, 2008).

Introduced species are another threat to biodiversity. Unplanned or poorly planned introduction of non-native ('exotic' or 'alien') species and genetic stocks is a major threat to terrestrial and aquatic biodiversity worldwide. There are hundreds, if not thousands, of new and foreign genes introduced with trees, shrubs, herbs, microbes and higher and lower animals each year (Sukopp and Sukopp, 1993; Kowarik, 2005). Many of those survive and can, after years and even many decades of adaptation, begin to be invasive. This might be misconstrued as increasing biodiversity, but the final effect is sometimes the opposite. The introduced species often displace native species such that many native species become extinct or severely limited.

Freshwater habitats worldwide are among the most modified by humans, especially in temperate regions. In most areas, introduction of non-native species is the most or second most important activity affecting inland aquatic areas, with significant and often irreversible impacts on biodiversity and ecosystem function.

A classic example is the extinction of one-half to two-thirds of the indigenous fish population in Lake Victoria after the introduction of the Nile perch, *Lates niloticus*, a top predator (Schofield and Chapman, 1999). Several species of free-floating aquatic plants able to spread by vegetative growth have dispersed widely over the globe and become major pests. Water hyacinth (*Eichhornia crassipes*) is a notable example in tropical waters as is *Anarchis canadensis* = *Elodea canadensis* in temperate waters of the northern hemisphere.

Biodiversity should still act as biological insurance for ecosystem processes, except when mean trophic interaction strength increases strongly with diversity (Thebault and Loreau, 2005). The conclusion, which needs to be tested against field studies, is that in tropical environments with a natural high biodiversity the interactions between potentially invasive hybrids of transgenic crops and their wild relatives should be buffered through the complexity of the surrounding ecosystems. This view is also confirmed by the results of Davis (Davis, 2003). Taken together, theory and data suggest that compared to intertrophic interaction and habitat loss, competition from introduced species is not likely to be a common cause of extinctions in long-term resident species at global, meta-community and even most community levels.

Two Case Studies on the Impact of Transgenic Crops on Biodiversity

Herbicide-tolerant crops; application of conservation tillage is easier with herbicide-tolerant crops

The soil in a given geographical area has played an important role in determining agricultural practices since the time of the origin of agriculture in the Fertile Crescent of the Middle East. Soil is a precious and finite resource; its composition, texture, nutrient levels, acidity, alkalinity and salinity are all determinants of productivity. Agricultural practices can lead to soil degradation and the loss in the ability of a soil to produce crops. Examples of soil degradation include erosion, salinization, nutrient loss and biological deterioration. It has been estimated that 67% of the world's agricultural soils have been degraded (World Resources Institute, 2000). It may also be worth noting that soil fertility is a renewable resource and that it can often be restored within several years of careful crop management.

In many parts of the developed and the developing world, tillage of soil is still an essential tool for the control of weeds. Unfortunately, tillage practices can lead to soil degradation by causing erosion, reducing soil quality and harming biological diversity. Tillage systems can be classified according to how much crop residue is left on the soil surface (Fawcett *et al.*, 1994; Trewavas, 2001, 2003; Fawcett and Towery, 2002). Conservation tillage is defined as 'any tillage and planting system that covers more than 30% of the soil surface with crop residue, after planting, to reduce soil erosion by water' (Fawcett and Towery, 2002). The value of reducing tillage was long recognized but the level of weed control a farmer required was viewed as a deterrent for adopting conservation tillage. Once effective herbicides were introduced in the latter half of the 20th century, farmers were able to reduce their dependence on tillage. The development of crop varieties tolerant to herbicides has provided new tools and practices for controlling weeds and has accelerated the adoption of conservation tillage practices, including the adoption of 'no-till' practices (Fawcett and Towery, 2002). Herbicide-tolerant cotton has been rapidly adopted since its introduction in 1994 (Fawcett *et al.*, 1994). In the USA, 80% of growers are making fewer tillage passes and 75% are leaving more crop residue (Cotton Council, 2003). In a farmer survey, 71% of the growers responded that herbicide-tolerant cotton had the greatest impact on soil fertility related to the adoption of reduced tillage or no-till practices (Cotton Council, 2003). In the case of soybean, the growers of glyphosate-tolerant soybean plant a higher percentage of their acreage using no-till or reduced tillage practices than growers of conventional soybeans (American Soybean Association, 2001). Fifty-eight per cent of gyphosate-tolerant soybean adopters reported making fewer tillage passes versus 5 years ago compared to only 20% of non-glyphosate-tolerant soybean users (American Soybean Association, 2001). Fifty-four per cent of growers cited the introduction of glyphosate-tolerant soybeans as the factor which had the greatest impact towards the adoption of reduced tillage or no-till (American Soybean Association, 2001). Today, the scientific literature on 'no-tillage' and 'conservation tillage' has grown to more than 6500 references, a selection of some 1200 references from the last 3 years are given at: http://www.botanischergarten.ch/Tillage/Bibliography-No-conservation-Tillage-2006-20080626.pdf.

Several important reviews have been published in recent months; they all report a positive story regarding the overall impact of herbicide-tolerant crops and the impact on the agricultural environment. A few examples and statements are cited below.

In a comprehensive review, Bonny (2008) describes the unprecedented success of the introduction of transgenic soybean in the USA. It is worthwhile to present one of the graphs on the statistics of glyphosate use (Fig. 11.1), thus correcting some of the legends spread by opponents, sometimes coming in seemingly sturdy statistics like those of Benbrook (2004) stating that the herbicide and pesticide use has increased ever since the introduction of transgenic crops. But a closer, more differentiated look reveals this to be an 'urban legend':

> A comparison of transgenic versus conventional soybean reveals that transgenic glyphosate-tolerant soybean allows both the simplification of weed control and greater work flexibility. Cropping transgenic soybean also fits well with conservation tillage. Transgenic soybean has an economic margin similar to conventional soybean, despite a higher seed cost. The next section describes the evolution of the use of herbicides with transgenic soybean, and some issues linked to the rapid increase in the use of glyphosate. At the beginning a smaller amount of herbicides was used, but this amount increased from 2002, though not steadily. None the less, the environmental and toxicological impacts of pesticides do not only depend on the amounts applied. They also depend on the conditions of use and the levels of toxicity and ecotoxicity. The levels of ecotoxicity seem to have somewhat

Percentage of total soybean

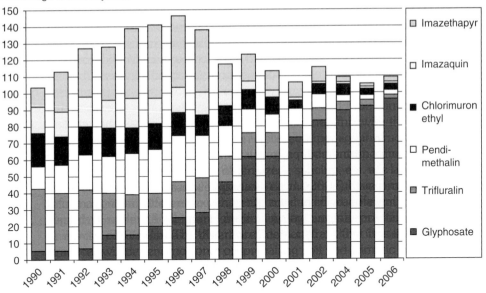

Fig. 11.1. Main herbicides used on total soybean acreage, 1990–2006 (as percentage of soybean surface treated by each herbicide; from USDA NASS, 1991–2007). With the development of glyphosate-tolerant soybean, this herbicide is used far more extensively. Indeed, it replaces the herbicides used previously; the Figure shows only a few of the latter. (From Bonny, 2008.)

> decreased. The success of transgenic soybeans for farmers has led to a higher use of glyphosate as a replacement for other herbicides, which has in turn led to a decline in its effectiveness. However, the issue here is not only genetic engineering in itself, but rather the management and governance of this innovation.
>
> (Carpenter and Gianessi, 2000)

Cerdeira *et al.* (2007) also emphasize the benefits, despite some green propaganda from Brazil and Argentina, but point also to some potential problems with the evolution of glyphosate-resistant weeds. Transgenic glyphosate-resistant soybeans (GRS) have been commercialized and grown extensively in the western hemisphere, including Brazil. Worldwide, several studies have shown that previous and potential effects of glyphosate on contamination of soil, water, and air are minimal, compared to those caused by the herbicides that they replace when GRS are adopted. In the USA and Argentina, the advent of glyphosate-resistant soybeans resulted in a significant shift to reduced- and no-tillage practices, thereby significantly reducing environmental degradation by agriculture. Similar shifts in tillage practiced with GRS might be expected in Brazil. Transgenes encoding glyphosate resistance in soybeans are highly unlikely to be a risk to wild plant species in Brazil; soybean is almost completely self-pollinated and is a non-native species in Brazil, without wild relatives, making introgression of transgenes from GRS virtually impossible. Probably the highest agricultural risk in adopting GRS in Brazil is related to weed resistance. Weed species in GRS fields have shifted in Brazil to those that can more successfully withstand glyphosate or to those that avoid the time of its application. These include *Chamaesyce hirta* (erva-de-Santa-Luzia), *Commelina benghalensis* (trapoeraba), *Spermacoce latifolia* (erva-quente), *Richardia brasiliensis* (poaia-branca) and *Ipomoea* spp. (corda-de-viola). Fourweed species, *Conyza bonariensis*, *Conyza canadensis* (*buva*), *Lolium multiflorum* (*azevem*) and *Euphorbia heterophylla* (*amendoim bravo*), have evolved resistance to glyphosate in GRS in Brazil and have great potential to become problems. These findings are also published in an earlier study with a worldwide scope looking at the herbicide tolerant crops of the western hemisphere by some of the same authors (Cerdeira and Duke, 2006) with the same outcome as above.

More pertinent review papers on soil erosion and other agronomic parameters have been published which are associated with the new agricultural management of herbicide-tolerant weeds: Bernoux *et al.* (2006), Beyer *et al.* (2006), Bolliger *et al.* (2006), Causarano *et al.* (2006), Chauhan *et al.* (2006), Etchevers *et al.* (2006), Wang *et al.* (2006), Anderson (2007), Gulvik (2007), Knapen *et al.* (2007), Knowler and Bradshaw (2007), Peigne *et al.* (2007), Raper and Bergtold (2007), Thomas *et al.* (2007), Thompson *et al.* (2008).

Impact of *Bt* maize on non-target organisms

In a recent study on environmental impact of *Bt* maize, the author included a commentary chapter on 180 scientific studies dealing with non-target organisms which could be harmed by the cultivation of *Bt* maize. Strictly observing the baseline comparison with non-*Bt* maize cultivation, it can be concluded that there is not a single publication pointing to detrimental effects of *Bt* maize compared to other maize traits. Four meta-studies have recently been published with more or less stringent selection of data published in scientific journals, and none show any sign of regulatory problems (Marvier *et al.*, 2007; Chen *et al.*, 2008; Duan *et al.*, 2008; Wolfenbarger *et al.*, 2008).

The work of Wolfenbarger *et al.* (2008) is singled out here since it is the best meta-analysis existing so far: the selection criteria are clearly defined on all levels and based on a carefully filtered data set which is a subset of the database published by Marvier *et al.* (2007; www.sciencemag.org/cgi/content/full/316/5830/1475/DC1). In total, the database used contained 2981 observations from 131 experiments reported in 47 published field studies on cotton, maize and potato. Maize was studied in the following two comparison categories (including also data on potato and cotton):

- The first set of studies contrasted *Bt* with non-*Bt* plots, neither of which received any additional insecticide treatments (Fig. 11.2). This comparison addresses the hypothesis that the toxins in the *Bt* plant directly or indirectly affect arthropod abundance. It can also be viewed as a comparison between the *Bt* crop and its associated unsprayed refuge (Gould, 2000).
- The second set of studies contrasted unsprayed *Bt* fields with non-*Bt* plots that received insecticides (Fig. 11.3). This comparison tests the hypothesis

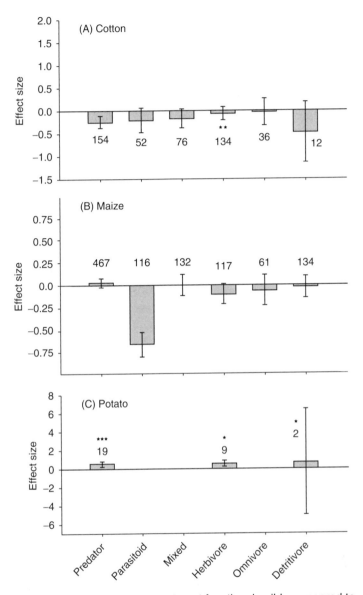

Fig. 11.2. The effect of *Bt* crops on non-target functional guilds compared to unsprayed, non-*Bt* control fields. Bars denote the 95% confidence intervals, asterisks denote significant heterogeneity in the observed effect sizes among the comparisons (*, 0.05; **, 0.01; ***, 0.001) and Arabic numbers indicate the number of observations included for each functional group; doi:10.1371/journal. pone.0002118.g001. (From Wolfenbarger *et al.*, 2008.)

that arthropod abundance is influenced by the method used to control the pest(s) targeted by the *Bt* crop. (The third set of studies contrasted fields of *Bt* crops and non-*Bt* crops both treated with insecticides, a category which did not occur in the maize studies cited above (Figs 11.4 and 11.5).)

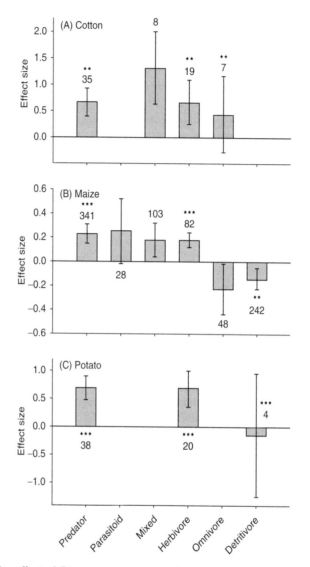

Fig. 11.3. The effect of *Bt* crops on non-target functional guilds compared to insecticide-treated, non-*Bt* control fields. Bars denote the 95% confidence intervals, asterisks denote significant heterogeneity in the observed effect sizes among the studies (*, 0.05; **, 0.01; ***, 0.001), and Arabic numbers indicate the number of observations included for each functional group. doi:10.1371/journal.pone.0002118. g002. (From Wolfenbarger *et al.*, 2008.)

In the above studies, great care was taken to eliminate redundant taxonomic units and multiple development stages of the same species, with a preference of the least mobile development stage; furthermore, the data sets were all derived from the same season.

In contrast to the following extract from a study by Marvier *et al.* (2007), the statistical analysis was not done with the original taxonomic units; rather

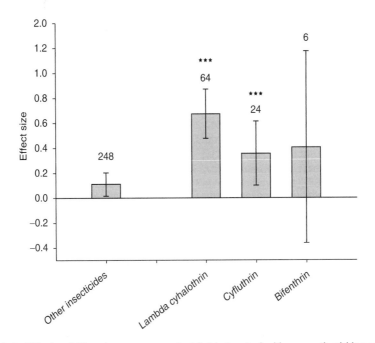

Fig. 11.4. Effects of *Bt* maize versus control fields treated with a pyrethroid insecticide on predatory arthropods. Bars denote the 95% confidence intervals, asterisks denote significant heterogeneity in the observed effect sizes among the studies (*, 0.05; **, 0.01; ***, 0.001), and Arabic numbers indicate the number of observations included for each functional group; doi:10.1371/journal.pone.0002118.g003.

the authors decided to use an additional descriptor, six '*functional guilds*' (herbivore, omnivore, predator, parasitoid, detritivore or mixed). More details can be read in the original publication, where database robustness and sensitivity of the data sets have been thoroughly discussed and careful decisions have been made in order to get maximum quality of the meta-analysis.

In maize, analyses revealed a large reduction of parasitoids in *Bt* fields. This effect stemmed from the lepidopteran-specific maize hybrids, and examining the 116 observations showed that most were conducted on *Macrocentrus grandii*, a specialist parasitoid of the *Bt*-target, *Ostrinia nubilalis*. There was no significant effect on other parasitoids, but *M. grandii* abundance was severely reduced by *Bt* maize. Higher numbers of the generalist predator, *Coleomegilla maculata*, were associated with *Bt* maize but numbers of other common predatory genera (*Orius*, *Geocoris*, *Hippodamia*, *Chrysoperla*) were similar in *Bt* and non-*Bt* maize.

In maize, the abundance of predators and members of the mixed functional guild were higher in *Bt* maize compared to insecticide-sprayed controls [Fig.11.2B]. Significant heterogeneity occurred in predators, indicating variation in the effects of *Bt* maize on this guild. For example, we detected no significant effect sizes for the common predator genera *Coleomegilla*, *Hippodamia* or *Chrysoperla*, but the predator *Orius* spp. and the parasitoid *Macrocentrus* were more abundant in *Bt* maize than in non-*Bt* maize plots

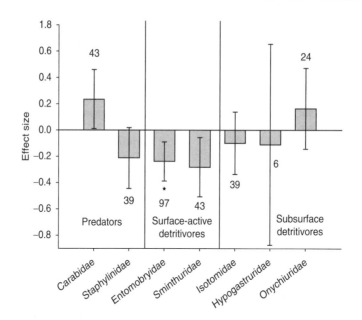

Fig. 11.5. Effect of *Bt* crops versus insecticide-treated, non-*Bt* control fields on soil-inhabiting predators and detritivores. Bars denote the 95% confidence intervals, asterisks denote significant heterogeneity in the observed effect sizes among the studies (*, 0.05; **, 0.01; ***, 0.001), and Arabic numbers indicate the number of observations included for each functional group; doi:10.1371/journal.pone.0002118.g004. (Wolfenbarger *et al.*, 2008.)

treated with insecticides. Partitioning by taxonomic groupings (Lepidoptera versus Coleoptera) or the target toxin did not reduce heterogeneity within predators. However, insecticides differentially affected predator populations. Specifically, application of the pyrethroid insecticides lambda-cyhalothrin, cyfluthrin, and bifenthrin in non-*Bt* control fields resulted in comparatively fewer predators within these treated control plots. Omitting studies involving these pyrethroids revealed a much smaller and homogeneous effect size. Predator abundance in *Bt* fields was still significantly higher compared with insecticide-treated plots, but the difference was less marked without the pyrethroids [Fig. 11.3]. Compared to the subset of controls using pyrethroids, *Bt* maize was particularly favorable to *Orius* spp.

Bt-maize favoured non-target herbivore populations relative to insecticide-treated controls, but there was also significant heterogeneity, some of which was explained by taxonomy. Aphididae were more abundant in insecticide sprayed fields and Cicadellidae occurred in higher abundance in the *Bt* maize. In contrast to patterns associated with predators and detritivores, type of insecticide did not explain the heterogeneity in herbivore responses. The pyrethroid-treated controls accounted for 85% of the herbivore records. Individual pyrethroids had variable effects on this group, and none yielded strong effects on the herbivores.

An underlying factor associated with the heterogeneity of the herbivore guild remained unidentified, but many possible factors were eliminated (e.g. Cry protein target, Cry protein, event, plot size, study duration, pesticide class, mechanism of pesticide delivery, sample method, and sample frequency).

The 'mixed' functional group was more abundant in *Bt* maize ($E = 0.1860.14$, $n = 103$) compared with non-*Bt* maize treated with insecticides. The majority of this functional group is comprised of carabids ($n = 33$), nitidulids ($n = 26$) and mites ($n = 23$).

For potatoes, the abundance of predators ($E = 0.6960.30$, $n = 38$), but not herbivores, was significantly higher in the *Bt* crop (Fig. 11.2C). Responses within each functional group were variable but sample sizes were too low to further partition this significant heterogeneity.

Predator–non-target herbivore ratio analyses

No significant change in predator:prey ratios was detected in cotton or potato; in maize there was a significantly higher predator/prey ratio in *Bt* maize plots than in the insecticide controls ($E = 0.6360.42$, $n = 15$). Significant heterogeneity for the predator:prey response existed in all three crops, but again sample sizes were too small to explore the cause of this variability.

Predator–detritivore analyses

The higher abundance of detritivores in sprayed non-*Bt* maize appeared to be driven primarily by two families of Collembola with a high proportion of surface-active species (Entomobryidae: $E = 20.2460.15$, $n = 97$; Sminthuridae: $E = 20.2860.23$, $n = 43$; Fig. 11.4). Three other families, Isotomidae, Hypogastruridae and Onychiuridae, with more sub-surface species, were similar in *Bt* and non-*Bt* fields. We would expect surface-active collembolans to be more vulnerable to surface-active predators, and we detected a significantly lower abundance in one predator of Collembola (Carabidae: $E = 0.2360.22$, $n = 43$), but not in another (Staphylinidae: $E = 20.2160.23$, $n = 39$; Fig. 11.4). The other two detritivore families occupy different niches than Collembola and responded differently to insecticide treatments. The abundance of Japygidae (Diplura) was unchanged ($E = 20.1160.35$, $n = 9$), but that for Lathridiidae (Coleoptera) was higher in *Bt* maize ($E = 0.7660.70$, $n = 6$), suggesting a direct negative effect of insecticides on this latter group. Lathridiid beetles, although being surface-active humusfeeders, are larger and more motile than Collembola and thus may be less vulnerable to predators and more vulnerable to insecticides.

As a whole, the study of Wolfenbarger *et al.* did not reveal any negative effects, confirming for a large amount of data and publications the environmental benefits of the *Bt* maize tested.

References

American Soybean Association (2001) *Homepage*. ASA. Available at: http://www.asa-europe.org/index.shtml

Anderson, P.C. (1998) History of harvesting and threshing techniques for cereals in the prehistoric Near East. In: Damania, A.B.

Valkoun, J. Willcox, G. and Qualset, C.O. (eds) *The Origins of Agriculture and Crop Domestication*. ICARDA, Aleppo, Syria, pp. 145–159.

Anderson, R.L. (2007) Managing weeds with a dualistic approach of prevention and control. A review. *Agronomy for Sustainable Development* 27(1), 13–18. Available at: < Go to ISI > ://WOS:000245743500002

Attaran, A. and Maharaj, R. (2000) Ethical debate – doctoring malaria, badly: the global campaign to ban DDT. *British Medical Journal* 321, 1403–1403.

Attaran, A., Roberts, D.R., Curtis, C.F., Kilama, W.L. (2000) Balancing risks on the backs of the poor. *Nature Medicine* 6, 729.

Batista, R., Saibo, N., Lourenco, T. and Oliveira, M.M. (2008) Microarray analyses reveal that plant mutagenesis may induce more transcriptomic changes than transgene insertion. *Proceedings of the National Academy of Sciences of the USA*, 105, 9, 3640–3645. Available at: < Go to ISI > ://WOS:000253846500082 and http://www.botanischergarten.ch/Genomics/Batista-Microarray-Analysis-2008.pdf

Baum, J.A., Bogaert, T., Clinton, W., Heck, G.R., Feldmann, P., Ilagan, O., Johnson, S., Plaetinck, G., Munyikwa, T., Pleau, M., Vaughn, T. and Roberts, J. (2007) Control of coleopteran insect pests through RNA interference. *Nature Biotechnology*, 25, 1322–1326. Available at: < Go to ISI > ://WOS:000251086500034 and http://www.botanischergarten.ch/Genomics/Baum-RNA-interference-2007.pdf

Bellon, M.R. and Berthaud, J. (2004) Transgenic maize and the evolution of landrace diversity in Mexico. The importance of farmers' behavior. *Plant Physiology* 134, 883–888.

Bellon, M.R. and Berthaud, J. (2006) Traditional Mexican agricultural systems and the potential impacts of transgenic varieties on maize diversity. *Agriculture and Human Values* 23(1), 3–14. Available at: < Go to ISI > ://000236273900002 and http://www.botanischergarten.ch/Maize/Bellon-Impacts-Landraces-2006.pdf

Bellon, M.R., Berthaud, J., Smale, M., Aguirre, J.A., Taba, S., Aragon, F., Diaz, J. and Castro, H. (2003) Participatory landrace selection for on-farm conservation: an example from the Central Valleys of Oaxaca, Mexico. *Genetic Resources and Crop Evolution* 50(4), 401–416. Available at: < Go to ISI > ://000183201600008 and

http://www.botanischergarten.ch/Maize/Bellon-Participatory-Maize-2003.pdf

Benbrook, C. (2004) Genetically Engineered Crops and Pesticide Use in the United States: The First Nine Years pp 53 Technical Paper Number 7 (Report) Available at: http://www.botanischergarten.ch/HerbizideTol/Benbrook-First-nine-2004.pdf

Berthaud, J. (2001) Maize Diversity in Oaxaca, Mexico: Simple Questions but No Easy Answers. CIMMYT-Report, p. 2. Available at: http://www.cimmyt.cgiar.org/whatiscimmyt/AR002001Spa/latinamerica/maize/diversity.htm#Maize

Bernoux, M., Cerri, C.C., Cerri, C.E.P., Neto, M.S., Metay, A., Perrin, A.S., Scopel, E., Razafimbelo, T., Blavet, D., Piccolo, M.D., Pavei, M. and Milne, E. (2006) Cropping systems, carbon sequestration and erosion in Brazil, a review. *Agronomy for sustainable development* 26(1), 1–8.

Beyer, M., Klix, M.B., Klink, H. and Verreet, J.A. (2006) Quantifying the effects of previous crop, tillage, cultivar and triazole fungicides on the deoxynivalenol content of wheat grain – a review. *Journal of Plant Diseases and Protection* 113(6), 241–246. Available at: < Go to ISI > ://WOS:000243738300001

Bolliger, A., Magid, J., Amado, T.J.C., Neto, F.S., Ribeiro, M.D.D., Calegari, A., Ralisch, R. and de Neergaard, A. (2006) Taking stock of the Brazilian zero-till revolution: A review of landmark research and farmers' practice. *Advances in Agronomy* 91, 47–110. Available at: < Go to ISI > ://WOS:000244524100002

Bonny, S. (2008) Genetically modified glyphosate-tolerant soybean in the USA: adoption factors, impacts and prospects. A review. *Agronomy for Sustainable Development* 28(1), 21–32. Available at: < Go to ISI > ://WOS:000253779900003 NEBIS 20080626

Carpenter, J. and Gianessi, L. (2000) Herbicide use on roundup ready crops. Science, 287, 5454, pp.503–804 <Go to ISI>://000085136400019 and http://www.botanischergarten.ch/HerbizideTol/CarpenterLetter-Herbicide-Use-2000.pdf

Cattivelli, L., Rizza, F., Badeck, F.W., Mazzucotelli, E., Mastrangelo, A.M., Francia, E., Mare, C., Tondelli, A. and Stanca, A.M. (2008) Drought tolerance improvement in crop plants: an integrated view from breeding to genomics. *Field Crops Research* 105, 1–14. Available at: < Go to ISI > ://WOS:000252464800001 and http://www.botanischergarten.ch/DroughtResistance/Cattivelli-Drought-Tolerance-2008.pdf

Causarano, H.J., Franzluebbers, A.J., Reeves, D.W. and Shaw, J.N. (2006) Soil organic carbon sequestration in cotton production systems of the southeastern United States: A review. *Journal of Environmental Quality* 35(4), 1374–1383. Available at: < Go to ISI > ://WOS: 000239189900044

CBD (1992) Convention on Biological Diversity. United Nations. Available at: http://www.biodiv.org/doc/publications/guide.asp

Cerdeira, A.L. and Duke, S.O. (2006) The current status and environmental impacts of glyphosate-resistant crops: a review. *Journal of Environmental Quality* 35(5), 1633–1658. Available at: < Go to ISI > ://000240924200001 and http://www.botanischergarten.ch/HerbizideTol/Cerdeira-Status-2006.pdf

Cerdeira, A.L., Gazziero, D.L.P., Duke, S.O., Matallo, M.B. and Spadotto, C.A. (2007) Review of potential environmental impacts of transgenic glyphosate-resistant soybean in Brazil. *Journal of Environmental Science and Health Part B-Pesticides Food Contaminants and Agricultural Wastes* 42(5), 539–549. Available at: < Go to ISI > ://WOS:000247652600009

Chauhan, B.S., Gill, G.S. and Preston, C. (2006) Tillage system effects on weed ecology, herbicide activity and persistence: a review. *Australian Journal of Experimental Agriculture* 46(12), 1557–1570. Available at: < Go to ISI > ://WOS:000241906000002

Chen, M., Zhao, J.-Z., Collins, H.L., Earle, E.D., Cao, J. and Shelton, A. (2008) A Critical Assessment of the Effects of Bt Transgenic Plants on Parasitoids. *PLoS ONE* 3, 5 pp. Published online 2008 May 28. doi: 10.1371 and http://www.physorg.com/news131726113.html, http://www.botanischergarten.ch/Bt/Chen-Critical-Assessment-2008.pdf, http://

www.botanischergarten.ch/Bt/Chen-Critical-TableS1-additional-2008.doc and http://www.botanischergarten.ch/Bt/Chen-Critical-Figure-Dose-additional-2008.doc

Cotton Council (2003) *National Cotton Council of America*. Cotton Council. Available at: http://www.cotton.org/

Curtis, C.F. (2002) Restoration of malaria control in the Madagascar highlands by DDT spraying. *American Journal of Tropical Medicine and Hygiene* 66(1), 1–1. Available at: < Go to ISI > ://000176672500001 and http://www.botanischergarten.ch/DDT/Curtis-Restoration-DDT-2002.pdf

Curtis, C.F. and Lines, J.D. (2000) Should DDT be banned by international treaty? *Parasitology Today* 16(3), 119–121. Available at: < Go to ISI > ://000085712500012 and http://www.botanischergarten.ch/DDT/Curtis-DDT-ban-2000.pdf

Cattivelli, L., Rizza, F., Badeck, F.W., Mazzucotelli, E., Mastrangelo, A.M., Francia, E., Mare, C., Tondelli, A. and Stanca, A.M. (2008) Drought tolerance improvement in crop plants: an integrated view from breeding to genomics. *Field Crops Research* 105, 1–14.

Darwin, C. (1845) *Journal of Researches into the Natural History and Geology of the Countries Visited During the Voyage of H.M.S. Beagle Round the World*, 2nd edn. John Murray, London. Available at: http://darwin-online.org.uk

Davis, M.A. (2003) Biotic globalization: does competition from introduced species threaten biodiversity? *Bioscience* 53(5), 481–489. Available at: < Go to ISI > ://000182833000011 and http://www.botanischergarten.ch/Africa-Harvest-Sorghum-Lit/Davis-Competition-2003.pdf

Dollaker, A. (2006) Conserving biodiversity alongside agricultural profitability through integrated R&D approaches and responsible use of crop protection products. *Pflanzenschutz-Nachrichten Bayer* 59(1), 117–134.

Dollaker, A. and Rhodes, C. (2007) Integrating crop productivity and biodiversity conservation pilot initiatives developed by Bayer CropScience, in Weed Science in Time of Transition. *Crop Science* 26(3), 408–416. Available at: http://www.sciencedirect.com/

science/article/B6T5T-4MV1NS0–1/2/
45ef7df1e5582 e65fbde44fa32651b26

Duan, J.J., Marvier, M., Huesing, J., Dively, G. and Huang, Z.Y. (2008) A meta-analysis of effects of Bt crops on honey bees (Hymenoptera: Apidae). *PLoS ONE* 3(1), e1415. Available at: http://dx.doi.org/ 10.1371%2Fjournal.pone.0001415 and http://www.botanischergarten.ch/Bt/Duan-Meta-Analysis-Effects-Bees-2008.pdf

Etchevers, J.D., Prat, C., Balbontin, C., Bravo, M. and Martinez, M. (2006) Influence of land use on carbon sequestration and erosion in Mexico, a review. *Agronomy for Sustainable Development* 26(1), 21–28. Available at: < Go to ISI > ://WOS: 000236003300003

Evans, L.T. (1998) *Feeding the Ten Billion: Plants and Population Growth.* Cambridge University Press, Cambridge.

Fawcett, R. and Towery, D. (2002) *Conservation Tillage and Plant Biotechnology: How New Technologies can Improve the Environment by Reducing the Need to Plow.* Purdue University, West Lafayette, Indiana. Available at: www.ctic.purdue.edu/CTIC/ CTIC.html or http://www.botanischergarten. ch/HerbizideTol/Fawcett-Biotech Paper.pdf

Fawcett, R., Christensen, B. and Tierney, D. (1994) The impact of conservation tillage on pesticide runoff into surface water, 49, 126–135. Available at: http://www. scientific-alliance.org/scientist_writes_ items/benefits_no_till.htm

Fedoroff, N.V. and Cohen, J.E. (1999) Plants and population: is there time? *Proceedings of the National Academy of Sciences of the USA* 96(11), 5903–5907. Available at: < Go to ISI > ://000080527100005 and http:// www.botanischergarten.ch/biodiversity/ Fedorova-Time-1999.pdf

Gagnon, P.R. and Platt, W.J. (2008) Multiple disturbances accelerate clonal growth in a potentially monodominant bamboo. *Ecology* 89(3), 612–618. Available at: < Go to ISI > ://WOS: 000254678200003

Gould, F. (2000) Testing Bt refuge strategies in the field. *Nature Biotechnology* 18(3), 266–267. Available at: < Go to ISI > ://000085771100017

Gulvik, M.E. (2007) Mites (Acari) as indicators of soil biodiversity and land use monitoring: A review. *Polish Journal of Ecology* 55, 415–440. Available at: < Go to ISI > ://WOS: 000249478600001

Harlan, J. and Zohary, D. (1966) Distribution of wild wheats and barley. *Science* 153, 1074–1080.

Harlan, J.R. (1971) Agricultural origins – centers and noncenters. *Science* 174(4008), 468. Available at: < Go to ISI > ://A1971K638 500005 and http://www.botanischergarten. ch/Africa-Harvest-Sorghum-Lit/Harlan-Centers-1971.pdf

Harlan, J.R. (1989) Wild-grass harvesting in the Sahara and Sub-Sahara of Africa. In: Harris, D.R. and Hillman, G.C. (eds) *Foraging and Farming: the Evolution of Plant Exploitation.* Unwin Hyman, London, pp. 79–98 and Figs 5.2–5.3.

Harlan, J.R. (1992) *Crops and Man,* 2nd edn. American Society of Agronomy, Madison, Wisconsin, p. 295.

Henneman, M.L. and Memmott, J. (2001) Infiltration of a Hawaiian community by introduced biological control agents. *Science* 293(5533), 1314–1316. Available at: http://www.sciencemag.org/cgi/content/ abstract/293/5533/1314 and http://www. botanischergarten.ch/BioControl/ Hennemann-Science-2001.pdf

Hillmann, G. (1996) Late Pleistocene changes in wild food plants available to huntergatherers of the northern Fertile Crescent: possible preludes to cereal cultivation. In: Harris, D.R. (ed.) *The Origin and Spread of Agriculture and Pastoralism in Eurasia.* University College Press, London, pp. 159–203.

Horton, J.E. (2000) Caution required with the precautionary principle. *The Lancet* 356(9226), 265–265. Available at: http:// www.sciencedirect.com/science/article/ B6T1B-40W5V2N-1/1/8f70aae7435c 15851abc3ee30f8d0572 and http://www. botanischergarten.ch/DDT/Horton-Caution-with-Precautionary-Principle-2000.pdf

Howarth, F.G. (1991) Environmental impacts of classical biological-control. *Annual Review of Entomology* 36, 485–509. Available at: < Go to ISI > ://A1991EQ80700021

Janzen, D. (1998) Gardenification of wildland nature and the human footprint. *Science* 279(5355), 1312–1313. Available at: < Go to ISI > ://000072251800035 and http://www.botanischergarten.ch/biodiversity/Janzen-Gardenification-SM-1998.pdf

Janzen, D. (1999) Gardenification of tropical conserved wildlands: multitasking, multi-cropping, and multiusers. *Proceedings of the National Academy of Sciences of the USA* 96(11), 5987–5994. Available at: < Go to ISI > ://000080527100018 and http://www.botanischergarten.ch/biodiversity/Janzen-Gardenification-1999.pdf

Kier, G., Mutke, J., Dinerstein, E., Ricketts, T.H., Kuper, W., Kreft, H. and Barthlott, W. (2005) Global patterns of plant diversity and floristic knowledge. *Journal of Biogeography* 32(7), 1107–1116. Available at: < Go to ISI > ://00022970590000

Knapen, A., Poesen, J., Govers, G., Gyssels, G. and Nachtergaele, J. (2007) Resistance of soils to concentrated flow erosion: a review. *Earth-Science Reviews* 80(1–2), 75–109. Available at: < Go to ISI > :// WOS: 000243646800003

Knowler, D. and Bradshaw, B. (2007) Farmers' adoption of conservation agriculture: review and synthesis of recent research. *Food Policy* 32(1), 25–48. Available at: < Go to ISI > ://WOS: 000243569800002

Kowarik, I. (2005) Urban ornamentals escaped from cultivation. In: Gressel, J. (ed.) *Crop Ferality and Volunteerism*. CRC Press, Boca Raton, Florida.

Kuper, W., Sommer, J.H., Lovett, J.C., Mutke, J., Linder, H.P., Beentje, H.J., Van Rompaey, R., Chatelain, C., Sosef, M. and Barthlott, W. (2004) Africa's hotspots of biodiversity redefined. *Annals of the Missouri Botanical Garden* 91(4), 525–535. Available at: < Go to ISI > ://000226362700001 and http://www.botanischergarten.ch/Africa-Harvest-Sorghum-Lit/Kueper-Hotspots-2004.pdf

Lee, T.M. and Jetz, W. (2008) Future battle-grounds for conservation under global change. *Proceedings of the Royal Society B-Biological Sciences* 275(1640), 1261–1270. Available at: < Go to ISI > :// WOS:000255056600005 and http://www.botanischergarten.ch/BiodiversityAgri/Lee-Future-Battlegrounds-2008.pdf

Lughadha, E.N., Baillie, J., Barthlott, W., Brummitt, N.A., Cheek, M.R., Farjon, A., Govaerts, R., Hardwick, K.A., Hilton-Taylor, C., Meagher, T.R., Moat, J., Mutke, J., Paton, A.J., Pleasants, L.J., Savolainen, V., Schatz, G.E., Smith, P., Turner, I., Wyse-Jackson, P. and Crane, P.R. (2005) Measuring the fate of plant diversity: towards a foundation for future monitoring and opportunities for urgent action. *Philosophical Transactions of the Royal Society B-Biological Sciences* 360(1454), 359–372. Available at: < Go to ISI > ://000228214600011

Mallory, A.C. and Vaucheret, H. (2006) Functions of microRNAs and related small RNAs in plants. *Nature Genetics* 38, S31–S36. Available at: < Go to ISI > ://000202983200007 and http://www.botanischergarten.ch/Genomics/Mallory-MicroRNAs-2006.pdf

Marvier, M., McCreedy, C., Regetz, J. and Kareiva, P. (2007) A meta-analysis of effects of Bt cotton and maize on nontarget inverte-brates. *Science* %R 10.1126/science. 1139208, 316(5830), 1475–1477. Available at: http://www.sciencemag.org/cgi/content/abstract/316/5830/1475 and http://www.botanischergarten.ch/Bt/Marvier-Meta-Analysis-2007.pdf and supporting data: http://www.botanischergarten.ch/Bt/Marvier-Meta-Analysis-Supporting-2007.pdf

Mattick, J.S. (2004) The hidden genetic program of complex organisms. *Scientific American* 291(4), 60–67. Available at: < Go to ISI > :// WOS:000223903500025 and http://www.botanischergarten.ch/Genomics/Mattick-Genome-Complexity-2004.pdf

May, R.M. (1990) How many species. *Philosophical Transactions of the Royal Society of London Series B-Biological Sciences* 330(1257), 293–304. Available at: < Go to ISI > ://WOS:A1990EM43100015 and http://www.botanischergarten.ch/Biodiver sityAgri/May-How-many-Species-1990.pdf

Morris, W.F., Kareiva, P.M. and Raymer, P.L. (1994) Do barren zones and pollen traps reduce gene escape from transgenic crops. *Ecological Applications* 4(1), 157–165. Available at: < Go to ISI > ://A1994MU60500017

and http://www.botanischergarten.ch/Africa-Harvest-Sorghum-Lit-1/Morris-Barrenzones-1994.pdf

Mucher, C.A., Steinnocher, K.T., Kressler, F.P. and Heunks, C. (2000) Land cover characterization and change detection for environmental monitoring of pan-Europe International. *Journal of Remote Sensing* 21(6–7), 1159–1181. Available at: http://www.ingentaconnect.com/content/tandf/tres/2000/00000021/F0020006/art00006 http://dx.doi.org/10.1080/014311600210128 and http://www.botanischergarten.ch/BiodiversityAgri/Mucher-Land-Cover-2000.pdf

Nevo, E. (1998) Genetic diversity in wild cereals: regional and local studies and their bearing on conservation *ex situ* and *in situ*. *Genetic Resources and Crop Evolution* 45(4), 355–370. Available at: < Go to ISI > :// 000075583700010

Peigne, J., Ball, B.C., Roger-Estrade, J. and David, C. (2007) Is conservation tillage suitable for organic farming? A review. *Soil Use and Management* 23(2), 129–144. Available at: < Go to ISI > ://WOS: 000247401500003

Prain, D. (1903) *Flora of the Sundribuns*, pp. 231–370.

Purvis, A. and Hector, A. (2000) Getting the measure of biodiversity. *Nature* 405(6783), 212–219. Available at: http://www.nature.com/cgi-taf/DynaPage.taf?file=/nature/journal/v405/n6783/full/ 405212a0_fs.html

Raikhel, N.V. and Minorsky, P.V. (2001) Celebrating plant diversity. *Plant Physiology* %R 10.1104/pp.900012, 127(4), 1325–1327. Available at: http://www.plantphysiol.org and http://www.botanischergarten.ch/Genomics/Raikhel-Celebrating-Plant-Diversity-2001.pdf

Raper, R.L. and Bergtold, J.S. (2007) In-row subsoiling: a review and suggestions for reducing cost of this conservation tillage operation. *Applied Engineering in Agriculture* 23(4), 463–471. Available at: < Go to ISI > ://WOS:000248928500009

Roberts, D.R., Manguin, S. and Mouchet, J. (2000) DDT house spraying and reemerging malaria. *The Lancet* 356(9226), 330–332. Available at: http://www.sciencedirect.com/science/article/B6T1B-40W5V2N-S/1/45b7aced36 db0dc7f42edeb587257897 and http://www.botanischergarten.ch/DDT/Roberts-DDT-House-spraying-2000.pdf

Robinson, R.A. and Sutherland, W.J. (2002) Post-war changes in arable farming and biodiversity in Great Britain. *Journal of Applied Ecology* 39(1), 157–176. Available at: http://www.botanischergarten.ch/Bt/Robinson-Post-war-Changes-Biodiversity-2002.pdf

Schlapfer, F., Pfisterer, A.B. and Schmid, B. (2005) Non-random species extinction and plant production: implications for ecosystem functioning. *Journal of Applied Ecology* 42(1), 13–24. Available at: < Go to ISI > ://000227175200003

Schofield, P.J. and Chapman, L.J. (1999) Interactions between Nile perch, *Lates niloticus*, and other fishes in Lake Nabugabo, Uganda. Available at: < Go to ISI > :// WOS: 000082086100001

Smith, A.G. (2000) How toxic is DDT? *The Lancet* 356(9226), 267–268. Available at: http://www.sciencedirect.com/science/article/B6T1B-40W5V2N-3/1/478cd64886 d364af58f6b73351114698 and http://www.botanischergarten.ch/DDT/Smith-How-toxic-DDT-2000.pdf

Sukopp, H. and Sukopp, U. (1993) Ecological long-term effects of cultigens becoming feral and of naturalization of nonnative species. *Experientia* 49(3), 210–218.

Taverne, J. (1999) DDT – to ban or not to ban? *Parasitology Today* 15(5), 180–181. Available at: http://www.sciencedirect.com/science/article/B6TB8-3WRB3N6-1K/1/01c87ce34b7286d71b07a1f2bc49c982 and http://www.botanischergarten.ch/DDT/Taverne-toban-nottoban-1999.pdf

Thebault, E. and Loreau, M. (2005) Trophic interactions and the relationship between species diversity and ecosystem stability. *American Naturalist* 166(4), E95–E114. Available at: < Go to ISI > ://000232270600002

Thomas, G.A., Titmarsh, G.W., Freebairn, D.M. and Radford, B.J. (2007) No-tillage and conservation farming practices in grain growing areas of Queensland – a review of 40 years of development. *Australian Journal of Experimental Agriculture* 47(8),

887–898. Available at: < Go to ISI > ://
WOS: 000248021900001

Thompson, J.P., Owen, K.J., Stirling, G.R. and
Bell, M.J. (2008) Root-lesion nematodes
(*Pratylenchus thornei* and *P. neglectus*): a
review of recent progress in managing a
significant pest of grain crops in northern
Australia. *Australasian Plant Pathology*
37(3), 235–242. Available at: < Go to ISI >
://WOS:000254549200004

Tilman, D. (2000) Causes, consequences and
ethics of biodiversity. *Nature* 405(6783),
208–211. http://www.3-Biodiv-Nature-Insight-
Causes-Tilman.pdf

Tilman, D., Polasky, S. and Lehman, C. (2005)
Diversity, productivity and temporal stabil-
ity in the economies of humans and nature.
*Journal of Environmental Economics and
Management* 49(3), 405–426. Available at:
< Go to ISI > ://000229345100001 and
http://www.botanischergarten.ch/Africa-
Harvest-Sorghum-Lit/Tilman-Diversity-
Stability-2005.pdf

Tren, R. and Bate, R. (2001) *Malaria and the
DDT Story*. The Institute of Economic
Affairs, London, IS: 0 255 36499 7, p. 112.
Available at: http://www.botanischergarten.
ch/DDT/Tren-Bate-IEA-DDT-story-2001.pdf

Trewavas, A.J. (2001) The population/biodi-
versity paradox. Agricultural efficiency to
save wilderness. *Plant Physiology* 125(1),
174–179. Available at: http://www.plant-
physiol.org/cgi/reprint/125/1/174.pdf

Trewavas, A. (2003) *Electronic Source:
Benefits To The Use Of Gm Herbicide
Tolerant Crops- The Challenge Of No-Till
Agriculture*. Scientific Alliance. Available at:
http://www.scientific-alliance.org/scientist_
writes_items/benefits_no_till.htm

Vareschi, V. (1980) *Vegetationsökologie der
Tropen Ulmer*. IS: 3800134233, Stuttgart,
Germany, p. 294. Available at: http://www.
payer.de/cifor/cif0203.htm

Wang, X.B., Cai, D.X., Hoogmoed, W.B.,
Oenema, O. and Perdok, U.D. (2006)
Potential effect of conservation tillage on
sustainable land use: a review of global
long-term studies. *Pedosphere* 16(5),
587–595 Available at: < Go to ISI > ://WOS:
000240820200006

Whitham, T.G., Martinsen, G.D., Floate, K.D.,
Dungey, H.S., Potts, B.M. and Keim, P.
(1999) Plant hybrid zones affect biodiversity: tools for a genetic-based understanding of community structure. *Ecology* 80(2),
416–428. Available at: < Go to ISI >
://000079036500007 and http://www.
botanischergarten.ch/Africa-Harvest-
Sorghum-Lit-1/Witham-biodiversiy-hybrids-
1999.pdf

WHO (2005) WHO Position On DDT Use In
Disease Vector Control Under The
Stockholm Convention On Persistent
Organic Pollutants, WHO, Geneva
(Report), p. 2. Available at: http://www.
botanischergarten.ch/DDT/WHO-Position-
on-DDT-2005.pdf

Witcombe, J.R., Hollington, P.A., Howarth, C.
J., Reader, S. and Steele, K.A. (2008)
Breeding for abiotic stresses for sustainable
agriculture. *Philosophical Transactions of
the Royal Society B-Biological Sciences*
363, 703–716 Available at: < Go to ISI > ://
WOS:000252663200003 and http://www.
botanischergarten.ch/DroughtResistance/
Witcombe-Breeding-abiotic-Stress-2008.
pdf

Wolfenbarger, L.L., Andow, D.A., Hilbeck, A.,
Nickson, T., Wu, F., Thompson, P.B. and
Ammann, K. (2004) GE crops: balancing
predictions of promise and peril. *Frontiers in
Ecology and the Environment* 2(3), 154–160.
Available at: < Go to ISI > ://000223960400
019 and www.frontiersinecology.org and
http://www.botanischergarten.ch/Frontiers-
Ecology/Ammann-Forum-def1.pdf

Wolfenbarger, L.L., Naranjo, S.E., Lundgren,
J.G., Bitzer, R.J. and Watrud, L.S. (2008) Bt
crop effects on functional guilds of non-
target arthropods: a meta-analysis. *PLoS
ONE* 3(5), e2118. Available at: http://dx.doi.
org/10.1371%2Fjournal.pone.0002118
and http://www.botanischergarten.ch/Bt/
LaReesa-Bt-crop-Meta-Analysis-2008.pdf
and http://www.botanischergarten.ch/Bt/
LaReesa-Meta-Analysis-Powerpoints-
2008.ppt

Wood, D. and Lenne, J. (1999) Agrobiodiversity
and natural biodiversity: some parallels. In:
Wood, D. and Lenne, J. (eds) *Agrobiodiversity,*

Characterization, Utilization and Management. CAB International, Wallingford, UK, pp. 425–445.

Wood, D. and Lenne, J. (2001) Nature's fields: a neglected model for increasing food production. *Outlook on Agriculture* 30(3), 161–170. Available at: < Go to ISI > ://000171396200003 and http://www.

botanischergarten.ch/Organic/Wood-Natures-Fields-2001.pdf

World Resources Institute (2000) People and Ecosystems, The Frayling Web of Life, World Resources Institute, UNDP, UNEP, World Bank, p. 36 Washington (Report). Available at: http://www.wri.org/wr2000/pdf/summary.pdf

12 Potential Wider Impact: Farmland Birds

M.J. WHITTINGHAM

School of Biology, Newcastle University, Newcastle upon Tyne, UK

Keywords: Sustainable agriculture, agroecology, biodiversity, farmland bird population declines

Summary

Birds are important biodiversity indicators because they depend on a range of invertebrates and plants for food. Farmland bird populations have declined dramatically, especially in Europe, in the latter half of the 20th century due to intensification in agricultural practice. It is important to assess the potential impacts of new technologies, such as genetically modified (GM) crops, on biodiversity given the large amount of funding spent on schemes designed to aid wildlife on farmland. There is not, to my knowledge, any published evidence of direct effects of GM crops on birds. However, there was considerable evidence of potential indirect effects of GM crops on farmland birds available from the recent UK Farm-scale Evaluation (FSE) trials (see Chapter 2, this volume). Results suggested that three out of four varieties of genetically modified herbicide-tolerant (GMHT) crops (spring-sown and winter-sown oilseed rape and sugarbeet) will support between two and three orders of magnitude lower weed abundances than conventionally managed crop varieties. These results were caused by differences in the type of pesticides sprayed on GMHT and conventional crops. For one crop (maize), there were more weeds on the GMHT crop than the conventional variety. Weeds provide key food resources for birds both directly, via seeds, and indirectly, via the invertebrate populations that they support. The declines in weed seed resources reported on the three GMHT crops suggest they have the potential to markedly reduce food resources for farmland birds, although the magnitude of how these changes in food could affect population levels is not currently known. If expensive schemes designed to enhance biodiversity on farmland are not to act in opposition to the environmental effects of GMHT crops then we need new ways of implementing GMHT crops to reduce their effects on weeds.

The Impact of Agricultural Practice on Farmland Bird Populations

Wildlife populations have been affected by clearance of habitats for agriculture ever since man first began to farm. However, for the context of this book I will focus on the effects of changing agricultural practice on birds over the latter half of the 20th century and the start of the 21st century.

Why concentrate on birds?

Undoubtedly the real interest in conservation ecology is preserving or enhancing the entire ecosystem (i.e. the community of organisms and their environment functioning as an ecological unit) on farmland or elsewhere, so why are birds so often singled out for study? The best option would be to have thorough data on all taxa, but for logistical reasons this is often not possible. Birds are an amenable study group because their populations are relatively easy to monitor and they are key indicators for the plants and animals on which they feed (Krebs *et al.*, 1999). Another reason why birds are often chosen as a study group is because, due to their appeal to many amateur birdwatchers and professional scientists alike, there is, arguably, simply a larger quantity and better quality of data available for birds than for any other taxa (Gaston and Blackburn, 2000; Orme *et al.*, 2006).

Have farmland birds declined more than populations of birds living in other habitats?

It is perhaps best to start with a caveat. Although farming is widespread around the world, a simple 'broad-brush' literature search revealed that 86% of studies of farmland birds were undertaken in European countries (Web of Science search on 24th January 2006 with search term 'Farmland Birds' yielded 568 hits and 487 of these related to studies carried out in European countries, the vast majority of which were empirical). Why is this so? Within Europe areas preserved for nature conservation are mixed into the landscape, thus the need to understand how to integrate the two. In other parts of the world (e.g. the USA), large areas of wilderness are largely set aside for nature, thus areas of farmland are generally considered primarily for agricultural production and not for landscape or wildlife needs. This is not a totally black-and-white situation, however. For example, in Canada the intensity of use of granular insecticides has been negatively correlated with population decline in a variety of species (Mineau and Downes, 2005); in North America grassland birds have been lost due to declines in the amount of grassland habitat per se (Brennan and Kuvlesky, 2005). It would be reasonable to suggest that there is stronger political will, up to now at least, to integrate farming and wildlife in Europe and so I concentrate on the situation in Europe primarily from here on. That is not to say that the principles applied within the European situation cannot be used elsewhere to help inform integration of wildlife and farming.

Farmland bird population trends in Europe have declined dramatically in the last few decades of the 20th century (Fuller *et al.*, 1995; Gregory *et al.*, 2004, 2005). Could these trends be due to the many changes happening in the world, such as climate change or increased pollution by modern technology?

There are three pieces of evidence which lend strong, albeit correlative, support to the idea that populations of birds which live on farmland have declined more than bird populations in different habitats (within the same geographical range). First, monitoring data for the UK has shown that populations of birds that live predominantly on farmland (e.g. skylark, *Alauda arvensis* L.; yellowhammer, *Emberiza citrinella* L.) have declined at a far greater rate over the same time period than those species which live in other habitats, such as generalist species (living in a range of habitats) or woodland species (Eaton *et al.*, 2006). This pattern was broadly similar for farmland bird specialists in Europe when compared with other European species living in other habitats (Gregory *et al.*, 2005). Second, Chamberlain *et al.* (2000) showed that the timing of changes in farmland bird populations matched, with a short time lag, the changes in agricultural practice (increases in the area of oilseed rape, autumn-sown cereals, and the use of pesticides and inorganic fertilizers). Third, Donald *et al.* (2001) showed that farmland bird population declines were greatest in European countries in which agricultural intensification, as measured by yield per unit of farmed land, was greatest. In other words, higher intensity farming leads to greater declines in farmland birds. These pieces of evidence are important because if global changes to climate or pollution were affecting species populations then why would these declines have focused with greater intensity on species living on farmland and why would the changes have matched changes in agricultural intensity both spatially and temporally? These studies therefore suggest that it is changes on farmland itself which have affected the group of bird species that live there and not other more general causes.

How has agricultural practice changed since the late 1950s and how have these changes affected farmland bird populations?

Within Europe there have been a multitude of changes in agricultural practice associated with intensification (see Table 12.1). These changes have had profound effects on bird populations on European farmland. An array of farmland bird studies has uncovered many different causal mechanisms underlying individual species population declines. For example, declines in skylark populations were attributed to changes in timing of sowing of cereal crops (Wilson *et al.*, 1997); the elimination of corncrakes as breeding birds throughout most of Britain was linked to a change from hand cutting of hay to horse-drawn mowing machines (Green, 1995); the rapid decline of cirl buntings, *Emberiza cirlus* L., in England was linked to declines in areas of mixed farming of low-intensity arable land and grassland (Aebischer *et al.*, 2000); the rapid decline in grey partridge populations was linked to increasing use of pesticides which reduced both invertebrates and the weed seeds on which the invertebrates feed – this led to reduced invertebrate availability for chicks which was identified as the key factor limiting populations (Potts 1980, 1986). The complex nature of changes in agricultural practice can however be neatly summed up by the term 'agricultural intensification'. Donald *et al.* (2001) eloquently show that on average farmland

Table 12.1. Examples of changes in British agricultural over recent decades likely to effect farmland birds. (From Krebs *et al.*, 1999; Fuller, 2000.)

Change in agriculture	Effect on birds
Land drainage (especially grasslands) results in drying out of fields	Reduces access to food for many species (drier soils mean birds that probe the ground for food have reduced food availability because prey are deeper in the soil thus reducing breeding productivity; M.J. Whittingham and C.L. Devereux, 2008, unpublished data)
Hedgerow removal	Less nesting sites, food and cover from predators (Whittingham and Evans, 2004)
'Improvement of pastures' (more fertilizers and monocultures)	Taller, faster growing swards are less suitable for many ground-nesting birds (Fuller, 2000) and it also reduces foraging rates and is likely to increase predation risk (Whittingham and Evans, 2004; Devereux *et al.*, 2006)
Increased agrochemical input	Pesticides reduce weeds and invertebrate populations and thus reduce breeding productivity and winter foraging habitat quality (e.g. Hart *et al.*, 2006)
Switch from spring to autumn sowing of cereals	Fewer nesting opportunities for species such as skylark (Wilson *et al.*, 1997)

bird population changes in European countries are correlated with intensification. Donald *et al.* used the metric of yield per unit of farmed land (e.g. the amount of cereal yield or milk yield) to indicate agricultural intensification.

The global view of agriculture

Given the focus of studies on farmland birds in Europe (see above), it is perhaps surprising that a recent study identified agriculture not only as the greatest threat to the extinction of bird species but also that the threat was greater in the developing world than in the developed world (Green *et al.*, 2005). The greatest difference between the developing and developed world is that in the former pristine habitats are cleared for farming and this results in the loss of many species, particularly those of conservation concern (Green *et al.*, 2005). In developed countries, on the other hand, there is very little pristine habitat left and so changes in agriculture, although clearly exerting strong negative effects on wildlife, are not of the magnitude shown in developing countries.

What is being done about these declines in farmland bird populations?

I hope it is clear from the above that populations of farmland birds in Europe have declined substantially over recent decades and that the overwhelming weight of evidence shows that these declines are due to changes in agricultural practice. If you are not convinced then so be it, but the governments of many European countries have been. This has been a strong driver in the formulation of agri-environment schemes (AES). Although the aims of these schemes include the enhancement of landscapes, the protection of the historic environment and promoting public access to the countryside, a major goal of these schemes is to benefit biodiversity (Whittingham, 2007). These schemes have used up a considerable amount of public money (€24 billion was spent by the European Union (EU) between 1992 and 2003 on these schemes) but the effects of these schemes on biodiversity have been mixed (Kleijn and Sutherland, 2003).

With the background of farmland birds and agricultural change now covered, I wish to explore the potential effects of genetically modified (GM) crops on farmland birds. Clearly any new technology that has the potential to affect biodiversity on farmland needs to be scrutinized in order to guide policy aimed at influencing biodiversity levels in agricultural systems. The key issue on farmland is that the effects of GM crops are currently part of a policy background (in Europe) in which increasing biodiversity on farmland is being promoted.

Impact of GM Herbicide-tolerant Crops on Farmland Birds

Direct effects

Prior to commercialization, GM crops must go through a rigorous screening process in order to demonstrate substantial equivalence to the non-GM comparator crop (Levidow *et al.*, 2007), the assumption being that any changes in the crop will be due to expression of the introduced transgene. No studies to date report direct effects of GM crops on wild birds and toxicity studies with GM feeds (e.g. *Bacillus thuringiensis* (*Bt*) maize) do not show any impact on farm birds (Aulrich *et al.*, 2002). Such non-target direct effects are recorded in arthropod species. For example, one study reported a deleterious effect of GM maize on the Monarch butterfly, *Danaus plexippus* L. (Losey *et al.*, 1999), although subsequent work has suggested that these claims are largely unfounded (Gatehouse *et al.*, 2002). These issues are covered in detail elsewhere in this book (see Chapters 8, 9 and 18, this volume).

Indirect effects: FSE trials

There is considerable evidence of the indirect effects of GM crops on birds, or more specifically the invertebrates and seed resources on which birds feed. The best evidence for this comes from the UK Farm-scale Evaluations (FSE) trials which investigated the effects of GM herbicide-tolerant (GMHT) crops on farmland biodiversity. The study was carried out over 3 years in 60 fields across England and Scotland (Freckleton *et al.*, 2003). Fields were divided into two: one-half was sown with a conventional crop and the other with a GMHT crop

(Firbank *et al.*, 2003). The crops grown were sugarbeet, maize and winter and spring oilseed rape. Below I illustrate the indirect effects of GMHT crops on birds based on the results from the FSE trials.

Both nesting and foraging behaviour can be affected by vegetation structure (Stephens and Krebs, 1986; Whittingham and Evans, 2004). What evidence is there to suggest that growing of GMHT crops could influence vegetation structure? GMHT crops in the FSE trials were not sprayed with a pre-emergence herbicide (because the broad-spectrum herbicides associated with GMHT crops are so strong that this is unnecessary); as a result weed densities were initially higher in GMHT crops. However, after herbicide application, the weeds were killed. Typically at the end of the growing season, there were fewer weeds in the GMHT oilseed rape and beet than in the conventionally grown crops. What does this mean for birds? Those ground-nesting species, such as yellow wagtail and skylark, which nest in dense weedy cover, could find that their nests become exposed following spraying in GMHT crops which leaves them especially vulnerable to predation (Donald *et al.*, 2002; Gilroy, 2007). Thus, GMHT crops have the potential to attract some species to nest in them but for these species to suffer from heavy nest predation late in the season. This area may make an interesting area for further study.

Food abundance and availability, as well as predation risk, largely determine the foraging efficiency of an animal (i.e. the amount of food eaten per unit time; Stephens and Krebs, 1986). Changes in vegetation structure caused by herbicide spraying could potentially influence foraging behaviour. Birds with a restricted field of view respond more slowly to an approaching predator and compensate by reducing intake rates of food and spending more time looking for predators (Whittingham *et al.*, 2004; Butler *et al.*, 2005). However, given that the crop is overshadowing the weeds beneath I would speculate that changes to vegetation structure due to GMHT crop management would make little difference to predation risk as the view of the surroundings is severely compromised anyway. However, food is less conspicuous on densely vegetated substrates, as is found earlier on in the season in some GMHT crops, which is likely to reduce foraging efficiency (Moorcroft *et al.*, 2002; Whittingham and Markland, 2002), although later in the season the opposite will apply.

The FSE trials showed that herbicide spraying used as part of the management of GMHT crops and conventional varieties had significant effects on weed populations and subsequently on seed abundance. Total seed counts from weeds in spring-sown wheat, *Beta vulgaris* L., and oilseed rape, *Brassica napus* L. (Heard *et al.*, 2003), and dicotyledonous weed seed rain in winter-sown oilseed rape (Bohan *et al.*, 2005), were significantly reduced in GMHT as compared with conventional varieties (Table 12.2). These effects persisted in the following year and estimates of 7% declines per annum in the seedbank were made for the two spring-sown crops in a typical cereal rotation (Heard *et al.*, 2003; Bohan *et al.*, 2005).

What effects are these changes in seeds likely to have on bird seed resources? The FSE trials were based at the field scale and because spraying occurred in the summer the spatial scale was limited in which to measure the response by breeding birds given that the fields used in the trial were widely

Table 12.2. The effect of GMHT crop management on weed seed and invertebrate resources likely to be important in the diet of farmland birds. These results are all derived from the UK Farm-scale Evaluation trials.

Crop	Differences in the abundance of key invertebrates in the diet of birds[a] between GM and conventional crop varieties	Differences in the abundance of key weed seeds[b] in the diet of birds between GM and conventional crop varieties[c]
Winter-sown oilseed rape	No significant difference between any group[d]	Approximately twice as many seeds present on conventional than GM crop variety[e]
Spring-sown oilseed rape	More spiders on conventional crops later in the year[f] and more bugs on conventional crops.[f] No significant difference for any other group[g]	Approximately three times as many seeds present on conventional than GM crop variety[e]
Sugarbeet	No significant difference between any group[g]	Approximately three times as many seeds present on conventional than GM crop variety[e]
Maize	No significant difference between any group[g]	More seeds present on GM crop variety than conventional crop[e]

[a]These include beetles (Carabidae species and Staphylinidae species), bugs (Heteroptera species) and spiders (Araneae species) (Wilson *et al.*, 1999).
[b]Key weed seeds in the diet of birds as defined by Gibbons *et al.* (2006) and Wilson *et al.* (1999).
[c]Based on averages across 17 species of farmland birds (Gibbons *et al.*, 2006).
[d]Bohan *et al.* (2005).
[e]Gibbons *et al.* (2006).
[f]Haughton *et al.* (2003).
[g]Brooks *et al.* (2003).

geographically separated. One attempt was made to measure bird abundance on a subset of FSE fields (of all four crop treatments). Yellowhammers and granivorous bird species collectively were more abundant on conventional than GMHT sugarbeet fields, and granivores were also more abundant on conventional than GMHT maize (Chamberlain *et al.*, 2007). This study also showed that in the winter several species were more abundant on maize stubbles following GMHT treatment. A major caveat of this work is that it was limited by the spatial scale on which the study was undertaken and so the statistical power to test for differences between treatments is likely to be weak.

However, another approach extrapolated the results on weed seeds from the FSE to potential effects on birds. Gibbons *et al.* (2006) predicted the effects of the reductions in seed supplies resulting from management of GM crops reported by studies of the FSE trials (Heard *et al.*, 2003; Perry *et al.*, 2003) on 17 granivorous farmland bird species (including bunting, *Emberiza* spp.; finch, *Fringilla* spp.; partridges; pigeons; sparrows, *Passer* spp.; and skylark, *A. arvensis* L.) whose diets were known from literature reviews (Wilson *et al.*, 1996, 1999). In all 17 species in both beet and spring oilseed rape, rain of weed seeds important in the diet were found to be approximately threefold less in the GMHT-managed crops as compared with the conventional crops, and in all but two cases these differences were significant (Table 12.2). Results were similar but slightly less emphatic for winter oilseed rape with a twofold difference on average across species, with 11 species showing significantly more seed resources in conventional crops (Gibbons *et al.*, 2006; Table 12.2). Results for GMHT maize contrasted with the other three crops, with increased seed rain for all 17 species, although in only seven species was this difference significant. Similar results were found when the energy content of seeds was analysed instead of seed abundance (Gibbons *et al.*, 2006). In summary, management of GMHT crops is likely to reduce weed seed resources for farmland birds in three of the crops trialled (beet, spring and winter oilseed rape) and increase seed resources in maize.

Differences in spraying regimes between GMHT and conventional crop varieties are also likely to affect invertebrate populations (via changes in weeds). Some invertebrate species form an important component of the diet of farmland birds and so the former is likely to covary with the latter. Here I will focus on invertebrates likely to be important in the diet of birds. There is a wide variety of invertebrates which have been shown to be important within farmland bird diets (Wilson *et al.*, 1999), but not all of these were measured in the FSE trials. Thus, I have concentrated on four invertebrate groups that have been shown to be important in the diet of farmland birds and were also measured in the trials namely: beetles (Carabidae species and Staphylinidae species), bugs (Heteroptera species) and spiders (Araneae species). Surprisingly, despite the clear differences in weed populations (see above), there were few differences in invertebrates sampled on GMHT crop varieties and on conventional crops (Table 12.2). There were significant differences in other invertebrate groups (e.g. butterflies, bees) with more being found on some conventional crops (Haughton *et al.*, 2003; Bohan *et al.*, 2005) but these groups are not generally food sources for birds.

GMHT maize contrasts with the other three crops in the FSE trial because it supports a greater abundance of weeds than conventional maize (Heard *et al.*, 2003; Gibbons *et al.*, 2006). The difference in GMHT maize was due to the fact that the broad-spectrum herbicide used on GMHT maize, glufosinate ammonium, was less effective at weed control than those used on conventional maize, mostly triazine herbicides such as atrazine (Heard *et al.*, 2003). At the time of the FSE trials, herbicide management of conventional maize crops included on the trial fields reflected accurately the standard practice within the UK (Champion *et al.*, 2003). However, triazine use will be prevented under future EU regulations (Brooks *et al.*, 2005). Perry *et al.* (2004) reanalysed the FSE data, by separating out sites not using triazines, and reported that although weed abundance would be likely to increase without triazines under conventional management, GMHT maize would still support a greater abundance of weeds.

In summary, GMHT crops are likely to reduce seed resources for many species of farmland birds, except on GMHT maize where a reverse effect is likely. The effects on invertebrates are substantially less obvious than for weeds.

However, given that weed populations are likely to be reduced substantially more in the long term, by a factor of 0.7–0.8 over 28 years on GMHT oilseed rape and beet (Heard *et al.*, 2005), it naturally follows that there may be, as yet undetected, long-term effects on invertebrates. At present, though, the FSE trials have provided little evidence of short-term changes in invertebrate populations due to the growing of GMHT crops.

What effect will GMHT crops have on bird populations?

The data presented in the previous section focus on how food resources for birds may be affected by GMHT crop management. However, it is not clear how these differences may affect bird populations. As noted by Freckleton *et al.* (2003), although the FSE is one of the most extensive and impressive ecological studies ever conducted it is not without limitations. One particular component highlighted is the lack of predictive power, especially at a landscape scale. Recently, however, one study has attempted to model the affects of GMHT crops at just such a scale (Butler *et al.*, 2007). Butler *et al.* used a model of the landscape split between cropped areas, margins and hedgerows. They used a simple matrix of how two changes in management will affect a range of components of the life history of bird species (e.g. foraging and nesting in both summer and winter) to estimate likely impacts. The study showed that species with narrower niches, notably those which are 'farmland' specialists (e.g. yellowhammer), are more at risk from agricultural change than more generalist species which also live on farmland (e.g. chaffinch, *F. coelebs* L.) and that the derived risk scores correlate well with population change. However, they go further and make predictions from their models, including predictions of future changes in GMHT crops. These predictions are perhaps a step too far for a number of reasons. First, although the risk assessment is significantly correlated with population change, the relationship is weak ($r^2 = 19\%$). Second, the model used in the study makes the prediction that GMHT crops reduce meadow pipit numbers based on the logic that GMHT crops reduce the biomass of invertebrates in cropped areas (via reducing weed populations) and that as meadow pipits feed on within-crop invertebrates they are likely to decline. However, this ignores the fact that meadow pipits make hardly any use of beet and rape (being mainly tied to grassland) and so are very unlikely to be affected by changes on these crops! In summary the effects on bird populations due to potential large-scale changes to GM crops remain poorly understood.

Although this chapter concentrates on farmland birds (and is therefore rather specific), I would like to make one further point about the FSE trials. As an environmental impact assessment, the FSE trials were rather one-dimensional in that they did not consider many additional factors like fossil fuel consumption, inputs, carbon dioxide, etc. In some of these areas GMHT crops are very efficient as they require less labour and input of active ingredient. Another advantage of GMHT crops is that glyphosate breaks down on contact with the soil and so contributes less to pollution. These areas are covered in detail elsewhere in the book (see Chapters 2 and 7, this volume), but it is important to consider that promotion of biodiversity (of which farmland birds is an indicator) includes a wide remit of the potential influences of crop production.

Conclusions

Growing of GMHT crops has been shown to reduce the abundance of weeds in three out of four crops studied in the FSE trials. The effects of these differences on invertebrate abundances were less obvious but in the long term may be negative. Results of the FSE trials suggest that the abundance of weed seeds, which are a key component of the diet of many farmland bird species, for three of the crops studied would be seriously reduced on GMHT crops as opposed to conventional crop varieties. Should these three GMHT crops replace their conventional counterparts wholesale then the resultant decline in food resources for farmland birds is likely to act in the opposite direction to expensive AES aimed to improve populations of farmland birds and other wildlife in agricultural systems.

There are potential solutions to this problem. The timing of spraying of herbicides is crucial to weed population dynamics and hence to invertebrates and birds at higher trophic levels. Freckleton *et al.* (2004) suggest that spraying early in the season may allow late emerging weeds to survive and thus positively influence weed populations on GMHT crops. If the agricultural landscape is filled with GMHT crops then it seems likely that these types of novel solutions will need to be explored so that agricultural policy aimed at production and that aimed at wildlife can be joined up to work together, and not in opposing directions.

Acknowledgements

Thanks to Rob Freckleton for helpful suggestions and to Guy Anderson and James Gilroy for comments about yellow wagtails.

References

Aebischer, N.J., Green, R.E. and Evans, A.D. (2000) From science to recovery: four cases of how research has been translated into conservation action in the UK. In: Aebischer, N.J., Evans, A.D., Grice, P.V. and Vickery, J.A. (eds) *Ecology and Conservation of Lowland Farmland Birds.* British Ornithologists Union, Norwich, UK.

Aulrich, K., Bohme, H., Daenicke, R., Halle, I. and Flachowsky, G. (2002) Novel feeds – a review of experiments at our institute. *Food Research International* 35, 285–293.

Bohan, D.A., Boffey, C.W.H., Brooks, D.R., Clark, S.J., Dewar, A.M., Firbank, L.G., Haughton, A.J., Hawes, C., Heard, M.S., May, M.J., Osborne, J.L., Perry, J.N., Rothery, P., Roy, D.B., Scott, R.J., Squire, G.R., Woiwood, I.P. and Champion, G.T. (2005) Effects on weed and invertebrate abundance and diversity of herbicide management in genetically modified herbicide-tolerant winter-sown oil-seed rape. *Proceedings of the Royal Society Series B* 272, 463–474.

Brennan, L.A. and Kuvlesky, W.P. (2005) North American grassland birds: an unfolding conservation crisis? *Journal of Wildlife Management* 69, 1–13.

Brooks, D.R., Bohan, D.A., Champion, G.T., Haughton, A.J., Hawes, C., Heard, M.S., Clark, S.J., Dewar, A.M., Firbank, L.G., Perry, J.N., Rothery, P., Scott, R.J., Woiwod, I.P., Birchall, C., Skellern, M.P., Walker, J.H., Baker, P., Bell, D., Browne, E.L., Dewar, A.J.G., Fairfax, C.M., Garner, B.H., Haylock, L.A., Horne, S.L., Hulmes, S.E., Mason, N.S., Norton, L.R., Nuttall, P., Randle, Z., Rossall, M.J., Sands, R.J.N., Singer, E.J. and Walker, M.J. (2003) Invertebrate responses to the management of genetically modified herbicide-tolerant and conventional spring crops. I. Soil-surface-active invertebrates. *Philosophical Transactions of the Royal Society of London* 358, 1847–1862.

Brooks, D.R., Clark, S.J., Perry, J.N., Bohan, D.A., Champion, G.T., Firbank, L.G., Haughton, A.J., Hawes, C., Heard, M.S. and Woiwod, I.P. (2005) Invertebrate biodiversity in maize following withdrawal of triazine herbicides. *Proceedings of the Royal Society Series B* 272, 1497–1502.

Butler, S.J., Bradbury, R.B. and Whittingham, M.J. (2005) Stubble height manipulation causes differential spatial use of stubble fields by farmland birds. *Journal of Applied Ecology* 42, 469–476.

Butler, S.J., Vickery, J.A. and Norris, K. (2007) Farmland biodiversity and the footprint of agriculture. *Science* 315, 381–384.

Chamberlain, D.E., Fuller, R.J., Bunce, R.G.H., Duckworth, J.C. and Shrubb, M. (2000) Changes in the abundance of farmland birds in relation to the timing of agricultural intensification in England and Wales. *Journal of Applied Ecology* 37, 771–788.

Chamberlain, D.E., Freeman, S.N. and Vickery, J.A. (2007) The effects of GMHT crops on bird abundance in arable fields in the UK. *Agriculture, Ecosystems and Environment* 118, 350–356.

Champion, G.T., May, M.J., Bennett, S., Brooks, D.R., Clark, S.J., Daniels, R.E., Firbank, L.G., Haughton, A.J., Hawes, C., Heard, M.S., Perry, J.N., Randle, Z.,

Rossall, M.J., Rothery, P., Skellern, M.P., Scott, R.J., Squire, G.R. and Thomas, M.R. (2003) Crop management and agronomic context of the Farm Scale Evaluations of genetically modified herbicide-tolerant crops. *Philosophical Transactions of the Royal Society of London* 358, 1801–1818.

Devereux, C.L., Whittingham, M.J., Fernandez-Juricic, E., Vickery, J.A. and Krebs, J.R. (2006) Foraging strategies and predator detection by starlings under differing levels of predation risk. *Behavioural Ecology* 17, 303–309.

Donald, P.F., Green, R.E. and Heath, M.F. (2001) Agricultural intensification and the collapse of Europe's farmland bird populations. *Proceedings of the Royal Society of London Series B* 268, 25–29.

Donald, P.F., Evans, A.D., Muirhead, L.B., Buckingham, D.L., Kirby, W.B. and Schmitt, S.I.A. (2002) Survival rates, causes of failure and productivity of skylark *Alauda arvensis* nests on lowland farmland. *Ibis* 144, 652–664.

Eaton, M.A., Ausden, M., Burton, N., Grice, P.V., Hearn, R.D., Hewson, C.M., Hilton, G.M., Noble, D.G., Ratcliffe, N. and Rehfisch, M.M. (2006) *The State of the UK's Birds 2005*. RSPB, BTO, WWT, CCW, EN, EHS and SNH, Sandy, Bedfordshire, UK.

Firbank, L.G., Heard, M.S., Woiwood, I.P., Hawes, C., Haughton, A.J., Champion, G.T., Scott, R.J., Hill, M.O., Dewar, A.M., Squires, G.R., May, M.J., Brooks, D.R., Bohan, D.A., Daniels, R.E., Osborne, J.L., Roy, D.B., Black, H.I.J., Rothery, P. and Perry, J.N. (2003) An introduction to the farm-scale evaluations of genetically-modified herbicide-tolerant crops. *Journal of Applied Ecology* 40, 2–16.

Freckleton, R.P., Sutherland, W.J. and Watkinson, A.R. (2003) Deciding the future of GM crops in Europe. *Science* 302, 994–996.

Freckleton, R.P., Stephens, P.A., Sutherland, W.J. and Watkinson, A.R. (2004) Amelioration of biodiversity impacts on genetically modified crops: predicting transient versus long-term effects. *Proceedings of the Royal Society Series B* 271, 325–331.

Fuller, R.J. (2000) Relationships between recent changes in lowland British agriculture and farmland bird populations: an overview. In: Aebischer, N.J., Evans, A.D., Grice, P.V. and Vickery, J.A. (eds) *Ecology and Conservation of Lowland Farmland Birds*. British Ornithologists Union, Norwich, UK.

Fuller, R.J., Gregory, R.D., Gibbons, D.W., Marchant, J.H., Wilson, J.D., Baillie, S.R. and Carter, N. (1995) Population declines and range contractions among lowland farmland birds in Britain. *Conservation Biology* 9, 1425–1441.

Gaston, K.J. and Blackburn, T.M. (2000) *Pattern and Process in Macroecology*. Blackwell Science, Oxford.

Gatehouse, A.M.R., Ferry, N. and Raemaekers, R.J.M. (2002) The case of the monarch butterfly: a verdict is returned. *Trends in Genetics* 18, 249–251.

Gibbons, D.W., Bohan, D.A., Rothery, P., Stuart, R.C., Haughton, A.J., Scott, R.J., Wilson, J.D., Perry, J.N., Clark, S.J., Dawson, R.J.G. and Firbank, L.G. (2006) Weed seed resources for birds in fields with conventional and genetically modified herbicide-tolerant crops. *Proceedings of the Royal Society Series B* 273, 1921–1928.

Gilroy, J.J. (2007) Breeding ecology and conservation of the yellow wagtail *Motacilla flava* in intensive arable farmland. PhD thesis, University of East Anglia.

Green, R.E. (1995) Diagnosing causes of bird population declines. *Ibis* 137, 47–55.

Green, R.E., Cornell, S.J., Scharlemann, J.P.W. and Balmford, A. (2005) Farming and the fate of wild nature. *Science* 307, 550–555.

Gregory, R.D., Noble, D.G. and Custance, J. (2004) The state of play of farmland birds: population trends and conservation status of lowland farmland birds in the United Kingdom. *Ibis* 146, 1–13.

Gregory, R.D., Van Strien, A., Vorisek, P., Gmelig Meyling, A.W., Noble, D.G., Foppen, R.P.B. and Gibbons, D.W. (2005) Developing indicators for European birds. *Philosophical Transactions of the Royal Society Series B* 360, 269–288.

Haughton, A.J., Champion, G.T., Hawes, C., Heard, M.S., Brooks, D.R., Bohan, D.A., Clark, S.J., Dewar, A.M., Firbank, L.G., Osborne, J.L., Perry, J.N., Rothery, P., Roy, D.B., Scott, R.J., Woiwod, I.P., Birchall, C., Skellern, M.P., Walker, J.H., Baker, P., Browne, E.L., Dewar, A.J.G., Garner, B.H., Haylock, L.A., Horne, S.L., Mason, N.S., Sands, R.J.N. and Walker, M.J. (2003) Invertebrate responses to the management of genetically modified herbicide-tolerant and conventional spring crops. II. Within-field and epigeal and aerial arthropods. *Philosophical Transactions of the Royal Society of London* 358, 1863–1877.

Hart, J.D., Milsom, T.P., Fisher, G., Wilkins, V., Moreby, S.J., Murray, A.W.A. and Robertson, P.A. (2006) The relationship between yellowhammer breeding performance, arthropod abundance and insecticide applications on arable farmland. *Journal of Applied Ecology* 43, 81–91.

Heard, M.S., Hawes, C., Champion, G.T., Clark, S.J., Firbank, L.G., Haughton, A.J., Parish, A.M., Perry, J.N., Rothery, P., Scott, R.J., Skellern, M.P., Squire, G.R. and Hill, M.O. (2003) Weeds in fields with contrasting conventional and genetically modified herbicide-tolerant crops. I. Effects on abundance and diversity. *Philosophical Transactions of the Royal Society Series B* 358, 1833–1846.

Heard, M.S., Rothery, P., Perry, J.N. and Firbank, L.G. (2005) Predicting longer-term changes in weed populations under GMHT crop management. *Weed Research* 45, 331–338.

Kleijn, D. and Sutherland, W.J. (2003) How effective are European agri-environment schemes on conserving and promoting biodiversity? *Journal of Applied Ecology* 40, 947–970.

Krebs, J.R., Wilson, J.D., Bradbury, R.B. and Siriwardena, G.M. (1999) The second silent spring? *Nature* 401, 611–612.

Levidow, L., Murphy, J. and Carr, S. (2007) Substantial equivalence – transatlantic governance of GM food. *Science Technology and Human Values* 32, 26–64.

Losey, J.E., *et al.* (1999) Transgenic pollen harms monarch larvae. *Nature* 399, 214.

Mineau, P. and Downes, M. (2005) Patterns of bird species abundance in relation to granular insecticide use in the Canadian prairies. *Ecoscience* 12, 267–278.

Moorcroft, D., Whittingham, M.J., Bradbury, R.B. and Wilson, J.D. (2002) The selection of stubble fields by wintering granivorous birds reflects vegetation cover and food abundance. *Journal of Applied Ecology* 39, 535–547.

Orme, C.D.L., Davies, R.G., Olson, V.A., Thomas, G.H., Ding, T.S., Rasmussen, P.C., Ridgely, R.S., Stattersfield, A.J., Bennett, P.M., Owens, I.P.F., Blackburn, T.M. and Gaston, K.J. (2006) Global patterns of geographic range size in birds. *PLOS Biology* 4, 1276–1283.

Perry, J.N., Rothery, P., Clark, S.J., Heard, M.S. and Hawes, C. (2003) Design, analysis and power of the Farm Scale Evaluations of genetically modified herbicide tolerant crops. *Journal of Applied Ecology* 40, 17–31.

Perry, J.N., Firbank, L.G., Champion, G.T., Clark, S.J., Heard, M.S., May, M.J., Hawes, C., Squire, G.R., Rothery, P., Woiwod, I.P. and Pidgeon, J.D. (2004) Ban on triazine herbicides likely to reduce but not negate relative benefits of GMHT maize cropping. *Nature* 428, 313–316.

Potts, G.R. (1980) The effects of modern agriculture, nest predation and game management on the population ecology of partridges (*Perdix perdix* and *Alectoris rufa*). *Advances in Ecological Research* 11, 1–79.

Potts, G.R. (1986) *The Partridge: Pesticides, Predation and Conservation.* Collins, London.

Stephens, D.W. and Krebs, J.R. (1986) *Foraging Theory.* Princeton University Press, Princeton, New Jersey.

Whittingham, M.J. (2007) Will agri-environment schemes deliver substantial biodiversity gain and if not why not? *Journal of Applied Ecology* 44, 1–5.

Whittingham, M.J. and Evans, K.L. (2004) The effects of habitat structure on predation risk of birds in agricultural landscapes. Ecology and Conservation of Farmland Birds II: the road to recovery. *Ibis* 146, 211–222.

Whittingham, M.J. and Markland, H.M. (2002) The influence of substrate on the functional response of an avian granivore and its implications for farmland bird conservation. *Oecologia* 130, 637–644.

Whittingham, M.J., Butler, S., Cresswell, W. and Quinn, J.L. (2004) The effect of limited visibility on vigilance behaviour and speed of predator detection: implications for the conservation of granivorous passerines. *Oikos* 106, 377–385.

Wilson, J.D., Arroyo, B.E. and Clark, S.C. (1996) *The Diet of Species of Lowland Farmland: A Literature Review.* Department of the Environment and English Nature, London.

Wilson, J.D., Evans, J., Browne, S.J. and King, J.R. (1997) Territory distribution and breeding success of skylarks *Alauda arvensis* on organic and intensive farmland in southern England. *Journal of Applied Ecology* 34, 1462–1478.

Wilson, J.D., Morris, A.J., Arroyo, B., Clark, S.C. and Bradbury, R.B. (1999) A review of the abundance and diversity of invertebrate and plant foods of granivorous birds in northern Europe in relation to agriculture. *Agriculture, Ecosystems and Environment* 75, 13–30.

13 Safety for Human Consumption

R.H. PHIPPS

School of Agriculture, Policy and Development, University of Reading, Reading, UK

Keywords: GM feeds, safety of milk, meat and eggs, tDNA and novel proteins

Summary

Since 1996, over 500 million ha of genetically modified (GM) crops have been grown worldwide. The principal GM crops are soybean, maize, cotton and canola which have been modified for herbicide tolerance (Ht) and/or insect resistance *Bacillus thuringiensis* (*Bt*). These crops are all used in monogastric and ruminant livestock production rations as energy and protein feed resources. GM feeds are included either as a whole crop (maize silage), a specific crop component (maize grain) or as co-products or crop residues such as oilseed meals, maize gluten feed, maize stover. GM crops with nutritionally enhanced characteristics for both food and feed are in various stages of development but are not dealt with in this chapter. The concept of a comparative safety assessment which is regarded by regulatory authorities as a robust starting point for the safety assessment for both GM food and feed is discussed. The questions posed by the use of GM feed in livestock production are: does their use influence animal health and productivity and is there any evidence that human health will be affected by the consumption of products such as milk, meat and eggs derived from livestock fed GM feed ingredients? Numerous studies have established that the chemical composition, nutritive value and animal performance of currently used GM feed ingredients are comparable to their near-isogenic non-GM counterpart and also conventional varieties. Although many organizations including the WHO do not consider the consumption of DNA from either conventional or GM crops as a human safety issue, since humans have consumed DNA from a wide variety of sources since evolution began, concern was raised that transgenic DNA (tDNA) and gene products (novel proteins) may accumulate in livestock products derived from animals receiving GM feed ingredients. To date no studies have reported the presence of tDNA that could encode for a gene or gene products in milk, meat and eggs produced by animals receiving GM feed ingredients. There is no evidence to suggest that food derived from animals fed GM products is anything other than as safe and as nutritious as that produced from conventional feed ingredients.

Introduction

As we enter the 21st century, world population is increasing, the area of land available for crop and livestock production is decreasing and the rate of crop improvement through conventional breeding is slowing and over 800 million people are malnourished. There is thus an urgent need for new technologies to increase crop yield, improve nutritional quality of food and feed and reduce crop losses. Societal pressure requires this to be achieved in a manner ensuring safety for the public and the environment. Tillman (1999) noted that this major challenge to decrease the environmental impact of agriculture while maintaining or improving its productivity and sustainability would have no single easy solution.

While it is recognized that there are many controversial issues associated with the introduction of genetically modified (GM) crops, they are considered by many as one of the possible ways forward to increase crop yields, and improve food and feed quality in an environmentally acceptable manner (Carpenter *et al.*, 2002; Phipps and Park, 2002; Bennett *et al.*, 2004; Brookes and Barfoot, 2006).

This chapter reviews data relating to the use of GM crops in animal production systems and considers some of the basic concepts associated with their safety and nutritional assessment and the likely implications for the safety of milk, meat and eggs derived from animals fed GM feed ingredients.

GM Feed Ingredients in Livestock Production

Between 1996 and 2007, the area of GM crops grown worldwide increased from 2 million to 114 million ha (James, 2007). In 2006, GM crops were grown by over 12 million farmers in 23 countries. The principal GM crops are soybean, maize, cotton and canola which are modified for agronomic input traits such as herbicide tolerance (Ht) and/or insect resistance (*Bt*; see Chapters 1, 5, 6, 7 and 18, this volume), and are all used in livestock production. These feeds may be included as a whole crop (maize silage), a specific crop component (maize grain) or as co-products or crop residues such as oilseed meals, maize gluten feed, maize stover. In many parts of the world, maize grain and soybean meal are the preferred choice of energy and protein supplements for use in both monogastrics and ruminant diets. Approximately 70 million t of GM maize grain and 115 million t of GM soybean meal are used annually in livestock production. Although GM crops with nutritionally enhanced characteristics, known as output traits, are in various stages of development, they are not considered in this chapter. Their safety and nutritional assessment, and examples of case studies have been published by the International Life Sciences Institute (ILSI, 2004, 2008).

Comparative Safety Assessment of GM Crops

Basic concept

The safety and nutritional assessment of GM crops has been the subject of several excellent reviews, including those by Kuiper *et al.* (2001), Chesson (2001), Cockburn (2002) and Kok and Kuiper (2003), and is discussed in the *Guidance Document of the Scientific Panel on Genetically Modified Organisms for the Risk Assessment of Genetically Modified Plants and Derived Food and Feed* published by the European Foods Standards Authority (EFSA, 2004). While this process will not be discussed in detail, it is important to describe the initial stage which is a comparative safety assessment in which the phenotypic, agronomic and compositional characteristics of the GM variety are compared with an appropriate comparator such as a near-isogenic non-GM counterpart and conventional varieties which provide a baseline being recognized as safe because of a history of long use. The aim of this initial step is to identify similarities and differences between the GM variety and closely related conventional counterparts and is not a safety assessment per se.

Once this initial phase has been completed, the focus of the safety assessment switches to addressing any differences that have been established and would include a detailed molecular characterization of the insert gene, the safety assessment of newly expressed protein(s) encoded by the gene which would include toxicity and allergenicity studies and, on a case-by-case basis, a nutritional evaluation with target species (EFSA, 2004).

Phenotypic, agronomic and compositional assessment

Cockburn (2002) has provided an excellent example of the measurements, taken by plant breeders, which are used in the case of maize to compare the phenotypic and agronomic characteristics of GM crops and their appropriate counterparts. These include measurements such as stand establishment, leaf orientation, plant height, ear height, ear tip fill, ear shape, silk date, tassel size, early plant vigour, leaf colour, silk colour, susceptibility to pests and diseases, reaction to pesticides and yield.

It is well known that geneticists and plant breeders have for at least 50 years used the results of compositional analyses as one of their main selection criteria in the development of new improved varieties produced through conventional breeding techniques such as radiation and chemical mutagenesis. The OECD has produced consensus documents that recommend which compositional analyses should be carried out for new varieties of a range of different crops (OECD, 2001a,b, 2002a–c) and are applicable to both conventional and biotech crops. However, it should be noted that even when statistically significant differences in compositional analyses are recorded between a GM crop and its appropriate comparator, these differences should be assessed carefully, because on their own they may not indicate the presence of an unintended effect arising from the inserted gene, nor have any implications for food or environmental safety. For example, the difference may fall within the natural and often wide variation that exists between currently available conventional varieties (ILSI, 2003). This emphasizes the importance of comparing a GM crop to both its near-isogenic parental line and also to a number of commercially relevant varieties. The range in composition within crop varieties is well illustrated in the ILSI (2003) crop composition database. If significant, biologically meaningful differences are noted then further investigation is needed in the safety and nutritional assessment of the new variety. Follow-up studies may include further analytical procedures and/or livestock feeding studies. The need for this additional work should be assessed on a case-by-case basis.

Numerous papers have now been published comparing the chemical composition of GM crops modified for herbicide tolerance and/or insect resistance (see Chapter 4, this volume) with their near-isogenic non-GM counterpart and commercial varieties. These studies have been summarized and reviewed by Clarke and Ipharraguerre (2000) and Flachowsky *et al.* (2005b) and showed no marked differences between the composition of the GM crop and their appropriate comparators. The work conducted by Ridley *et al.* (2002) provides an excellent example of the extensive compositional analyses conducted when comparing the grain and forage component of Ht maize with the near-isogenic counterpart and a number of commercially grown varieties. The material analysed was obtained over two seasons from different geographical zones and from both replicated and non-replicated studies. Parameters measured included: key nutrients (protein, fat, ash, acid and neutral detergent fibre and non-structural carbohydrates), minerals (calcium, magnesium, phosphorus, potassium, copper, iron, zinc and manganese), 18 amino acids, eight fatty acids and anti-nutrients and secondary metabolites (phytic acid, trypsin inhibitor, ferulic and p-courmeric acids and raffinose). Compositional comparability was clearly demonstrated and even though some small differences between the GM material and its near-isogenic counterpart were statistically significant, the values noted fell within the range of the currently available varieties and that noted in historical literature (ILSI, 2003).

Nutritional assessment of first generation GM crops

Feeding studies with target livestock species including chickens (broilers and laying hens), pigs, sheep, dairy cows, beef cattle, rabbits and a range of fish species have now been conducted as part of the nutritional assessment of a range of first-generation GM crops to establish their effect on animal performance. Examples of these studies are discussed below.

Nutrient bioavailability

Although compositional analyses provide a cornerstone in the nutritional evaluation of feeds, they cannot provide information on nutrient digestion which is an important parameter. Numerous livestock feeding studies have now compared the *in vivo* bioavailability of nutrients from a range of crops with their near-isogenic non-GM counterpart and commercial varieties (Hammond *et al.*,

1996 (broilers, lactating dairy cattle, catfish); Maertens *et al.*, 1996 (rabbits); Daenicke *et al.*, 1999 (sheep); Boehme *et al.*, 2001 (pigs and sheep); Aulrich *et al.*, 2001 (broilers); Barriere *et al.*, 2001 (sheep); Gaines *et al.*, 2001b (pigs); Reuter *et al.*, 2002a,b (pigs); Stanford *et al.*, 2003 (sheep); Hartnell *et al.*, 2005 (sheep)). The results all showed that the bioavailability of a wide range of nutrients from a range of GM crops modified for agronomic input traits was comparable with those for near-isogenic non-GM and conventional varieties. While some statistically significant differences were noted these were generally small, inconsistent and not considered to be biologically meaningful.

Production studies with monogastric livestock

POULTRY Numerous feeding studies with day-old broiler chicks and laying hens have now been reported (Brake and Vlachos, 1998; Gaines *et al.*, 2001a; Stanisiewski *et al.*, 2002; Taylor *et al.*, 2002, 2003, 2004, 2005; Brake *et al.*, 2003; Elangovan *et al.*, 2003; Kan and Hartnell, 2004a,b). The diets were all formulated to contain a high proportion of the test material which included GM varieties of *Bt* and Ht maize, soybean, canola and wheat, and appropriate counterparts. In each study, the chemical composition of the feed ingredient produced from the GM varieties, the near-isogenic non-GM varieties and the commercial hybrids was determined and found to be comparable. Furthermore, the results established that the nutritional value was also comparable as no biologically meaningful differences in the production parameters were measured.

Recent m ıltigenerational studies comparing diets with non-GM and GM insect-resistant maize (expressing *Bt*) with quail and laying hens for ten and four generations, respectively, have been reported by Flachowsky *et al.* (2005b) and Halle *et al.* (2006). The authors reported that the GM maize did not significantly affect hatchability, health or performance of poultry, nor quality of meat and eggs when compared with the non-GM isogenic comparator.

The conclusion from these recent publications is that once the chemical composition of the GM feed is shown to be comparable with its appropriate non-GM counterpart, the nutritional value can be assumed to be similar and further animal feeding studies will add little to their nutritional assessment.

PIGS Numerous comparative feeding studies have now also been conducted with growing and finishing pigs (Bohme *et al.*, 2001; Gaines *et al.*, 2001b; Stanisiewski *et al.*, 2001; Weber and Richert, 2001; Reuter *et al.*, 2002a,b; Cromwell *et al.*, 2004; Hyun *et al.*, 2004; Stein *et al.*, 2004). In these studies, a range of feeds, including maize grain, sugarbeet, soybean meal, canola meal, rice and wheat, modified for agronomic input traits such as Ht and *Bt* were compared with near-isogenic non-GM and commercial varieties. With few exceptions these studies contained data on both the compositional analysis of the feeds and the results of nutritional assessment using a range of end points for the feeding study.

As with poultry studies, trials with pigs have also shown that when compositional analyses of GM varieties and the near-isogenic non-GM and commercial varieties were comparable, then nutritional value was also similar.

Production studies with ruminant livestock
Ruminant livestock may consume both forages, which form 20–100% of the diet and consist of fresh (e.g. grass and lucerne) or ensiled forage (e.g. grass, lucerne or maize silage) or crop residues (e.g. maize stover or cereal straw) and supplements to provide additional energy (e.g. cereal grain), and protein (oilseed meals such as soybean meal and canola meal). Many of these feed resources are now obtained from GM crops. Sheep, beef cattle and dairy cows have all been used in studies to compare feed ingredients derived from a range of GM crops and their near-isogenic non-GM counterpart and commercial hybrids.

BEEF CATTLE Studies with beef cattle including those reported by Daenicke *et al.* (1999), Petty *et al.* (2001a,b), Berger *et al.* (2002, 2003) and Folmer *et al.* (2002) are among those reviewed by Flachowsky *et al.* (2005a) who reported that the performance of beef cattle fed maize grain, maize silage or stover from GM crops was comparable to those recorded for conventional crops. In addition, they noted that when compositional analyses of GM varieties and the near-isogenic non-GM and commercial varieties were comparable, then the nutritional value was also similar when used in beef cattle.

DAIRY COWS Between 1996 and 2004, over 20 studies were conducted in which the performance of lactating dairy cows which received feed ingredients derived from GM crops modified for agronomic input traits has been compared with their near-isogenic non-GM control and commercial hybrids. An extensive range of GM feed ingredients have been used in these studies and include *Bt* maize silage and maize grain, derived from crops modified to be protected against European corn borer (Barriere *et al.*, 2001; Donkin *et al.*, 2003) and corn root worm (Grant *et al.*, 2003), Ht and *Bt* cotton seed (Castillo *et al.*, 2004), Ht soybeans (Hammond *et al.*, 1996), Ht maize silage (Ipharraguerre *et al.*, 2003) and Ht fodder beet (Weisbjerg *et al.*, 2001). These studies demonstrated that the important end points of feed intake, milk yield and composition of lactating dairy cows were unaffected by the inclusion of feed ingredients derived from this wide range of GM crops.

As with other livestock species, studies with lactating dairy cows also showed that once the chemical composition of the GM feed is shown to be comparable with its appropriate non-GM counterpart, their nutritional value can be assumed to be similar (Table 13.1). While these studies provided little further relevant information on the nutritional assessment of these GM feed ingredients, they did provide valuable public reassurance with the introduction of a new technology.

Production studies with fish and rabbits
Production studies have also been carried out with catfish (Hammond *et al.*, 1996), rainbow trout (Brown *et al.*, 2003), salmon (Sanden *et al.*, 2004) and rabbits (Maertens *et al.*, 1996; Chrastinova *et al.*, 2002), and provided similar conclusions to those drawn from studies conducted with other livestock species.

In conclusion, it has been established that GM crops currently used have similar chemical composition to appropriate comparator varieties, and it is therefore not surprising that in the numerous feeding studies that have now

Table 13.1. Chemical composition of whole crop maize silage and maize grain from GM *Bt* MON810, GM Ht (GA 21 glyphosate tolerance) and the non-GM controls, and the milk production of lactating dairy cattle receiving diets containing either the GM silage and grain compared with conventional feed ingredients (after Donkin *et al.*, 2003).

g/kg DM[a]	GM *Bt* (MON810)	Control	GM Ht (GA 21)	Control
	Chemical composition			
Dry matter (g/kg)	43.3 (86.4)§	41.4 (86.4)	37.3 (86.7)	38.9 (85.6)
CP	7.8 (8.6)	7.9 (9.15)	8.8 (9.7)	8.7 (9.9)
NDF	41.5 (8.7)	43.2 (9.3)	43.1 (8.6)	41.0 (8.9)
WSC	42.5 (76.7)	40.5 (75.6)	40.5 (76.0)	42.3 (75.1)
NE_L (Mcal/kg)	1.61 (2.11)	1.58 (2.09)	1.54 (2.09)	1.58 (2.11)
Calcium	0.24 (0.01)	0.23 (0.01)	0.24 (0.04)	0.23 (0.02)
Phosphorus	0.21 (0.36)	0.25 (0.32)	0.23 (0.31)	0.23 (0.42)
	Milk production			
DM I (kg/day)	24.4	25.4	21.8	21.5
Milk yield (kg/day)	34.9	35.7	27.8	27.5
Fat (%)	3.46	3.42	3.61	3.55
Protein (%)	3.00	2.97	3.24	3.25
Lactose (%)	4.66	4.70	4.72	4.70

[a]Unless otherwise stated § grain data in brackets.

been conducted with a wide range of target livestock species, there has been no evidence to suggest that their performance differed in any respect from those fed the non-GM counterpart (Clarke and Ipharraguerre, 2000; OECD, 2003; Flachowsky *et al.*, 2005a; Phipps *et al.*, 2006). It is also concluded that routine-feeding studies with target species generally added little to a safety and nutritional assessment of GM crops, but provides public reassurance as to the wholesomeness of the food produced.

Fate of Transgenic DNA and Encoded Proteins in GM Feed Ingredients

Conservation and processing of GM feeds

The ensiling process with chopping of plant tissue and the subsequent lowering of pH by lactic acid fermentation produces a harsh environment for DNA and will accelerate its degradation. In a recent review, Jonas *et al.* (2001) noted that studies conducted by Hupfer *et al.* (1999) indicate that, while the origin of the silage made from GM crops could be confirmed, it showed that the ensiling process resulted in major fragmentation of the transgenic DNA (tDNA) and that the presence of intact, functional genes after an extended time of ensiling was highly unlikely.

Feed processing also results in DNA fragmentation. While grinding and milling has little effect on DNA fragment size, mechanical expulsion or chemical extraction of oil from seeds can cause extensive fragmentation. However, Chiter *et al.* (2000) have shown that processed feed samples may contain DNA fragments large enough to contain functional genes. Thus, it is accepted that farm livestock are consuming both transgenic DNA and novel protein, in addition to endogenous DNA and protein.

Consumption and digestion of DNA

The United Nations Food and Agriculture Organization, the WHO, the US Food and Drug Administration and the US Environmental Protection Agency have each stated very clearly that the consumption of DNA from all sources – including plants improved through biotechnology – is safe and does not produce a risk to human health, given the long history of safe consumption of DNA. Jonas *et al.* (2001) have reviewed the issues surrounding the safety of DNA in food. They noted using data from Austria that the total per capita intake of tDNA that may arise from GM maize, soya and potatoes is $0.38\,\mu g\ day^{-1}$, which represents 0.00006% of a typical daily intake of 0.6 g. They emphasized that tDNA was equivalent to that from existing foods and concluded that information reviewed did not indicate any safety concerns associated with the ingestion of DNA per se from GMOs and considered it to be as safe as any other DNA in food.

In the case of farm livestock, the extensive and aggressive digestion processes to which protein and DNA are subjected to in both ruminants and monogastrics have been described in detail in a number of reviews (Beever and Kemp, 2000; Beever and Phipps, 2001; Beever, *et al.*, 2002). They concluded that the processes of nucleic acid digestion in ruminants and non-ruminants provides substantial evidence that the chances of intact DNA (either transgenic or native) being absorbed and incorporated into the host animal's genome are extremely remote. In addition, Beever *et al.* (2002) noted that the limited amount of research on the digestive fate of novel proteins encoded by the transgene also shows that the normal processes of protein digestion in both ruminants and non-ruminants appear to be more than adequate to prevent any intact proteins being absorbed across the intestinal wall.

Detection of Endogenous and Transgenic DNA and Novel Proteins in the Intestinal Tract of Livestock Species and Subsequent Presence in Food Products

The search for fragments of endogenous and tDNA and novel proteins encoded by the transgene in the intestinal tract of livestock and their presence in animal-derived foods such as milk, meat and eggs, have been a major research focus with studies conducted in a wide range of target livestock species.

Polymerase chain reaction (PCR) analyses have been used to determine the presence or absence of DNA fragments of multi- (e.g. rubisco) or single-copy endogenous genes (e.g. lectin) and the single-copy transgene inserted into the plant. In the case of multi-copy genes, there are up to 10,000 copies in every cell while in the case of single-copy genes there is only one copy of that gene per cell. It has been suggested by a number of authors, including Weber and Richert (2001) and Jennings *et al.* (2003a), that the greater abundance of DNA from endogenous multi-copy genes is likely to make it easier to detect than DNA from single-copy genes, irrespective of whether they are endogenous or transgenic genes. Enzyme-linked immunosorbent assay (ELISA) kits have been developed and used in the detection of novel proteins present in GM crops and which may be present in a range of samples from the gastrointestinal (GI) tract and food derived from animals fed GM feed ingredients.

Presence of endogenous and transgenic DNA and novel proteins in the gastrointestinal tract of livestock

Several workers have now reported the presence of DNA fragments from endogenous multi-copy plant genes in the intestinal tract of poultry (Aeschbacher *et al.*, 2004; Deaville and Maddison, 2005), pigs (Klotz *et al.* 2002; Chowdhury *et al.*, 2003a; Reuter and Aulrich, 2003) and dairy cows (Phipps *et al.*, 2003) and that the size and frequency of the fragments tended to decrease as digesta moved through the intestinal tract having been subjected to natural digestive processes. These studies showed, however, that the process of DNA digestion was not complete, indicating the opportunity for DNA to transfer across the intestinal wall and thus into peripheral tissues and subsequent animal-derived food.

Feed proteins are subjected to extensive and aggressive digestion processes in the GI tract of both monogastrics and ruminants and this topic has been reviewed by a number of authors including Beever and Kemp (2000). *In vitro* assays which have been developed to simulate monogastric digestion also indicate the extremely rapid degradation of feed proteins including novel proteins encoded by transgenes; such studies suggest that these proteins are unlikely to be found intact in the GI tract and are thus most unlikely to be absorbed and accumulated in animal-derived foods.

However, recent studies by Chowdhury *et al.* (2003a,b) and Einspanier *et al.* (2004) who used commercially available ELISA kits, reported an unexpected persistence of CRY1(A)b protein immunoactivity in the bovine GI tract when cows were fed *Bt* maize. In response to these findings, Lutz *et al.* (2005) from Technical University of Munich conducted further studies in which digesta samples were analysed by both ELISA and immunoblotting methods to determine if intact CRY1(A)b protein could in fact be detected in digesta samples. While the ELISA method again indicated that the concentration of CRY1(A)b protein increased during passage down the GI tract, results from the immunoblotting assays showed significant degradation of the protein in the GI tract. This led the authors to conclude that CRY1(A)b protein is rapidly and extensively degraded during digestion in cattle. These results, however, draw attention to the need to ensure that the most appropriate analytical methods are used and to interpret results with caution.

Presence of endogenous and transgenic DNA and novel proteins in animal-derived food products

Small fragments from the multi-copy endogenous plant gene rubisco have been found in milk (Phipps *et al.*, 2003) and poultry meat (Einspanier *et al.*, 2001; Deaville and Maddison, 2003), while Klotz *et al.* (2002) found similar plant DNA fragments in poultry meat obtained off supermarket shelves. While DNA fragments from single-copy endogenous genes have occasionally been found in the upper reaches of the intestinal tract of both monogastric and ruminant livestock, they have not been detected in milk or poultry meat (Deaville and Maddison, 2003; Phipps *et al.*, 2003).

The recent excellent review by Flachowsky *et al.* (2005b) of the role of GM plants in animal nutrition reported on the results of 23 studies with dairy cows, pigs and poultry when fed diets containing either GM or conventional feed ingredients. The GM feed ingredients used in these studies were from a range of different crops including soybean, maize and cotton which had been modified with either Ht and/or *Bt*. Results from these studies are presented in Table 13.2 and show that tDNA was not detected in milk, or poultry and pig meat produced in any of these 23 studies. A further study with Atlantic salmon has produced similar results as the authors (Sanden *et al.*, 2004) were also unable to detect small fragments of tDNA in fish muscle when fed Ht soybean meal.

However, in 2006, a study using an assay with greatly enhanced analytical sensitivity identified very small fragments (106–146 base pairs (bp)) of the transgenes *cry1Ab* and *cp4 epsps* in conventional and 'organic' milk samples in Italy (Agodi *et al.*, 2006). The tDNA fragments reported were so small that the results must be interpreted in the context of the minimal size for a functional gene. For example, the size of the intact *cry1Ab* gene is 3500 bp, and its minimal functional unit is encoded on 1800 bp of DNA, both of which are considerably larger than the 106–146 bp DNA fragments detected in milk.

A number of studies have now been conducted showing that novel proteins have not been detected in either milk (Calsamiglia *et al.*, 2003; Jennings *et al.*, 2003c; Phipps *et al.*, 2003; Yonemochi *et al.*, 2003) or pig and poultry meat (Weber and Richert, 2001; Ash *et al.*, 2003; Jennings *et al.*, 2003a,b; Yonemochi *et al.*, 2002).

To conclude, the detection of DNA in milk, meat and eggs is likely to be a function of the abundance of the gene, the size of the fragment of DNA being tested for, and the sensitivity of the analytical methods used, such that detection of fragments of tDNA would be much rarer and more difficult than detection of high copy number endogenous genes. Given the long history of safe consumption of meat, milk and eggs from animals in which endogenous plant and animal gene fragments have now been shown to be detectable, and given that the DNA of a transgene is identical to all other types of DNA, then products from animals fed GM crops would not differ from foods already deemed safe (Jonas *et al.*, 2001). In addition, novel proteins have not been detected in any animal-derived foods.

Table 13.2. Studies with livestock species to determine the presence or absence of transgenic DNA in animal-derived products.

Animal category	GM crop	Tissues examined	Detection of tDNA	Reference
Ruminants				
Dairy cow	*Bt* maize (silage)	Milk	Not detected	Faust and Miller (1997)
Dairy cow	Ht soybean	Milk	Not detected	Klotz and Einspanier (1998)
Dairy cow	*Bt* maize grain	Milk	Not detected	Einspanier *et al.* (2001)
Beef steer	*Bt* maize grain	Muscle	Not detected	Einspanier *et al.* (2001)
Dairy cow	Ht soybean	Milk	Not detected	Phipps *et al.* (2001)
Dairy cow	*Bt* maize grain	Milk	Not detected	Phipps *et al.* (2002)
Dairy cow	*Bt* cottonseed	Milk	Not detected	Jennings *et al.* (2003c)
Dairy cow	Ht/*Bt* silage	Milk	Not detected	Calsamiglia *et al.* (2003)
Dairy cow	Ht soya/*Bt* maize grain	Milk	Not detected	Phipps *et al.* (2003)
Dairy cow	*Bt* maize grain	Milk	Not detected	Yonemochi *et al.* (2003)
Dairy cow	Ht/*Bt* cottonseed	Milk	Not detected	Castillo *et al.* (2004)
Dairy cow	Ht maize silage	Milk	Not detected	Phipps *et al.* (2005)
Monogastrics				
Broiler	*Bt* maize	Muscle, eggs	Not detected	Einspanier *et al.* (2001)
Laying hen	Ht soybean	Eggs, liver	Not detected	Ash *et al.* (2000)
Broiler	Ht soybean	Muscle, skin, liver	Not detected	Khumnirdpetch *et al.* (2001)
Broiler	*Bt* maize	Muscle, eggs	Not detected	Aeschbacher *et al.* (2001)
Broiler	*Bt* maize	Muscle	Not detected	Yonemochi *et al.* (2002)
Broiler	*Bt* maize	Muscle	Not detected	Chowdhury *et al.* (2002)
Broiler	*Bt* maize	Muscle	Not detected	Jennings *et al.* (2003a)
Broiler	*Bt* maize	Muscle	Not detected	Tony *et al.* (2003)
Broiler		Muscle	Not detected	El Sanhorty (2004)
Pig	*Bt* maize	Muscle	Not detected	Weber and Richert (2001)
Pig	*Bt* maize	Muscle	Not detected	Einspanier *et al.* (2001)
Pig	*Bt* maize	Muscle	Not detected	Klotz *et al.* (2002)
Pig	*Bt* maize	Muscle	Not detected	Chowdhury *et al.* (2003a)

Continued

Table 13.2. Continued.

Animal category	GM crop	Tissues examined	Detection of tDNA	Reference
Pig	*Bt* maize	Muscle	Not detected	Jennings *et al.* (2003b)
Pig	*Bt* maize	Muscle	Not detected	Reuter and Aulrich (2003)
Pig	gdhA maize	Muscle	Not detected	Beagle *et al.* (2004)
Pig	gdhA maize	Muscle	Not detected	Qiu *et al.* (2004)
Other species				
Atlantic salmon	Ht soybean	Muscle	Not detected	Sanden *et al.* (2004)

Conclusions

The uptake of GM crops such as soybean, maize, cotton and oilseed rape, modified for herbicide tolerance and/or insect protection, has been dramatic and they are now used extensively in both ruminant and monogastrics diets. These crops have been subjected to a robust and rigorous safety assessment with a comparative safety assessment as a starting point. Numerous studies have established that once the chemical composition of the GM crops was established as similar to their appropriate counterpart it could be assumed that the GM crop had a similar nutritional value and thus further feeding studies with livestock added little to the overall nutritional evaluation.

While concern had been expressed over the possible accumulation of tDNA and novel proteins encoded by the transgene in animal-derived foods, studies to date have failed to detect either the presence of tDNA fragments that could retain any biological activity or genetic functionality, or novel protein.

The current paper finds no evidence to suggest that food derived from animals fed GM feed ingredients is anything other than as safe as that produced by conventional feed ingredients.

References

Aeschbacher, K., Messikommer, R. and Wenk, C. (2001) Physiological characteristics of Bt 176-corn in poultry and destiny of recombinant plant DNA in poultry products. *Annals of Nutrition Metabolism*, 45 (Suppl. 1): 376 (Abstract).

Aeschbacher, K., Messikommer, R. Meile, L., and Wenk, C. (2004) Bt 176 corn in poultry nutrition: physiological characteristics and fate of recombinant plant DNA in chickens. *Poultry Science* 84, 385–394.

Agodi, A., Barchitta, M., Grillo, A. and Sciacca, S. (2006) Detection of genetically modified DNA sequences in milk in the Italian market. *International Journal of Hygiene and Environmental Health* 209, 81–88.

Ash, J.A., Novak, C.L. and Scheideler, S.E. (2000) The fate of genetically modified protein from Roundup Ready soybeans in the laying hen. *Journal of Applied Poultry Research* 12, 242–245.

Ash, J.A., Novak, C.L. and Scheideler, E. (2003) The fate of genetically modified protein from Roundup Ready® soybeans in the laying hen. *Journal of Applied Poultry Research* 12, 242–245.

Aulrich, K., Böhme, H., Daenicke, R., Halle, I. and Flachowsky, G. (2001) Genetically modified feeds in animal nutrition. 1st communication: *Bacillus thuringiensis* (Bt) corn in poultry, pig and ruminant nutrition. *Archive Animal Nutrition* 54, 183–195.

Barriere, Y., Verite, R., Brunschwig, P., Surault, F. and Emile, J.C. (2001) Feeding value of corn silage estimated with sheep and dairy cows is not altered by genetic incorporation of Bt176 resistance to *Ostrinia nubilalis*. *Journal of Dairy Science* 84, 1863–1871.

Beagle, J.M., Apgar, G.A., Jones, K.L., Griswold, K.E., Qui, X. and Martin, M.P. (2004) The digestive fate of the *gdh A* transgene in corn diets fed to weanling swine. *Journal of Animal Science* 82, 457.

Beever, D.E. and Kemp, C.F. (2000) Safety issues associated with the DNA in animal feed derived from genetically modified crops. A review of scientific and regulatory procedures. *Nutrition Abstracts and Reviews* 70, 197–204.

Beever, D.E. and Phipps, R. (2001) The fate of plant DNA and novel proteins in feeds for farm livestock: a United Kingdom perspective. *Journal of Animal Science* 79, E290–E295.

Beever, D.E., Glenn, K. and Phipps, R.H. (2002) A safety evaluation of genetically modified feedstuffs for livestock production; the fate of transgenic DNA and proteins. Xth Asian Australian Association of Animal Production Societies Congress India, Delhi, September 2002.

Bennett, R.M., Phipps, R.H., Strange, A.M. and Grey, P.T. (2004) Environmental and human health impacts of growing genetically modified herbicide-tolerant sugar beet: a life-cycle assessment. *Plant Biotechnology Journal* 2, 273–278.

Berger, L.L., Robbins, N.D., Sewell, J.R and Stanisiewski, E.P. (2002) Effect of feeding diets containing corn grain with Roundup (event GA21 or NK603), control, or conventional varieties on steer feedlot performance and carcass characteristics. *Journal of Animal Science* 80: Suppl. 1, 270 (Abstract 1080).

Berger, L.L., Robbins, N.D., Sewell, J.R., Stanisiewski, E.P. and Hartnell, G.F. (2003) Effect of feeding diets containing corn grain with corn rootworm protection (event MON863), control, or conventional varieties on steer feedlot performance and carcass characteristics. *Journal of Animal Science* 81, Suppl. 1, 214 (Abstract M150).

Böhme, H., Aulrich, K., Daenicke, R. and Flachowsky, G. (2001) Genetically modified feeds in animal nutrition 2nd communication: Glufosinate tolerant sugar beets (roots and silage) and maize grains for ruminants and pigs. *Archives of Animal Nutrition* 54, 197–207.

Brake, J. and Vlachos, D. (1998) Evaluation of event 176 'Bt' corn in broiler chickens. *Journal Poultry Science* 77, 648–653.

Brake, J., Faust, M.A. and Stein, J. (2003) Evaluation of transgenic event Bt 11 hybrid corn in broiler chickens. *Poultry Science* 82, 551–559.

Brookes, G. and Barfoot, P. (2006) GM Crops: The First Ten Years – Global Socio-Economic and Environmental Impacts. International Service for the Acquisition of Ag-Biotech Applications, Brief 36.

Brown, P.B., Wilson, K.A., Jonker, Y. and Nickson, T.E. (2003) Glyphosate tolerant canola meal is equivalent to the parental line in diets fed to rainbow trout. *Journal of Agriculture and Food Chemistry* 51, 4268–4272.

Calsamiglia, S., Hernandez, B., Hartnell, G.F. and Phipps, R.H. (2003) Effect of feeding corn silage produced from corn containing MON810 and GA21 genes on feed intake, milk production and composition in lactating dairy cows. *Journal of Dairy Science* 86(Suppl 1) (Abstract 247).

Carpenter, J., Felsot, A., Goode, T., Onstad, D. and Sankula, S. (2002) Comparative environmental impacts of biotechnology-derived and traditional soybean, corn and cotton crops. Council for Agricultural Science and Technology, Ames, Iowa.

Castillo, A.R., Gallardo, M.R., Maciel, M., Giordano, J.M., Conti, G.A., Gaggiotti, M.C.,

Quaino, O., Gianni, C. and Hartnell, G.F. (2004) Effect of feeding dairy cows with either BollGard, BollGard II, Roundup Ready or control cottonseeds on feed intake, milk yield and milk composition. *Journal of Dairy Science* 84(Suppl. 1), 413 (Abstract 1712).

Castillo, A.R., Gallardo, M.R., Maciel, M., Giordano, J.M., Conti, G.A., Gaggiotti, M.C., Quaino, O., Gianni, C. and Hartnell, G.F. (2004) Effects of feeding rations with genetically modified whole cottonseed to lactating holstein cows. *Journal of Dairy Science* 87, 1778–1785.

Chesson, A. (2001) Assessing the safety of GM food crops. In: R.E. Hester and R.M. Harrison (ed.) *Issues in Environmental Science and Technology* 15, 1–24. (Reprinted from *Food Safety* and *Food Quality*.)

Chiter, A., Forbes, J.M. and Blair, G.E. (2000) DNA stability in plant tissues: implications for the possible transfer of genes from genetically modified food. *FEBS Letters* 481, 164–168.

Chowdhury, E.H., Kuribara, H., Hino, A., Sultana, P., Mikami, O., Shimada, N., Guruge, K.S., Saito, M. and Nakajima, Y. (2003a) Detection of corn intrinsic and recombinant DNA fragments and Cry1Ab protein in the gastrointestinal contents of pigs fed genetically modified corn Bt11[1]. *Journal of Animal Science* 81, 2546–2551.

Chowdhury, E.H., Mikami, O., Nakajima, Y., Kuribara, H., Hino, A., Suga, K., Hanazumi, M. and Yomemochi, C. (2003b) Detection of genetically modified maize DNA fragments in the intestinal contents of pigs fed StarLink CBH351. *Veterinary and Human Toxicology* 45, 95–96.

Chrastinova, L., Sommer, A.I., Rafay, J., Caniga, R. and Prostredna, M. (2002) Genetically modified maize in diets for rabbits – influence on performance and product quality. In: *Proceedings of the Society of Nutritional Physiology* 11, 195 (Abstract).

Clarke, J.H. and Ipharraguerre, I.R. (2000) Livestock performance: feeding biotech crops. In: *Proceedings of Symposium – Agriculture, Biotechnology, Market*. ADAS-ASAS, Baltimore, Maryland.

Cockburn, A. (2002) Assuring the safety of genetically modified (GM) foods: the impor-tance of an holistic, integrative approach. *Journal of Biotechnology* 98, 79–106.

Cromwell, G.L., Henry, J. and Fletcher, D.W. (2004) Herbicide-tolerant rice versus conventional rice in diets for growing-finishing pigs. *Journal of Animal Science* 82, 329 (Abstract W67).

Daenicke, R., Aulrich, K. and Flachowsky, G. (1999) GVO in der Fütterung. *Mais* 27, 135–137.

Deaville, E.R. and Maddison, B.C. (2005) Detection of transgenic and endogenous plant DNA fragments in blood, tissues and digesta of broilers. *Journal of Agricultural and Food Chemistry* 53, 10268–10275.

Donkin, S.S., Velez, J.L., Totten, A.K., Stanisiewski, E.P. and Hartnell, G.F. (2003) Effects of feeding silage and grain from glyphosate-tolerant or insect-protected corn hybrids on feed intake, ruminal diges-tion, and milk production in dairy cattle. *Journal of Dairy Science* 86, 1780–1788.

Einspanier, R., Klotz, A., Kraft, J., Aulrich, K., Poser, R., Schwagele, F., Jahreis, G. and Flachowsky, G. (2001) The fate of forage DNA in farm animals: a collaborative case-study investigating cattle and chicken fed recombinant plant material. *European Food Research and Technology* 212, 129–134.

Einspanier, R., Lutz, B., Rief, S., Berezina, O., Zerlov, V., Schwarz, W. and Mayer, J. (2004) Tracing residual recombinant feed molecules during digestion and rumen diversity in cattle. *European Food Research Technology* 218, 269–273.

European Food Standards Authority (2004) Guidance document on the scientific panel on genetically modified organisms for the risk assessment of genetically modified plants and derived food and feed. *EFSA Journal* 99, 1–94.

Elangovan, A., Mandal, A. and Johri, T. (2003) Comparative performance of broilers fed diets containing processed meals of *Bt*, parental non-*Bt* line or commercial cotton seeds. *Asian-Australasian Journal of Animal Sciences* 16, 57–62.

El Sanhorty, R.M.E. (2004) Quality control for foods produced by genetic engineering. Disseration, Technische Universität Berlin, Berlin.

Faust, M. and Miller, L. (1997) *Study finds no Bt in milk*. IC-478. Fall Special Livestock Edition. Iowa State University Extension, Ames, Iowa, pp. 6–7.

Flachowsky, G., Chesson, A. and Aulrich, K. (2005a) Animal nutrition with feeds from genetically modified plants. *Archive Animal Nutrition* 59, 1–40.

Flachowsky, G., Halle, I. and Aulrich, K. (2005b) Long term feeding of Bt-corn – a ten-generation study with quail. *Archive Animal Nutrition* 59, 449–451.

Folmer, J.D., Grant, R.J., Milton, C.T. and Beck, J. (2002) Utilization of *Bt* corn residues by grazing beef steers and *Bt* corn silage and grain by growing beef cattle and lactating dairy cows. *Journal of Animal Science* 80, 1352–1361.

Gaines, A.M., Allee, G.L. and Ratliff, B.W. (2001a) Nutritional evaluation of *Bt* (MON810) and Roundup Ready corn compared with commercial hybrids in broilers. *Poultry Science* 80, 51 (Abstract 214).

Gaines, A.M., Allee, G.L. and Ratliff, B.W. (2001b) Swine digestible energy evaluations of *Bt* (MON810) and Roundup corn compared with commercial varieties. *Journal of Animal Science* 79, 109 (Abstract 453).

Grant, R.J., Fanning, K.C., Kleinschmit, D., Stanisiewski, E.P. and Hartnell, G.F. (2003) Influence of glyphosate-tolerant (event NK603) and corn rootworm protected (event MON863) corn silage and grain on feed consumption and milk production in Holstein cattle. *Journal of Dairy Science* 86, 1707–1715.

Halle, I., Aulrich, K. and Flachowsky, G. (2006) Four generations feeding GMO-corn to laying hens. In: *Proceedings of the Society of Nutritional Physiology* 15, 114 (Abstract).

Hammond, B.G., Vicini, J.L., Hartnell, G.F., Naylor, M.W., Knigth, C.D., Robinson, E.H., Fuchs, R.L. and Padgette, S.R. (1996) The feeding value of soybeans fed to rats, chickens, catfish and dairy cattle is not altered by genetic incorporation of glyphosate tolerance. *Journal of Nutrition* 126, 717–727.

Hartnell, G.F., Hvelplund, T. and Weisbjerg, M.R. (2005) Nutrient digestibility in sheep fed diets containing Roundup Ready or conventional fodder beet, sugar beet, and beet pulp. *Journal of Animal Science* 83, 400–407.

Hupfer, C., Hotzel, H., Sachse, K. and Engel, K.H. (1999) The effect of ensiling on PCR based detection of genetically modified *Bt* maize. *European Food Research and Technology* 209, 301–304.

Hyun, Y., Bressner, G.E., Ellis, M., Lewis, A.J., Fischer, R., Stanisiewski, E.P. and Hartnell, G.F. (2004) Performance of growing-finishing pigs fed diets containing Roundup Ready corn (event NK603), a nontransgenic genetically similar corn, or conventional corn lines. *Journal of Animal Science* 82, 571–580.

ILSI (2003) ILSI crop composition database. Washington, DC, International Life Sciences Institute. Available at: http://www.cropcomposition.org

ILSI (2004) Nutritional and safety assessments of foods and feeds nutritionally improved through biotechnology. *Comprehensive Reviews of Food Science and Food Safety* 3, 35–104.

ILSI (2008) Nutritional and safety assessments of foods and feeds nutritionally improved through biotechnology: case studies. *Comprehensive Reviews of Food Science and Food Safety,* 7(1), 50–115.

Ipharraguerre, I.R., Younker, R.S., Clark, J.H., Stanisiewski, E.P. and Hartnell, G.F. (2003) Performance of lactating dairy cows fed corn as whole plant silage and grain produced from a glyphosate tolerant hybrid (event NK 603). *Journal of Dairy Science* 86, 1734–1741.

James, C. (2007) *Global Review of Commercialised Transgenic Crops*. International Service for the Acquisition of Agri-biotech Applications Ithaca, New York.

Jennings, J.C., Albee, L.D., Kolwyck, D.C., Surber, J.B., Taylor, M.L., Hartnell, G.F., Lirette, R.P. and Glenn, K.C. (2003a) Attempts to detect transgenic and endogenous plant DNA and transgenic protein in muscle from broilers fed YieldGard Corn Borer corn. *Poultry Science* 82, 371–380.

Jennings, J.C., Kolwyck, D.C., Kays, S.B., Whetsell, A.J., Surber, J.B., Cromwell, G.L., Lirette, R.P. and Glenn, K.C. (2003b)

Determining whether transgenic and endogenous plant DNA and transgenic protein are detectable in muscle from swine fed Roundup Ready soybean meal. *Journal of Animal Science* 81, 1447–1455.

Jennings, J.C., Whetsell, A.J., Nicholas, N.R., Sweeney, B.M., Klaften, M.B., Kays, S.B., Hartnell, G.F., Lirette, R.P. and Glenn, K.C. (2003c) Determining whether transgenic or endogenous plant DNA is detectable in dairy milk or beef organs. *Bulletin 383/2003 of the International Dairy Federation*, 41–46.

Jonas, D.A., Elmadfa, I., Engel, K.-H., Heller, K.J., Kozianowski, G., Konig, A., Muller, D., Narbonne, J.-F., Wackernagel, W. and Kleiner, J. (2001) Safety considerations of DNA in Food. *Annals of Nutrition and Metabolism* 45, 1–20.

Kan, C.A. and Hartnell, G.F. (2004a) Evaluation of broiler performance when fed insect-protected, control, or commercial varieties of dehulled soybean meal. *Poultry Science* 83, 2029–2038.

Kan, C.A. and Hartnell, G.F. (2004b) Evaluation of broiler performance when fed Roundup Ready wheat (event MON 71800), control, and commercial wheat varieties. *Poultry Science* 83, 1325–1334.

Khumnirdpetch, V., Udormsri Intarchote, A., Treemanee, S., Tragoonroong, S. and Thummabood, S. (2001) *Detection of GMOs in the Broilers that Utilised Genetically Modified Soyabeanmeals as a Feed Ingredient.* Plant and Animal Genome IX Conference, San Diego, California, p. 585.

Klotz, A. and Einspanier, R. (1998) Detection of 'novel-feed' in animals? Injury of consumers of meat or milk is not expected. *Mais* 3, 109–111.

Klotz, A., Meyer, J. and Einspanier, R. (2002) Degradation and possible carry over effects of feed DNA monitored in pigs and poultry. *European Food Research and Technology* 214, 271–275.

Kok, E.J. and Kuiper, H.A. (2003) Comparative safety assessment of biotech crops. *Trends in Biotechnology* 10, 439–444.

Kuiper, H.A., Kleter, G.A., Noteborn, H.P.J.M. and Kok, E.J. (2001) Assessment of the food

safety issues related to genetically modified foods. *The Plant Journal* 27, 503–528.

Lutz, B., Weidemann, S., Einspanier, R., Mayer, J. and Albrecht, C. (2005) Degradation of Cry1Ab protein from genetically modified maize in the bovine gastrointestinal tract. *Journal of Agricultural and Food Chemistry* 53, 1453–1456.

Maertens, L., Luzi, F. and Huybechts, I. (1996) Digestibility of non-transgenic and transgenic oilseed rape in rabbits. In: *Proceedings of 6th World Rabbit Congress,* Toulouse, France, pp. 231–235.

OECD (2001a) Series on the Safety of Novel Foods and Feeds No. 1. Consensus document on key nutrients and key toxicants in low erucic acid rapeseed (canola). Organisation for Economic Co-operation and Development, Paris.

OECD (2001b) Series on the Safety of Novel Foods and Feeds No. 2. Consensus document on compositional considerations for new varieties of soybean: key food and feed nutrients and anti-nutrients. Organisation for Economic Co-operation and Development, Paris.

OECD (2002a) Series on the Safety of Novel Foods and Feeds No. 3. Consensus document on compositional considerations for new varieties of sugar beet: key food and feed nutrients and anti-nutrients. Organisation for Economic Co-operation and Development, Paris.

OECD (2002b) Series on the Safety of Novel Foods and Feeds No. 4. Consensus document on compositional considerations for new varieties of soybean: key food and feed nutrients and anti-nutrients and toxicants. Organisation for Economic Co-operation and Development, Paris.

OECD (2002c) Series on the Safety of Novel Foods and Feeds No. 5. Consensus document on compositional considerations for new varieties of maize (*Zea mays*): key food and feed nutrients and anti-nutrients and secondary metabolites. Organisation for Economic Co-operation and Development, Paris.

OECD (2003) Series on the Safety of Novel Foods and Feeds No. 9. Considerations for the safety assessment of animal feedstuffs derived from genetically modified plants. Organisation for Economic Co-operation and Development, Paris.

Petty, A.T., Hendrix, K.S., Stanisiewski, E.P. and Hartnell, G.F. (2001a) Feeding value of *Bt* corn grain compared with its parental hybrid when fed in beef cattle finishing diets. Abstract 320 presented at the Midwester Section ASAS and Midwest Branch ADSA 2001 Meeting, Des Moines, Iowa.

Petty, A.T., Hendrix, K.S. Stanisiewski, E.P. and Hartnell, G.F. (2001b) Performance of beef cattle fed Roundup Ready corn harvested as whole plant silage or grain. Abstract 321 presented at the Midwester Section ASAS and Midwest Branch ADSA 2001 Meeting, Des Moines, Iowa.

Phipps, R.H. and Park, J.R. (2002) Environmental benefits of genetically modified crops: global and European perspectives on their ability to reduce pesticide use. *Journal of Animal Feed Sciences* 11, 1–18.

Phipps, R.H., Beever, D.E. and Tingey, A. (2001) Detection of transgenic DNA in bovine milk: results for cows receiving a TMR containing maize grain modified for insect protection (MON 810). *Journal of Animal Science* 79, 114–115.

Phipps, R.H., Beever, D.E. and Humphries, D.J. (2002) Detection of transgenic DNA in milk from cows receiving herbicide tolerant (CP4EPSPS) soyabean meal. *Livestock Production Science* 73, 269–273.

Phipps, R.H., Deaville, E.R. and Maddison, B.C. (2003) Detection of transgenic DNA and protein in rumen fluid, duodenal digesta, milk, blood and faeces of lactating dairy cows. *Journal of Dairy Science* 86, 4070–4078.

Phipps, R.H., Jones, A.K., Tingey, A.P. and Abeyasekera, S. (2005) Effect of corn silage from an herbicide tolerant genetically modified variety on milk production and absence of transgenic DNA in milk. *Journal of Dairy Science* 88, 2870–2878.

Phipps, R.H., Einspanier, R. and Faust, M. (2006) Safety of meat, milk, and eggs from animals fed crops derived from modern biotechnology. Council for Agricultural

Science and Technology (CAST). Issue Paper 33. Ames, Iowa.

Qiu, G.A., Apgar, G.A., Griswold, K.E., Beagle, J.M., Martin, M.P., Jones, K.L., Iqbal, M.J., Lightfoot, D.A. (2004) Digestive fate of a *gdhA* gene from a genetically modified corn fed to growing pigs. *Journal of Animal Science* 82, 329 (Abstract W66).

Reuter, T. and Aulrich, K. (2003) Investigations on genetically modified maize (*Bt*-maize) in pig nutrition: fate of feed ingested foreign DNA in pig bodies. *European Food Research Technology* 216, 185–192.

Reuter, T., Aulrich, K., Berk, A. and Flachowsky, G. (2002a) Investigations on genetically modified maize (*Bt*-maize) in pig nutrition. Chemical composition and nutritional evaluation. *Archives of Animal Nutrition* 56, 23–31.

Reuter, T., Aulrich, K. and Berk, A. (2002b) Investigations on genetically modified maize (Bt-maize) in pig nutrition: fattening performance and slaughtering results. *Archives of Animal Nutrition* 56, 319–326.

Ridley, W.P., Sidhu, R.S., Pyla, P.D., Nemeth, M.A., Breeze, M.L. and Astwood, J.D. (2002) Comparison of the nutritional profile of glyphosate-tolerant corn event NK 603 with that of conventional corn (*Zea mays* L.). *Journal of Agricultural and Food Chemistry* 50, 7235–7243.

Sanden, M., Bruce, I.J., Rahman, M.A. and Hemre, G.I. (2004) The fate of transgenic sequences present in genetically modified plant products in fish feed, investigating the survival of GM soybean DNA fragments during feeding trials in Atlantic salmon, *Salmo salar* L. *Aquaculture* 237, 391–405.

Stanford, K., Aalhus, J.L., Dugan, M.E.R., Wallins, G.L., Sharma, R. and McAllister, T.A. (2003) Effects of feeding transgenic canola on apparent digestibility, growth performance and carcass characteristics of lambs. *Canadian Journal of Animal Science* 83, 299–305.

Stanisiewski, E.P., Hartnell, G.F. and Cook, D.R. (2001) Comparison of swine performance when fed diets containing Roundup Ready corn (GA21), parental line or conventional corn. *Journal of Animal Science* 79, 319–320.

Stanisiewski, E.P., Taylor, M.L., Hartnell, G.F., Riordan, S.G., Nemeth, M.A., George, B. and Astwood, J.D. (2002) Broiler performance when fed RR (event RT 73) or conventional canola meal. *Poultry Science* 81, 95 (Abstract 408).

Stein, H.H., Sauber, T., Rice, D., Hinds, M., Peters, D., Dana, G. and Hunst, P. (2004) Comparison of corn grain from biotech and non-biotech counterparts for grow-finish pig performance. *Journal of Animal Science* 82, 328 (Abstract W65).

Taylor, M.L., Hartnell, G.F., Nemeth, M.A., George, B. and Astwood, J.D. (2002) Comparison of broiler performance when fed diets containing YieldGard (MON810) x Roundup Ready (NK603), nontransgenic control, or commercial corn. *Poultry Science* 81, 95 (Abstract 405).

Taylor, M.L., Hyun, Y., Hartnell, G.F., Riordan, S.G., Nemeth, M.A., Karunanandaa, K., George, B. and Astwood, J.D. (2003) Comparison of broiler performance when fed diets containing grain from YieldGard Rootworm (MON863), YieldGard Plus (MON810 x MON863), nontransgenic control, or commercial reference corn hybrids. *Poultry Science* 82, 1948–1956.

Taylor, M.L., Stanisiewski, E.P., Riordan, S.G., Nemeth, M.A., George, B. and Hartnell, G.F. (2004) Comparison of broiler performance when fed diets containing Roundup Ready (event RT73), nontransgenic control, or commercial canola meal. *Poultry Science* 83, 456–461.

Taylor, M.L., Hartnell, G., Nemeth, M., Karunanandaa, K. and George, B. (2005) Comparison of broiler performance when fed diets containing corn grain with insect-protected (corn rootworm and European corn borer) and herbicide-tolerant (glyphosate) traits, control corn, or commercial reference corn. *Poultry Science* 84, 587–593.

Tillman, D. (1999) Global environmental impacts of agricultural expansion: the need for sustainable and efficient practices. In: *Proceedings of the National Academy of Science of the USA.* 96, 5995–6000.

Tony, M.A., Butschke, A., Broll, H., Grohmann, L., Zagon, J., Halle, I., Dänicke, S., Schauzu, M., Hafez, H.M. and Flachowsky, G. (2003) Safety assessement of *Bt* 176 maize in broiler nutrition: degradation of maize-DNA and its metabolic fate. *Archives of Animal. Nutrition* 57, 235–252.

Weber, T.E. and Richert, B.T. (2001) Grower-finisher growth performance and carcass characteristics including attempts to detect transgenic plant DNA and protein in muscle from pigs fed genetically modified '*Bt*' corn. *Journal of Animal Science* 79, 67 (Abstract 162).

Weisbjerg, M.R., Purup, S., Vestergaard, M., Hvelplund, T. and Sejrsen, K. (2001) Undersøgelser af genmodificerede foderroer til malkekøer. DJF Rapport Husdyrbrug, Nr. 25, Maj 2001, 39.

Yonemochi, C., Fujisaki, H., Harada, C., Kusama, T. and Hanazumi, M. (2002) Evaluation of transgenic event CBH 351 (Starlink) corn in broiler chicks. *Animal Science* 73, 221–228.

Yonemochi, C., Ikeda, T., Harada, C., Kusama, T. and Hanazum, M. (2003) Influence of transgenic corn (CBH 351, named Starlink) on health condition of dairy cows and transfer of Cry9C protein and *cry9c* gene to milk, blood, liver and muscle. *Animal Science* 74, 81–88.

14 Biofuels: *Jatropha curcas* as a Novel, Non-edible Oilseed Plant for Biodiesel

A. Kohli,[1,2] M. Raorane,[1] S. Popluechai,[1,3] U. Kannan,[1] J.K. Syers[3] and A.G. O'Donnell[1,4]

[1]*School of Biology, Institute for Research on Environment and Sustainability, Newcastle University, Newcastle upon Tyne, UK;* [2]*Plant Breeding Genetics and Biotechnology, International Rice Research Institute, Metro Manila, The Philippines;* [3]*School of Science, Mae Fah Luang University, Chiang Rai, Thailand;* [4]*Faculty of Natural and Agricultural Sciences, The University of Western Australia, Perth, Australia*

Keywords: Renewable energy, biofuels, non-edible oilseed, *Jatropha curcas*, fatty acid methyl esters, sustainable development

Summary

The negative environmental impacts, the limited sources and rising prices of fossil fuels pose significant environmental and socio-economic challenges. Globally, major national and international initiatives are under way to identify, revive, research and recommend renewable sources of energy. One such renewable source is esterified vegetable oil, i.e. biodiesel. Crop plants yielding edible oilseeds can be diverted to the biodiesel market only to a limited extent due to their value in the food sector. One route to meeting the gap between the demand for food oils and the need for alternative fuel oils is the use of non-edible oilseed plants such as *Jatropha curcas*. *J. curcas* or physic nut is a member of the Euphorbiaceae family and has been the subject of much interest as a source of biodiesel due to a number of perceived advantages. For example, the by-products of *J. curcas*-based biodiesel production have potential as a nutritious seed cake for fodder, as a soil amendment or as a biogas feedstock. Glycerol can be used in a variety of industrial applications and *J. curcas* leaf, stem and bark extracts have uses in the medicinal, cosmetics, plastics and insecticide/pesticide industries. As an aid to sustainable rural development, *J. curcas* grows on marginal and wastelands promoting effective land use, gender empowerment and soil rehabilitation. However, neither *J. curcas* nor any other potentially useful non-edible oilseed plant is currently grown commercially. In fact, such plants are generally undomesticated and have yet to be subject to any genetic improvement with respect to yield quality or quantity. Also, many *J. curcas* accessions can be toxic to humans and animals due to the presence of toxic compounds such as curcins and phorbol esters. Thus, despite the enthusiasm in some countries for widespread plantation cropping, *J. curcas* is currently not commercially viable as a biodiesel feedstock without genetic improvement either through conventional breeding or molecular engineering because of unpredictable yield patterns, varying, but often low, oil content, the presence of toxic and carcinogenic compounds, high male to female flower ratio, asynchronous and multiple flowering flushes, low seed germination frequency, plant height and its susceptibility to biotic and abiotic stresses. This paper reviews the potential of *Jatropha* as a model, non-edible, oilseed plant and the research needed to realize its potential as a bioenergy crop.

The Need for Renewable Sources of Energy

Depletion of non-renewable resources

Non-renewable fossil fuels are a limited resource that supplies nearly 90% of the world's energy demand. Sustained economic growth in some countries, and a desire for similar trends in others, has led to an exponential increase in global energy consumption. Although it is much debated as to whether substantial oil reserves lie undiscovered, inaccessible or environmentally hazardous to recover, it is widely accepted that the rate of consumption will continue to increase. Considering anticipated energy demands, it is expected that oil reserves will last many years less than projections based on the current rate of consumption and there is a general consensus that non-renewable energy resources will become limited sooner rather than later. Predictions for world population of up to 8 billion by 2025 foresee most of this increased energy demand coming from developing and transition economies including China, India and Brazil (Focacci, 2005; Mathews, 2007).

Environmental damage

Fossil fuel combustion contributes to global warming and acid rain. Sun and Hansen (2003) have used surface air temperature change data from 1951 to 2000 to compute anticipated changes for 2050 and suggest that surface air temperatures will increase by an average of 0.3–0.4°C. This is close to the average surface air temperature rise of 0.5°C in the last century (Jones and Moberg, 2003; Jones *et al.*, 2006). By the end of the 21st century, temperatures are expected to rise by 6.4°C (Friedlingstein, 2008). The increases in surface air temperature are expected to continue even if carbon dioxide (CO_2) emissions were stopped today. The reason for current and expected temperature rise is the increasing concentration of 'greenhouse gases'. Affects such as rise in sea levels, unpredictable drought-flood water cycles, quantitative and habit change in flora and fauna have all been related to climate change. The two most important components of greenhouse gases are CO_2 and methane (CH_4). Fossil fuel combustion in power plants, transport and industry is the major source of CO_2 emissions and accounts for nearly 80% of worldwide anthropogenic CO_2 emissions. CO_2 levels are also affected by deforestation where the loss of plants that trap CO_2 leads to higher atmospheric levels and thus to a rise in temperature. Similarly, the release of CH_4 from agricultural enterprises, including animal rearing and rice farming, along with the consumption of natural gas also contributes to global warming.

Rising prices

Fossil fuels encompass a range of commodities such as crude oil, natural gas and coal. Although prices for each of these commodities are driven by different factors, as a global average they follow similar trends. Since crude oil is the source of a number of petroleum products feeding the transportation and other industries, the economic impact of oil prices are easily visible through rising prices of other commodities. Oil prices have risen sixfold over the last century but over the period 2002–2008 oil prices have increased fourfold from US$30 to more than US$130 per barrel. Higher oil prices are becoming restrictive to economic growth, hence energy efficiency and energy diversification have become a top priority in terms of both energy generation and energy use in several countries.

The case for renewables

Current political and economic scenarios surrounding energy reserves and their utilization encourage investment in the discovery and mining of new oil and gas reserves while simultaneously developing highly efficient conversion technologies and considering alternative, renewable energy sources. At a regional scale, renewable fuels can be very cost-effective if considering ancillary benefits such as rural development, land and soil reclamation and environmental amelioration through the use of biomass-based fuels. Wind, water, solar and geothermal-based energy can also be highly suitable and cost-effective for selective areas. However, the highly centralized and largely efficient fossil fuels-based supply and distribution system currently puts renewables at a disadvantage. This is because the localized, domestic nature of renewables threatens those countries and corporations that control the source, production and distribution of fossil fuels. However, over the next 20 years developing countries will account for a large part of the projected energy demand and as such are major sources and markets for renewable energy plants. Renewables must therefore become an integral part of a diversified energy portfolio. Global awareness of the limited supply and environmental damage caused by fossil fuels combined with their increasing prices has raised government, industry and researcher interests in renewable energy. Currently, because of its importance to the transport sector, biodiesel is regarded as one of the main solutions.

The Need for Replacement Transport Fuels

Why the transport sector?

The total energy consumed in any country is primarily divided into the following types: electricity, heat, power and fuel. These are in turn used mainly by the residential, commercial, industrial and transport sectors. Although the transport sector globally uses 22% of the total energy and accounts for 27% of the global CO_2 emissions (deLa Rue du Can and Price, 2008), it has become the immediate target for renewables. This is because the transport sector is expected to see the greatest and most immediate surge in energy demand. Although this is likely to be the case in developing and transition economies such as Brazil, China and India, where vehicle fitness and emission regulations are not strictly imposed, it may also impact on cities of more developed countries where transportation remains the largest emitter of CO_2 (deLa Rue du Can and Price, 2008) since energy conversion efficiency and emission control in transportation is generally lower than in other energy sectors.

The options

Almost 97% of the world's transportation-related energy demand whether road, rail, air or sea runs either on petroleum-based products or natural gas (deLa Rue du Can and Price, 2008). The major non-fossil alternative fuels are biofuels that include biodiesel, bioethanol, biogas and bioelectricity produced using hydrogen fuel cells. With the exception of bioethanol produced directly from plant sugars or starches (e.g. sugarcane, sugarbeet, cassava or cereal grains), both biogas and bioalcohols (including biopropanol and biobutanol being investigated by some companies) remain restrictively expensive for use as vehicle fuels. Brazil has set an example for the use of bioethanol, but for most other countries biodiesel is the biofuel of choice.

Biodiesel

A mix of fatty acid methyl esters (FAMEs) made from a biological source such as vegetable oil or unused or used animal fat is called biodiesel. FAMEs can be produced by different methods of esterification, though most follow a similar basic approach (Fig. 14.1) in which the oil is first filtered to remove water and contaminants and then pretreated to remove, or esterified to transform, free fatty acids (FFAs) into biodiesel. Generally, only recycled animal fat containing more than 4% w/v FFAs needs pretreatment. Following pretreatment, the oils are mixed with methanol and a catalyst (usually sodium or potassium hydroxide) to trans-esterify the triglycerides into FAMEs and glycerol. The glycerol can then be separated and the FAMEs used as biodiesel. The trans-esterification of vegetable oils is not new and has been practised since the 1800s when it was used to produce glycerin for soaps. In soap production, however, the FAMEs

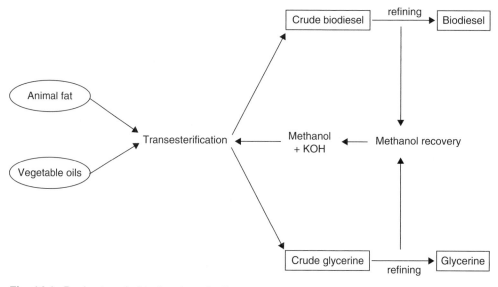

Fig. 14.1. Basic steps in biodiesel production.

are the by-products. The use of vegetable oils in transport is also not new. More than a century ago Rudolph Diesel, the inventor of the diesel engine, first powered his engine with groundnut oil. It was not till the 1920s when petro-diesel became readily available that biodiesel went out of favour.

Of the different biofuel options currently available, biodiesel is the only one that is economically and technologically feasible and is expected, along with the Brazilian model of bioethanol, to remain competitive at least in the short term (Johnston and Holloway, 2007). Within the European Union, biodiesel production and consumption over the last decade has continued to increase. The total EU27 biodiesel production for 2006 was over 4.8 million t, an increase of 54% over the 2005 figures (EBB, 2008). While the USA has been producing biodiesel from soya oil to a limited extent for many years, it has in recent years dramatically increased production from 20 million gal (1 gal = 4.54609 l) in 2003 to over 250 million gal by 2006 (NBB, 2008).

Biodiesel is the only biofuel for which there is a strong market demand at present. The temporal and spatial flexibility in biodiesel production due to the variety of feedstocks available is a major strength. Most of Germany and France's biodiesel is made from rapeseed oil, while most of that from the USA is derived from soybean. Other vegetable oils have also been used for producing biodiesel, for example, oil from groundnut, sunflower, safflower, canola, linseed and palm. However, these generally cater to local small-scale consumption. Heterotrophic algae (Chisti, 2007) and bacteria (Kalscheuer *et al.*, 2006) have also been used following genetic engineering to produce oil for conversion to biodiesel. Blending biodiesel with petro-diesel is more popular than the use of B100 (100% biodiesel). The advantages of biodiesel over petro-diesel depend on several factors but mostly on the composition of the blend used, for example, a 20% blend of biodiesel (B20) with petro-diesel works well in terms of economies of scale and allied benefits. Based on the average diesel consumption of the USA and on soya-based biodiesel production, B100 yields 3.2 units of fuel energy compared to the 0.83 unit yield of petro-diesel per unit of fossil energy consumed (Mixon *et al.*, 2003). In countries with lower transportation and feedstock production costs, this yield comparison is higher.

Table 14.1 compares some critical properties of biodiesel and petro-diesel. Based on these properties there are obvious environmental benefits in using biodiesel. For example, in B100 there is zero sulfur, so sulfate and sulfur dioxide (SO_2; contributors to acid rain) are eliminated. It also has 67% less unburned hydrocarbons reducing the carcinogen concentrations and smog-forming capacity in air. There is 48% less carbon monoxide (CO) and 47% less particulate matter (PM) than is found in low-sulfur petro-diesel. Biodiesel does, however, increase nitrous oxide (NOx) by 10%, but the total smog-forming potential is much lower and the absence of SO_2 should allow the introduction of NOx control technologies such as exhaust gas recirculation (EGR) in diesel engines to be employed (Agarwal *et al.*, 2006). The high cetane number (CN) of biodiesel leads to similar power, torque and fuel economy in terms of performance while higher lubricity provides yet another advantage (Agarwal and Das, 2001). Biodiesel is inferior to petro-diesel in engine performance in cold temperatures and its corrosive effects on rubber components. However, both these issues can easily be addressed using antifreeze and alternative tubing materials,

Table 14.1. Comparison of properties between biodiesel and petro-diesel. (Adapted from Mixon *et al.*, 2003.)

Property	Biodiesel	Petroleum diesel (CARB low-sulfur diesel)
Cetane number	51–62	44–49
Lubricity	More than diesel, comparable to oil lubricants	Low-sulfur fuel has very low lubricity factor
Biodegradability	Readily biodegradable	Poorly biodegradable
Toxicity	Essentially non-toxic	Highly toxic
Oxygen	Up to 11% free oxygen	Very low
Aromatics	No aromatic compounds	18–22%
Sulfur	None	0.05%
Cloud point	Slightly worse than diesel	NA
Flash point	300–400°F	125°F
Spill hazard	None	High
Material compatibility	Degrades natural rubber	No effect on natural rubber
Shipping	Non-hazardous and non-flammable	Hazardous
Heating value	2–3% higher than diesel	1
Renewable supply	Renewable fuel	Non-renewable
Supply	USA estimated 2 billion gal/year	Limited
Energy security	Domestic raw material	Mix of domestic and imports
Alternative fuel	Yes	No
Production process	Chemical reaction	Reaction + fractionation

respectively. None the less, more research is needed to address the issue of cold temperature performance if biodiesel is to be used more extensively in colder regions.

Despite these minor technical drawbacks and policy frameworks that are still not so supportive of biodiesel compared to petro-diesel, the demand for biodiesel continues to rise. EU production and consumption is rising and there is a clear road map for EU countries to replace 5.75% of all its transport fuel with biofuels by 2010 with a planned increase to 10% by 2020 (BRAC, 2006). Total energy supplied by renewables is planned to be 20% by 2020. Similarly, countries such as China and India are also moving towards renewable fuels. India has an ambitious target of replacing 20% of its transport fuel with biodiesel by 2010 (Biodiesel 2020, 2008). Therefore, biofuels in general and specifically biodiesel is expected to be the fuel of choice over the next 5–10 years (Biodiesel 2020, 2008). It is argued, however, that a coherent vision on the global role of biofuels in energy, economic and environmental policy is still missing, but that for long-term efficiency biofuels industry should develop on a 'level trade policy playing field' and it is important that the sector does not receive inordinate 'shelter' (Motaal, 2008). However, at present biofuel operations are heavily subsidized globally and in 2007 alone tax credits in the developed world totaled €10 billion (Anonymous, 2008), prompting a UK legislation requiring biofuels to show significantly smaller carbon footprints than their petroleum-based equivalents in order to keep their government subsidies.

Oils for Food or Fuel?

Non-edible oilseed plants for biodiesel

Current and future demands for biodiesel are encouraging the diversion of food and feed crops into fuel. Under present tax and subsidy regimes the price increases in US maize are predicted to increase the world maize price by 20% by 2010 (Fairless, 2007; *Nature* Editorial, 2007). According to a World Bank report, a 1% increase in the price for staple food leads to a 0.5% drop in calorific consumption (Fairless, 2007; *Nature* Editorial, 2007), which is likely to impact most significantly on some of the world's poorest people. Using edible crops for fuel will also drastically alter land-use pattern because the need for food supplies will have to be maintained. While there are growing concerns over the use of edible crops such as rape, soy and palm for fuel oil, the use of non-edible seed oils or the use of direct bioconversions from waste (Demirbas, 2007, 2008a,b) is now considered a major alternative. In principle, any oil-rich plant seed can be used as a source of biodiesel. Azam *et al.* (2005) conducted an extensive study that compared 75 non-traditional oilseed plant species containing at least 30% w/w seed oil. The FAME composition, iodine value (IV) and CN were compared to assess the suitability as a feedstock for biodiesel production. Their analysis revealed that nearly one-third of the plants analysed contained suitable seed oils and made a strong case against using edible oils for biodiesel. In fact the seed oils of 26 potentially useful species actually meet the diesel standards of the USA, Germany and the European Standards Organization (ESO; Azam *et al.*, 2005). The list of alternative oilseed plants can be narrowed further to a few useful plants according to their oil productivity per hectare; potential economically useful by-products; growth habit (tree or shrub); habitat (arid, semi-arid, tropical) and cultivation requirements such as fertilizers, water and the need for plant protection. Taking all of these criteria into consideration, *J. curcas* was recommended as one of the most suitable sources of non-edible oilseeds for biodiesel feedstock (Azam *et al.*, 2005; Chhetri *et al.*, 2008) without comparisons to the edible oilseed-based biodiesel sector for yield of oil per unit area. However, the conclusions of an international conference on *J. curcas* in 2007 stated that 'the positive claims on *J. curcas* are numerous, but that only a few of them can be scientifically substantiated (Jongschaap *et al.*, 2007). The claims that have led to the popularity of the crop are based on the incorrect combination of positive characteristics which are not necessarily present in all *J. curcas* accessions and have certainly not been proven beyond doubt in combination with its oil production' (Jongschaap *et al.*, 2007). Nevertheless the popularity of *J. curcas* has been pushing research on claimed benefits and a number of these are being increasingly documented through peer-reviewed publications as is partly obvious with regard to the co-product utilization (see Table 14.2) and *J. curcas* life-cycle sustainability assessment for rural development potential (Achten *et al.*, 2007a). If most 'traditional' claims can be validated then the improved elite varieties of *J. curcas* may live up to projected promises. Francis (2008) argued that the success of *J. curcas* as a biofuel crop would largely depend on a 'refined set of agronomic plantation management practices' and realization of the value of principal co-products, especially the seed cake. Recently, methods for detoxifying the seed cake have been proposed and also a number of uses for toxic seed cake have been proposed. These are discussed below.

The Case for *Jatropha curcas*

From the data presented by Azam *et al.* (2005), it is clear that the two plants most suitable for biodiesel are *J. curcas* and *Pongamia pinnata*. Of the two, *P. pinnata* is a bigger tree and not readily amenable to pruning, whereas *J. curcas* is a shrub that can be checked from growing tall and a pruning regime standardized to balance the vegetative and the reproductive growth for maximum yields (Jongschaap *et al.*, 2007). Therefore, the ease of harvesting the fruit from *J. curcas* in comparison to *P. pinnata* makes the former a more acceptable source of seed oil when grown as a large-scale, plantation crop. Furthermore, Modi *et al.* (2006) demonstrated that among *J. curcas*, *P. pinnata* and sunflower, the maximum conversion of 93% oil to biodiesel was achieved with *Jatropha*.

The origin of *J. curcas* is still unclear but there is some evidence that it originated from Central America (Heller, 1996). *J. curcas*, also known as 'physic nut', 'pignut', 'vomit nut' or 'fig nut', is a perennial, monoecious shrub and Heller (1996) has provided a detailed morphological description of the plant. The

Table 14.2. Possible uses of different parts of *Jatropha*.

Plant Part	Uses	References
Seeds	Wastewater copper biosorption by seedcoat	Jain *et al.* (2008)
	Seed diterpenes and toxins as antimicrobial	Goel *et al.* (2007)
	Antitumour, molluscicidal and insecticidal	Luo *et al.* (2007); Liu *et al.* (1997)
	Jatropholone used to prevent gastric lesions	Pertino *et al.* (2007)
Seed oil	Biodiesel	Kaushik *et al.* (2007)
	Resins and varnishes	Patel *et al.* (2008)
	Soap, lubricants and illuminants	Roy (1998)
	Antimicrobial	Eshilokun *et al.* (2007)
	Biopesticide	Goel *et al.* (2007)
Seed cake	Protein-rich feed	Devappa and Swamylingappa (2008)
	Organic fertilizer	Mendoza *et al.* (2007)
	Lipase and protease production	Mahanta *et al.* (2008)
	Activated carbon	Sricharoenchaikul *et al.* (2008)
Fruit husk	Used as fuel through combustion	Singh *et al.* (2008)
	Pyrolysed for biogas	Vyas and Singh (2007)
Leaf/extract	Feed for silkworms in silviculture	Grimm *et al.* (1997)
Oil	Larvicidal	Rahuman *et al.* (2008)
	Bactericidal	Eshilokun *et al.* (2007)
	Fungicidal	Onuh *et al.* (2008)
Stem/bark	Dark blue dye	Srivastava *et al.* (2008)
	Waxes and tannins	Burkill (1985)
	Jatrophone is antineoplastic	Biehl and Hecker (1986)
Subterranean stems	Jatrophone used against snake bite	Brum *et al.* (2006)
Root	Antibacterial diterpenoids	Alyelaagbe *et al.* (2007)

inflorescence is a complex cyme formed terminally and the unisexual male flowers contain ten stamens arranged in two distinct whorls of five each while the female flowers contain the gynoecium with three slender styles dilating to a massive bifurcate stigma. The fruit is an ellipsoid capsule 2.5–3 cm long, 2–3 cm in diameter, yellow, turning black that contains three black seeds per fruit. Seeds are ellipsoid, triangular-convex in shape and measure 1.5–2 × 1–1.1 cm. Recently, a number of physical and mechanical properties of the fruits, nuts and the kernels have been described in terms of their importance in harvesting, handling and processing the oil for biodiesel production (Sirisomboon *et al.*, 2007).

Seed oil and biodiesel characteristics of *J. curcas*

The physical characteristics of *J. curcas* seeds vary depending on their geographical origin. Generally, seed weight varies from 0.4 to 0.7 g and seed dimensions vary with length and width from 15 to 17 mm and 7 to 10 mm, respectively (Martinez-Herrera *et al.*, 2006). The kernel to shell ratio is usually around 60:40 and the seed oil content is in the range of 25–40% w/w. Investigating the physico-chemical properties of *J. curcas* from Benin, Kpoviessi *et al.* (2004) reported a higher range with an upper limit close to 50%. Kaushik *et al.* (2007) observed significant differences in seed size, 100-seed weight and oil content between 24 accessions from different agro-climatic zones of the northern state of Haryana in India. They concluded that habitat and prevailing environmental conditions were more important than genotype in determining phenotypic variation (higher coefficient of variation) in *J. curcas* and its suitability as a fuel crop (see below).

Oil samples of different accessions mainly contain unsaturated fatty acids in the form of oleic acid (18:1; 43–53%) and linoleic acid (18:2; 20–32%). Akintayo (2004) characterized *J. curcas* seed oil from Nigeria and found oleic acid as the most abundant fatty acid. Martinez-Herrera *et al.* (2006) have characterized four provenances of *J. curcas* from different agro-climatic regions of Mexico (Castillo de Teayo, Pueblillo, Coatzacoalcos and Yautepec) and showed that seed kernels were rich in crude protein (31–34.5%) and lipid (55–58%). Here also the major fatty acids found in the oil samples were oleic (41.5–48.8%) and linoleic acid (34.6–44.4%) though palmitic and stearic acids were also reported in lower amounts (10–13% and 2.3–2.8%, respectively). Work in our own laboratories (Popluechai *et al.*, 2008a) shows that the oil yield of *J. curcas* seeds from six provenances in Thailand varies from 20 to 35% w/w. The major fatty acids found in the oil samples were oleic (36–44%), linoleic (29–35%), palmitic (12–14%) and stearic (8–10%); these values are in general agreement with the *J. curcas* oil compositions reported elsewhere. The elevated levels of oleic and linoleic acids make the respective FAMEs suitable for biodiesel. Reksowardojo *et al.* (2006) compared five diesel types – petro-diesel, *J. curcas* B10, B100 and palm oil B10 and B100 and found the biodiesels to be more efficient in direct injection (DI) engines. A comparison of properties of petroleum diesel with *J. curcas* biodiesel (Table 14.3) shows that it does provide a suitable replacement for diesel. Furthermore, Foidl *et al.* (1996) showed that both methyl and ethyl esters of *J. curcas* fatty acids could be used without engine modification. However, the use of pure *J. curcas* biodiesel (B100) is contested (Wood, 2008) due to high NOx emissions (Reksowardojo *et al.*, 2006) and it has been shown that engine performance can be improved with petro-diesel and biodiesel blends (Pramanik, 2003). Blending with mixes of biodiesel, e.g. from *J. curcas* and palm oil, showed better stability at low temperature and also improved oxidation stability compared to the use of either *J. curcas* or palm oil biodiesel alone (Sarin *et al.*, 2007). Corrosion tests on engine parts and emissions analysis showed that both *J. curcas* and palm oil biodiesel make acceptable substitutes for petro-diesel (Reksowardojo *et al.*, 2006; Kaul *et al.*, 2007). The effects of using biofuels in internal combustion engines have recently been reviewed by Agarwal (2006), while Rao *et al.* (2007) specifically compared diesel to *J. curcas* biodiesel and its blends with diesel. Their results indicated that ignition delay, maximum heat release rate and combustion duration were lower for *J. curcas* biodiesel and its blends

Table 14.3. Comparison of properties of diesel and *Jatropha* biodiesel.

Parameter	Diesel	*Jatropha* oil
Energy content (MJ/kg)	42.6–45.0	39.6–41.8
Specific weight (15/40 C)	0.84–0.85	0.91–0.92
Solidifying point	14	2
Flash point	80	110–240
Cetane value	47.8	51
Sulfur (%)	1.0–1.2	0.13

compared to diesel. *J. curcas* had lower tailpipe emissions than diesel except for nitrogen oxides (Reksowardojo *et al.*, 2006). Pradeep and Sharma (2007) recently addressed the latter with *J. curcas* biodiesel and concluded that hot EGR was an effective solution to reducing NOx. Research has also demonstrated that effective routes can be found to obtaining *J. curcas* biodiesel with minimal FFAs (Tiwari *et al.*, 2007; Berchmans and Hirata, 2008) so that seed oil from varieties with relatively large FFAs can also be used as long as the C16:C18 ratio is acceptable.

Potential *J. curcas* co-products

Co-products produced during and after the extraction and conversion of the oil, provide added value to the use of *J. curcas* as a bioenergy crop. Table 14.2 lists some of the products that can be obtained from different parts of the plant and are identified as useful in the energy, chemical, medical, cosmetics or other industries. An array of low- to high-value products obtained from the plant would help to realize the concept of a 'biorefinery' whereby the entire plant is used to ensure maximum commercial, social and environmental impact.

The potential of *J. curcas* seed oil as a source of biodiesel, and the expected availability of residue materials for exploitation as co-products, is encouraging research on their processing and economic validation. For example, biodiesel production generates fruit husk that can be used as feedstock for open-core downdraught gasifiers (Vyas and Singh, 2007) for wood gas. Singh *et al.* (2008) reported on using combustion, gasification and oil and biodiesel extraction from *J. curcas* husk, shells and seeds for a holistic approach to optimal utilization of all parts of the fruits towards obtaining maximum possible energy. Recently, Sricharoenchaikul *et al.* (2008) demonstrated that *J. curcas* waste pyrolysed at 800°C followed by potassium hydroxide (KOH) activation could generate a low-cost-activated carbon as adsorbent with desirable surface properties for application in various industries. The glycerin generated as a by-product of trans-esterification of the oil can be used to convert it to a number of value-added products (Pagliaro *et al.*, 2007).

The seed cake provides a rich source of protein when detoxified of its toxins, co-carcinogens and antinutrients factors providing an excellent animal feed (Devappa and Swamylingappa, 2008). Also, Mahanta *et al.* (2008) have demonstrated a more high-value use of the seed cake, without the need for detoxification, using solid-state fermentation with *Pseudomonas aeruginosa* to produce enzymes such as proteases and lipases. The seed cake has also been used as a nutrient-rich substrate to grow *Fusarium monoliforme* for gibberellin production (U. Kannan, Newcastle, 2008, personal communication). In the seed cake, the co-carcinogenic compounds, diterpene phorbol esters (PEs), are a major cause of concern limiting the widespread commercialization of *J. curcas*. However, some diterpenes are known for their antimicrobial and antitumour activities. For example, *J. curcas* was shown to contain jatropholone A and B (Ravindranath *et al.*, 2004) which have been recently shown to have gastro-protective and cytotoxic effects (Pertino *et al.*, 2007). Other terpenoids such as *Jatropha* trione and acetylaleuritolic acid obtained from other species of *Jatropha* have also been shown to have antitumour activities (Torrance *et al.*, 1976, 1977). The terpenoid compounds can be used as biopesticides and bioinsecticides. Goel *et al.* (2007) have recently reviewed the documented beneficial effects of PEs from *Jatropha* spp. and concluded that any such uses must also consider the fate of the terpenoids in the soil, water, plants and human health following application. Burkill (1985) has provided an extensively referenced list of uses of various parts of *J. curcas* in the Kew Royal Botanic Gardens entry for the plant. Table 14.2 provides a list of some of those and additional new uses that have recently been documented. Min and Yao (2007) have strongly recommended the use of *J. curcas* for high-value compounds along with the biodiesel industry. These high-value co-products paint a rather rosy scenario for the economic feasibility of *J. curcas*-based biodiesel enterprise, but it must be noted that a number of high-value uses of the seed cake studied (Table 14.2) were on unmodified, toxic seed cake. It remains to be seen if seed cake that has been detoxified or modified for no/minimal toxins/ PE can still be a source of such compounds. If the answer is negative, then does the economics of 'toxin-less' seed cake as a source of animal feed make *J. curcas*-based biodiesel enterprise economically feasible? Use of *J. curcas* for biodiesel or high-value products will require specialized varieties for each of the two sectors, but in either case dealing with plant toxins and irritants is an issue and must be addressed.

Rural, socio-economic and ecological sustainability

Traditionally, *J. curcas* is used as a biofence in India and other countries in Africa and Asia. Although its primary purpose is to keep cattle away from crop plants, planting *J. curcas* also serves as a means of reducing soil erosion. It is claimed that as a drought-tolerant perennial in which the root system helps to hold the soil structure together in poor soil types, *J. curcas* helps maintain valuable topsoil, which might otherwise be lost due to erosion. Recently, Achten *et al.* (2007b) have reported on studies on *J. curcas* root architecture and its relation to erosion control. Their preliminary results exhibited 'promising erosion control potential' but they stressed that further research was needed to optimize the agroforestry and plantation systems for *J. curcas*. In another study, the same group (Achten *et al.*, 2007a) described the rural, socio-economic development potential and soil rehabilitation potential of *J. curcas* plantations and highlighted the need to consider the local land, water and infrastructure conditions before undertaking *J. curcas* plantations. Their qualitative sustainability assessment, focusing on environmental impacts and socio-economic issues, suggested that *J. curcas*-based biodiesel enterprises could help support rural development provided wastelands or degraded lands were planted. Francis *et al.* (2005) described the potential of *J. curcas* in simultaneous wasteland reclamation, fuel production and socio-economic development in areas with degraded land in India. Similar but also preliminary studies with examples from different countries in Asia and Africa have been described in the proceedings of the FACT Foundation symposium on *J. curcas* (Jongschaap *et al.*, 2007).These studies supported the potential of *J. curcas* for land/soil rehabilitation and for contributing to rural socio-economic development in a sustainable manner. However, the same report warned on extrapolating results from small-scale episodic planting of *J. curcas* to plantation-scale because as a crop the plant would indeed have its requirements of water, nutrients and fertilizers for optimal yield even from unimproved varieties.

An economic feasibility study for *J. curcas* undertaken by the business school of the University of California, Berkeley, for opportunities in India considered both the rural development business model and the large-scale industrial model (Khan *et al.*, 2006). Identifying the cost drivers to crop production, the study showed that the cost required during the first 3 years was substantially higher compared to the ongoing marginal costs of maintaining the crop once established with these low marginal cost contributing to the high profit margins. Investments in the first 3 years could be recovered by the sixth year and profits thereafter were seen to be high. Forecasting cash flows over 50 years – the productive lifetime of a shrub – positive net present value and high internal rate of return were reported. It was concluded that due to immense interest in its potentialities for holistic and sustainable development on different fronts, *Jatropha*-based business was viable provided appropriate tax, subsidy and biofuels policies created enough demand for the biodiesel. In both models, the critical point was minimizing the risk and uncertainty of feedstock supply through research and development.

Research and Development Issues for *Jatropha*

The projected socio-economic, environmental and political advantages of *J. curcas* as an alternate energy crop have attracted interest from both businesses and governments. Countries in Asia, Africa and Latin America with vast tracts of non-agrarian marginal land are rapidly putting policies in place to promote biofuels. Starting with India and Thailand (Bhasabutra and Sutiponpeibun, 1982; Takeda, 1982; Banerji *et al.*, 1985), *J. curcas* is becoming increasingly popular not only with other developing and transition economies in the region but even with Brazil where *J. curcas* biodiesel potential is being explored for trucks and lorries (http://www.abelard.org/news/archive-oil1–2.htm#oi151202), since their bioethanol markets cater mainly for cars. It is even claimed that in Parana state of Brazil there are some *J. curcas* varieties that are frost-tolerant/resistant (http://www.abelard.org/briefings/energy-economics.asp). If true, this is expected to increase interest in *J. curcas* as the more temperate countries of Europe and the North America might also adopt it as a potential alternative to edible oils. Molecular responses of *J. curcas* to cold stress have been recently reported to better understand routes to generating cold-tolerant varieties (Liang *et al.*, 2007). Although Thailand is now increasingly exploring the bioethanol from cassava route for biofuels, *J. curcas* remains highly popular as a source of biodiesel in India. Worryingly, however, the increasing popularity of *J. curcas* is fast outpacing research efforts to improve it for delivering on its potential at a large scale – whether through small farm cooperatives or through extensive plantations. There are major issues surrounding the large-scale adoption of *J. curcas* for biodiesel and all its ancillary benefits. Despite the proposed benefits of *J. curcas* for biodiesel, biogas, biofertilizer, biopesticide, medicine, cosmetics, rural development and soil and environmental amelioration, *J. curcas* remains an undomesticated plant with variable, often low and usually unpredictable yield patterns. To obtain optimal benefits from a potentially useful plant, *J. curcas* needs to be improved with respect to the following key research areas.

Levels of toxins and co-carcinogens

J. curcas seeds contain elevated levels of toxic and anti-nutritive factors (ANFs) such as saponins, lectins, curcins, phytate, protease inhibitors and PEs (Martinez-Herrera *et al.*, 2006) and are generally considered toxic to humans and animals when ingested. Other studies also report on the toxicity of seeds, oil and press cake (Gubitz *et al.*, 1999) and these were recently reviewed by Gressel (2008). Methods to detoxify the seed cake have been proposed (Devappa and Swamylingappa, 2008); however, the co-carcinogenic PEs and curcins are not easily and cost-effectively removed.

The difficulty of removing PEs not only from the seed cake but also from the oil and even from the biodiesel itself, currently limits the commercial exploitation of *J. curcas* and its co-products, despite their potential added value. Gressel (2008) recently proposed that transgenic strategies would be the most effective route to obtaining toxin- and PE-free varieties of *J. curcas*. Use of transgenic approaches to achieve reduction/elimination of PE would require the identification of the relevant PE synthesis genes in *Jatropha* and the availability of appropriate plant transformation protocols. Although transformation of *J. curcas* using *Agrobacterium* has recently been reported (Li *et al.*, 2008), it is not a routine procedure in different laboratories because replicating it has been difficult. Moreover, transforming *J. curcas* to remove or minimize PE contents might not be straightforward because the diterpenoid biosyntheis pathway used for PE synthesis is the same as that for key plant hormones such as gibberellins and abscisic acid. It is important to note that PEs found in high concentrations in the *J. curcas* kernels of different accessions were not detected in the samples from Castillo de Teayo, Pueblillo and Yautepec in Mexico

(Makkar *et al.*, 1998) where the seeds are traditionally eaten as roasted nuts. This offers the possibility of using conventional plant breeding to generate cultivars lacking PEs. However, as we discuss below, the limited genetic variability among *J. curcas* accessions from across the world did not result in reduced PE cultivars when toxic and non-toxic accessions were crossed (Sujatha *et al.*, 2005). However, more extensive breeding efforts may offer the prospect of reducing PE content in the plant by conventional or marker-assisted breeding. We have successfully obtained hybrids with markedly lower PE levels using interspecific breeding programmes with *J. curcas* and *Jatropha integerrima* (Popluechai *et al.*, 2008b). This approach has been corroborated by recent work in Dr Hong Yan's group (Singapore, 2008, personal communication), who also found lower PE levels in some hybrids of *J. curcas* and *J. integerrima*. Recently, a cost-effective method for the reduction of PE to less than 15 parts per million (ppm) in the seed cake has been reported (H.P.S. Makkar, Brussels, 2008, personal communication), but the method is currently in the process of being patented. The method uses the carp fish as a model, which is sensitive to more than 15 ppm PE, however, for other animal and human exposures much lower ppm of PE is desirable. It is not clear to us if the method at this stage is suitable for generating seed cake with the desired reduced levels of PE.

Limited genetic variability

Genetic diversity among *J. curcas* accessions was compared by Reddy *et al.* (2007) using amplified fragment length polymorphism/random amplified polymorphic DNA (AFLP/RAPD) on 20 Indian accessions for identification of drought and salinity tolerance. An 8–10% AFLP-mediated and 14–16% RAPD-mediated polymorphism was found among accessions suggesting that intraspecific genetic variation was limited in *J. curcas*. Recently, Basha and Sujatha (2007) used 42 accessions of *J. curcas* from different regions in India to identify genetically divergent materials for breeding programmes. They also included a non-toxic genotype from Mexico and reported inter-accession molecular polymorphism of 42.0% with 400 RAPD primers and 33.5% with 100 ISSR primers indicating only modest levels of genetic variation among Indian cultivars. However, the Mexican variety could easily be differentiated from the 42 Indian accessions suggesting that almost all Indian accessions had a similar ancestry and that using accessions from different parts of the world may reveal genetic variation suitable for breeding programmes. Ranade *et al.* (2008) used single-primer amplification reaction (SPAR) to compare 21 *J. curcas* accessions from different parts of India using wild unknown accessions and classified accessions held in research institutes. Three accessions from the North-east states were clearly different among themselves and from other accessions analysed, while most other accessions were highly similar. Sudheer *et al.* (2008) studied differences among seven species of *Jatropha* from India and found *J. integerrima* to be most closely related to *J. curcas*. We have compared 17 *J. curcas* and one *Jatropha podagrica* accession using RADP/AFLP (Fig. 14.2). The 17 accessions were obtained from Thailand

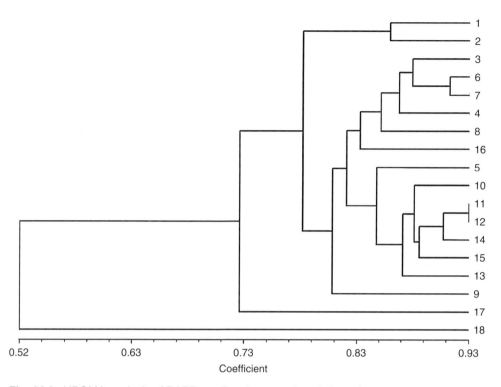

Fig. 14.2. UPGMA analysis of RAPD-mediated genotyping of *Jatropha curcas* accessions: 1–17 represent *J. curcas* accessions from India (1 and 2); Nigeria (3) and Thailand (4 –17); while 18 represents *Jatropha podagrica* from Thailand. All *J. curcas* accessions fall into one cluster while *Jatropha podagrica* is separated from *J. curcas*.

(14; six provenances), India (two) and Nigeria (one), while the *J. podagrica* was also obtained from Thailand. UPGMA-mediated cluster analysis revealed two major clusters one containing all of the 17 *J. curcas* accessions and the second containing *J. podagrica*, which showed a genetic similarity coefficient of 52% with *J. curcas*. This was the same as recently reported using 26 RAPD primers by Ram *et al.* (2007). We also showed that the non-toxic accession from Mexico clustered separately from other *J. curcas* accessions and that the six Thai provenances could not be separated. However, the genetic similarity coefficient between the Thai and the Mexican accessions was high (0.76) as was also noted by Basha and Sujatha (2007). These results show the importance of testing accessions from wider eco-geographic regions and including assessments of other *Jatropha* spp. in addition to *J. curcas*. We have recently extended these studies to 38 *J. curcas* accessions from 13 countries on three continents, along with six different species of *Jatropha* from India using a novel approach of combinatorial tubulin-based polymorphism (cTBP; Breviario *et al.*, 2007). Once again, except for the accessions from Mexico and Costa Rica, which separated into independent groups, all other *J. curcas* accessions from different countries clustered together reiterating a narrow genetic base for

J. curcas (Popluechai *et al.*, 2008b). Similarly, Ram *et al.* (2007) obtained three clusters, one cluster contained all five of the *J. curcas* accessions from India; the second contained six of the seven different species investigated while the third contained a single species, *Jatropha glandulifera*. *J. curcas* accessions exhibit monomorphism even with microsatellite markers (G.H. Yue, Singapore, 2008, personal communication). These studies indicate the limitations of using intraspecific breeding for *J. curcas* improvement; as exemplified by the failure to obtain low-PE hybrids in crosses between *J. curcas* toxic accessions from India and the non-toxic accession from Mexico (Sujatha *et al.*, 2005). In our laboratories in Thailand, India and the UK, we have employed interspecific breeding and obtained hybrids with low PE and improved agronomic traits (Popluechai *et al.*, 2008b). For most future efforts targeting *J. curcas* improvement, we would recommend using interspecific hybridization. Therefore, characterization of different species of *Jatropha* with a view to selecting the hybridizing parent carrying the desirable trait becomes even more important. Recently, Sunil *et al.* (2008) recorded the phenotypic traits of plants *in situ* for plants from four different eco-geographical regions of India, to develop a method for identification of superior lines of *J. curcas*.

Seed germination

J. curcas is propagated through cuttings and seeds. Cuttings are useful because they maintain the clonal nature and hence the plant's original characteristics. However, the vigour, vitality and yield of the plant suffer as a consequence of successive propagation (Goleniowski *et al.*, 2003; Clark and Hoy, 2006); thus seed germination of identified stock is preferred. We have observed that the germination efficiency of fresh *J. curcas* seeds is good for up to 6 months but that after this period germination rate is markedly reduced and can reach as low as 30% in seeds stored for more than a year. Due to its thick seedcoat, *J. curcas* germination needs good moisture conditions while the temperature, light and oxygen conditions are, as for other seeds, also important. *J. curcas* seed germination is similar to the epigeal germination of the castor seed where the *hypocotyl* elongates and forms a hook, pulling rather than pushing the cotyledons and apical meristem through the soil. Germination rates in raised beds were found to be higher than in sunken beds (Sharma, 2007). Thus, the development of standardized plantation-scale seed germination practices, founded on detailed analyses of seed metabolites in relation to varying germination rates, are needed. Studies on seed germination of *Ricinus*, a related Euphorbiaceous plant, have provided a valuable insight into seed oil content, composition and hydrolysis (Kermode and Bewley, 1985a,b, 1986; Kermode *et al.*,1985, 1988, 1989a,b). We used the 2D-PAGE analysis approach and identified six patatin like lipases that were up-regulated in germinating seeds. Given that during castor seed development lipases have also been shown to be up-regulated (Eastmond, 2004), we have cloned these genes to investigate their role in determining seed oil content, composition and susceptibility to hydrolysis.

Reproduction

The plant produces unisexual flowers in cymose inflorescences with the male and female flowers produced in the same inflorescence. Female flowers are different to the male flowers in shape, are relatively larger and are borne on central axes. In *Jatropha*, various issues of flower development directly affect yield. For example, a high ratio of male to female flowers in the inflorescence leads to few female flowers setting seeds. Solomon Raju and Ezradanam (2002) reported that one inflorescence can produce 1–5 female and 25–93 male flowers. The average male to female flower ratio is 29:1 (Solomon Raju and Ezradanam, 2002). However, this ratio might be expected to vary among the different provenances. Prakash *et al.* (2007) have reported that the male to female flower ratio increased in the second year crop and that flowering flushes numbered from one to three per year. Aker (1997) monitored the episodic flowering dependent on precipitation in Cabo Verde in 1994. First flowering started in early May within 2–5 days of the first rains at the end of April and lasted nearly 1 month, being the largest of the four flowering episodes. The second flush resulted from rain on 17 May and started on 9 June with few flowers. Wet season started with rains again on 3 August lasting till mid-November during which time two flowering flushes were observed, one in mid-August and another in the third week of September. Solomon Raju and Ezradanam (2002), working in south-east India, reported a single flowering spread along the wet season from late July to late October, while Sukarin *et al.* (1987), working in Thailand, observed two flowering peaks, one in May and the other in November. The flowering depends on the location and agro-climatic conditions. Normally in North India flowering occurs once a year; however, in the south-eastern state of Tamil Nadu fruiting occurs almost throughout the year. A few lines are bimodal and flower twice a year. The flowering is mostly continuous in such types in the presence of optimum moisture (Gour, 2006). Although synchronous development of the male and female flowers has been reported by Solomon Raju and Ezradanam (2002), the asynchronous maturity of fruits makes the harvesting of *J. curcas* fruits difficult because a certain stage of fruit development is needed for picking; manual harvesting becomes highly time-consuming and problematic due to the number and variability of cycles during the season. At each cycle, picking at the correct stage also requires the retention of skilled pickers. Pollination from male flowers of other inflorescences effectively reduces the chances of success. Hence, studies on genetic and environmental factors affecting flower ratio, timing and cycles are an important area of research and development. For mechanical harvesting to be feasible, and in a commercial operation this will be essential, it will be necessary to synchronize flowering/fruit set as well as reduce plant height. There is substantial scope here for plant breeding initiatives to address and resolve these critical issues. Interestingly, *J. curcas* accessions that exhibit synchronous flowering and low male to female flower ratio have now been recorded (S. Mulpuri, Hyderabad, 2008, personal communication).

Propagation

There are two major methods used to propagate *Jatropha*, generative and vegetative propagation. Generative propagation is normally through direct seedling to plantation or transplanting from the nursery. The seeding method, seeding depth, seasonal timing, quality of the seed and spacing are all important factors effecting growth and yield in *Jatropha*. Chikara *et al.* (2007) have reported on the effect of plantation spacing on yield and showed that seed yield per plant was increased significantly by increasing the spacing between plants; obviously this reduced the seed yield per hectare. The optimum spacing could vary in different climatic and soil types. For vegetative propagation, rooting of stem cuttings is a crucial step in the propagation of woody plants and there is great variability in the rooting ability of different species. Kochhar *et al.* (2005) reported on the effect of auxins on the rooting and sprouting behaviour of *J. curcas* and *J. glandulifera* and on the physiological and biochemical changes accompanying germination. The results showed that the sprouting of buds took place much earlier than rooting in both species. Application of indole butyric acid (IBA) and naphthalene acetic acid (NAA) increased survival percentages with IBA more effective in *J. curcas* and NAA in *J. glandulifera*. The number and type of roots per cutting is also an important factor. Roots of *J. curcas* seedlings are of the taproot system and as such are characteristically drought-tolerant. However, roots of stem cuttings lack the main taproot and are more fibrous in nature affecting their water-harvesting potential. Propagation using *in vitro* manipulations to induce strong and deep roots and the development of suitable de-differentiation and regeneration protocols is therefore important for micropropagation and prospective genetic engineering.

Biotic stress assessment and management

Although *J. curcas* was not widely cropped until the early 2000s, its potential as an energy crop was identified earlier through a project in Mali (Henning, 1996). Consequently, the International Plant Genetic Resources Institute (IPGRI) commissioned a monograph by Joachim Heller that was published in 1996. With regard to the various pests and pathogens that affect *J. curcas*, the monograph lists the following diseases and their causal organisms: damping off caused by *Phytophthora* spp., *Phythium* spp. and *Fusarium* spp.; leaf spot caused by *Helminthosporium tetramera*, *Pestalotiopsis* spp., *Cercospora jatrophae-curces*; dieback due to infection with *Ferrisia virgata* and *Pinnapsis strachani*; fruit-sucking damage caused by insects such as *Calidea dregei* and *Nezara viridula*; leaf damage by *Spodoptera litura* and *Oedaleua senegalensis*; and seedling loss due to *Julus* spp. (Heller, 1996). Despite the plants richness in pest and pathogen-repellant toxins, this is a limited list and one that is likely to grow as more plantations are developed in different parts of the world. For example, following the recent establishment of a number of *J. curcas* plantations in India, Narayana *et al.* (2006) have reported an increased incidence of the *Jatropha* mosaic virus disease. *Bemisia tabaci*, the carrier of mosaic virus disease and a known pest for cassava in Uganda, was reported to infect *J. curcas* (Sseruwagi *et al.*, 2006). Thus, a careful assessment under different agro-climatic conditions is necessary to determine the potential biotic stressors and their management in *J. curcas* plantations, a challenge that may be confounded by manipulating through breeding and genetic engineering of toxin production in *Jatropha*.

Agronomic requirements

Work on agronomic issues such as water requirement, the timing of its application, and the rate and timing of fertilizers, spacing, etc. are largely unknown and poorly characterized. Jongschaap *et al.* (2007) have provided one-stop information on different experiments conducted in different countries with regard to understanding and optimizing agronomic practices for optimal *J. curcas* cultivation. The differences demonstrate that it is likely that these practices will depend not only on prevailing soil and climatic conditions, but also on any socio-economic and environmental impact expected from *Jatropha* cultivation in different countries. This will in turn be guided by wasteland availability along with other economic factors. Depending on the range of soils capable of supporting *J. curcas* plantations, soil science and agronomy have a major role to play in delivering on biodiesel per se and on the soil conservation and rehabilitation potential of *Jatropha*. Fundamental studies on plant physiology of *J. curcas* will also help to define the correct agronomic practices for maximal yields. In our own work, we have noted a clear difference in the water-use efficiency between four accessions tested in the field and under glasshouse conditions (Popluechai *et al.*, 2008b) indicating one accession to be much superior compared to others.

Gene mining and expression analysis

In a recent article, Gressel (2008) stressed the need to identify and understand the expression patterns of different genes that needed to be up-regulated or silenced as part of a transgenic approach to *Jatropha* improvement. It is only since its popularity as a biodiesel feedstock since the early 2000s that basic information on properties such as genome, genes and genotypes of *J. curcas* has become available. We have recently estimated that the genome size of *J. curcas* is in the region of 980 megabase (Mb) and have obtained pictures for chromosome spreads of the pro-metaphase of the root-tip cells (Popluechai *et al.*, 2008b). This work has also confirmed the presence of 11 pairs of chromosomes, as reported earlier (Soontornchainaksaeng and Jenjittikul, 2003). However, Carvalho *et al.* (2008) have reported the genome size to be 416 Mb while a value of 240 Mb has also been described (G.H. Yue, Singapore, 2008, personal communication). It is not clear why there is such variability in genome size or why there seems to be no consensus in plant materials that all contained $2n = 22$ chromosomes.

Given the importance of the seed oil, the fatty acid biosynthesis (FAB) genes in *J. curcas* were among the first to be investigated and identified. DNA sequences of some FAB genes and others of *J. curcas* have only recently become available. Important among these are those for steroyl acyl destaurase (Tong *et al.*, 2006), curcin (Luo *et al.*, 2007), DRE-binding ERF3 genes (Tang *et al.*, 2007), beta-ketoacyl-ACP synthase (Li *et al.*, 2008), aquaporin (Zhang *et al.*, 2007) and betaine-aldehyde dehydrogenase (Zhang *et al.*, 2008). Additional gene sequences now available from the European Molecular Biology Laboratory (EMBL) and the National Centre for Biotechnology Information (NCBI) are UDP-glycosyltransfersase (EMBL AAL40272.1), phosphoenolpyruvate carboxylase (EMBL AAS99555.1), acetyl-CoA carboxylase (EMBL ABJ90471.1), omega-3 and omega-6 fatty acid desaturase (EMBL ABG49414.1), diacylglycerol acyltransferase (EMBL ABB84383.1), acyl-ACP thioesterase (EU106891), ribulose-1,5-bisphosphate carboxylase/oxygenase (EU395776) and phenylalanine ammonia lyase (DQ883805).

We have recently identified three oleosin genes and the upstream promoter sequence for one of them along with the gene for grain softness protein (Popluechai *et al.*, 2008b) and have used these, together with those of published sequences, to study the expression patterns in different accessions and in different tissues of the plant. These studies are now being extended using expressed sequence tag (EST), complementary DNA (cDNA) and bacterial artificial chromosome (BAC) libraries to identify other genes of potential for the improvement of *Jatropha* as a fuel crop.

Conclusions

A number of options are now available as potential sources of renewable bioenergy including a variety of routes to the production of biodiesel, bioethanol and biogas. Apart from the Brazilian model of bioethanol production, only biodiesel is applicable and cost-effective over the short to medium term. For biodiesel, the non-edible oilseed plants are favoured due to questions of whether good agricultural land should be used for food or fuel. As a non-edible oil crop, *J. curcas* has attracted much attention in this respect. However, the unsubstantiated claims and known facts must be balanced to have an objective view on the chances of this plant delivering its perceived potential (Jongschaap *et al.*, 2007). Continued evaluation of *J. curcas* accessions in terms of their agronomic performance suggests that the optimization of simple agronomic practices can help deliver quantifiable increases in yields. However, to realize the full potential of *J. curcas* as a bioenergy crop requires the development of elite lines and elimination/reduction of the toxins and co-carcinogens such as curcin and PE. Currently, the transgenic approaches needed to achieve this in *J. curcas* are limited by lack of knowledge on relevant genes, and although transformation of *J. curcas* has been reported, it is not a routine procedure and replicating the transformation has proven difficult. Additionally, conventional plant-breeding approaches are hampered by lack of knowledge on the genetic variability in global accessions of *J. curcas*. Vegetative and apomictic reproduction (Datta *et al.*, 2005) may have contributed to maintaining and propagating a limited stock of *J. curcas* over time, thereby limiting genetic variability. The capacity for apomixis in *J. curcas* can be used to stabilize hybrid vigour and to help maintain high-yielding stocks. Recent descriptions of fixing apomixis in *Arabidopsis* (Ravi *et al.*, 2008) may be transferrable to *Jatropha*. The importance of genetic variability is evident from the work on using interspecific hybrid lines to breed for desirable traits despite the fact that *J. integerrima* itself lacks the desirable range of targeted traits. However, although genetically polymorphic, it is sufficiently closely related (Sudheer *et al.*, 2008) to produce viable hybrids.

Much has yet to be done if *Jatropha* is to become established as a commercially viable source of biodiesel and little should be expected from the widespread planting of existing varieties. Sujatha *et al.* (2008) recently reviewed the role of biotechnology in the improvement of *J. curcas*. As indicated above, work is starting to appear that suggests that through a combination of breeding and transgenic approaches, elite varieties capable of delivering the necessary yields and free of toxins could be developed. Whether or not this happens depends very much on global commodity markets and the competition from other energy technologies and energy crops.

References

Achten, W.M.J., Aerts, R., Mathijs, E., Verchot, L., Singh, V.P. and Muys, B. (2007a) *Jatropha* biodiesel fueling sustainability? *Biofuels Bioproducts and Biorefining* 1, 283–291.

Achten, W.M., Reubens, B., Maes, W., Mathijs, E.,Verchot, L., Singh, V.P., Poesens, J. and Muys, B. (2007b) Root architecture of the promising bio-diesel plant *Jatropha*. *Communications in Agricultural and Applied Biological Sciences* 72, 81–85.

Agarwal, D., Sinha, S., Kumar, A. and Agarwal, A. (2006) Experimental investigation of control of NOx emissions in biodiesel-fueled compression ignition engine. *Renewable Energy* 31, 2356–2369.

Agarwal, A.K. (2006) Biofuels (alcohols and biodiesel) applications as fuels for internal combustion engines. *Progress in Energy and Combustion Science* 33, 233–271.

Agarwal, A.K. and Das, L.M. (2001) Biodiesel development and characterization for use as a fuel in compression ignition engine. *Journal of Engineering for Gas Turbines Power* 123, 440–447.

Aker, C.L. (1997) Growth and reproduction of *Jatropha* curcas. Biofuels and industrial products from *Jatropha* curcas. Developed from the Symposium 'Jatropha 97' Managua, Nicaragua, 23–27 February.

Akintayo, E.T. (2004) Characteristics and composition of Parkia biglobbossa and *Jatropha* curcas oils and cakes. *Bioresource Technology* 92, 307–310.

Alyelaagbe, O.O., Adesogan, K., Ekundayo, O. and Gloer, J.B. (2007) Antibacterial diterpenoids from *Jatropha podagrica* hook. *Phytochemistry* 68, 2420–2425.

Anonymous (2008) Subsidy shake up to roil biofuel market. *Chemistry and Industry*. 6, 15.

Azam, M.M., Waris, A. and Nahar, N.M. (2005) Prospects and potential of fatty acid methyl esters of some non-traditional seed oils for use as biodiesel in India. *Biomass and Bioenergy* 29, 293–302.

Banerji, R., Chowdhury, A.R., Misra, G., Sudarsanam, G., Verma, S.C. and Srivastava, G.S. (1985) *Jatropha* curcas seed oils for energy. *Biomass* 8, 277–282.

Basha, S.D. and Sujatha, M. (2007) Inter and intra-population variability of *Jatropha curcas* (L.) characterized by RAPD and ISSR markers and development of population-specific SCAR markers. *Euphytica* 156, 375–386.

Berchmans, H.J. and Hirata, S. (2008) Biodiesel production from crude *Jatropha curcas* L. seed oil with a high content of free fatty acids (reduced via a 2-step process). *Bioresource Technology* 99, 1716–1721.

Bhasabutra, R. and Sutiponpeibun, S. (1982) *Jatropha curcas* oil as a substitute for diesel engine oil. *Renewable Energy Review Journal* 4, 56–60.

Biehl, J. and Hecker, E. (1986) Irritant and antineoplastic principles of some species of the genus *Jatropha* (Euphorbiaceae). *Planta Medica* 52, 430–430.

Biodiesel 2020: Global Market Survey, Feedstock Trends and Forecasts (2008 release). *Multi-Client Study*, 2nd edn. Emerging Markets Online, 685 pp.

Biofuels Research Advisory Council Report (BRAC) (2006) A Vision for 2030 and Beyond.

Breviario, D., Baird, Wm.V., Sangoi, S., Hilu, K., Blumetti, P. and Gianì, S. (2007) High polymorphism and resolution in targeted fingerprinting with combined beta-tubulin introns. *Molecular Breeding* 20, 249–259.

Brum, R.L., Cardoso, C.A.L., Honda, N.K., Andreu, M.P. and Viana, L.L.D.S. (2006) Quantitative determination of Jatrophone in 'Cachaqa' prepared with *Jatropha elliptica*. *Chemical and Pharmaceutical Bulletin* 54, 754–757.

Burkill, H.M. (1985) *The Useful Plants Of West Tropical Africa, Vol 2, 1. Families A–D.* Royal Botanical Gardens, Kew, UK. 960 pp.

Carvalho, C.R., Clarindo, W.R., Praca, M.M., Araujo, F.S. and Carels, N. (2008) Genome size, base composition and karyotype of *Jatropha curcas* L., an important biofuel plant. *Plant Science* 174, 613–617.

Chhetri, A.B., Tango, M.S., Budge, S.M., Watts, K.C. and Islam, M.R. (2008) Non-edible plant oils as new sources for biodiesel production. *International Journal of Molecular Sciences* 9, 169–180.

Chikara, J., Ghosh, A., Patolia, J.S., Chaudhary, D.R. and Zala, A. (2007) Productivity of *Jatropha curcas* L. under different spacing. FACT Seminar on *Jatropha curcas* L.: Agronomy and Genetics, Wageningen, The Netherlands, 26–28 March 2007, FACT Foundation.

Chisti, Y. (2007) Biodiesel from microalgae. *Biotechnology Advances* 25, 294–306.

Clark, C.A. and Hoy, M.W. (2006) Effects of common viruses on yield and quality of Beauregard sweetpotato in Louisiana. *Plant Disease* 90, 83–88.

Datta, S.K., Bhattacharya, A. and Datta, K. (2005) Floral biology, floral resource constraints and pollination limitation in *Jatropha curcas* L. *Pakistan-Journal-of-Biological-Sciences* 8, 456–460.

de la Rue du Can, S. and Price, L. (2008) Sectoral trends in global energy use and greenhouse gas emissions. *Energy Policy* 36, 1386–1403.

Demirbas, A. (2007) Progress and recent trends in biofuels. *Progress in Energy and Combustion Science* 33, 1–18.

Demirbas, A. (2008a) Biomethanol production from organic waste materials. *Energy Sources Part A – Recovery Utilization and Environmental Effects* 30, 565–572.

Demirbas, A. (2008b) The importance of bioethanol and biodiesel from biomass. *Energy Sources Part B – Economics Planning and Policy* 3, 177–185.

Devappa, R.K. and Swamylingappa, B. (2008) Biochemical and nutritional evaluation of *Jatropha* protein isolate prepared by steam injection heating for reduction of toxic and antinutritional factors. *Journal of the Science of Food and Agriculture* 88, 911–919.

Eastmond, P.J. (2004) Cloning and characterization of the acid lipase from castor beans. *Journal of Biological Chemistry* 279, 45540–45545.

Eshilokun, A.O., Kasali, A.A., Ogunwande, I.A., Walker, T.M. and Setzer, W.N. (2007) Chemical composition and antimicrobial studies of the essential oils of *Jatropha integerrima* Jacq (leaf and seeds). *Natural Producs Communications* 2, 853–855.

European Biodiesel Board (EBB) (2008) Statistics.

Fairless, D. (2007) Biofuel: the little shrub that could-maybe. *Nature* 449, 652–655.

Focacci, A. (2005) Empirical analysis of the environmental and energy policies in some developing countries using widely employed macroeconomic indicators: the cases of Brazil, China and India. *Energy Policy* 33, 543–554.

Foidl, N., Foidl, G., Sanchez, M., Mittelbach, M. and Hackel, S. (1996) *Jatropha curcas* L. as a source for the production of biofuel in Nicaragua. *Bioresource Technology* 58, 77–82.

Francis, G. (2008) *Jatropha* as a biofuel crop: potential and issues. In: Syers, K.J., Wood, D. and Thongbai, P. (eds) *Proceedings International Technical Workshop on the Feasibility of Non-edible Oil Seed Crops for Biofuel Production 25–27 May 2007*, Chaing Rai, Thailand, pp. 38–48.

Francis, G., Edinger, R. and Becker, K. (2005) A concept for simultaneous wasteland reclamation, fuel production, and socio-economic development in degraded areas in India: need, potential and perspectives of *Jatropha* plantations. *Natural Resources Forum* 29, 12–24.

Friedlingstein, P. (2008) A steep road to climate stabilization. *Nature* 451, 297–298.

Goel, G., Makkar, H.P.S., Francis, G. and Becker, K. (2007) Phorbol esters: structure, biological activity and toxicity in animals. *International journal of Toxicology* 26, 279–288.

Goleniowski, M.E., Flamarique, C. and Bima, P. (2003) Micropropagation of oregano (*Origanum vulgare* x *applii*) from meristem tips. In vitro *Cell and Developmental Biology* 39, 125–128.

Gour, V.K. (2006) Production practices including post-harvest management of *Jatropha curcas*. In: Singh, B., Swaminathan, R. and Ponraj, V. (eds) *Biodiesel Conference Towards Energy Independence – Focus on Jatropha on 9–10 June 2006*, pp. 223–251.

Gressel, J. (2008) Transgenics are imperative for biofuel crops. *Plant Science* 174, 246–263.

Grimm, C., Somarriba, A. and Foidl, N. (1997) Development of eri silkworm *Samia cynthia ricini* (Boisd.) (Lepidoptera: Saturniidae) on different provenances of *Jatropha* curcas leaves. Biofuels and Industrial Products from *Jatropha curcas*. Symposium '*Jatropha* 97' Managua, Nicaragua. 23–27 February.

Gubitz, G.M., Mittelbach, M. and Trabi, M. (1999) Exploitation of the tropical oil seed plant *Jatropha curcas* L. *Bioresource Technology* 67(1), 73–82.

Heller, J. (1996) *Physic Nut* Jatropha curcas *L.: Promoting the Conservation and Use of Underutilised and Neglected Crops 1.* International Plant Genetic Resources Institute, Rome.

Henning, R. (1996) Combating desertification: the *Jatropha* project of Mali, West Africa. Aridlands Newsletter 40. Available at: http://ag.arizona.edu/OALS/ALN/aln40/Jatropha.html

Jain, N., Joshi, H.C., Dutta, S.C., Kumar, S. and Pathak, H. (2008) Biosorption of copper from wastewater using *Jatropha* seed coat. *Journal of Scientific and Industrial Research* 67, 154–160.

Johnston, M. and Holloway, T. (2007) A global comparison of national biodiesel production potentials. *Environmental Science and Technology* 41, 7967–7973.

Jones, P.D. and Moberg, A. (2003) Hemispheric and large-scale surface air temperature variations: an extensive revision and an update to 2001. *Journal of Climate* 16, 206–223.

Jones, P.D., Parker, D.E., Osborn, T.J. and Briffa, K.R. (2006) Global and hemispheric temperature anomalies – land and marine instrumental records. In: *Trends: A Compendium of Data on Global Change*. Carbon Dioxide Information Analysis Center, Tennessee.

Jongschaap, R.E.E., Corre, W.J., Bindraban, P.S. and Brandenburg, W.A. (2007) Claims and Facts on *Jatropha curcas* L: Global *Jatropha curcas* Evaluation Breeding and Propagation Programme. Plant Research International Wageningen Report 158, Wageningen, The Netherlands.

Kalscheuer, R., Stolting, T. and Alexander, S.A. (2006) Microdiesel: *Escherichia coli* engineered for fuel production. *Microbiology* 152, 2529–2536.

Kaul, S., Saxena, R.C., Kumar, A., Negi, M.S., Bhatnagar, A.K., Goyal, H.B. and Gupta, A.K. (2007) Corrosion behavior of biodiesel from seed oils of Indian origin on diesel engine parts. *Fuel Processing Technology* 88, 303–307.

Kaushik, N., Kumar, Kumar, K., Kumar, S., Kaushik, N. and Roy, S. (2007) Genetic variability and divergence studies in seed traits and oil content of Jatropha (*Jatropha curcas* L.) accessions. *Biomass and Bioenergy* 31, 497–502.

Kermode, A.R. and Bewley, J.D. (1985a) The role of maturation drying in the transition from seed development to germination. *Journal of Experimental Botany* 36, 1906–1915.

Kermode, A.R. and Bewley, J.D. (1985b) The role of maturation drying in the transition from seed development to germination.2. postgerminative enzyme-production and soluble-protein synthetic pattern changes within the endosperm of *Ricinus communis* seeds. *Journal of Experimental Botany* 36, 1916–1927.

Kermode, A.R. and Bewley, J.D. (1986) The role of maturation drying in the transition from seed development to germination.4. protein-synthesis and enzyme-activity changes within the cotyledons of *Ricinus communis* L. seeds. *Journal of Experimental Botany* 37, 1887–1898.

Kermode, A.R., Gifford, D.J. and Bewley, J.D. (1985) The role of maturation drying in the transition from seed development to germination.3. insoluble protein synthetic pattern changes within the endosperm of *Ricinus communis* L. seeds. *Journal of Experimental Botany* 36, 1928–1936.

Kermode, A.R., Pramanik, S.K. and Bewley, J.D. (1988) The role of maturation drying in the transition from seed development to germination.5. responses of the immature castor bean embryo to isolation from the whole seed – a comparison with premature desiccation. *Journal of Experimental Botany* 39, 487–497.

Kermode, A.R., Pramanik, S.K. and Bewley, J.D. (1989a) The role of maturation drying in the transition from seed development to germination.6. desiccation-induced changes in messenger-rna populations within the endosperm of *Ricinus communis* L. seeds. *Journal of Experimental Botany* 40, 33–41.

Kermode, A.R., Pramanik, S.K. and Bewley, J.D. (1989b) The role of maturation drying in the transition from seed development to germination.7. effects of partial and complete desiccation on abscisic-acid levels and sensitivity in *Ricinus communis* L. seeds. *Journal of Experimental Botany* 40, 303–313.

Khan, A., Rajagopal, D. and Yoo, K.J. (2006) Plant Oils – Entrepreneurial Opportunities in Alternative Fuel Production and Rural Development in India. Report – MOT-UNIDO Bridging the Divide 2005–2006. University of California, Berkeley, California.

Kochhar, S., Kochhar, V.K., Singh, S.P., Katiyar, R.S. and Pushpangadan, P. (2005) Differential rooting and sprouting behaviour of two *Jatropha* species and associated physiological and biochemical changes. *Current Science* 89, 936–939.

Kpoviessi, D.S.S., Accrombessi, G.C., Kossouoh, C., Soumanou, M.M. and Moudachirou, M. (2004) Physicochemical properties and composition of pignut (*Jatropha curcas*) non-conventional oil from different regions of Benin. C.R. *Chimie* 7, 1007–1012.

Li, J., Li, M.R., Wu, P.Z., Tian, C.E., Jiang, H.W. and Wu, G.J. (2008) Molecular cloning and expression analysis of a gene encoding a putative – ketoacyl-acyl carrier protein (ACP) synthase III (KAS III) from *Jatropha* curcas. *Tree Physiology* 28, 921–927.

Liang, Y., Chen, H., Tang, M.J., Yang, P.F. and Shen, S.H. (2007) Responses of *Jatropha curcas* seedlings to cold stress: photosynthesis-related proteins and chlorophyll fluorescence characteristics. *Phyiologia Plantarum* 131, 508–517.

Liu, S.Y., Sporer, F., Wink, M., Jourdane, J., Henning, R., Li, Y.L. and Ruppel, A. (1997) Anthraquinones in *Rheum palmatum* and *Rumex dentatus* (Polygonaceae) and phorbolesters from *Jatropha curcas* (Euphorbiaceae) with molluscicidal activity against the schistosomias vector snails *Oncomelania, Biomphalaria* and *Bulinus. Tropical Medicine and International Health* 2, 179–188.

Luo, M.J., Liu, W.X., Yang, X.Y., Xu, Y., Yan, F., Huang, P. and Chen, F. (2007) Cloning, expression, and antitumor activity of recombinant protein of curcin. *Russian Journal of Plant Physiology* 54, 202–206.

Mahanta, N., Gupta, A. and Khare, S.K. (2008) Production of protease and lipase by solvent tolerant *Pseudomonas aeruginosa* PseA in solid state fermentation using *Jatropha curcas* seed cake as substrate. *Bioresource technology* 99, 1729–1735.

Makkar, H.P.S., Becker, K. and Schmook, B. (1998) Edible provenances of *Jatropha curcas* from Quintana Roo state of Mexico and effect of roasting on antinutrient and toxic factors in seeds. *Plant Foods for Human Nutrition* 52, 31–36.

Martínez-Herrera, J., Siddhuraju, P., Francis, G., Dávila-Ortíz, G. and Becker, K. (2006) Chemical composition, toxic/antimetabolic constituents, and effects of different treatments on their levels, in four provenances of *Jatropha curcas* L. from Mexico. *Food Chemistry* 96, 80–89.

Mathews, J. (2007) Seven steps to curb global warming. *Energy Policy* 35, 4247–4259.

Mendoza, T.C., Castillo, E.T. and Aquino, A.L. (2007) Towards making *Jatropha curcas* (tubang bakod) a viable source of biodiesel oil in the Philippines. *Philippine Journal of Crop Science* 32, 29–43.

Min, E. and Yao, Z.L. (2007) The development of biodiesel industry in recent years – Peculiarity, predicament and countermeasures. *Progress in Chemistry* 19, 1050–1059.

Mixon, J., Dack, J., Kraukunas, I. and Feng, J. (2003) A case for biodiesel submitted by the biodiesel team. Available at: http://depts.washington.edu/poeweb/grad programs/envmgt/2003symposium/biodiesel_finalpaper.pdf

Modi, M.K., Reddy, J.R.C., Rao, B.V.S.K. and Prasad, R.B.N. (2006) Lipase-mediated transformation of vegetable oils into biodiesel using propan-2-ol as acyl acceptor. *Biotechnology Letters* 28, 637–640.

Motaal, D.A. (2008) The biofuels landscape: is there a role for WTO? *Journal of World Trade* 42, 61–86.

Narayana, D.S.A., Shankarappa, K.S., Govindappa, M.R., Prameela, H.A., Rao, M.R.G. and Rangaswamy, K.T. (2006) Natural occurrence of *Jatropha* mosaic virus disease in India. *Current Science* 91, 584–586.

National Biodiesel Board – USA (NBB) (2008) Statistics.

Nature Editorial (2007) Kill king corn. *Nature* 449, 637.

Onuh, M.O., Ohazurike, N.C. and Emeribe, E.O. (2008) Efficacy of *Jatropha curcas* leaf extract in the control of brown blotch disease of cowpea (*Vigna unguiculata*). *Biological Agriculture and Horticulture* 25, 201–207.

Pagliaro, M., Ciriminna, R., Kimura, H., Rossi, M. and Della Pina, C. (2007) From Glycerol to value-added products. *Angwandte Chemie-International edition* 46, 4434–4440.

Patel, V.C., Varughese, J., Krishnamoorthy, P.A., Jain, R.C., Singh, A.K. and Ramamoorty, M. (2008) Synthesis of alkyd resin from *Jatropha* and rapeseed oils and their applications in electrical insulation. *Journal of Applied Polymer Science* 107, 1724–1729.

Pertino, M., Schmeda-Hirschmann, G., Rodriguez, J.A. and Theoduloz, C. (2007) Gastroprotective effect and cytotoxicity of semisynthetic jatropholone derivatives. *Planta Medica* 73, 1095–1100.

Popluechai, S., Raorane, M., O'Donnell, A.G. and Kohli, A. (2008a) Research progress towards *Jatropha* as alternate oilseed for biodiesel. In: Syers, K.J., Wood, D. and Thongbai, P. (eds) *Proceedings of the International Technical Workshop on the Feasibility of Non-edible Oil Seed Crops for Biofuel Production May 25–27, 2007*, Chaing Rai, Thailand, pp. 136–145.

Popluechai, S., Breviario, D., Mulpuri, S., Makkar, H.P.S., Raorane, M., Emami, K., Reddy, A.R., Gatehouse, A.M.R., Syers, J.K., O'Donnell, A.G. and Kohli, A. (2008b)

Limited genotypic diversity in *J. curcas*: implications and solutions for improved varieties. Submitted.

Pradeep, V. and Sharma, R.P. (2007) Use of HOT EGR for NOx control in a compression ignition engine fuelled with bio-diesel from *Jatropha* oil. *Renewable Energy* 32, 1136–1154.

Prakash, A.R., Patolia, J.S., Chikara, J. and Boricha, G.N. (2007) Floral biology and flowering behaviour of *Jatropha curcas*. FACT Seminar on *Jatropha curcas* L.: Agronomy and Genetics, Wageningen, The Netherland, 26–28 March 2007, FACT Foundation.

Pramanik, K. (2003) Properties and use of *Jatropha curcas* oil and diesel fuel blends in compression ignition engine. *Renewable Energy* 28, 239–248.

Rahuman, A.A., Gopalakrishnan, G., Venkatesan, P. and Geetha, K. (2008) Larvicidal activity of some Euphorbiaceae plant extracts against *Aedes aegypti* and *Culex quinquefasciatus* (Diptera: Culicidae). *Parasitology Research* 102, 867–873.

Ram, S.G., Parthiban, K.T., Kumar, R.S., Thiruvengadam, V. and Paramathma, M. (2007) Genetic diversity among *Jatropha* species as revealed by RAPD markers. *Genetic Resources and Crop Evolution*. Doi 10.1007/s10722-007-9285-7

Ranade, S.A., Srivastava, A.P., Rana, T.S., Srivastava, J. and Tuli, R. (2008) Easy assessment of diversity in *Jatropha curcas* L. plants using two single-primer amplification reaction (SPAR). *Biomass and Bioenergy*. Doi:10.1016/ j.biombioe.2007.11.006.

Rao, G.L.N., Prasad, B.D., Sampath, S. and Rajagopal, K. (2007) Combustion analysis of diesel engine fueled with *Jatropha* oil methyl-ester diesel blends. *International Journal for Green Energy* 4, 645–658.

Ravi, M., Marimuthu, M.P.A. and Siddiqi, I. (2008) Gamete formation without meiosis in Arabidopsis. *Nature* 451, 1121–1124.

Ravindranath, N., Ravireddy, M., Ramesh, C., Ramu, R., Prabhakar, A., Jagdeesh, B. and Das, B. (2004) New lathyrane and podocarpane diterpenoids from *Jatropha*

curcas. *Chemical and Pharmaceutical Bulletin* 52, 608–611.

Reddy, M.P., Chikara, J., Patolia, J.S. and Ghosh, A. (2007) Genetic improvement of *Jatropha curcas* adaptability and oil yield. FACT Seminar on *Jatropha curcas* L.: Agronomy and Genetics, Wageningen, The Netherland, 26–28 March 2007, FACT Foundation.

Reksowardojo, I.K., Dung, N.N., Tuyen, T.Q., Sopheak, R., Brodjonegoro, T.P., Soerwidjaja, T.H., Ogawa, I.H. and Arismunandar, W. (2006) The comparison of effect of biodiesel fuel from palm oil and physic nut oil (*Jatropha curcas*) on an direct injection (DI) diesel engine. *Proceedings of FISITA 2006 Conference in Yokohama, Japan*.

Roy, S. (1998) Biofuel production from *Jatropha curcas*. *10th European Conference and Technology Exhibition on Biomass for Energy and Industry, 08–11 June 1998, Wurzburg, Germany. Biomass for Energy and Industry*, pp. 613–615.

Sarin, R., Sharma, M., Sinharay, S. and Malhotra, R.K. (2007) *Jatropha*-palm biodiesel blends: an optimum mix for Asia. *Fuel* 86, 1365–1371.

Sharma, N. (2007) Effect on germination on raised bed and sunken bed nursery in different provenances of *Jatropha curcas* L. Fact Seminar on *Jatropha curcas* L.: Agronomy and Genetics, Wageningen, The Netherland, 26–28 March 2007, FACT Foundation.

Singh, R.N., Vyas, D.K., Srivastava, N.S.L. and Narra, M. (2008) SPRERI experience on holistic approach to utilize all parts of *Jatropha curcas* fruit for energy. *Renewable Energy* 33, 1868–1873.

Sirisomboon, P., Kitchaiya, P., Pholpho, T. and Mahuttanyavanitch, W. (2007) Physical and mechanical properties of *Jatropha curcas* L. fruits, nuts and kernels. *Biosystems Engineering* 97, 201–207.

Solomon Raju, A.J. and Ezradanam, V. (2002) Pollination ecology and fruiting behaviour in a monoecious species, *Jatropha curcas* L. (Euphorbiaceae). *Current Science* 83, 1395–1398.

Soontornchainaksaeng, P. and Jenjittikul, T. (2003) Karyology of *Jatropha* (Euphorbiaceae) in Thailand. *Thai Forest Bulletin* (Bot) 31, 105–112.

Sricharoenchaikul, V., Pechyen, C., Aht-ong, D. and Atong, D. (2008) Preparation and characterization of activated carbon from the pyrolysis of physic nut (*Jatropha curcas* L.) waste. *Energy and Fuels* 22, 31–37.

Srivastava, S.K., Tewari, J.P. and Shukla, D.S. (2008) A folk dye from leaves and stem of *Jatropha curcas* L. used by Tharu tribes of Devipatan division. *Indian Journal of Traditional Knowledge* 7, 77–78.

Sseruwagi, P., Maruthi, M.N., Colvin, J., Rey, M.E.C., Brown, J.K. and Legg, J.P. (2006) Colonization of non-cassava plant species by cassava whiteflies (*Bemisia tabaci*) in Uganda. *Entomologia Experimentalis et Applicata* 119, 145–153.

Sudheer Pamidiamarri, D.V.N., Pandya, N., Reddy, M.P. and Radhakrishnan, T. (2008) Comparative study of interspecific genetic divergence and phylogenic analysis of genus *Jatropha* by RAPD and AFLP: genetic divergence and phylogenic analysis of genus *Jatropha*. *Molecular Biology Reporter*. Doi 10.1007/s11033-008-9261-0.

Sujatha, M., Makkar, H.P.S. and Becker, K. (2005) Shoot bud proliferation from auxillary nodes and leaf sections of non-toxic *Jatropha curcas* L. *Plant Growth Regulation* 47, 83–90.

Sujatha, M., Reddy, T.P. and Mahasi, M.J. (2008) Role of biotechnological interventions in the improvement of castor (*Ricinus communis* L.) and *Jatropha curcas* L. *Biotechnology Advances*. Doi: 10.1016/j.biotechadv.2008.05.004.

Sukarin, W., Yamada, Y. and Sakaguchi, S. (1987) Characteristics of physic nut, *Jatropha curcas* L. as a new biomass crop in the tropics. *Japanese Agricultural Research Quarterly, Japan* 20, 302–303.

Sun, S. and Hansen, J.E. (2003) Climate simulations for 1951–2050 with a coupled atmosphere-ocean model. *Journal of Climate* 16, 2807–2826.

Sunil, N., Varaprasad, K.S., Sivaraj, N., Kumar, T.S., Abraham, B. and Prasad, R.B.N. (2008) Assessing *Jatropha curcas* L. germplasm *in situ* – A case study. *Biomass and Bioenergy* 32, 198–202.

Takeda, Y. (1982) Development study on *Jatropha curcas* (sabu dum) oil as a substitute for diesel engine oil in Thailand. *Journal of Agriculrural Association, China* 120, 1–8.

Tang, M., Sun, J., Liu, Y., Chen, F. and Shen, S. (2007) Isolation and functional characterization of the JcERF gene, a putative AP2/EREBP domain-containing transcription factor, in the woody oil plant *Jatropha curcas*. *Plant Molecular Biology* 63, 419–428.

Tiwari, A.K., Kumar, A. and Raheman, H. (2007) Biodiesel production from *Jatropha* (*Jatropha curcas*) with high free fatty acids: an optimized process. *Biomass and Bioenergy* 31, 569–575.

Tong, L., Shu-Ming, P., Wu-Yuan, D., Dan-Wei, M., Ying, X., Meng, X. and Fang, C. (2006) Characterization of a new stearoyl-acyl carrier protein desaturase gene from *Jatropha curcas*. *Biotechnology Letters* 28, 657–662.

Torrance, S.J., Wiedhopf, R.M. and Cole, J.R. (1977) Antitumor agents from *Jatropha macrorhiza* (Euphorbiaceae). III: acetyla-leuritolic acid. *Journal of Pharmaceutical Sciences* 66(9), 1348–1349.

Torrance, S.J., Wieldhopf, R.M., Cole, J.R., Arora, S.K., Bates, R.B., Beavers, W.A. and Cutler, R.S. (1976) Antitumor agents from *Jatropha macrorhiza* (Euphorbiaceae). II. Isolation and Characterization of *Jatropha* trione. *Journal of Organic Chemistry* 41(10), 1855–1857.

Vyas, D.K. and Singh, R.N. (2007) Feasibility study of *Jatropha* seed husk as an open core gasifier feedstock. *Renewable Energy* 32, 512–517.

Wood, D. (2008) Target properties for biofuels in Thailand. In: Syers, K.J., Wood, D. and Thongbai, P. (eds) *Proceedings International Technical Workshop on the Feasibility of Non-edible Oil Seed Crops for Biofuel Production 25–27 May 2007,* Chaing Rai, Thailand, pp. 50–61.

Zhang, F.L., Niu, B., Wang, W.C., Chen, F., Wang, S.H., Xu, Y., Jiang, L.D., Gao, S., Wu, J., Tang, L. and Jia, Y.R. (2008) A novel betaine aldehyde dehydrogenase gene from *Jatropha curcas*, encoding an enzyme implicated in adaptation to environmental stress. *Plant Science* 174, 510–518.

Zhang, Y., Wang, Y.X., Jiang, L.D., Xu, Y., Wang, Y.C., Lu, D.H. and Chen, F. (2007) Aquaporin JcPIP2 is involved in drought responses in *Jatropha curcas. Acta Biochimicaet Biophysica Sinica* 39, 787–794.

III Global Perspectives

15 European Commercial Genetically Modified Plantings and Field Trials

F. Ortego,[1] X. Pons,[2] R. Albajes[2] and P. Castañera[1]

[1]Departamento de Biología de Plantas, Centro de Investigaciones Biológicas, Madrid, Spain; [2]Centre UdL-IRTA, Universitat de Lleida, Lleida, Spain

Keywords: Bt maize, post-market monitoring, resistance, non-target arthropods, field trials

Summary

Genetically modified (GM) maize crops expressing the Cry1Ab toxin from *Bacillus thuringiensis* (*Bt* maize) have been cultivated in Spain on a commercial scale since 1998, reaching an area of about 75,000 ha (around 21% of total maize-growing area) in 2007. A post-market environmental monitoring plan for the *Bt* maize varieties derived from Event *Bt*176 was carried out during the period 1998–2005, and a second ongoing monitoring plan for *Bt* maize varieties derived from Event MON810 was initiated in 2003. Both of these consider case-specific monitoring for the evolution of resistance in target insects, as well as for the potential effects on non-target arthropods. Monitoring for field resistance in target insects, *Sesamia nonagrioides* and *Ostrinia nubilalis*, is being assessed by changes in susceptibility from baseline levels. To date, no changes in their susceptibility to the Cry1Ab toxin have been found, which is consistent with the fact that there have been no control failures reported in transgenic *Bt* maize fields. Field trials to determine potential effects of *Bt* maize on non-target arthropods have focused on three functional groups: herbivores, predators and parasitoids. No detrimental effects have been observed on any of the main predatory groups that are common in maize fields. Nevertheless, field trials in the same areas for longer periods are necessary to discard long-term potential cumulative effects.

Commercial Planting of *Bt* Maize in the European Union

Genetically modified (GM) maize plants expressing the Cry1Ab toxin from *Bacillus thuringiensis* (*Bt* maize) is the only commercial GM crop grown in the European Union (EU), and Spain is the only member state where *Bt* maize varieties have been cultivated at a commercial scale since 1998. A mean

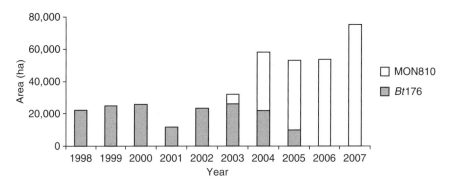

Fig. 15.1. Area (ha) of *Bt* maize cultivated in Spain between 1998 and 2007. (Data from the Spanish Ministry of Agriculture, available at: http://www.mapa.es/es/agricultura/pags/semillas/estadisticas.htm.)

growing area of about 20,000–25,000 ha of *Bt* maize (5% of the total maize-growing area) was cultivated between 1998 and 2002, increasing to 50,000–75,000 ha (around 12–21%) in the following years (Fig. 15.1). Germany, France, Portugal, the Czech Republic and Slovakia also grew some *Bt* maize during this period but reported very small areas, usually under 5000 ha (James, 2007). In 2007, 21,200 ha of *Bt* maize were cultivated in France, and Romania and Poland planted *Bt* maize for the first time, bringing the total number of countries planting biotech crops in the EU to eight (James, 2007).

The first commercial GM maize hybrids registered in Spain were two varieties, Jordi and Compa CB (Syngenta Seeds), carrying the genetic modification identified as Event *Bt*176, approved for placing on the market under Directive 90/220/EEC in 1997 (Commission Decision 97/98/EC of 23 January 1997, http://eur-lex.europa.eu/en/index.htm). Two new varieties derived from the Event *Bt*176 were registered in 2003 and 2004, although Compa CB was essentially the only commercial variety grown between 1998 and 2003. In 2004, the European Food Safety Authority (EFSA, http://www.efsa.europa.eu/en.html) declared that the antibiotic-resistant marker gene present in the *Bt*176 maize should not be present in commercial GM crops (EFSA, 2004), and in August of 2005, the Spanish Ministry of Agriculture prohibited further planting of all *Bt*176 varieties. The commercial planting of MON810 maize was authorized under Directive 90/220/EEC in 1998 (Commission Decision 98/294/CE of 22 April 1998). Four varieties derived from this genetic modification and commercialized by different seed companies (Monsanto Agricultura, Pioneer Hi-Bred and Nickerson Sur) were registered in Spain in 2003. New varieties were approved in 2004, 2005 and 2006, including those derived from the inscription of MON810 varieties authorized in France and Germany in the Common EU Catalogue of Varieties. A total of 51 varieties offered by nine companies (Monsanto Agricultura, Pioneer Hi-Bred, Limagrain Ibérica, Semillas Fito, Arlesa Semillas, Koipesol, Agrar Semillas, Corn States International, Coop de Pau) were allowed to be cultivated in Spain in 2007.

Bt maize provides an effective control of two key lepidopteran pests: the Mediterranean corn borer (MCB), *Sesamia nonagrioides* (Lefèbvre; Lepidoptera:

Noctuidae) and the European corn borer (ECB), *Ostrinia nubilalis* (Hübner; Lepidoptera: Crambidae). In 2007, 75,150 ha of *Bt* maize was grown, representing approximately 21% of the total Spanish maize-growing area (http:// www.mapa.es/es/agricultura/pags/semillas/estadisticas.htm). Regional rates of *Bt* maize adoption are quite variable, but it seems to be related to corn borer damage. Thus, in Catalonia and Aragon, where corn borer pressure is high, *Bt* maize represents 64% and 54%, respectively, of the maize-growing area. The average yield advantage of *Bt* maize over conventional maize in Spain for the 2002–2004 period was 4.7% (Gómez-Barbero and Rodríguez-Cerezo, 2006). This represents an increase of 13% over the average gross margin obtained by a maize farmer in Spain, including subsidies. These benefits, however, vary widely in the three regions studied, ranging from the high gross margin differences in Aragon (€122/ha) to just €9/ha in Albacete (Gómez-Barbero *et al.*, 2008).

Post-market Monitoring Programmes in Spain

EU legislation for the cultivation of GM plants is regulated by Directive 2001/18/EC, and by Regulation (EC) 1829/2003 on GM food and feed. This legislation requires a case-by-case assessment of the risk to human health and the environment before GM plants can be authorized for placing on the market. Data on composition, toxicity, allergenicity, nutritional value and environmental impact must be evaluated during the environmental risk assessment (ERA) process (also see Chapters 4 and 13, this volume). The identification of areas of uncertainty or risk, including the impact of exposure over long periods of time and cumulative long-term effects when GM plants are commercialized, may give rise to the need for further specific activities, including post-market monitoring. Thus, current EU legislation requires the obligation for applicants to implement, if appropriate, an environmental monitoring programme. According to Annex VII of the Directive and the accompanying guidance notes (Council Decision 2002/811/EC of 3 October 2002), the goals of environmental monitoring are: (i) to confirm any hypothesis relative to the possible adverse effects of the GM plant identified by the ERA; and (ii) to identify the occurrence of adverse effects not considered in the ERA. With respect to their general principles, monitoring plans will be developed on a case-by-case basis during the approved period for the commercialization of the product, usually 10 years, although this period of time may differ according to the GMO and the information relating to its subsequent use. It should include general surveillance to detect unforeseen adverse effects, and if necessary, case-specific monitoring to elucidate the impact of identified risks in both the medium and long term. If changes in the environment are observed, it will be necessary to perform a new risk evaluation to establish if these are a consequence of the GMO itself, or a consequence of its use.

The Council Decision 2002/811/EC establishing guidance notes supplementing Annex VII to Directive 2001/18/EC expand upon the objectives and general principles for post-market monitoring of GM plants, but explicitly do not attempt to provide details for their development. Instead, this Decision refers to the possible need to complement the existing 'framework with more specific, supplementary guidance on monitoring plans or checklists with regard to particular traits, crops or groups of GMOs'. EFSA has provided further guidance to applicants on the preparation and presentation of general surveillance plans as part of post-market monitoring (EFSA, 2006). In addition, the European Commission has set up a 'Working Group on Guidance Notes on Monitoring Supplementing Annex VII of directive 2001/18/EC' to elaborate EU-level monitoring concepts, plans, methods and parameters for case-specific monitoring and general surveillance, as well as to address the issues of EU-wide harmonization and coordination of data resulting from post-market monitoring. However, no EU-wide consensus on how to design post-market monitoring programmes has been defined (Sanvido et al., 2005). Indeed, there is still controversy on the definition of the specific functions of general surveillance and case-specific monitoring (Den Nijs and Bartsch, 2004; Sanvido et al., 2005), since the Council Decision 2002/811/EC provides no clear differentiation for the conceptual differences between them. According to EFSA (2006), the difference is related to the predictable or unpredictable nature of the effect. Where there is scientific evidence of a potential adverse effect linked to the genetic modification, then case-specific monitoring should be carried out in order to confirm the assumptions of the environmental risk assessment. Consequently, case-specific monitoring is not obligatory and is only required to verify the risk assessment, whereas general surveillance is always required in the environmental monitoring plan (EFSA, 2006).

Although post-market monitoring plans are only required in the EU since Directive 2001/18/EC, they have been a requirement of Spanish legislation for the registration of commercial varieties since 1998 (Order of the Ministry of Agriculture of 23 March 1998, later actualized by the 'Real Decretos' 323/2000 of 3 March 2000 and 178/2004 of 30 January 2004, http://www.boe.es/g/eng/index.php). The Spanish authorities, taking into consideration the precautionary principle, required a monitoring plan for all transgenic varieties registered in the Spanish Commercial Variety Register, in advance of what was later included in the Directive 2001/18/EC. A post-market environmental monitoring plan for the Bt maize varieties derived from Event Bt176 was carried out during the period 1998–2005. A second ongoing monitoring plan for Bt maize varieties derived from Event MON810 was initiated in 2003. These plans were designed by the notifiers of the events in collaboration with experts of the Spanish National Biosafety Commission (Ministry of Environment), and were approved by the Ministry of Agriculture in the corresponding Ministerial Orders (Roda, 2006). Both plans were designed to evaluate the efficacy of the genetic modification incorporated into each particular variety and consider case-specific monitoring for the evolution of resistance in target insects (see below), and for the potential effects on non-target arthropods (see below) and soil microorganisms (Badosa et al., 2004). An additional study to determine the potential effects on nutrition and gut flora of animals fed with Bt maize was required for those varieties derived from Event Bt176 that contained a marker gene that conferred resistance to ampicillin. Furthermore, general surveillance on MON810 varieties is also being conducted, based on farmer questionnaires. Companies must implement these monitoring plans and report the annual results to the Spanish Authorities for a period of at least 5 years. These studies have been funded in part by the Spanish Ministry of Environment and the Ministry of Education. The Spanish National Biovigilance Commission, responsible to the Ministry of Agriculture, was created for the supervision of post-market monitoring programmes and for the adoption of the measures to be taken if adverse effects were detected.

Monitoring for Field Resistance in Target Insect Pests

The Spanish research programme to monitor corn borer resistance focuses on similar issues to those considered in the draft protocol proposed by the Expert Group of the European Commission (Draft Protocol for the monitoring of the European Corn Borer and Mediterranean Corn Borer Resistance to *Bt* Maize, Document XI/157/98, submitted by DGXI 29/04/98 [SCP/ GMO/014]). According to this protocol, evolution of resistance to *Bt* maize in target pest populations is being assessed by changes in susceptibility from baseline levels (Farinós *et al.*, 2004a). Coincident with the commercial release of *Bt* maize in Spain in 1998, González-Núñez *et al.* (2000) established the baseline susceptibility to the insecticidal protein Cry1Ab in ECB and MCB populations from representative *Bt* maize-growing regions. This study revealed small differences in susceptibility among populations, which can be attributed to natural variation, since they were collected from non-transgenic fields with no records of *Bt* products being used. Annual monitoring has then been conducted on *Bt* maize fields from the same maize-growing areas during the period 1999–2007 (Farinós *et al.*, 2004b; F. Ortego and P. Castañera, unpublished results). To date, no changes in the susceptibility to the Cry1Ab toxin have been found, which is consistent with the fact that there have been no control failures reported in transgenic *Bt* maize fields. Moreover, monitoring of populations of MCB and ECB with a history of high exposure to Cry1Ab toxin versus conspecific populations from non-*Bt* maize is being performed as a complementary method to annual monitoring, with similar results (F. Ortego and P. Castañera, unpublished results). Recently, Saeglitz *et al.* (2006) established the baseline susceptibility to Cry1Ab of ECB populations in Germany. Overall, the data suggest little differentiation among European populations in terms of their susceptibility to Cry1Ab. However, it is important that the same toxin batches will be available throughout the life of the monitoring programmes, since susceptibility may vary considerably between different batches and formulations of *Bt* toxins (Farinós *et al.*, 2004a; Saeglitz *et al.*, 2006).

The relevance of laboratory assays to forecast the evolution of insect resistance in the field has been questioned, because the selection pressure is lower than in the field (Chaufaux *et al.*, 2001). In Spain, where only the variety Compa CB was cultivated until 2003, MCB and ECB larvae of the second and third generation were exposed to sublethal doses of the toxin, since toxin titre decreases after anthesis, and therefore laboratory selection might be relevant. Laboratory selection assays for eight generations yielded selected strains of MCB and ECB, that were 21-fold and tenfold more tolerant to Cry1Ab than the corresponding unselected strains (Farinós *et al*., 2004b), suggesting that both species have the potential to develop low to moderate levels of tolerance to the toxin. However, none of the laboratory-selected larvae was able to survive on *Bt* maize, probably because laboratory colonies usually start with limited genetic variation and thus may not contain the rare resistant alleles present in field populations.

The high-dose/refuge (HDR) strategy is considered the most effective for delaying resistance in target pests to *Bt* maize and it is the one recommended in Spain; it being mandatory to deploy refuges for *Bt* maize fields over 5 ha. This strategy is based on the key assumptions that resistance is functionally recessive, resistance alleles are initially rare ($<10^{-3}$), and random mating occurs within the typical dispersal distances of the adults. To test the first assumption, an F2 screen was conducted on natural populations of MCB from Spain and Greece (Andreadis *et al*., 2007) and ECB from France (Bourguet *et al*., 2003). No major resistance alleles were found; the frequency of resistance alleles being $<8.6 \times 10^{-3}$ for the MCB Spanish population, $<9.7 \times 10^{-3}$ for the MCB Greek population and $<9.2 \times 10^{-4}$ for the ECB French population. These results suggest that the frequency of alleles conferring resistance to Cry1Ab in both MCB and ECB populations may be sufficiently rare so that the HDR strategy could be applied with success for resistance management. On the other hand, the knowledge of genetic differentiation between populations is also of critical importance to predict the efficiency of HDR strategy, since the potential to evolve *Bt*-resistance is strongly related to the migration between populations. Bourguet *et al*. (2000) showed that the genetic structure of ECB populations in maize fields was compatible with random mating and a high gene flow between fields across the whole of France. However, a significant host-plant effect on ECB genetic differentiation was evident when wild and cultivated host plants were considered (Leniaud *et al*., 2006). Genetic differentiation of MCB populations analysed by random amplified polymorphic DNA (RAPD) markers (De la Poza *et al*., 2006) supports previous studies on allozyme analysis (Buès *et al*., 1996; Leniaud *et al*., 2006) suggesting a limited dispersive behaviour in MCB in comparison with that found for ECB. In addition, behavioural studies on MCB adults suggest that males and females from adjacent maize fields could mate at random, and that refuges can be located at least 400 m from *Bt* maize fields with no decrease in random mating (Eizaguirre *et al*., 2004, 2006).

Field Trials and Non-target Arthropods

As explained elsewhere (Chapter 8, this volume), risk assessment trials are usually conducted with a sequential process from early tests in the laboratory to the most complex and realistic trials in the field, through intermediate testing in semi-field or glasshouse environments. However, post-market monitoring must focus on field trials that consider the different local and commercial conditions under which transgenic crops are grown, since these can vary depending on the area, growers' habits, commercial variety and many environmental variables (e.g. climate, diseases, insect pests, soil fertility). Field trials allow us to test the effects of arthropod exposure to *Bt* toxins via multiple and complex trophic pathways, and also to study the consequences, particularly for specialist natural enemies, of removing or altering certain hosts or prey; a situation that is difficult to reproduce in the laboratory or even under semi-field conditions. Nevertheless, laboratory tests under worst-case scenarios may help to interpret field results that may be difficult to unambiguously attribute to the transgenic trait under field conditions. In addition, to measure the effects of GM crops on non-target arthropods we need to previously define those organisms that are ecologically more relevant and to elucidate their ways of exposure to *Bt* maize toxins.

Arthropod fauna in maize fields

ECB and MCB, and more recently the Western corn rootworm (WCR), *Diabrotica virgifera virgifera* LeConte, are the most harmful maize pest in Europe. The ECB is distributed in all Western European maize-growing areas, whereas MCB is restricted to areas located under the parallel 45°N. The WCR was introduced into Europe in the early 1990s and has subsequently become an important pest in eastern and south-eastern areas, including Serbia, Hungary, Romania and northern Italy. Other herbivores that can reach high densities in maize in Spain, but whose incidence is sporadic, are cutworms (*Agrotis segetum* Denis and Schiffemüller), armyworms (*Pseudaletia unipuncta* Haworth), bollworms (*Helicoverpa armigera* Hübner), wireworms (*Agriotes lineatus* L.), spider mites (*Tetranychus urticae* Koch), aphids (*Rhopalosiphum padi* L., *Sitobion avenae* F. and *Metopolophium dirhodum* Walker) and leafhoppers (*Zyginidia scutellaris* Herrich-Schäfer) (Castañera, 1986; Pons and Albajes, 2003).

Predatory fauna in maize has been particularly well studied in Spain (Asín and Pons, 1998; Albajes *et al.*, 2003; De la Poza *et al.*, 2005; Farinós *et al.*, 2008). An intensive survey conducted over several years in Lleida (north-eastern Spain) has provided a sound picture of the most important insect predatory groups (Table 15.1). More than 25 species or genera of plant-dwelling predators that belonged to 15 families of Insecta and five Arachnida taxons were identified. However, only ten species/groups were regularly the most common: the anthocorid *Orius* spp. (especially *O. majusculus* (Reuter) and *O. niger* (Wolff.)), the nabid *Nabis provencalis* Remane, the thrips *Aeolothrips tenuicornis* (Bagnall), the green lacewing *Chrysoperla carnea* (Stephens), the carabid *Demetrias atricapillus* (L.), the coccinellids *Coccinella septempunctata* L., *Hippodamia variegata* (Goeze) and *Stethorus punctillum* Weise, the staphylinid *Tachyporus* sp., several species of Syrphidae, the cecidomyid *Aphidoletes aphidimyza* (Rondani) and several unidentified species of spiders. Among soil-dwelling predators caught by pitfall traps, carabids (mainly *Agonum dorsale* (Pontoppidan), *Poecilus cupreus* L., and *Harpalus rufipes* (De Geer)), dermapterans (mainly *Labiduria riparia* (Pallas)) and spiders (unidentified species) were the prevalent groups in Lleida. Surveys performed in

Table 15.1. Plant-dwelling and soil-dwelling insect predators recorded in visual samplings and caught in pitfall traps, respectively, in Lleida, Spain (Asín and Pons, 1998; Albajes *et al.*, 2003; De la Poza *et al.*, 2005; X. Pons and R. Albajes unpublished data). Only species or genera which represented more than 5% within their family are listed.

		Species or genus	
Order	Family	Plant-dwelling predators	Ground-dwelling predators
Dermaptera	Forficulidae	*Forficula auricularia* L.	
	Labiduridae		*Labiduria riparia* (Pallas)
Heteroptera	Anthocoridae	*Orius majusculus* (Reuter), *Orius niger* (Wolffer), *Orius laevigatus* (Fieber), *Orius laticollis* (Reuter)	
	Nabidae	*Nabis provencalis* Ramane	*N. provencalis*
	Miridae	*Trigonotylus caelestiallum* (Kirkaldy), *Adelphocoris* sp., *Lygus rugulipennis* (Poppius) *Lygus* spp., *Creontiades pallidus* Ramlur	
	Geocoridae	Unidentified	
Thysanoptera	Aeolothripidae	*Aeolothrips tenuicornis* (Bagnall), *Aeolothrips fasciatus* (L.)	
Neuroptera	Chrsyopidae	*Chrysoperla carnea* (Stephens)	
	Hemerobiidae	Unidentified spp.	
Coleoptera	Carabidae	*Demetrias atricapillus* (L.), other unidentified spp.	*Agonum dorsale* (Pontoppidan), *Amara anthobia* Villa, *Angoleus nitidus* (Dejean), *Bembidion lampros* Herbst., *Brachynidius scopleta* L.F., *Brachinus crepitans* L., *Calathus circumseptus* German, *Campalita maderae* ssp. *indagator* F., *Harpalus rufipes* (De Geer), *Harpalus distinguendus* Duftschmid, *Metallina properans* Stephens, *Phyla tethis* Netolitzky, *Poecilus cupreus* (L.), *Poecilus kugelani* Niger Letz, *Poecilus purpurascens* (Dejean), *Testedium bipunctulatum* L.

Continued

Table 15.1. Continued.

| Order | Family | Species or genus | |
		Plant-dwelling predators	Ground-dwelling predators
	Coccinellidae	*Coccinella septempunctata* L., *Hippodamia variegata* (Goeze), *Propylea quatuordecimpunctata* (L.), *Scymnus* spp., *Stethorus punctillum* Weise	
	Cantharidae	*Rhagonycha* sp.	
	Staphylinidae	*Tachyporus* spp.,	Unidentified species
Diptera	Cecidomyiidae	*Aphidoletes aphidimyza* (Rondani)	
	Syrphidae	*Episyrphus balteatus* (De Geer), *Sphaerophoria scripta* (L.), *Scaeva pyrastri* (L.)	

Madrid (central Spain) revealed that *Orius* spp. were also the most abundant plant-dwelling arthropods, followed by *S. punctillum* and spiders. Likewise, carabids and spiders were the most abundant ground-dwelling predators, followed by staphylinids.

Unfortunately, to date, there is little information available regarding the parasitoids associated with maize field ecosystems in Europe. The most abundant parasitoid of MCB and ECB in Spain is the tachinid *Lydella thompsoni* Herting, although ichneumonids (*Ichneumon* spp.) have also been reported (Castañera, 1986). Aphid parasitoids are also abundant when aphids are at high densities. The most commonly identified species are: *Aphidius ervi* Haliday, *Aphidius rhopalosiphi* DeStefani-Perez, *Praon volucre* (Haliday) and *Lysiphlebus testaceipes* (Cresson) (Pons and Stáry, 2003).

Routes of exposure of non-target arthropods to *Bt* maize toxins

Risk has been classically defined as the result of hazard and exposure. Exposure of non-target arthropods to *Bt* maize may occur through several ways. Non-target herbivores may ingest *Bt* toxin when feeding on toxin-expressing tissues. In a recent survey of *Bt* toxin movement via the trophic web in commercial *Bt*-maize fields in Spain, Obrist *et al.* (2006) recorded remarkably high levels of Cry1Ab in the spider mite *T. urticae*, the leaf beetle *Oulema melanopus* (L.) and the zoophytophagous mirid bug *Trigonotylus caelestialium* (Kirkaldy), whereas no, or extremely low, levels of the toxin were detected in Homoptera and Thysanoptera. It is well known that aphids do not acquire the toxin when feeding in *Bt* maize because of the very low levels of toxin in the phloem sap (Raps *et al.*, 2001). From herbivores, toxins may move to predators and parasitoids that consume contaminated prey. *T. urticae* appears to be the better pest candidate for transferring the toxin to predators, since *O. melanopus* was only found to contain the *Bt* toxin when feeding on the early growth stages of maize, and *T. caelestiallum* is not very abundant in Spanish maize (Obrist *et al.*, 2006). Natural enemies may also be exposed to *Bt* toxins when they feed on plant tissues or materials (i.e. pollen). Thus, field trials confirmed that the *Bt* toxin could be transferred to predators, for example, *Orius* spp., *Chrysoperla* spp. and *S. punctillum*, when *Bt* maize pollen or spider mites were available (Obrist *et al.*, 2006). The passage of Cry1Ab to *O. majusculus* via these two food sources was confirmed in the laboratory using maize cultivars containing the Event *Bt*176 that expresses the toxin in pollen (Obrist *et al.*, 2006). Likewise, it has been possible to confirm the transmission of Cry1Ab toxin from MON810 varieties to *S. punctillum* via *T. urticae* (Álvarez-Alfageme *et al.*, 2008). Exposure of predators and parasitoids via the environment may also occur when toxins from plant residues persist in the soil.

In addition to those direct effects, natural enemies may also be affected indirectly by *Bt* maize due to changes in the availability or nutritional quality of preys and hosts. A reduction in their quantity is more likely to impact parasitoids and specialist predators, such as aphidophagous coccinellids, lacewings, syrphids and the spider mite predator *S. punctillum*. Low nutritional quality in prey could primarily affect predators feeding on lepidopterans exposed to sublethal doses of the toxin, although this kind of effect is not likely to be observed in the field because most predators that potentially prey on such larvae are generalists.

Field trials to assess potential effects of *Bt* maize on non-target arthropods

Since the initial steps of *Bt* maize deployment in Spain, field trials have focused on monitoring their potential effects on three groups of non-target arthropods: herbivores, predators and parasitoids. These studies have been based on comparing fauna composition and abundance on *Bt* and near-isogenic non-*Bt* maize. Comparisons have sometimes included an insecticide treatment as a baseline, because *Bt* maize is intended to replace chemical treatments.

Herbivores
The impact of transgenic *Bt* maize (Compa CB, Event 176), expressing the Cry1Ab protein, on aphids, leafhoppers, cutworms and wireworms, was evaluated at the farm scale by comparing their abundance on *Bt* plots, and on plots of the near-isogenic variety, over three consecutive growing seasons (Eizaguirre *et al.*, 2006). Soil-borne insect pests (cutworms and wireworms) were consistently not affected by the transgenic crop. However, effects on aphidophauna were unexpected, as more aphids were found on *Bt* maize than on the near-isogenic counterparts (Lumbierres *et al.*, 2004; Pons *et al.*, 2005).

The analysis of aphid age structure showed more individuals on *Bt* plots for alates, apterous adults and young nymphs of *R. padi*, apterous adults and apterous fourth instar nymphs of *S. avenae*, and alates, apterous adults and apterous fourth instar nymphs of *M. dirhodum*. Leafhoppers, particularly mature nymphs of *Z. scutellaris*, were also more abundant on *Bt* maize. Differences in aphid and leafhopper densities, an effect that cannot be directly attributed to *Bt* toxins, were not sufficiently high to affect yield, but they may have affected the availability of prey for predators.

Other field trials have been conducted in Europe for measuring impacts of *Bt* maize on non-target maize herbivores. Gathmann *et al.* (2006) reported no adverse effects of *Bt* maize to non-target lepidopteran larvae living on accompanying weeds in maize field margins. No differences in the abundance of several species of aphids (Lozzia and Rigamonti, 1998; Lozzia, 1999; Bourguet *et al.*, 2002) and leafhoppers (Rauschen *et al.* 2004) were found, presumably because the intensity and duration of these field trials were not sufficient to display those differences that were observed in long-term studies conducted in Spain.

Predators

A similar scheme as to the one mentioned above for herbivores was used to assess the potential impact of *Bt* maize on predatory arthropods. Field trials were conducted at two different agronomic areas in north-eastern (Lleida) and central (Madrid) Spain over a period of 3 years (De la Poza *et al.*, 2005; Farinós *et al.*, 2008). Treatments were also *Bt* versus non-*Bt* maize (Compa CB, Event 176), but with the addition of a third treatment consisting of the non-transgenic cultivar treated with imidacloprid as a seed dressing. Plant predators were monitored by visual plant scouting, while ground-dwelling predators were monitored using pitfall traps. In general, the abundance varied from year to year and between locations, but no detrimental effects of *Bt* maize on predators were found. *Orius* species are a priori among the predators most likely exposed to *Bt* maize, since they can acquire the toxin through contaminated prey and feeding on pollen (Obrist *et al.*, 2006), and they are a major predator of ECB that is normally absent from *Bt* maize during most of the season. Field trials showed that their abundance was not negatively affected by *Bt* maize in any of the six combinations at the two locations over the 3-year period (Fig. 15.2); these findings are consistent with the lack of negative effects reported for *Bt* maize on *O. majusculus* under laboratory conditions (Zwallen *et al.*, 2000; Pons *et al.*, 2004). Moreover, the wide range of prey that can be consumed by *Orius* spp. in addition to ECB eggs and young larvae may overcome the absence of this particular prey species. The abundance of the coccinelid *S. punctillum*, a rather specific predator of spider mites, varied among years and localities depending on the presence of *T. urticae* in the maize fields, but no differences were found between *Bt* and non-*Bt* plots (Fig. 15.2). Although the passage of *Bt* toxin to *S. punctillum* via its prey has been confirmed (Obrist *et al.*, 2006; Álvarez-Alfageme *et al.*, 2008), no subsequent negative effects on larval development, and adult longevity and fecundity were reported (Álvarez-Alfageme *et al.*, 2008); No differences in abundance of the three main predatory groups recorded in pitfall traps – ground beetles, spiders and earwigs – were found between *Bt* and

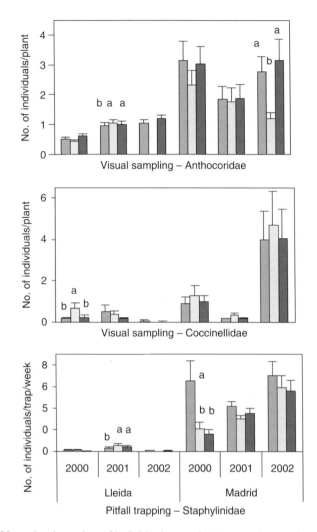

Fig. 15.2. Mean (±SE) number of individuals per plant, or per trap and week, in each treatment, location and year for anthocorids and coccinelids recorded in visual sampling and staphylinids caught in pitfall traps. Treatments represented are: transgenic *Bt* maize (*Bt+*), ■; isogenic maize with insecticide-seed treatment (*Bt–/I*), ▨; and isogenic maize with no insecticide (*Bt–*), ■; Within each year and location, values marked with different letters were significantly different (*P* < 0.05). (Reproduced from De la Poza *et al.*, 2005.)

non-*Bt* plots (De la Poza *et al.*, 2005; Farinós *et al.*, 2008). The only group of ground-dwelling predators that showed a significant reduction in abundance in *Bt* plots were the staphylinids in field trials carried out in Madrid in 2000. However, differences varied from one year to another and according to the location, with an increase in abundance in *Bt* plots in Lleida in 2001 (Fig. 15.2).

The results obtained from the Spanish farm-scale studies agree with data available from other field trials conducted in Europe and elsewhere (reviews in Romeis *et al.*, 2006; Marvier *et al.*, 2007). Habustova *et al.* (2006) did not find differences between MON810 *Bt* maize and non-transgenic maize plots (about 0.5 ha) in a 3-year field trial in the Czech Republic, when the abundance of plant and soil-dwelling predators were compared. A 3-year study conducted in maize fields in Germany found that, except for higher abundance of the mycetophagous/saprophagous beetle species *Cortinicara gibbosa* (Herbst) in MON810 *Bt* maize than in the isogenic control in the first year of the study, no other significant differences in arthropod communities between *Bt* maize and the control were observed (Eckert *et al.*, 2006). Likewise, Freier *et al.* (2004) conducted field trials to assess the abundance of selected arthropod taxa, including relevant herbivores and predators, during a 4-year period in paired (transgenic *Bt* versus non-*Bt*) fields of commercial size in Germany. This design, in addition to the low number of samples, introduced a large amount of variability in the results that, together with year variability, led the authors to conclude that differences between *Bt* and non-*Bt* fields were much smaller than variability of environmental factors. Field studies conducted in Italy indicated no effects of *Bt* maize on ground beetle assemblages (Lozzia, 1999). A short-term (1 year) study carried out in France found no effects of MON810 *Bt* maize on soil fauna (Naïbo, 2003). Furthermore, no differences in the abundance of the main plant-dwelling predators were found by Güllü *et al.* (2004) during a 2-year trial in Turkey. A similar conclusion was reported by Bourguet *et al.* (2002) for the predators *Orius insidiosus*, *Syrphus corollae* (F.), *Coccinella septempunctata* and *C. carnea* at two sites in France. Specific field trials on spiders carried out in comparative commercial fields in Germany during a 3-year period could not detect differences between the spider communities of transgenic and non-transgenic maize (Volkmar *et al.*, 2004; Ludy and Lang, 2006).

Parasitoids
Much less work has been devoted to study the potential effects of *Bt* maize on insect parasitoids. In field trials in France, Bourguet *et al.* (2002) found that ECB larvae collected from *Bt*176 maize displayed a lower level of parasitism by the tachinids *L. thompsoni* and *Pseudoperichaeta nigrolineata* (Walker) than did larvae collected from non-*Bt* maize. The authors argue that, since a direct effect of Cry1Ab toxin on tachinids seems unlikely, the parasitoids may have been indirectly affected by a reduction in the quantity and quality of ECB populations. No differences were found in parasitization rate and species composition when aphids were caught in the field in *Bt* and non-*Bt* plots at the peak of the aphid population (Pons and Stáry, 2003). Likewise, no differences in the number of aphid parasitoids on *Bt* versus non-*Bt* plots were reported by Bourguet *et al.* (2002).

Conclusions

Field trials are essential to enable a science-based discussion on the risks and safety of GM plants; hence, it is considered a key topic in the design of post-market monitoring plans in Europe. In the case of Spain, the only member state where Bt maize varieties have been cultivated at a commercial scale for the last 9 years, monitoring plans for Bt176 and MON810 varieties consider case-specific monitoring for the development of resistance in target insects and for the potential effects on non-target arthropods.

No consistent shifts in susceptibility for field populations of MCB and ECB have been reported after 9 years of Bt maize cultivation in Spain. Furthermore, different studies in Europe have shown that gene flow and the frequency of resistance alleles of ECB and MCB populations are compatible with the high-dose/refuge strategy. Nevertheless, monitoring should be continued to ensure early detection of resistance, and additional ecological, physiological and behavioural studies are required for the implementation of appropriate management decisions.

No detrimental effects of farm-scale Bt maize have been observed on the main predator taxa or on the whole functional group, suggesting that Bt maize could be compatible with natural enemies that are common in maize fields in Europe. Nevertheless, field trials in the same areas for longer periods are necessary to discard long-term potential cumulative effects.

References

Albajes, R., López, C. and Pons, X. (2003) Predatory fauna in corn fields and response to imidacloprid seed-treatment. *Journal of Economic Entomology* 96, 1805–1813.

Álvarez-Alfageme, F., Ferry, N., Castañera, P., Ortego, F. and Gatehouse, A.M.R. (2008) Prey mediated effects of Bt maize on fitness and digestive physiology of the red spider mite predator *Stethorus punctillum* Weise (Coleoptera: Coccinellidae) *Transgenic Research* 17, 943–954.

Andreadis, S.S., Álvarez-Alfageme, F., Sánchez-Ramos, I., Stodola, T.J., Andow, D.A., Milonas, P.G., Savopoulou-Soultani, M. and Castañera, P. (2007) Frequency of resistance to *Bacillus thuringiensis* toxin Cry1Ab in Greek and Spanish population of *Sesamia nonagrioides* (Lepidoptera: Noctuidae). *Journal of Economic Entomology* 100, 195–201.

Asín, L. and Pons, X. (1998) Aphid predators in maize fields. *IOBC/WPRS Bulletin* 21, 163–170.

Badosa, E., Moreno, C. and Montesinos, E. (2004) Lack of detection of ampicillin resistance gene transfer from BT176 transgenic corn to culturable bacteria under field conditions. *FEMS Microbiology Ecology* 48, 169–178.

Bourguet, D., Bethenod, M.T., Trouvé, C. and Viard, F. (2000) Gene flow in the European corn borer *Ostrinia nubilalis*: implications for the sustainability of transgenic insecticidal maize. *Proceedings of the Royal Society of London* 267, 1177–1184.

Bourguet, D., Chaufaux, J., Micoud, A., Delos, M., Naibo, B., Bombarde, F., Marque, G., Eychenne, N. and Pagliari, C. (2002) *Ostrinia nubilalis* parasitism and the field abundance of non-target insects in transgenic *Bacillus thuringiesis* corn (*Zea mays*). *Environmental and Biosafety Research* 1, 49–60.

Bourguet, D., Chaufaux, J., Seguin, M., Buisson, C., Hinton, J.L., Stodola, T.J., Porter, P., Cronholm, G., Buschman, L.L. and Andow, D.A. (2003) Frequency of alleles conferring resistance to Bt maize in French and US corn belt populations of *Ostrinia nubilalis. Theoretical and Applied Genetics* 106, 1225–1233.

Buès, R., Eizaguirre, M., Toubon, J.F. and Albajes, R. (1996) Différences enzymatiques et écophysiologiques entre populations de *Sesamia nonagrioides* Lefèbvre

(Lepidoptera: Noctuidae) originaires de l'ouest du Bassin Méditerranéen. *Canadian Entomologist* 128, 849–858.

Castañera, P. (1986) Plagas del maíz. *Proceedings of the IV Jornadas Técnicas de Maíz, Plagas*, Lleida, Spain, pp. 1–24.

Chaufaux, J., Segui, M., Swanson, J.J., Bourguet, D. and Siegfried, B.D. (2001) Chronic exposure of the European corn borer (Lepidoptera: Crambidae) to Cry1Ab *Bacillus thuringiensis* toxin. *Journal of Economic Entomology* 94, 1564–1570.

De la Poza, M., Pons, X., Farinós, G.P., López, C., Ortego, F., Eizaguirre, M., Castañera, P. and Albajes, R. (2005) Impact of farm-scale Bt maize on abundance of predatory arthropods in Spain. *Crop Protection* 24, 677–684.

De la Poza, M., Farinós, G.P., Ortego, F., Hernández-Crespo, P. and Castañera, P. (2006) Genetic structure of *Sesamia nonagrioides* populations: implications for Bt-maize resistance management. *IOBC/WPRS Bulletin* 29, 119–121.

Den Nijs, H. and Bartsch, D. (2004) Introgression of GM plants and the EU Guidance Note for monitoring. In: Den Nijs, H., Bartsch, D. and Sweet, J. (eds) *Introgression from Genetically Modified Plants into Wild Relatives and Its Consequences*. CAB International, Wallingford, UK, pp. 365–390.

Eckert, J., Schupham, I., Hothorn, L.A. and Gathmann, A. (2006) Arthropods on maize ears for detecting impacts of Bt maize on nontarget organisms. *Environmental Entomology* 35, 554–560.

EFSA (2004) Opinion of the scientific panel on genetically modified organisms on the use of antibiotic resistance genes as marker genes in genetically modified plants. *The EFSA Journal* 48, 1–18.

EFSA (2006) Guidance document of the scientific panel on genetically modified organisms for the risk assessment of genetically modified plants and derived food and feed. *The EFSA Journal* 99, 1–100.

Eizaguirre, M., López, C. and Albajes, R. (2004) Dispersal capacity in the Mediterranean corn borer *Sesamia nonagrioides*. *Entomologia Experimentalis et Applicata* 113, 25–34.

Eizaguirre, M., Albajes, R., López, C., Eras, J., Lumbierres, B. and Pons, X. (2006) Six years after the commercial introduction of Bt maize in Spain: field evaluation, impact and future prospects. *Transgenic Research* 15, 1–12.

Farinós, G.P., De La Poza, M., González-Núñez, M., Hernández-Crespo, P., Ortego, F. and Castañera, P. (2004a) Research programme to monitor corn borers resistance to Bt-maize in Spain. *IOBC/WPRS Bulletin* 27, 73–77.

Farinós, G.P., De la Poza, M., Hernández-Crespo, P., Ortego, F. and Castañera, P. (2004b) Resistance monitoring of field populations of the corn borers *Sesamia nonagrioides* and *Ostrinia nubilalis* after five years of Bt maize cultivation in Spain. *Entomologia Experimentalis et Applicata* 110, 23–30.

Farinós, G.P., de la Poza, M., Hernández-Crespo, P., Ortego, F. and Castañera, P. (2008) Diversity and seasonal phenology of aboveground arthropods in conventional and transgenic maize crops in Central Spain. *Biological Control* 44, 362–371.

Freier, B., Schorling, M., Traugott, M., Juen, A. and Volkman, C. (2004) Results of a 4-year survey and pitfall trapping in Bt maize and conventional maize fields regarding the occurrence of selected arthropod taxa. *IOBC/WPRS Bulletin* 27, 79–84.

Gathmann, A., Wirooks, L., Hothorn, L.A., Bartsch, D. and Schuphan, I. (2006) Impact of Bt maize pollen (MON810) on lepidopteran larvae on accompanying weeds. *Molecular Ecology* 15, 2677–2685.

Gómez-Barbero, M. and Rodríguez-Cerezo, E. (2006) *Economic Impact of Dominant GM Crops Worldwide: A Review*. Joint Research Centre, Institute for Prospective Technological Studies, Scientific and Technical Research series, Seville, Spain.

Gómez-Barbero, M., Berbel, J. and Rodríguez-Cerezo, E. (2008) *Bt* corn in Spain–the performance of the EU's first GM crop. *Nature Biotechnology* 26, 384–386.

González-Núñez, M., Ortego, F. and Castañera, P. (2000) Susceptibility of Spanish populations of the corn borers *Sesamia nonagrioides* (Lepidoptera: Noctuidae) and *Ostrinia nubilalis* (Lepidoptera: Crambidae) to a

Bacillus thuringiensis endotoxin. *Journal of Economic Entomology* 93, 459–463.

Güllü, M., Tatli, F., Kanat, A.D. and Islamoglu, M. (2004) Population development of some predatory insects on Bt and non-Bt maize hybrids in Turkey. *IOBC/WPRS Bulletin* 27, 85–91.

Habustova, O., Turanli, F., Dolezai, P., Ruzicka, V., Spitzer, L. and Hussein, H. (2006) Environmental impact of Bt maize: three years of experience. *IOBC/WPRS Bulletin* 29, 57–63.

James, C. (2007) *Global Status of Commercialized Biotech/GM Crops: 2007.* ISAAA Brief No. 37. ISAAA, Ithaca, New York.

Leniaud, L., Audiot, P., Bourguet, D., Frerot, B., Genestier, G., Lee, S.F., Malausa, T., Le Pallec, A.H., Souqual, M.C. and Ponsard, S. (2006) Genetic structure of European and Mediterranean maize borer populations on several wild and cultivated host plants. *Entomologia Experimentalis et Applicata* 120, 51-U4.

Lozzia, G.C. (1999) Biodiversity and structure of ground beetle assemblages (Coleoptera: Carabidae) in Bt corn and its effects on non target insects. *Bolletino di Zoologia Agraria e di Bachicoltura* Ser. II 31, 37–58.

Lozzia, G.C. and Rigamonti, I.E. (1998) Preliminary study on the effects of transgenic maize on non-target species. *IOBC/WPRS Bulletin* 21, 171–180.

Ludy, C. and Lang, A. (2006) A 3-year field-scale monitoring of foliage-dwelling spiders (Araneae) in transgenic *Bt* maize fields and adjacent field margins. *Biological Control* 38, 314–324.

Lumbierres, B., Albajes, R. and Pons, X. (2004) Transgenic Bt maize and *Rhopalosphum padi* (Homoptera: Aphididae) performance. *Ecological Entomology* 29, 309–317.

Marvier, M., McCreedy, C., Regetz, J. and Kareiva, P. (2007) A meta-analysis of effects of Bt cotton and maize on nontarget invertebrates. *Science* 36, 1475–1477.

Naïbo, B. (2003) Le maïs Bt préserve la faune utile. *Perspectives Agricoles* 293, 14–17.

Obrist, L.B., Dutton, A., Albajes, R. and Bigler, F. (2006) Exposure of arthropod predators to Cry1Ab toxin in Bt maize fields. *Ecological Entomology* 31, 143–154.

Pons, X. and Albajes, R. (2003) Control of maize pests with imidacloprid seed dressing treatment in Catalonia (NE Iberian Peninsula) under traditional crop conditions. *Crop Protection* 21, 943–950.

Pons, X. and Stáry, P. (2003) Spring aphid-parasitoid (Hom.; Aphididae, Hym., Braconidae) associations and interactions in a Mediterranean arable crop ecosystem, including Bt maize. *Journal of Pesticide Science* 76, 133–138.

Pons, X., Lumbierres, X., López, C. and Albajes, R. (2004) No effects of Bt maize on the development of *Orius majusculus*. *IOBC/WPRS Bulletin* 27, 131–136.

Pons, X., Lumbierres, B., López, C. and Albajes, R. (2005) Abundance of non-target pests in transgenic Bt-maize: a farm scale study. *European Journal of Entomology* 102, 73–79.

Raps, A., Kehr, J., Gugerli, P., Moar, W.J., Bigler, F. and Hilbeck, A. (2001) Detection of Cry1Ab in phloem sap of *Bacillus thuringiensis* corn in selected herbivore *Rhopalosiphum padi* (Homoptera: Aphididae) and *Spodoptera littoralis* (Lepidoptera: Noctuidae). *Molecular Ecology* 10, 525–533.

Rauschen, S., Ecker, J., Gathman, A. and Schuphan, I. (2004) Impact of growing Bt maize on cicadas: diversity, abundance and methods. *IOBC/WPRS Bulletin* 27, 137–142.

Roda, L. (2006) Evaluación del riesgo de los organismos modificados geneticamente y planes de seguimiento. In: Muñoz, E. (ed.) *Organismos Modificados Geneticamente.* Editorial Ephemera, Madrid, pp. 175–191.

Romeis, J., Meissle, M. and Bigler, F. (2006) Transgenic crops expressing *Bacillus thuringiensis* toxins and biological control. *Nature Biotechnology* 24, 63–71.

Saeglitz, C. Bartsch, D., Eber, S., Gathmann, A., Priesnitz, K.U. and Schupham, I. (2006) Monitoring the Cry1Ab sysceptibility of European corn borer in Germany. *Journal of Economic Entomology* 99, 1768–1773.

Sanvido, O., Widmer, F., Winzeler, M. and Bigler, F. (2005) A conceptual framework for the design of environmental post-market monitoring of genetically modified plants. *Environmental and Biosafety Research* 4, 13–27.

Volkmar, C., Traugott, M., Juen, A., Schorling, M. and Freier, B. (2004) Spider communities in Bt maize and conventional maize fields. *IOBC/WPRS Bulletin* 27, 165–170.

Zwallen, C., Nentwig, W., Bigler, F. and Hilbeck, A. (2000) Tritrophic interactions of transgenic *Bacillus thuringiensis* corn, *Anaphothrips obscurus* (Thysanoptera: Thripidae), and on the predator *Orius majusculus* (Heteroptera: Anthocoridae). *Environmental Entomology* 29, 846–850.

16 Monitoring *Bt* Resistance in the Field: China as a Case Study

K.L. He, Z.Y. Wang and Y.J. Zhang

The State Key Laboratory for Biology of Plant Disease and Insect Pests, Institute of Plant Protection, Chinese Academy of Agricultural Sciences, Beijing, China

Keywords: *Bt* crops, commercialization of Biotech crops, cotton, benefit, insect resistance, insect resistance management (IRM)

Summary

China was the first country to commercialize biotech crops with the commercialization of tobacco in the early 1990s. In 1997, it formally approved the commercialization of *Bt* cotton. Despite the availability of local cotton varieties expressing *Bacillus thuringiensis* (*Bt*), Monsanto's *Bt* cotton varieties were also introduced. The adoption of *Bt* cotton increased consecutively for the first 7 years and from 2004 to 2007 occupied more than 66% of the national total cotton acreage. More than 8 million smallholder, resource-poor cotton farmers derived significant productivity, economic, environmental, health and social benefits, including a substantial contribution to the alleviation of poverty in some areas, as a result of higher incomes from *Bt* cotton. In most regions, typical cotton farms are on a small scale. Cotton bollworm (CBW) *Helicoverpa armigera* (Hübner) is the major target insect of *Bt* cotton. Due to the growing of *Bt* cotton, in 2003 there was a reduction of more than 95,000t of pesticide. As an important component of the monitoring programme, baseline Cry1A(c) susceptibility data for the CBW were established; the timing of this coincided with the commercialization of *Bt* cotton. A systematic monitoring programme for CBW resistance to *Bt* cotton has been carried out and, to date, no field resistance has been detected. A refuge-based strategy has been employed in resistance management. Natural refuges have successfully been adopted in the Yellow River region, the largest cotton cropping area where mixed cropping with a wide variety of crops, such as maize, soybean, groundnut, oilseed rape, legumes, etc., is generally practised by the small farms. Non-*Bt* cotton refugia have been recommended in the Changjiang (Yangtse) River region and the North-western region since cotton is the sole host plant for pink bollworm (*Pectinophora gossypiella* (Saunders)) in the Changjiang River region.

Bt Crops: Acreage in China

China is one of the pioneers in the commercial cultivation of biotech crops. Transgenic tobacco, expressing coat protein (CP) genes from tobacco mosaic virus (TMV) for viral resistance, was the first commercial biotech crop to be planted in China during the early 1990s (Wang *et al.*, 2005). Transgenic tobacco variety NC89 that contains two CP genes from TMV and cucumber mosaic virus (CMV) was grown on approximately 10,000 ha during 1992 (Zhou and Fang, 1992). Two lines of transgenic tobacco, PK863 and PK893, both expressing insecticidal genes from *Bacillus thuringiensis* (Bt), were the first Bt crops globally to be commercialized and were commercialized in China during 1992 (Pray, 1999). These transgenic varieties of tobacco were grown on 1.6 million ha during 1997, and dominated the world production of biotech crops. However, during 1998, their cultivation was significantly decreased because of a decline in the market from international tobacco importers.

Prior to 1996, individual organizations were responsible for conducting and managing their own biotech crop field trials. Subsequently, guidelines (Safety Administration Implementation Regulation on Agricultural Biological Genetic Engineering, issued by the Ministry of Agriculture on 10 July 1996) were issued by the Chinese Ministry of Agriculture regulating all activities relating to field trials and commercialization of biotech crops in China; new biotech crop varieties were commercialized through the normal varietal release procedures. During the 1990s, field trials were conducted on a range of different biotech crops (such as tobacco, tomato, sweet pepper, chilli pepper, cotton, rice, wheat, maize, potato, cabbage, soybean, papaya, groundnut, melon, rapeseed, etc.) with a diversity of traits (such as viral and bacterial diseases and insect resistance, herbicide tolerance, improved shelf life, etc.; Jia, 1990, 1995, 1997; Huang *et al.*, 2002b). The biotech tobacco varieties listed above were not officially approved by the National Genetically Modified Organism (GMO) Biosafety Committee of China though commercialized for several years. Although numerous field trials have been conducted on insect-resistant transgenic rice and maize, such as *sck* rice, *sck* + *cry*1Ac rice, *cry*1Ab rice, and *cry*1Ab maize and *cry*1A maize since 1997 (Wu *et al.*, 2001; He *et al.*, 2003; Liu *et al.*, 2003, 2004; Zhu, 2006), as yet, none of these have been approved for commercialization. Therefore, in this chapter we will focus more on Bt cotton.

The first commercialized Bt cotton was the GK series of varieties expressing the *cry*1A gene for control of the cotton bollworm (CBW; *Helicoverpa armigera* (Hübner)); this variety was developed by the Biotechnology Research Institute of Chinese Academy of Agricultural Sciences (CAAS). Although Bt cotton has been developed in China using different transformation methods (i.e. pollen-tube pathway) to that used in the USA, the introduction of US Bt cotton was permitted while commercializing local Bt cotton. Bollgard Nuctn33B, an American Bt cotton variety expressing the *cry*1Ac gene that possess excellent efficacy against target insects including the CBW in field trials in northern China, was approved by the National GMO Biosafety Committee of China for commercialization in 1997 and was available to farmers in Hebei Province during 1998. Since then, the cropping areas of these Bt cotton varieties have spread rapidly throughout the cotton-growing regions of China (Table 16.1).

Initially, the extension of Bt cotton relied on varieties introduced by public research organizations, the extension system itself, and seeds sold by the state-run seed network. Based on biosafety issues, prior to 1999, Bt cotton could

Table 16.1. Adoption of *Bt* cotton in China. (From NSBC; Song and Wang, 2001; Statistic for National extension of dominating crop varieties 2002–2005; and Report on China Cotton Production Prosperity by Mao and Wang, 2007.)

	1997	1998	1999	2000	2001	2002	2003	2004	2005	2006	2007
Total cotton area	4.5	4.4	3.7	4.0	4.8	4.2	5.1	5.6	5.0	5.3	5.7
Local *Bt* cotton	0.034	0.09	0.1	0.23	0.39	0.96	1.43	2.47	2.41	n	n
Bt + CpTi	n	n	n	0.024	0.21	0.23	0.45	0.53	0.52	n	n
Monsanto	n	0.17	0.55	0.94	1.00	0.91	1.12	0.70	0.47	n	0.30
Total *Bt* cotton	0.034	0.26	0.65	1.2	1.6	2.1	3.0	3.7	3.4	3.5	3.8

n means there were no data or data were not available.

only be planted within the limits of Hebei, Henan, Shandong, Anhui and Shanxi provinces. However, because of the promising efficacy of *Bt* cotton for control of CBW, the adoption of *Bt* varieties has been the result of decisions by millions of small-scale resource-poor cotton farmers in other regions. This eliminates any doubt that biotech crops are able to play an important role for helping poor farmers in developing countries. After being introduced during 1999, the cropping areas for these *Bt* cotton varieties have increased substantially in all cotton cropping regions of China. The acreage of *Bt* cotton has continuously increased over the last decade, from almost nothing in 1996, to about 3.4 million ha in 2005, thus representing 67.2% of China's cotton area of 5.0 million ha. During the first 5 years of *Bt* cotton commercialization, Monsanto's *Bt* cotton was the dominant variety (Table 16.1). However, the domestic *Bt* cotton, especially after SGK cotton (containing *Bt* + CpTI), was approved for commercialization in 1999, increased significantly and became the dominant variety (Table 16.1).

On average, each household growing cotton accounts for 0.42 ha (Pray *et al.*, 2002), and it is estimated that about 8 million households have now adopted *Bt* cotton. By comparison to the non-*Bt* cotton varieties with good management and intensive pesticide spraying regimes, the yield increase and the price of *Bt* varieties were marginal. However, the cost saving and reduction in labour enjoyed by *Bt* cotton growers reduced the cost of producing 1 kg of cotton by 28%, from US$2.23 to US$1.61 (Huang *et al.*, 2002b; Dong *et al.*, 2004). On average, *Bt* cotton farmers apply pesticide only 6.6 times for non-bollworm pests per season compared to nearly 20 times for bollworm and non-bollworm pests per season by non-*Bt* cotton farmers. On a per hectare basis, the pesticide use for non-*Bt* cotton production is more than five times higher than for *Bt* cotton in terms of both quantity and expenditure (Huang *et al.*, 2001). More than 8 million small resource-poor cotton farmers have derived significant productivity, economic, environmental, health and social benefits, including a substantial contribution to the alleviation of poverty in some areas, as a result of higher incomes from *Bt* cotton.

Farming Practice

China is one of the leading producers and consumers of cotton in the world. The cotton cropping area, after the widespread adoption of *Bt* cotton, was at around 5 million ha for the last decade. This produces 4.5–6.5 million t of cotton per year, which meets about 20% of the annual global demand for cotton. In 2005, China utilized 9.5 million t of cotton with about 2.57 million t having to be imported to meet this demand.

Cotton production in China can be divided into five agroecological regions on the basis of growing-season variables crucial to cotton production, including rainfall, temperature and duration of the growing season. These regions are: Southern China Region (SCR), the Changjiang (Yangtse) River Region (CRR), the Yellow River Region (YRR), the Precocious Region (PR) and the North-western Region (NWR; Table 16.2). From the long history of cotton cultivation,

Table 16.2. Cotton agroecological production regions and associated climatic conditions in China. (Available at: http://www.cricaas.com.cn/digit/data/vall.htm)

	SCR[a]	CRR[b]	YRR[c]	NWR[d]	PR[e]
Latitude and longitude	<25°N 97–120°E	25°N–34°N 103–122°E	34°N–40°N 105–122°E	>35°N 76–105°E	>35°N 105–124°E
Annual rainfall (mm)	1200–2000	800–1200	500–800	~200	500–700
Monthly average temperature (°C)	14–26 (annually)	21–24 April–October	19–22 April–October	17–25 April–October	16–24 May–September
Accumulated temperature (≥10°C) (degree-day)	5500–9000	4800–5500	4200–4800	3450–4500	3300–3600
Frost-free days	300–365	220–300	180–230	170–230	150–170
Annual sunshine hours	2400–2600	1200–2400	2200–2900	2700–3300	1200–1300 (May–September)
Percentage of national cotton area[f]	<0.1	24	54	22	<0.1

[a]SCR including Guandong, Guangxi, Hainan, Taiwan, most of Yunnan, Xichang of Sichuan and southern parts of Guizhou and Fujian.
[b]CRR including the provinces of Zhejiang, Shanghai, Jiangxi, Hunan, Hubei, southern valley of Huai River in Jiangsu and Anhui, Sichuan (excluding Xichang), Nanyang and Xinyang of Henan, Southern Shaanxi, Northern Yunnan, Northern Guizhou and Northern Fujian.
[c]YRR including south of the Great Wall of Hebei, Shandong, Henan (excluding Nanyang and Xinyang), southern Shaanxi, middle Shaanxi, southern Gansu, northern Valley of Huai River in Jiangsu and Anhui, Beijing and Tianjin.
[d]NWR including Xinjiang Uygur Autonomous Region and north-western Gansu province.
[e]PR including Liaoning, middle Shaanxi, Northern Hebei (north of Great Wall), Northern Shaanxi, east Gansu.
[f]Data calculated from NSBC (2003).

distinct cotton cultivation systems have been adopted to meet the specific climatic and socio-economic requirements.

Following the rural reformation, smallholder family farming dominates China's agriculture. In SCR, CRR, YRR and PR, in particular, the average farm size of a typical cotton farmer is less than 1 ha, of which the cotton area is less than 0.5 ha (Pray et al., 2002). Collectively they cultivate 78% of China's cotton area and produce 65% of China's cotton. In contrast, large-scale farming is carried out in the NWR, which cultivates around 22% of China's cotton area and produces about 35% of China's cotton (NSBC, 2003). However, cotton farming is still labour-intensive in this region.

In CRR, YRR, PR and SCR, small-scale farms generally practise mixed crop farming, with a wide variety of crops, such as maize, soybean, groundnut, oilseed rape, legumes, etc. Typically, double cropping systems are adopted by cotton farmers in CRR, which are usually wheat and cotton, with a small amount of oilseed rape. A wheat–cotton intercropping system is usually practised for optimal use of the limited land available. The wheat–cotton intercropping system is also effective for biological control of the cotton aphid in this region since large populations of predators such as ladybirds on the wheat will move to the seedlings of cotton after wheat harvest. In YRR, cotton is often grown in monoculture, or, in some areas, is seeded after the harvest of wheat or as an intercrop with wheat.

Throughout the cotton-growing regions of China, improved cotton varieties are being planted. The National and/or Provincial Committees of Crop Varietal Testing and Certification are responsible for testing, certification, licensing and releasing of new varieties. During the 1990s, the public agricultural technical extension system played an important role in extension of new varieties. About 50% of seeds are usually sold by the state-run seed network and provincial seed companies. Foreign firms entered into China's seed market mainly in the form of joint ventures with organizations such as the provincial seed companies, the China National Seed Industry Group and the provincial seed-testing stations. For example, in 1998 Monsanto and Delta and Pine Land established a joint venture with the Hebei Provincial Seed Company to sell transgenic cotton seed (in 2007 Monsanto acquired Delta and Pine Land). However, private companies are now becoming more and more important in the seeds market, which contributed almost 70% of the seed supply in 2006, compared to less than 30% in 2001 (Fig. 16.1). As a consequence, the supply of seeds from the public sector has consecutively decreased over the last 6 years, by more than 40% to less than 13.0%. It is noteworthy that approximately 20% of cotton farmers still save seed for replanting, although this has been decreased from 30%. However, in 2007 there was a reversal in this trend with seed supply from the public sector increasing to 32.3% due to a new policy of governmental subsidy for planting improved and high-quality seed in eight provinces (planting 1 ha cotton with improved seed from the public sectors, farmers can get RMB ¥225 ≈ US$32 in the form of a direct subsidy; Mao and Wang, 2007).

Due to the diverse agroecological planting zones, a wide range of different varieties of cotton have been adopted across the country. This was driven by

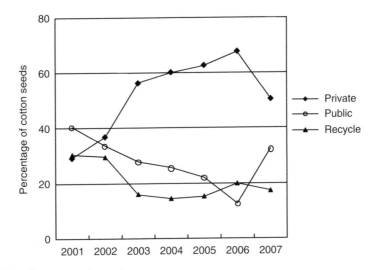

Fig. 16.1. Cotton seed suppliers.

farmer demand as they recognized the benefits of this cost-saving technology when it was introduced into YRR. As a result, new *Bt* cotton varieties that grow and seed well in these different environments are being developed in response to this demand. The National GMO Biosafety Committee of China has sped up the accessing and approving of new *Bt* cotton varieties. From only two varieties approved during 1997, 41 varieties were certified by 2003. However, to keep pace with this growing demand, the number of varieties certified by the National GMO Biosafety Committee of China increased to 112 in 2004 and 212 in 2005 (Mao and Wang, 2006). Many other varieties from provincial institutes are being grown, but some of these local varieties did not go through the official approval procedure set by the National GMO Biosafety Committee of China.

As the use of *Bt* cotton has spread, some research institutes at provincial and prefectural levels have also produced new *Bt* varieties by backcrossing CAAS and the Monsanto varieties into their own local varieties (Pray *et al.*, 2002). These varieties are also widespread in almost every province where they could be adopted. A research report on cotton seed inspection by the Cotton Research Institute, CAAS, showed that 194 and 215 insect-resistant varieties were planted in 2006 and 2007, respectively (Mao and Wang, 2006, 2007). Among them, 49 and 58 varieties have been approved by the National GMO Biosafety Committee of China and certified by National or Provincial Committees of Crop Varietal Testing and Certification; these occupy about 32.1% and 34.1% of the nation's total cotton acreage in 2006 and 2007, respectively. The remaining have the trait for bollworm resistance, but they have gone through the procedure of varietal testing and certification as non-*Bt* cotton varieties although some of them have been approved by the National GMO Biosafety Committee of China. These varieties account for approximately 32.4% and 32.0% of the nation's total cotton acreage in 2006 and 2007, respectively.

Pest Problems

Cotton-associated arthropod pests in China have been reviewed by Wu and Guo (2005). The main pest problems ranked by their economic importance are lepidopteran pest complex, spider mites (*Tetranychus cinnabarinus* (Boisduval), *T. truncates* Ehara, *T. turkestani* Ugavov et Nikolski and *T. dunhuangensis* Wang), cotton aphids (*Aphis gossypii* Glover, *A. atrata* Zhang, *A. medicaginis* Koch and *Acyrthosiphon gossypii* Mordviiko), mirids (*Adelphocoris suturalis* Jakovlve, *A. lineolatus* Geore, *A. fasciaticollis* Rueter, *Lygus lucorum* Meyer-Dür and *L. pratensis* L.), thrips (*Thrips tabaci* Lindemen, *T. flavus* Schrank and *Frankliniella intonsa* (Trybom)), whiteflies (*Bemisia argentifolii* Bellows & Perring and *B. tabaci* (Gennadius)) and the leafhoppers (*Empoasca biguttula* (Ishida) and *E. flavescens* (Fabricius)). The lepidopteran pest complex normally includes the CBW, the pink bollworm (*Pectinophora gossypiella* (Saunders)), the beet armyworm (*Spodoptera exigua* (Hüebner)), the Asian corn borer (*Ostrinia furnacalis* (Guenée)), the spiny bollworm (*Earias cupreoviridis* Walker, *E. fabil* Stoll and *E. insulana* (Boisduval)) and the cotton looper (*Anomis flava* (Fabricius)). Yield losses caused by major arthropod pests are detailed in Table 16.3. The CBW is the most widespread pest problem of cotton in China. Since CBW is the target pest of *Bt* cotton, we will focus more on this particular species.

The CBW occurs in all cotton-growing regions of China and infests cotton, wheat, maize, tomato, pea, etc. The optimal conditions for CBW development are at a temperature of 25–28°C and 70–90% relative humidity (RH). To adapt to these diverse geographic conditions of the different cotton-growing regions, the CBW can be distinguished into four genotypes, i.e. the tropical, subtropical, temperate and Xinjiang genotypes which are adapted to SCR, CRR, YRR and NWR, respectively.

Based on evidence from phenomenological studies, including the capture of CBW moths at high altitudes or over the Bohai Sea (Wu *et al.*, 1998), the analysis of pollen attached to the proboscis of moths (Xu *et al.*, 2000a), as well as gene exchanges between populations among different ecological

Table 16.3. Major arthropod pests of China cotton and their impact on estimated crop loss before and after *Bt* cotton commercialization. (From Wu and Guo, 2005.)

Pest complex	Estimated percentage of crop loss, 1994				Estimated percentage of crop loss, 2001			
	CRR	YRR	NWR	Total	CRR	YRR	NWR	Total
Cotton bollworm	5.97	10.13	0.32	6.60	3.91	2.92	0.32	2.47
Cotton aphids	0.44	1.46	1.27	1.03	0.33	0.84	0.25	0.52
Pink bollworm	2.02	0.14	0.00	0.85	0.90	0.01	0.00	0.27
Spider mites	0.58	0.61	0.76	0.63	1.22	0.39	0.22	0.59
Mirids	0.26	0.32	0.09	0.15	0.54	0.23	0.08	0.28
Other pests	0.34	0.30	0.18	0.29	0.23	0.23	0.28	0.24
Total	9.60	12.96	2.62	9.65	7.13	4.62	1.15	4.38

regions (Wu and Guo, 2000; Xu *et al.*, 2000b), it is clear that CBW is a long-distance migratory insect (Wu *et al.*, 2002a). Recently, radar observations have provided detailed parameters of their flight behaviour during this long-distance wind-borne nocturnal migration (Feng *et al.*, 2004, 2005). Although the diapausing pupae of CBW cannot overwinter in areas with cold winters, long-distance facultative migration during the summer East Asia monsoon of larvae belonging to the temperate genotype can occur, and so cause damage to cotton, maize, other crops and wild grass grown in these cold regions.

The CBW goes through three to four generations per year in YRR and NWR and four to six generations in CRR. During the first generation when adults are present, the cotton either has not been planted or the plant is too small to infest. Therefore, the CBW usually infests flax, pea, tomato in NWR, and wheat, pea, tomato, legumes, verbena grass, etc. in YYR and CRR. In general, the most damaging generations on cotton are the second and third generations in YRR, third and fourth generations in CRR, second generation in NWR and PR, and third to fifth generations in SCR. Late in the season, CBW usually moves out of the cotton fields and infests other host crops and wild grass. Generally this pest moves to the maize fields during the last generation; this is particularly so in YRR (Wang *et al.*, 2002). The population densities are highly dynamic during a given season. Rainfall during the growing season is an important climatic factor that impacts on regional population densities. High levels of rainfall in the early part of the season will significantly restrain early generation population development, which usually reaches outbreak levels in later generations that occur during the dry season (Wu *et al.*, 1993; Yang *et al.*, 1998, 2001a,b).

Reductions in Traditional Pesticide Use

In China, for control of CBW and pink bollworm, farmers used pesticides. Initially, they applied chlorinated hydrocarbons (such as dichlorodiphenyl-trichloroethane DDT)) until they were banned for environmental and health reasons in the early 1980s (Stone, 1988). In the mid-1980s farmers began to use organophosphates, but they were neither effective nor safe. In the early 1990s, these were replaced by the use of pyrethroids, which proved to be more effective and safer than organophosphates. However, resistance to this pesticide evolved very quickly. With rising pest resistance and pest populations, the application of different types of pesticides by cotton farmers in China rose sharply, even though some of the pesticides had little effect on the target pests (Huang *et al.*, 2002a). Due to the over-application and abuse of pesticides, a lot of problems gradually began to surface. First, the cost of pesticide applications was much higher. Chinese cotton farmers expend nearly US$500 million on pesticides annually (Pray *et al.*, 2002), which reduced the farmers' net income. Second, farmers spent a longer time in spraying pesticides. In the past, large numbers of farmers became sick from pesticide application each year (Qiao *et al.*, 2000); therefore, the farmers' health was greatly affected. Third, pesticide application seriously polluted the environment, particularly drinking water for local farmers in cotton-planting regions, where farmers depend on groundwater both for living and irrigation uses. Also, the biodiversities in the cotton-producing region were greatly affected.

In China, *Bt* cotton has been deployed for targeting CBW and pink bollworm since 1997 (Wu and Guo, 2004). Farmers are receiving the greatest benefits from *Bt* cotton's reduced pesticide need. Although farmers still have to spray in the early part of the season, pesticide spraying is substantially reduced or eliminated during the middle and late seasons due to the high expression of *Bt* toxin in *Bt* cottons (Zhang *et al.*, 2001). Therefore, the times for pesticides spraying were greatly decreased. Some Chinese cotton farmers reduced the number of the times they sprayed from 30 to three times, although, more often, the reduction was from 12 to three or four times (Pray *et al.*, 2001). The total decline in pesticide application has been impressive in China. In 1999, the reduction in pesticide use was about 20,000t. While in 2001, due to an increased area for *Bt* cotton, the reduction in pesticide use was about 78,000t (Pray *et al.*, 2002). However, due to the continuously increased *Bt* cotton-growing area, the reduction of pesticide application reached 95,000t in 2003 (Huang *et al.*, 2004). On average, compared to non-*Bt* cotton farmers, *Bt* cotton farmers reduced pesticide use by 49.9 kg ha^{-1} per person, equating to a reduction of 60% of pesticide use in cotton-producing regions (Huang *et al.*, 2002b). Reduction rates varied among provinces, and ranged from 20–50% in the Lower Reach of Changjiang River Basin to 70–80% in the Northern China cotton production region (Huang *et al.*, 2002a,c). Assuming 320,000 ha of *Bt* cotton, its uptake reduced pesticide use by at least 15,000t (Pray *et al.*, 2001).

With the rapid spread of transgenic *Bt* cotton, the net income of the cotton farmer was significantly improved. *Bt* cotton farmers increased their income by being able to reduce the use of both pesticides and labour. Due to the greatly reduced application of pesticides, while farmers averagely paid additional RMB ¥410 ha^{-1} for the higher *Bt* cotton seed price, the net income gains from *Bt* cotton production were RMB ¥1378 ha^{-1} (about US$166). Meanwhile, the cotton farmer also saved 41 days ha^{-1} as a direct result of decreased labour (Huang *et al.*, 2004).

By reducing the application of the pesticides, transgenic *Bt* cotton has also reduced the number of farmers poisoned. Pray *et al.* (2002) found that for non-*Bt* cotton producing farmers, the percentages of reported farmers being poisoned were particularly high, about 22% and 29% in the first 2 years. In contrast, for *Bt* cotton producing farmers, only 5–8% of farmers were reported to become sick from spraying pesticides.

Adoption of *Bt* cotton reduced pesticide application greatly, which also had a positive and significant impact on the environment, such as less air pollution, less water pollution, and greater biodiversity, including that of natural enemies (see Chapters 8 and 11, this volume).

Monitoring Resistance

Although resistance to *Bt* cotton in the field has not yet been detected in CBW in China, a *Bt*-resistant strain has been developed under laboratory selection, both in China (Liang *et al.*, 2000) as well as in other countries (Bird and Akhurst, 2004, 2005; Kranthi *et al.*, 2006). Field studies on CBW in China have also demonstrated that about 5–20% of naturally occurring larvae could survive on *Bt* cotton in the late season (Wu *et al.*, 2003). This indicates that *Bt* cotton does not produce a sufficiently high dose relative to the tolerance of this major pest.

The importance of baseline information in monitoring target insects evolving resistance to *Bt* crops, as well as how to establish this baseline have been well-documented in the literature (Stone and Sims, 1993; Wu *et al.*, 1999, 2002c; Bentur *et al.*, 2000; Gujar *et al.*, 2000; Glaser and Matten, 2003; Huang, 2006). In China, Wu *et al.* (1999) established baseline Cry1A(c) susceptibility data for CBW in 1997; this coincided with the time when *Bt* cotton had just began to be commercialized. They also measured geographic variability in the response of the CBW to Cry1A(c). Resistance monitoring has subsequently been conducted on a yearly basis. During 1998–2000, a total of 41 different CBW strains were sampled, and most of them were collected from *Bt* cotton planting regions. The ranges of IC_{50} values (concentration producing 50% inhibition of larval development to third instar) among different populations in 1998, 1999 and 2000 were 0.020–0.105, 0.016–0.099 and 0.016–0.080 µg ml^{-1}, respectively. Diagnostic concentration studies (IC_{99}) showed that the percentage of individuals reaching third instar ranged from 0% to 4.35%, with only eight of the 41 tested populations showing values above 0%. During 2001–2004, 53 different CBW strains from the *Bt* cotton planting regions were sampled. The range of concentration producing 50% inhibition of larval development to third instar ($IC_{50)}$ values) among different populations in 2001, 2002, 2003 and 2004 was 0.014–0.046, 0.010–0.062, 0.005–0.062 and 0.005–0.035 µg ml^{-1}, respectively. Diagnostic concentration studies (IC_{99}) showed that the percentage of individuals reaching third instar ranged from 0% to 9.09%, with only four among 53 tested strains showing values above 0%. Those data indicate that the susceptibility to Cry1Ac of the field populations sampled is not different from the baseline in 1997, and no movement towards resistance among the CBW strains is apparent (Wu *et al.*, 2002c, 2006).

Although a resistance monitoring system is already in place in China, it is not efficient for detecting frequencies of resistance as low as 0.005. In 2003–2005, the sensitivities of isofemale lines F_1/F_2 of the CBW collected from Anci County (Hebei Province, a complex cropping system with maize, soybean, groundnut and *Bt* cotton) and Xiajin County (Shandong Province, a concentrative cropping system mainly with maize and *Bt* cotton) in northern China to Cry1Ac toxin protein were monitored systematically by the diagnostic concentration method, and simulation models of adaptation of the CBW to *Bt* cotton were established (Li *et al.*, 2004). A conservative estimation was carried out and showed that the resistance gene frequencies to Cry1Ac in Xiajin population in 2003, 2004 and 2005 were 0, 0.00068 and 0.00233, respectively, which presented an obvious enhanced tendency, while for those in Anci population, the increase was negligible. It was suggested that the resistance gene frequency of the CBW increased in the region where the *Bt* cotton was cultivated in large acreages. The results of 2002–2005 showed that the relative average development rate (RADR) of the CBW larvae in F_1 test has increased significantly year by year ($P = 0.0001$), which was higher than those of two susceptible populations. Although these two field populations (Anci and Xiajin) still had not reached the mean RADR value of the resistant population, the results suggested that the tolerance level of CBW to Cry1Ac increased with the increasing length of time of *Bt* cotton cultivation. Moreover, the mean RADR value of the CBW larvae in F_1 test in Xiajin population was significantly higher than that of the Anci population, suggesting that the tolerance level to Cry1Ac increased more quickly in the region where larger acreages of *Bt* cotton were being cultivated. The totals of 99 and 83 isofemale lines from the Anci population and the Xiajin population, respectively, were tested respectively in F_1 and F_2 generations during 2002–2005. There was a significant positive correlation between RADR of CBW larvae in F_1 test and their offspring (F_2 test) in these two regions, which indicated that tolerance to Cry1Ac was heritable. Based on the results from Xiajin and Anci during 2002–2005, the values of heritability were 0.43 and 0.62 in Anci and Xiajin, respectively. The means proportion of selection were 20% and 30% in Anci and Xiajin, respectively, and the means of RADR of three genotypes were 0.30, 0.53 and 0.8. Based on a population size of 2000 moths, and a fecundity of 100 eggs per female, the simulation results of the QuCim model suggested that in Anci it would take 15 years for the resistant allele frequency to increase from the mean of 0.000357 to 0.5 and in Xiajin population it would take 11 years to increase from 0.0009 to 0.5, suggesting that the rate of evolution to *Bt* is quicker in an intensive *Bt* cotton-growing region than in a multiple-cropping system.

Other baselines have also been established for other insect pests, for example, the striped stem borer, *Chlio suppressalis*, to Cry1Ac and Cry1Ab toxins, and the intrapopulation variation in susceptibility to Cry1Ac in the main rice-growing areas; and Asian corn borer in main spring- and summer-maize-growing areas of China to Cry1Ab (Meng *et al.*, 2003; He *et al.*, 2005; Han *et al.*, 2006). As yet, however, *Bt* rice and *Bt* maize have not been approved for commercialization.

Implementation of Integrated Resistance Management

Since the adoption of *Bt* crops, various resistance management strategies have been proposed and debated in an attempt to address concerns over the potential for resistance development and to preserve the utility of *Bt* crops (McGaughey and Whalon, 1992; Tabashnik, 1994; Alstad and Andow, 1995). The refuge plus high-dose strategy is currently the most widely accepted approach for *Bt* crop insect resistance management. Although the high-dose/refuge strategy, adopted in the USA and some other countries, seems reasonable and effective, it is difficult to implement in small farm and cropping systems in YRR and CRR, since farming in these regions is quite different from the large-scale farming in the USA and Australia. Mixed plantings of cotton, maize, soybean and groundnut are common (Wu and Guo, 2005). Because of the complexity of managing individual fields in the same region, educating and monitoring more than 10 million farmers made the use of the refuge strategy difficult to implement (Ru *et al.*, 2002; Wu and Guo, 2005). In consideration of the fact that, among others, the small farm and cotton plot sizes may allow non-cotton crops to function as refugia, China does not mandate the conventional cotton refuge strategy for insect-resistant management, as occurs in the USA and Australia (Wu *et al.*, 2002b). A preliminary field survey of functional refuge for CBW within the *Bt* cotton-growing areas in North China has been conducted (Wu *et al.*, 2002b, 2004). Field trials indicate that both soybean and groundnut can supply refuges for the second and third generations of the CBW. As the most abundant crop, maize is planted widely and has a long sowing date, from April to June in YRR. The varied planting time for maize in YRR increases the overlap between the moth oviposition period and the occurrence of maize in the silk stage, which could serve as the refuge for the third and fourth generations of the CBW (Wang *et al.*, 2002). A planting system consisting of wheat, soybean and/or groundnut, maize and *Bt* cotton, which can supply susceptible refugees for the CBW all season long, is considered to be a good low-risk farming model (Wu, 2007). A planting system consisting of wheat, soybean and/or groundnut, and *Bt* cotton, which can supply enough susceptible refuges for the second and third generations, but could not supply sufficient susceptible refuges for the fourth generation, is considered as the middle-risk farming model. A planting system consisting of wheat, *Bt* cotton and maize, which can supply enough susceptible refuges for the fourth generation, but could not supply sufficient susceptible refuges for the second and third generations, is considered to be a high-risk farming model. Therefore, the low-risk farming model is recommended in YRR. Continuous monitoring showed this strategy works well because the resistance level of the CBW is still very low in China (Wu *et al.*, 2002c, 2006; Li *et al.*, 2004).Thus, this low-risk farming strategy has been recommended for areas where farmers exclusively grow cotton without natural refuge from other crops (Wu *et al.*, 2002b, 2004).

In addition, gene flow derived from migration of the CBW over a large area is also an important factor that delays the evolution of *Bt* resistance. It is possible that immigrant individuals from non-*Bt*-cotton-growing areas account for a large proportion of the population during some years owing to the decrease of moths from the *Bt* cotton in the local area (Wu *et al.*, 2002a, 2004).

In the NWR of China, the growing of *Bt* cotton is expected to significantly increase in the near future. Since the cropping system is similar to that in the USA, the growing of non-*Bt* cotton to serve as refugia will be required. In CRR, another important target pest of *Bt* cotton is pink bollworm; in contrast to CBW, cotton is the sole host plant in the region for pink bollworm, and monocultures of lines that express *Bt* toxins continuously are likely to select intensely for resistance. The lower efficacy of *Bt* cotton against pink bollworm in China indicates that further studies on *Bt* cotton deployment and resistance management strategies in CRR are necessary (Wan *et al.*, 2004).

A challenge in China is the commercialization of *Bt* maize, which is an important issue related to the pest management of cotton. Although *Bt* maize is at present not grown in China, its commercialization is currently under consideration by the Chinese government. Some scientists suggest that the commercialization of *Bt* maize, a key refuge for CBW will be lost and resistance to *Bt* cotton may evolve more rapidly in China (Wu and Guo, 2005).Thus, a mutual refuge for *Bt* cotton and *Bt* maize should be considered. In reality, in northern China, small farms grow a variety of crops at any one time, for example, soybean, groundnut and vegetables, plus some wild grass – this type of cropping system will thus provide a natural refuge for the CBW, especially for the first generation of the CBW which mainly feed on winter wheat, for the susceptible population of overwintering adults from non-*Bt* hosts mating with surviving adults from *Bt* crops, and producing heterozygotes on winter wheat which still cannot survive on *Bt* crops. Thus, it is anticipated that this type of cropping system will delay the evolution of resistance within the pest population.

Extensive laboratory and field trials have been conducted for evaluation of the efficiency of transgenic rice and maize on target lepidopteran pests and potential ecological risks on non-target arthropods (He *et al.*, 2003; Wang *et al.*, 2004; Chen *et al.*, 2006). Studies for developing a proactive insect resistance management programme for transgenic rice and maize in the future are under consideration to ensure the sustainable use of transgenic rice and maize in China. This information is essential to managing resistance in pest populations and especially in assessing whether a field control failure was due to actual resistance or other factors affecting expression of the *Bt* protein.

References

Alstad, D.N. and Andow, D.A. (1995) Managing the evolution of resistance to transgenic plants. *Science* 268,, 1894–1896.

Bentur, J.S., Cohen, M.B. and Gould, F. (2000) Variation in performance on Cry1Ab-transformed and non-transgenic rice among populations of *Scirpophaga incertulas* (Lepidoptera: Pyralidae) from Luzon Island, Philippines. Journal of Economic Entomology 96, 1773–1778.

Bird, L. and Akhurst, R.J. (2004) Relative fitness of Cry1A resistant and susceptible *Helicoverpa armigera* (Lepidoptera: Noctuidae) on conventional and transgenic cotton. *Journal of Economic Entomology* 97, 1699–1709.

Bird, L. and Akhurst, R.J. (2005) Fitness of Cry1A-resistant and susceptible *Helicoverpa armigera* (Lepidoptera: Noctuidae) on transgenic cotton with reduced levels of Cry1Ac. *Journal of Economic Entomology* 98, 1311–1319.

Chen, M., Ye, G.Y., Yao, H.W., Chen, X.X., Shen, Z.C., Hu, C. and Datta, S.K. (2006)

Field assessment of the effects of transgenic rice expressing a fused gene of cry1Ab and cry1Ac from *Bacillus thuringiensis* Berliner on nontarget planthopper and leafhopper populations. *Environmental Entomology* 35, 127–134.

Dong, H.Z., Li, W.J., Tang, W. and Zhang, D.M. (2004) Development of hybrid Bt cotton in China – a successful integration of transgenic technology and conventional techniques. *Current Science* 86, 778–782.

Feng, H.Q., Wu, K.M., Cheng, D.F. and Guo, Y.Y. (2004) Northward migration of *Helicoverpa armigera* (Lepidoptera: Noctuidae) and other moths in early summer observed with radar in northern China. *Journal of Economic Entomology* 97, 1874–1883.

Feng, H.Q., Wu, K.M., Ni, Y.X., Cheng, D.F. and Guo, Y.Y. (2005) High-altitude windborne transport of *Helicoverpa armigera* (Lepidoptera: Noctuidae) and other moths in mid-summer in northern China. *Journal of Insect Behavior* 18, 335–349.

Glaser, J.A. and Matten, S.R. (2003) Sustainability of insect resistance management strategies for transgenic Bt corn. *Biotechnology Advances* 22, 45–69.

Gujar, G.T., Kumari, A., Kalia, V. and Chandrashekar, K. (2000) Spatial and temporal variation in susceptibility of the American bollworm, *Helicoverpa armigera* (Hübner) to *Bacillus thuringiensis var. kurstaki* in India. *Current Science* 78, 995–1001.

Han, L.Z., Wu, K.M., Peng, Y.F., Wang, F. and Guo, Y.Y. (2006) Evaluation of transgenic rice expressing Cry1Ac and CpTI against *Chilo suppressalis* and intrapopulation variation in susceptibility to Cry1Ac. *Environmental Entomology* 35, 1453–1459.

He, K.L., Wang, Z.Y., Zhou, D.R., Wen, L.P., Song, Y.Y. and Yao, Z.Y. (2003) Evaluation of transgenic Bt corn for resistance to the Asian corn borer (Lepidoptera: Pyralidae). *Journal of Economic Entomology* 96, 935–940.

He, K.L., Wang, Z.Y., Wen, L., Bai, S.X., Ma, X. and Yao, Z.Y. (2005) Determination of baseline susceptibility to Cry1Abprotein for Asian corn borer (Lepidoptera: Crambidae). *Journal of Applied Entomology* 129, 407–412.

Huang, F.N. (2006) Detection and monitoring of insect resistance to transgenic Bt crops. *Insect Science* 13, 73–84.

Huang, J.K., Hu, R.F., Rozelle, S., Qiao, F.B. and Pray, C.E. (2001) Small holders, transgenic varieties, and production efficiency: the case of cotton farmers in China. Working Paper No. 01-015, University of California, Davis, California. March 2001, p. 30.

Huang, J., Hu, R., Rozelle, S., Qiao, F. and Pray, C.E. (2002a) Small holders, transgenic varieties, and production efficiency: the case of cotton farmers in China. In: Evenson, R.E., Santaniello, V. and Zilberman, D. (eds) *Economic and Social Issues in Agricultural Biotechnology.* CAB International, Wallingford, UK, pp. 393–407.

Huang, J.K., Rozelle, S., Pray, C.E. and Wang, Q.F. (2002b) Plant biotechnology in China. *Science* 295, 674–677.

Huang, J.K., Hu, R.F., Rozelle, S., Qiao, F.B. and Pray, C.E. (2002c) Transgenic varieties and productivity of small holder cotton farmers in China. Australian *Journal of Agricultural and Resource Economics,* 46, 367–387.

Huang, J.K., Hu, R.F., Meijl, H. van and Tongeren, F. van. (2004) Plant biotechnology in China: public investments and impacts on farmers. In: *New Directions for a Diverse Planet.* Proceedings for the 4th International Crop Science Congress, Brisbane, Australia, 26 Sep - 1 Oct, 2004. Available at: http://www.cropscience.org.au

Jia, S.R. (1990) Progress of genetic transformation in higher plants. Jiangsu. *Journal of Agricultural Sciences* 6, 44–47.

Jia, S.R. (1995) Progress in plant genetic engineering. *News Letter of High-tech* 5, 1–2.

Jia, S.R. (1997) Current development of field tests on GMOs and guidelines in China. In: Matsui, S., Miyazaki, S. and Kasamo, K. (eds) *The Biosafety Results of Field Tests of Genetically Modified Plants and Microorganisms.* Japan International Research Center for Agricultural Sciences, Tsukuba, Ibaraki, Japan.

Kranthi, K.R., Dhawad, C.S., Naidu, S.R., Mate, K., Behere, G.T., Wadaskar, R.M. and Kranthi, S. (2006) Inheritance of resistance in Indian *Helicoverpa armigera* (Hübner) to Cry1Ac toxin of *Bacillus thuringiensis.* *Crop Protection* 25, 119–124.

Li, G., Wu, K., Gould, F., Feng, H., He, Y. and Guo, Y. (2004) Frequency of Bt resistance genes in *Helicoverpa armigera* populations from the Yellow River cotton-farming region of China. *Entomologia Experimentalis et Applicata* 112, 135–143.

Liang, G., Tan, W. and Guo, Y. (2000) Study on screening and mode of inheritance to Bt transgenic cotton in *Helicoverpa armigera.* *Acta Entomologica Sinica* 43, 57–62.

Liu, Z.C., Ye, G.Y., Hu, C. and Datta, S.K. (2003) Impact of transgenic indica rice with a fused gene of cry1Ab/cry1Ac on the rice paddy arthropod community. *Acta Entomologica Sinica* 46, 454–465.

Liu, Z.C., Ye, G.Y. and Hu, C. (2004) Effects of *Bacillus thuringiensis* transgenic rice and chemical insecticides on arthropod communities in paddy fields. *Chinese Journal of Applied Ecology* 15, 2309–2314.

McGaughey, W.H. and Whalon, M.E. (1992) Managing insect resistance to *Bacillus thuringiensis* toxins. *Science* 258, 1451–1455.

Mao, S.C. and Wang, X.H. (2006) Report on inspection for national cotton varieties planted in 2006. Report on China Cotton Production Prosperity. No. 105. Available at: http://www.cricaas.com.cn/jingqibaogao/ccppi105.htm

Mao, S.C. and Wang, X.H. (2007) Report on inspection for national cotton varieties planted in 2007. Report on China Cotton Production Prosperity. No. 128. Available at: http://www.cricaas.com.cn/jingqibaogao/ccppi128.htm

Meng, F., Wu, K., Gao, X., Peng, Y. and Guo, Y. (2003) Geographic variation in susceptibility of *Chilo suppressalis* (Lepidoptera: Pyralidae) to *Bacillus thuringiensis* toxins in China. *Journal of Economic Entomology* 96, 1838–1942.

National Statistic Bureau of China (NSBC) (2003) National cotton acreage since 1994. Available at: http://www.cricaas.com.cn/frame/digit.htm

Pray, C.E. (1999) Public and private collaboration on plant biotechnology in China. *AgBioForum* 2, 48–53.

Pray, C.E., Ma, D.M., Huang, J.K. and Qiao, F.B. (2001) Impact of Bt cotton in China. *World Development* 29, 813–825.

Pray, C.E., Huang, J.K., Hu, R.F. and Rozelle, S. (2002) Five years of Bt cotton in China – the benefits continue. *The Plant Journal* 31, 423–430.

Qiao, F., Huang, J. and Rozelle, S. (2000) Pesticide and Human Health: The Story of Rice in China. Working Paper. Department of Agricultural and Resource Economics, University of California, Davis, California.

Ru, L.J., Zhao, J.Z. and Rui, C.H. (2002) A simulation model for adaptation of cotton bollworm to transgenic Bt cotton in northern China. *Acta Entomologica Sinica* 45, 153–159.

Song, X.X. and Wang, S.M. (2001) Status and evaluation on the extension of cotton varieties in the production in China in the past 20 years. *Cotton Science* 13, 315–320.

Stone, B. (1988) Developments in agricultural technology. *The China Quarterly* 116, 767–822.

Stone, T.B. and Sims, S.R. (1993) Geographic susceptibility of *Heliothis virescens* and *Helicoverpa zea* (Lepidoptera: Noctuidae) to *Bacillus thuringiensis*. *Journal of Economic Entomology* 86, 989–994.

Tabashnik, B.E. (1994) Evolution of resistance to *Bacillus thuringiensis*. *Annual Review Entomology* 39, 47–79.

Wan, P., Wu, K., Huang, M. and Wu, J. (2004) Seasonal pattern of infestation by pink bollworm *Pectinophora gossypiella* (Saunders) in field plots of Bt transgenic cotton in the Yangtze River Valley of China. *Crop Protection* 23, 463–467.

Wang, D.Y., Wang, Z.Y., He, K.L., Cong, B., Bai, S.X. and Wen, L.P. (2004) Temporal and spatial expression of Cry1Ab toxin in transgenic Bt corn and its effects on Asian corn borer, *Ostrinia furnacalis* (Guenée). *Scientia Agricultura Sinica* 37, 1155–1159.

Wang, H.G., Ma, H.J., An, D.C. and Zhang, M. (2005) *Developing Biotechnology and Leading Bioeconomy*. China Medical Science and Technology Press, Beijing, China, pp. 181–184.

Wang, Z.Y., He, K.L., Wen, L.P., Zhang, G.Y. and Zheng, L. (2002) Spatial and temporal distributions of the fourth generation cotton bollworm eggs on summer corn seeded at different times in north China. *Agricultural Sciences in China* 1, 96–100.

Wu, G., Cui, H.R., Shu, Q.Y., Ye, G.Y., Xie, X.B., Xia, Y.W., Gao, M.W. and Altosaar, I. (2001) Expression patterns of *cry*1Ab gene in progenies of 'Kemingdao' and the resistance to striped stem borer. *Scientia Agricultura Sinica* 34, 496–501.

Wu, K., Guo, Y. and Wu, Y. (2002a) Ovarian development of adult females of cotton bollworm and its relation to migratory behavior around Bohai Bay of China. *Acta Ecologica Sinica* 22, 1075–1078.

Wu, K., Guo, Y. and Gao, S. (2002b) Evaluation of the natural refuge function for *Helicoverpa armigera* (Lepidoptera: Noctuidae) within

Bacillus thuringiensis transgenic cotton growing areas in northern China. *Journal of Economic Entomology* 95, 832–837.

Wu, K., Guo, Y., Lv, N., Greenplate, J.T. and Deaton, R. (2002c) Resistance monitoring of *Helicoverpa armigera* (Lepidoptera: Noctuidae) to Bt insecticidal protein in China. *Journal of Economic Entomology* 95, 826–831.

Wu, K., Guo, Y., Lu, N., Greenplate, J.T. and Deaton, R. (2003) Efficacy of transgenic cotton containing a cry1Ac gene from *Bacillus thuringiensis* against *Helicoverpa armigera* (Lepidoptera: Noctuidae) in northern China. *Journal of Economic Entomology* 96, 1322–1328.

Wu, K., Guo, Y. and Head, G. (2006) Resistance monitoring of *Helicoverpa armigera* (Lepidoptera: Noctuidae) to Bt insecticidal protein during 2001–2004 in China. *Journal of Economic Entomology* 99, 893–898.

Wu, K.M. (2007) Environmental impacts of Bt cotton commercialization and strategy for risk management in China. *Journal of Agricultural Biotechnology* 15, 1–5.

Wu, K.M. and Guo, Y.Y. (2000) The coordinated development and analysis of contributing factors of cotton bollworm resistance to insecticides in Round-Bohai Bay Region. *Acta Phytophylacica Sinica* 27, 173–178.

Wu, K.M. and Guo, Y.Y. (2004) Changes in susceptibility to conventional insecticides of a Cry1Ac-selected population of *Helicoverpa armigera* (Hübner) (Lepidoptera: Noctuidae). *Pest Management Science* 60, 680–684.

Wu, K.M. and Guo, Y.Y. (2005) The evolution of cotton pest management practices in China. *Annual Review of Entomology* 50, 31–52.

Wu, K.M., Xu, G. and Guo, Y.Y. (1998) Observations on migratory activity of cotton bollworm moths across the Bohai Gulf in China. *Acta Phytophylacica Sinica* 25, 337–340.

Wu, K.M., Guo, Y.Y. and Lv, N. (1999) Geographic variation in susceptibility of *Helicoverpa armigera* (Lepidoptera: Noctuidae) to *Bacillus thuringiensis* insec-

ticidal protein in China. *Journal of Economic Entomology* 92, 273–278.

Wu, K.M., Feng, H.Q. and Guo, Y.Y. (2004) Evaluation of maize as a refuge for management of resistance to Bt cotton by *Helicoverpa armigera* in the Yellow River cotton farming region of China. *Crop Protection* 23, 523–530.

Wu, Z., Xue, Z. and Zhang, Z. (1993) A study on the nature population life table of cotton bollworm and its application. *Acta Ecologica Sinica* 13, 185–193.

Xu, G., Guo, Y., Wu, K. and Jiang, J. (2000a) On the mark-release techniques of *Helicoverpa armigera* (Hübner). *Acta Gossypii Sinica* 12, 245–250.

Xu, G., Guo, Y. and Wu, K. (2000b) Allozyme variations within and among five geographic populations of *Helicoverpa armigera* (Hübner). *Acta Entomologica Sinica* 43, 63–69.

Yang, Y.T., Wang, D.H. Zhu, M.H. and Yi, H.J. (1998) Studies on the relationship between soil moisture and the occurrence of cotton bollworm. *Acta Gossypii Sinica* 10, 210–215.

Yang, Y.T., Wang, D.H. and Zhu, M.H. (2001a) Damage and economic threshold of the third generation of cotton bollworm in the Yangtze River cotton region of Jiangsu province. Chinese. *Journal of Applied Ecology* 12, 86–90.

Yang, Y.T., Wang, D.H and Zhu, M.H. (2001b) Effects of soil characteristics on the occurrence on *Helicoverpa armigera* (Hübner) and its region division. *Acta Ecologica Sinica* 21, 959–963.

Zhang, Y., Wu, K. and Guo, Y. (2001) On the spatial-temporal expression of the contents of Bt insecticidal protein and the resistance of Bt transgenic cotton to cotton bollworm. *Acta Phytophylacica Sinica* 28, 1–6.

Zhou, R.H. and Fang, R.X. (1992) Field testing and applying of novel virus resistant tobacco variety-transgenic NC89. *Crops* 1, 25.

Zhu, Z. (2006) Development and biosafety assessment of insect-resistant transgenic rice. In: *Development of Agricultural GMOs. Proceedings of the first International Furum of Agricultural GMOs*. September 20, 2006, Beijing, China, pp. 5–6.

17 Current Status of Crop Biotechnology in Africa

D. GEORGE

School of Biology, Institute for Research on Environment and Sustainability, Newcastle University, Newcastle upon Tyne, UK and Ministry of Agriculture, Department of Agricultural Research, Gaborone, Botswana

Keywords: Gene revolution, drought, politics, infrastructure, famine, biotechnology, yield, population

Summary

The continent of Africa as a whole is faced with a number of critical challenges as Africa's level of crop production has not kept pace with population growth. Hunger and malnutrition are prevalent. Good agricultural land is declining as a result of poor farming practices, and pests and diseases cause high yield losses. Water is limited and is a major constraint on agricultural production in Africa. In the light of these problems, there is much optimism regarding the potential role that agricultural biotechnology (genetically modified (GM) crops) could play in reviving agriculture in Africa. Proponents of this revolution, also known as the 'Gene Revolution', contend that the technology has the potential to address problems not solved by conventional means. Conversely critics of the technology claim that GM crops will affect human health, non-target organisms and also damage the environment. Despite this controversy GM crops are been grown by many countries in the world to address agricultural challenges; however, South Africa is the only country in Africa commercially growing GM crops although field trials on GM crops have been conducted in Egypt, Kenya and Zimbabwe. It is imperative that Africa is not left behind in the new Agricultural Revolution, the Gene Revolution. While it is not suggested that GM crops are a panacea for Africa's problems, it may still bring immense befits to the people. Unfortunately, there are a number of constraints to biotechnological development in Africa that need to be resolved. These include a lack of resources, political instability, lack of networks, intellectual property right law, trade imbalances, the current legislative framework, the actual crops chosen for modification and biosafety issues.

Introduction

The role of agriculture is to produce food for people. In Africa, food crises of catastrophic proportions have been a permanent feature of many countries for a number of years. This has been exacerbated by drought, flood and political instability. Africa lost out on the Green Revolution that brought unprecedented benefits to Asia, and now there is hope that the 'Gene Revolution' or agricultural biotechnology could help alleviate some of its problems.

Agricultural biotechnology encompasses not only the production of genetically modified (GM) crops, but also micropropagation, marker-assisted breeding, genomics and bioinformatics (De Vries and Toenniessen, 2001). There is considerable optimism regarding the potential role that biotechnology could play in reviving agriculture in Africa. Proponents of this revolution contend that the technology has the potential to address problems currently not solved by conventional means (Huang *et al.*, 2002, http://www.ids.ac.uk). It is claimed that GM crops have the potential to be healthier, for example, cancer-fighting tomatoes (Science daily, 2002), more nutritious (golden rice) and more productive (insect-resistant crops) than organisms derived by conventional means. Conversely, critics of the technology claim that GM crops will affect human health and non-target organisms and also damage the environment. It is also claimed that transgenic crops will only benefit large multinational corporations. However, the economic evidence available suggests that GM crops do not only support large firms but that the benefits created by transgenic crops are shared by consumers, technology suppliers and adopting farmers (Raney, 2006; James, 2008). In 2007, 12 million farmers benefited from biotech crops, and 11 million of these farmers were resource-poor farmers from developing countries (James, 2008).

GM crop uptake is more rapid than any other agricultural technology in history (Raney, 2006) suggesting that farmers clearly do see obvious benefits. The area planted with transgenic crops in their first year of commercialization was 1.7 million ha, it increased significantly year on year to reach some 114.3 million ha in 2007 (Table 17.1 and Fig. 17.1). The dominant biotech crops are soybean, maize, cotton and canola (Fig. 17.2). The dominant trait of biotech crops is herbicide tolerance (HT) which constitutes 69% of the 102 million ha of the total transgenic area, followed by insect resistance (IR) with 19%, and both HT and IR stacked in the same plant with 13.1% (James, 2007). The estimated value of transgenic crops grown globally in 2006 was US$6.15 billion. This was comprised of US$2.68 billion for biotech soybean, US$2.39 billion for biotech maize, US$0.87 billion for biotech cotton and US$0.21 billion for biotech canola (James, 2007). Clearly these crops have delivered economic benefits to farmers.

In Africa, biotechnology is still in the embryonic stage. South Africa (SA) is the only country on the continent commercially growing transgenic crops and it has a strong private sector involved in biotechnological research. The dominant biotech crops grown in South Africa are the insecticidal *Bacillus thuringiensis* (*Bt*) Cry protein expressing cotton and *Bt* maize. Apart from South Africa countries such as Egypt, Kenya, Zimbabwe and Nigeria are currently conducting trials on transgenic crops but biotechnological development in Africa is constrained by a number of factors. Problems include a lack of resources, choice of commercialized GM crops suited to African agriculture, political instability, lack of research networks and lack of intellectual property rights (IPR) legislation.

The aim of this chapter is to give a brief overview of the status of biotechnology in Africa. It is divided into a number of sections looking at challenges

Table 17.1. Global area of biotech crops in 2007: by country.

Rank	Country	Area (million hectares)	Biotech crops
1	USA	57.7	Soybean, maize, cotton, squash, papaya, lucerne
2	Argentina	19.1	Soybean, maize, cotton
3	Brazil	15.0	Soybean, cotton
4	Canada	7.0	Canola, maize, soybean
5	India	6.2	Cotton
6	China	3.8	Cotton, tomato, poplar, petunia, papaya, sweet pepper
7	Paraguay	2.6	Soybean
8	South Africa	1.8	Maize, soybean, cotton
9	Uruguay	0.5	Soybean, maize
10	The Philippines	0.3	Maize
11	Australia	0.1	Cotton
12	Spain	0.1	Maize
13	Mexico	0.1	Cotton, soybean
14	Colombia	<0.1	Cotton, carnation
15	Chile	<0.1	Maize, soybean, canola
16	France	<0.1	Maize
17	Honduras	<0.1	Maize
18	Czech Republic	<0.1	Maize
19	Portugal	<0.1	Maize
20	German	<0.1	Maize
21	Slovakia	<0.1	Maize
22	Romania	<0.1	Maize
23	Poland	<0.1	Maize

facing agriculture in Africa, the status of biotechnology in Africa, constraints to biotechnological development in Africa and concerns raised regarding the development and deployment of transgenic crops. Finally, it considers the future for this technology in the context of African agriculture.

Agricultural Challenges in Africa: Can Biotechnology Help to Solve Them?

There are a number of critical challenges facing Africa today. Primarily there is increasing demand for food as a result of an increasing population. Crop production in Africa has fallen to about 1 t ha^{-1}; this is similar to the average productivity of British farmers during the reign of the Roman Empire! (55 BC to AD 500; Conway, 2003). Current annual production shortfalls of cereals are met by imports of approximately 21.5 million t and it is predicted that these short falls will increase tenfold by 2050 (Persley, 2000). Africa's low productivity is due to a number of reasons including erratic rainfall across the continent, often leading to severe drought resulting in food crises, and severe

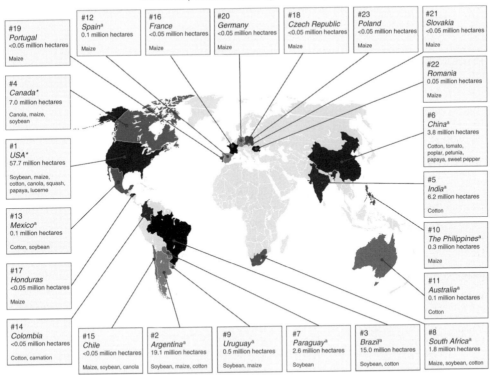

Biotech crop countries and Mega-countries[a], 2007

[a]Thirteen biotech mega-countries growing 50,000 h, or more, of biotech crops.

Fig. 17.1. Biotech crop countries and mega-countries. (From James, 2007.)

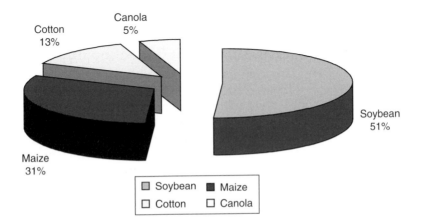

Fig. 17.2. Global biotech area of different crops. (From James, 2008.)

damage by pests and diseases causing considerable crop losses; furthermore, there is increasing pressure on land from urbanization and industrialization. These challenges are discussed below in detail.

Production output/levels

In sub-Saharan Africa 50–75% of the population and labour force are engaged in agriculture (Ndiritu, http://www.cgiar.org/biotech/rep0100/Ndiritu.pdf). However, despite this large labour force African agricultural growth has actually been declining over the last two decades. In fact the cereal yield has been flat since 1960 (Eicher *et al.*, 2006) with the annual growth rate falling from 2.3% in the 1970s to 2% in the years 1980–1992. Chetsanga reported that in the growing seasons for 1997–1998 US farmers planted 32.6 million ha maize and obtained a total production of 263 million t this equates to an average yield of $8.1\,t\,ha^{-1}$. (www.cgiar.org/biotech/rep0100/chetsanga.pdf). In comparison, during the same period, the whole of sub-Saharan Africa planted only about 22 million ha maize, with an average yield of $1.2\,t\,ha^{-1}$. Of the major developing regions of the world, only sub-Saharan Africa has seen a decline in the per capita output for food cereals in the last 30 years (Ndiritu, http://www.cgiar.org/biotech/rep0100/Ndiritu.pdf). These low production levels frequently result in food insecurity and have led to acute shortages of food in many parts of the continent. Biotechnology may offer the best opportunity to increase yield not only by controlling pests, but also by the development of stress-tolerant crops, such as drought-tolerant varieties.

Erratic rainfall/drought

The majority of the farming systems in Africa are rain-fed, with only a very small area irrigated. In 1995, 96% of cereals in sub-Saharan Africa were sown in rain-fed agricultural systems (Rosegrant *et al.*, 2002). The Food and Agriculture Organization of the United Nations (FAO) crop prospects for 2008 paint a bleak picture for food production in Southern Africa. It estimates that food production in Zimbabwe, Lesotho and Swaziland will decline by approximately 44%, 51% and 60%, respectively. This decline is mainly attributed to dry spells and erratic rainfall. There was severe drought in Africa in the period 2002–2004 and again in 2006. In 2006, the Horn of Africa was severely affected by drought – the worst hit countries were Ethiopia, Kenya, Somalia and Djibouti. The total number of people who required emergency food aid was 7.9 million (FAO, 2006). To compound this situation further, recent reports have indicated that water shortages are likely to result in civil strife in the 21st century (Vidal, 2006).

There is therefore a clear and urgent need to develop drought-tolerant crops. Unfortunately, conventional breeding through pollen transfer is very time-consuming, it generally takes 10–15 years to fully develop an improved hybrid maize variety (Chetsanga, www.cgiar.org/biotech/rep0100/chetsanga.pdf) and the remaining genetic potential for improvement of the major crops is limited (Gressel, 2008). Biotechnology thus has the potential to offer viable alternatives to enhancing the breeding process.

Pests and diseases

Pests and disease are major constraints to agricultural production in Africa. Pests cause significant damage to crops; estimates of crop losses vary considerably from country to country, and crop to crop (Gouse *et al.*, 2005). The witchweed (*Striga* sp.) is a major agricultural problem. It is a parasite of maize in sub-Saharan Africa, with yield losses of 65–92% having been recorded (http://www.interacademycouncil.net/CMS/Reports/AfricanAgriculture/6967/7014/708.aspx; http://www.interacademycouncil.net/CMS/Reports/AfricanAgriculture/6967/7014/7056.aspx; http://www.interacademycouncil.net; accessed 01/04/2008). This parasitic plant is difficult and expensive to control and poses a significant problem in many regions of Africa. Insect damage is another major constraint on crop productivity. Serious insect pest damage not only leads to direct losses but also often leads to high incidence of diseases as well. For example, cassava mosaic disease, which can completely destroy a crop in heavily infested regions is vectored by the whitefly (*Bemisia tabaci*; http://www.interacademycouncil.net/CMS/Reports/AfricanAgriculture/6967/7014/708.aspx; http://www.interacademycouncil.net/CMS/Reports/AfricanAgriculture/6967/7014/7056.aspx) and again control of this pest is problematic without expensive chemicals. Furthermore, some parts of Africa experience periodic explosions of insect populations that can devastate whole areas of crops, e.g. locusts.

Poor soils

A large portion of Africa is covered by desert and good agricultural land is in decline. Deforestation is rife and soils are becoming poorer and poorer in nutrients as a result of over-utilization without being fertilized. It is estimated that 8 million t of nutrients are depleted annually in Africa (Fleshman, 2006). Overgrazing is the most important cause of soil degradation, accounting for 49% of the area lost, followed by agricultural activities (24%), deforestation (14%) and overexploitation of vegetative cover (13%; http://www.interacademycouncil.net/CMS/Reports/AfricanAgriculture/6967/7014/708.aspx; http://www.interacademycouncil.net/CMS/Reports/AfricanAgriculture/6967/7014/7056.aspx). Furthermore, levels of fertilizer use in Africa are among the lowest in the world (Fleshman, 2006).

It is apparent that the challenge facing agriculture is to produce more from less resources (Persley, 2000). Only by increasing yield on lands currently under cultivation can hundreds of millions of hectares of tropical forest and other natural environments be saved from conversion to agriculture (Toenniessen *et al.*, 2003). Biotechnological innovations could go a long way in solving some of the problems presented above. Biotechnology, however, is not a panacea to Africa's agricultural problems as political instability and trade with other nations (particularly those that refuse GM imports) present significant problems.

Current Status of Plant Biotechnology in Africa

Biotechnology research in Africa is predominantly conducted by, and funded by, the public sector. To date, South Africa is the only country that has commercialized GM crops, although Egypt, Kenya and Zimbabwe are currently conducting field trials. Unfortunately, these countries all have different agricultural goals and priorities. Ruttan (1999) suggests a three-stage classification of the goals of agricultural biotechnological development. Stage one, the goal is to lift the yield ceiling of cereals (cereal yield in Africa has stagnated over the last three decades). Stage two, the goal is to enhance the nutritive value of cereals with products such as golden rice, which increases vitamin A levels and reduces cases of child blindness. Stage three focuses on the development of plants as nutrient factories to supply food, feed and fibre. Byerlee and Fischer (2002) have laid out a three-stage model of the abilities of countries to develop biotechnology:

- Type 1 countries have weak National Agricultural Research Systems (NARS) with capacity for tissue culture but little private sector activity. Many African countries fall into this stage.
- Type 2 countries have medium to strong NARS with strong national commodity research programmes and have some/limited capacity in molecular biology.
- Type 3 countries have very strong NARS with considerable research being conducted on transgenic crops.

Biotechnology in South Africa (Type 3)

The type of country not only influences its ability to conduct biotechnology research but also, critically, its ability to legislate adequately.

Laws and regulations

South Africa passed the Genetically Modified Organisms Act in 1997. The act applies to the: genetic modification of organisms, development, production, release and application of genetically modified organisms (GMOs) and the use of gene therapy. (www.info.gov.za/acts/1997/act15.htm). It applies only to living GMOs and not to their products (Cloete *et al.*, 2006); furthermore, it does not cover techniques involving human gene therapy (www.info.gov.za/acts/1997/act15.htm).

The South African Ministry of Health in conjunction with the Department of Agriculture has established two advisory groups on GM food labelling. One group, run by the Bureau of Standards, has the responsibility to develop an identity preservation system to track food ingredients and check label claims. The other group, under the Centre for Scientific and Industrial Research, reviews sampling and detection methods. South Africa has ratified the Cartagena Protocol on Biosafety and is party to the Convention on Biological Diversity.

Institutes involved in biotechnology in South Africa

South Africa is the leader in GMO research in Africa. To date there are over 110 biotechnology groups, with over 160 projects. More than 150 research trials have been conducted (Mulder and Henschel, 2003) and there are a number of leading research institutes involved in Biotechnology in South Africa. These, together with some of their major projects, are outlined below:

AGRICULTURAL RESEARCH COUNCIL (ARC) The council has a number of units conducting research on important aspects of agriculture; the Grain Crops Institute (ARC-GCI); the Small Grain Institute (ARC-SGI); the Animal Improvement Institute (ARC-AII); and the Institute of Tropical and Subtropical Crops (ARC, http://www.arc.agric.za/home.asp?pid=272).

UNIVERSITY OF PRETORIA The Forestry and Agricultural Biotechnology Institute (FABI) was established in 1997 (FABI, 2005/2006, http://www.fabinet.up. ac.za/). The primary objective of the institute is to promote the broad field of plant biotechnology through an interdisciplinary approach. The institute is involved in the sequencing of *Eucalyptus*.

UNIVERSITY OF CAPE TOWN Research in biotechnology was initiated in the 1980s and the university is currently conducting a number of cutting-edge research activities with direct relevance to African agriculture. For example, it has generated tobacco that is resistant to cucumber mosaic and tobacco neocrosis viruses (http://www.mcb.uct.ac.za/powerpoints/engineering.htm#Regeneration) and has developed a transgenic potato that is resistant to potato virus Y and leaf roll virus (in collaboration with ARC). In collaboration with a local company (PANNAR Ltd), the university has developed techniques for the regeneration and transformation of local maize varieties. Significantly scientists at the University of Cape Town are developing maize resistant to maize streak virus and tolerant to drought and other forms of abiotic stress (Thompson, 2004). Interestingly the South African plant *Xerophyta viscosa* is used as a source of genes for abiotic stress resistance.

In addition to research of agricultural relevance, transgenic crops such as tomato and tobacco are also being used for production of medically important vaccines (http://www.mcb.uct.ac.za/powerpoints/engineering.htm#Regeneration). For example, tobacco is used for the production of vaccines against papilloma virus (this is the biggest cause of cervical cancer among women in Africa) and the South African HIV subgroup (Thompson, 2004).

Field trials

More than 150 research trials have been conducted in South Africa (Mulder and Henschel, 2003). These include trials on multiple resistance, insect resistance and herbicide tolerance in cotton and maize and single glyphosate tolerance in eucalyptus. In 2007, more field trials on different crops including potatoes, groundnuts, sugarcane, maize and cotton were conducted.

Commercial release

GM crops grown in South Africa by both small- and large-scale farmers include *Bt* maize, Ht soybean and *Bt* cotton. In 2007, 57% of the total maize crop, 90% of the total cotton and 80% of the total soybean acreage was planted to GM varieties (Bosman, http://www.africabio.com).

It is clear that South Africa is well positioned to benefit directly from the current commercial GM crops, and uniquely for Africa it has infrastructure in place to develop its own varieties of crops, suited to local conditions and resistant to relevant pests and abiotic stresses.

Biotechnology in Egypt (Type 2)

While not as advanced as the situation in South Africa, Egypt has developed biosafety guidelines and the main institute involved in biotechnology research is the Agricultural Genetic Engineering Research Institute (AGERI).

Plant biotechnology research at AGERI includes conferring:

1. Virus resistance in crops such as tomato, faba bean, cucurbits and banana;
2. Insect resistance in crops such as potato, tomato, cotton and maize;
3. Stress tolerance in crops such as tomato, cotton, faba bean and wheat;
4. Fungal resistance in crops such as tomato, maize and faba beans (Madkour, http://www.cgiar.org/biotech/rep0100/Madkour.pdf).

In addition to agribiotech research, AGERI (in partnership with the University of Wyoming) has developed a pesticide based on a highly potent strain of *Bt* isolated from the Nile Delta. It is highly effective against a broad range of insects including Lepidoptera, Coleoptera and Diptera (Madkour, http://www.cgiar.org). AGERI has also established BIOGRO International, a company responsible for the commercialization of research. It has several business interests including an agreement with an American private company (Piooner Hi-Bred) and under this partnership scientists from AGERI can be trained at Pioneer in agricultural biotechnology. In fact, a USA–Egypt joint Science and Technology Joint Fund was established in 1995 to strengthen scientific and technological capabilities between the two countries (http://www.ageri.sci.eg/topic4/SCIENCETEC.HTM). These two countries are also involved in bilateral research on *Bt* genes (Madkour, http://www.cgiar.org). Further funded projects include Genomic Characterization of Stress Related genes form Wild Barley; Genomic Markers for Salt Tolerance in Sugar Beets; and Generation of Genetically Modified Corn Expressing an Insect Chitinase Gene (http://www.ageri.sci.eg/topic4/SCIENCETEC.HTM).

Biotechnology in Kenya (Type 2)

Institutions involved in biotechnology in Kenya

The institutions involved in agricultural biotechnology research in Kenya are the Kenya Agricultural Research Institute (KARI); Kenyatta University; Jomo Kenyatta University of Agriculture and Technology; the National Potato Research Centre (NPRC); and the Faculty of Agriculture at Moi University.

The International Livestock Research Institute (ILRI) also carries out biotechnological research on livestock diseases. ILRI is a CGIAR centre based in Nairobi; it is a world-leading institute in tropical livestock disease research and develops techniques for their management. A number of regional and international organizations are also based in Kenya, among which are: the African Biotech Stakeholders Forum (ABSF), African Agricultural Technology Foundation (AAFT) and African Harvest Biotechnology Foundation International (AHFBI).

The Kenya Agricultural Research Institute (KARI) and the International Centre for Maize and Wheat Research, in collaboration with the Syngenta Foundation, are currently involved in a project known as Insect-Resistant Maize for Africa (http://www.cimmyt.org/ABC/InvestIn-InsectResist/pdf/IRMAPartners.pdf). The aim of this programme is to develop maize resistant to insect pest stem borers (see below). KARI and in this case CIMMYT (International Maize and Wheat Improvement Centre) are also involved in a project to develop GM herbicide resistance in maize to combat the *Striga* witchweed.

Insect-resistant maize for Africa (IRMA)

With funding from the World Bank and Monsanto, The Kenya Agricultural Research Institute and the International Maize and Wheat Improvement Centre (Mexico) are developing GM maize resistant to stem borers. This is being carried out by incorporating modified genes derived from *Bacillus thuringiensis* (*Bt*) into the crop genome (De Groote *et al.*, 2003). Maize transformed with different *Bt* genes were transferred from Mexico to Kenya for leaf bioassay trials. These trials evaluated the efficacy against five species of stem borer, representing major economic pests (*Chilo partellus, C. orichaociliellus, Busseola fusca, Eldana saccharina* and *Sesamia calamistis*). Unfortunately, no event provided complete control against *B. fusca* (De Groote, 2003). Field trials were started in 2005 (http://www.scidev.net/en/news/kenya-begins-first-open-field-trials-of-gm-maize.html) and are currently ongoing.

Biotechnology in other African countries

The three countries discussed above are the current leaders in biotechnology research in Africa, although some limited work is being conducted elsewhere. The current status of this work in Zimbabwe and Burkina Faso will be addressed in brief.

Biotechnology in Zimbabwe (Type 2)

The Biotechnology Research Institute (BRI) is one of the 11 institutes of the Scientific and Industrial Research and Development Centre (SIRDC, http://www.sirdc.ac.zw/institutes/bri/index.htm; http://www.sirdc.ac.zw/institutes/bri/research.htm). The centre carries out research in animal and plant genetics with a view to producing animal/crop varieties that will boost agricultural production. It provides advisory services on GMOs and food crops.

The Biotechnology Research Institute (BRI) projects include:

- *Maize improvement*: Zimbabwe was the first country in the world to produce a single cross-hybrid of maize. This was officially released in 1960 (SIRDC, http://www.sirdc.ac.zw/institutes/bri/index.htm; http://www.sirdc.ac.zw/institutes/bri/research.htm).

The institute is currently involved in Research and Development of drought-tolerant and insect-resistant maize hybrids.

- *Mushroom project*: The institute aims to develop, produce and sell high-quality oyster, button and chanterelle mushroom spawn (SIRDC, http://www.sirdc.ac.zw/institutes/bri/index.htm; http://www.sirdc.ac.zw/institutes/bri/research.htm).

Biotechnology in Burkina Faso (Type 1)

Although field trials on *Bt* cotton have been carried out, as yet there have been no commercial releases of any GM crops. Other major projects currently being carried out include the development of GM cowpeas tolerant to drought, and resistant to insects and viruses; these are collaborative projects between INERA (Environment and Agricultural Research Institute) and The International Institute of Tropical Agriculture (IITA) in Nigeria.

Major Goals for African Agriculture

It is clear from the examples cited above that despite a considerable number of research programmes in existence, there appears (despite published works; Ruttan, 1999) to be a complete lack of a single common policy detailing the major goals for the development of African agriculture. While Africa is a huge continent and each country will have specific needs, some common goals should be identified, even if initially done on a regional basis.

Biotechnology products for Africa: two examples

1. Fortified Sorghum

A consortium of institutions from Africa, Japan and the USA aim to produce sorghum that is fortified with amino acids, proteins, iron, zinc and vitamin E for improved human health. This consortium, funded by the Bill and Melinda Gates Foundation, includes: Africa Harvest, The Council for Scientific and Industrial research of South Africa, The African Agricultural Technology Foundation, The Forum for Agricultural Research in Africa and the Agricultural Research Council of South Africa (http://www.supersorghum.org/donor.htm). It is critical that such consortia are formed to address the nutritional deficiencies that continue to create major health crises across the continent.

2. New Rice for Africa (NERICA)
Scientists at the West Africa Rice Development Association (WARDA) in Benin have developed a hybrid rice called New Rice For Africa (NERICA) by crossing Asian rice (*Oryza sativa*) with African rice (*O. glaberrima*). Asian rice is high yielding but poorly adapted to African conditions, whereas the African rice is well adapted to African conditions but is prone to lodging and grain shattering. The new rice varieties have qualities such as higher yields, shorter-growing seasons, resistance to local stresses and higher protein content (http://www.warda.org/NERICA%20flyer/advantage.htm). The new varieties have been released in countries such as Nigeria, Uganda and Cote d'Ivoire. While this has been achieved by conventional methods, further research, using molecular tools, is being conducted particularly with respect to nutritional content.

Constraints to Biotechnological Development in Africa

As we can see from the various examples above, considerable potential exists for GM technology in Africa. However, there are also many significant constraints that have to be addressed. Many African governments are sceptical about GM foods (Eicher *et al.*, 2006). In 2002, Zambia rejected food aid from the USA as the grain was genetically modified and Angola requires food aid grain to be milled before it is distributed (Eicher *et al.*, 2006).

Only a few countries have well-defined national goals and priorities in the area of biotechnology (South Africa, Kenya and Egypt). Other countries do not have such well-defined biotechnology policies and priorities. Many countries are however signatories to Cartagena Protocol on Biosafety.

A number of policy challenges have been identified as a hindrance to the development and application of biotechnology in many African countries. They include, significantly, a lack of clear priorities and investment strategies; further, many African countries have not identified specific areas of biotechnology in which to invest to meet specific goals and this makes it difficult to make long-term policies. Many African countries spread thinly their limited financial and human resources across sectors and research agencies. This results in duplication, and places a heavy strain on available resources (Kasota, 1999). As can be seen from the previous section, the countries actively conducting GM and biosafety research often have competing institutes.

Lack of resources

Human, capital, infrastructure
There is an acute shortage of skilled personnel in the area of biotechnology in Africa. According to a UNESCO Science Report South Africa has 13,000 full-time researchers and Egypt has 10,000. No other African country has more than 4000 researchers. In some cases trained scientists are lost due to HIV/

AIDS, this is a problem particularly in Southern Africa. Working conditions are poor in many African countries. Scientists are paid meagre salaries and there are no well-defined policies in place for rewarding outstanding scientists. This often leads to nepotism in promotions and outstanding scientists end up being demoralized, leading to poor quality of work. Some scientists go to the extent of leaving the continent to seek greener pastures elsewhere. The problem of the brain drain in Africa is very serious and needs immediate attention.

Biotechnology is a capital-intensive venture. In Africa, this capital has to be provided by the government because of the weak private sector in most countries, apart from South Africa. The role of governments is therefore important. Governments in Africa are operating on a shoestring budget and, therefore, find it difficult to fund research and development (R&D) in biotechnology. Competing for funding with R&D in Biotechnology are immediate national concerns like education, health and even defence. This leads to low levels of funding for research. In fact funds for public sector agricultural research are declining (Njobe-Mbuli, http://www.cgiar.org/biotech/rep0100/Njobe.pdf). According to the UNESCO Science Report (2005), more than half of science research funds in Africa are from International organizations.

Africa is characterized by an underdeveloped infrastructure. Very few countries in Africa have laboratories with the capacity to produce transgenic plants. Basic facilities and equipment are lacking. Communication networks and transport facilities are substandard and this imparts negatively on the quality of the research. In some cases it is difficult to get laboratory consumables on time.

Political instability

Scientists, like all other professionals, need an enabling environment to perform to their optimum. In many parts of the African continent there are conflicts which place an intolerable stress on the people. Lives are lost, people are displaced and infrastructure destroyed. This places an intolerable burden on many scientists, most of whom are left with no option but to flee these areas. Currently there are conflicts in Sudan, Uganda, Chad and Côte d'Ivore. In Uganda 1.4 million people have been displaced by conflict between the Lord Resistance Army (LRA) and the Government of Uganda (FAO, 2007).

Lack of networks

Networks are critical for technological development in Africa. Africa lags behind in technological development in the field of biotechnology. The gap between technologically proficient continents and Africa can be bridged only by developing linkages and networks. Strong ties should be forged between Africa and the rest of the world to facilitate exchange of knowledge. Seconding of African scientists to leading institutes and vice versa is one option that needs to be fully explored. Strong electronic communicative networks between Africa and the rest of the world should also be established.

Intellectual property rights

Patent laws are still being introduced in many African countries. There is demand from the World Trade Organization for African countries to revise their Intellectual property laws to meet the requirements of the Agreement on Trade Related Aspects of Intellectual Property Rights (TRIPS).

Patent issues

There are two organizations involved in patent law in Africa (Kameri-Mbote, 1991):

1. Organisation Africaine de la Propriete Intellecttuelle (OAPI), to which Francophone countries belong;
2. African Regional Intellectual Property Organization (ARIPO) based in Harare, to which most of the anglophone countries belong.

ARIPO members are categorized into three groups:

1. Botswana, Lesotho and Swaziland, which confer automatic protection to patents registered in South Africa;
2. Gambia, Ghana, Seychelles, Sierra Leone and Uganda, which require that patents be first registered in the UK before registration in the home country;
3. Kenya, Liberia, Mauritius, Nigeria, Somalia, Sudan, Tanzania, Zambia and Zimbabwe, which have established independent patent laws based on guidelines provided by the World Intellectual Property Organization (WIPO) or utilize the UK as a model.

Until there is a single standard for IPR they present a hindrance to national and international cooperation.

Trade imbalances

Trade liberalization and tariff barriers have been identified as areas that have been detrimental to the African farmer (http://www.pambazuka.org/en/category/gcap/30139; accessed 1/09/07). European and American farmers are highly subsidized and African farmers are unable to compete with them both in the domestic and export markets. Structural programmes put in place by the International Monetary Fund (IMF) and the World Bank eliminated subsidies and reduced tariffs for the African countries; this has created many problems for some countries. For example, four countries, Benin, Chad, Mali and Burkina Faso, rely on cotton for 40% of their annual export market. However, the US domestic and export subsidy for cotton farmers is estimated at US$5 billion, which they simply cannot compete with. While this situation exists the trade agreements that began in 1994 have generated inequalities in agricultural trade.

This is critical to Africa. In 2001, agriculture provided US$20.7 billion to Africa's economy. Farming employs at least 70% of the workforce in sub-Saharan Africa, and generates about 30% of the continent's gross domestic product (http://www.pambazuka.org/en/category/gcap/30139). Access to fair and equitable agricultural trade policies is crucial for Africa in terms of food security, economic development and poverty eradication (http://www. pambazuka.org/en/category/gcap/30139). The Uruguay Round of trade agreements, which began in 1994 and eliminated subsidies to the continent, has had detrimental effects on African farmers while farmers in America and Europe are highly subsidized.

Political will/legislative framework

There is a general consensus in some quarters that some African governments are not doing enough to promote the development of science on the continent. The feeling is that governments should take the initiative in science innovations because the continent lacks strong and vibrant private science companies. However, to date, very few countries have well-defined policies on the development of biotechnology. Regulations on GMOs are also non-existent in some countries. Research programmes are often isolated and not need-driven (Brink *et al.*, 1998). There are, however, regional programmes like the Southern Africa Regional Biosafety (SARB) programme aimed at providing technical training in biosafety regulatory implementation. SARB is made up of the following countries: Malawi, Mauritius, Mozambique, Namibia, South Africa, Zambia and Zimbabwe (Environmental News Service, 2002).

In view of such huge constraints on the development of biotechnology in Africa, it is disheartening to know that the major crops commercially developed for use in the Americas and Europe are of limited use in Africa.

Choice of crops

The majority of the commercialized GM crops are not important crops in Africa. These crops are not well adapted to the local conditions and are therefore susceptible to diseases and pests. Africa's major crops have been left off the biotechnology bandwagon. Major staple crops like cassava have not been thoroughly researched by the multinational corporations. These crops are well adapted to local conditions and farmers know how to cultivate them. For biotechnology to have far-reaching benefits, it is imperative that crops that are important to African farmers are put on the agenda of the large multinationals so that farmers can readily identify with this technology. But could African farmers afford a product produced in this way?

With biotechnology so little developed in most African Nations, it is surprising that the biosafety of crops is such a major issue. However, given all the problems listed above (e.g. lack of political will, inadequate legislation, etc.), it will no doubt heavily impact on some African governments' ability to effectively regulate and monitor GM crops.

Biosafety

Cohen and Paarlberg (2004) concluded that biosafety procedures for GM crops in developing countries are not working well. The authors pointed out that it is time-consuming to make and enforce regulatory decisions because decisions must be applied at three points: approval for trials; approval for larger location trials; and finally approval for commercial use.

Cohen (2005) found that there is a high cost of compliance with regulation as the cost of approval of a single transformation event ranges from US$700,000 for virus-resistant papaya to US$4 million for herbicide-resistant soybean. This is obviously restrictive in Africa and any other developing nations.

There is also concern that many legal and regulatory processes in Africa take a 'broad-brush approach' to regulation, i.e. that different uses of biotechnology are regulated in the same way. For example, biotechnology products that are used in research and development and those that are used as food are actually regulated in the same way (http://www.nepadst.org/doclibrary/pdfs/biotech_africa_2007.pdf). It is imperative that these issues must be addressed across countries in a consistent manner if biotechnology is to be adopted.

Concerns About Biotechnology in Africa

Safety issues related to biotechnology

Safety issues relating to toxicity, allergenicity and feed safety (Chapter 13, this volume), effects on non-target beneficial organisms including biological control agents (Chapters 8 and 9, this volume), cross-pollination (Chapter 7, this volume), the fate of the protein in the soil (Chapter 10, this volume) and evolution of pest resistance (Chapters 5 and 6, this volume) are all major concerns and they are discussed in brief below and in detail in the preceding chapters of this book.

Safety of *Bt* Cry proteins

Cry proteins are considered insect specific and therefore are safe to non-target organisms (Romeis *et al.*, 2008).

Cross-pollination and gene transfer

Gene transfer

There are two mechanisms in which genes move from one organism to another.

First, transfer of genes between two related species, through reproduction, such as cross-pollination of plants and interbreeding of animals (referred to as vertical gene transfer). Genes can also be transferred between related or unrelated organisms without the process of reproduction. This method of gene transfer is referred to as horizontal gene transfer (HGT), although no experimental evidence exists for this happening in the field. However, concern has been raised that the planting of GM crops will lead to an increase in fitness and invasiveness of weedy related species (Warwick *et al.*, 2008). Currently, the dominant trait of biotech crops is herbicide tolerance and since many crops express this trait there is thus a risk of gene escape.

Gene flow from GM crops to their wild relatives has been reported in many crops. In Africa, this presents a serious problem because the continent is a centre of origin of many crops such as African rice, pearl millet, sorghum and cowpeas (Gepts, 2003). Precautions that can be taken to minimize cross-pollination include allowing space between fields, strips of traditional crops surrounding GM plantings (the refugia concept; see Chapters 5, 15 and 16, this volume) and cleaning of harvesting equipment. However, minimizing gene flow can prove to be problematic in certain regions of Africa given the economic situations and educational standards of the farmers. Future transgene mitigation strategies may help this situation (Gressel, 2008).

Fate of *Bt* proteins in soil

It is feared that soil organisms may be affected on being exposed to Cry proteins that may be incorporated into the soil through crop residues. Studies have been conducted to determine the amount of *Bt* protein leached by roots and also from other plants parts incorporated in the soil and its effect on soil rhizosphere and non-rhizosphere miocroflora, soil collembola and earthworms. Most studies have found that there were no adverse effects (see Chapter 10, this volume).

Insect resistance management

Pest populations continuously exposed to *Bt*-crops for several years have the potential to evolve resistance to Cry proteins. This is currently dealt with using gene staking and refuges (see Chapter 5, this volume) and presents no significant risk greater than that already experienced with pesticides.

GM crops receive a level of regulatory and scientific scrutiny on a scale that no other novel agricultural product has ever been subjected to. While some risks may exist, for example Africa is the centre of origin of certain crops and landraces and wild relatives may be at risk from gene flow from GM crops, this must be balanced against the potential benefits of GM.

Potential for Biotechnology

In addition to examples already cited above, a number of other interesting developments are currently underway that have the potential to improve African agricultural production:

- *Disease-resistant sweet potato*: A new variety of sweet potato is being field-tested in Kenya that is resistant to viruses. The sweet potato is energy-rich, vitamin-packed and drought-resistant and, as such, an ideal crop for Africa (http://www.nuffieldbioethics.org/go/ourwork/gmcrops/pressrelease_ 183.html).

- *Drought-resistant rice*: This variety has been developed by scientists from the USA (Cornell University) jointly with researchers from Korea. Researchers took the genes that synthesize trehalose, a simple sugar that is produced naturally in a wide variety of organisms – and inserted it into rice (http:// news.bbc.co.uk/1/low/sci/tech/2512195.stm). This is a particularly important trait for Africa due to its erratic rainfall patterns.

- *Flood-resistant rice*: Scientists at the International Rice Research Institute in Philippines have developed rice that is resistant to waterlogging (http:// news.ninemsn.com.au/article.aspx?id=173754). This will prove useful in some areas and shows the challenges of improving agriculture in a climatically diverse continent.

- *Cancer-fighting tomatoes*: It is claimed that tomatoes with high levels of antioxidants can help combat certain cancers. These are currently being field-tested by Purdue University and the US Department of agriculture's Agricultural Research Services. The new variety offers three times the amount of the anti-oxidant lycopene compared to that present in conventional varieties (Science Daily, 2002; http://www.sciencedaily.com/releases/2002/06/02061907 3727.htm). This particular example stresses the importance of the nutritional improvement of crops for Africa.

- *Tastier tomatoes*: Researchers at the US department of Agriculture (USDA) and the Boyce Thompson Institute for Plant Research (BTI) at Cornell University have discovered a gene (*rin*) in tomato. By silencing this gene, USDA and BTI researchers have developed a way for tomatoes to stay on the vine longer so that they develop more nutrients, colour and taste, which is important when many people are malnourished (http:// www.ars.usda.gov/is/pr/2002/020411.htm).

Conclusions

Agricultural production in Africa has not kept pace with population growth. The continent is faced with frequent food crises. The situation is exacerbated by a number of factors such as drought, flood, pests, diseases and poor soils. Africa did not benefit from the Green Revolution that brought unprecedented benefits to Asia. It is imperative that Africa is not left behind in this new Agricultural Revolution, the Gene Revolution.

Farmers around the world are already reaping the benefits of biotechno-logy, but biotechnology in Africa is still in an embryonic stage. To date South Africa is the only country in the continent commercially growing GM crops, while other countries such as Egypt, Kenya and Zimbabwe are conducting field trials. Biotechnology, however, cannot solve all of Africa's agricultural problems. A number of impediments hamper biotechnological innovations in Africa, among these are political instability, lack of resources, lack of net-works, intellectual property rights, trade imbalances, crops chosen for modi-fication and biosafety issues. These issues need to be resolved for biotechnology to have far-reaching benefits to African farmers and the pop-ulation at large.

References

Bosman, L. (2008) Available at: http://www.africabio.com/pdf/MEDIARELEASE%20ISAAA.pdf

Brink, J.A., Woodward, B.R. and Da Silva, E.J. (1998) Plant biotechnology: a tool for development in Africa. *EJB Electronic Journal of Biotechnology* 1, 142–151.

Byerlee, D. and Fischer, K. (2002) Accessing modern science: policy and institutional options for agricultural biotechnology in developing countries. *World Development* 30, 931–948.

Chetsanga, C.J. (1999) Zimbabwe: exploitation of biotechnology in agricultural research. In: *Agricultural Biotechnology and the Poor* Consultative Group an International Agricultural Research (CGIAR) and World Bank. Available at: http://www.cgiar.org/biotech/rep0100/chetsanga.pdf

Cloete, T.E., Nel, L.H. and Theron, J. (2006) Biotechnology in South Africa. *Trends in Biotechnology* 24, 557–562.

Cohen, J.I. (2005) Poor nations turn to publicly developed GM crops. *Nature Biotechnology* 23, 27–33.

Cohen, J.I. and Paarlberg, R. (2004) Unlocking crop biotechnology in developing countries – a report from the field. *World Development* 32, 1563–1577.

Conway, D. (2003) Biotechnology and Hunger. Available at: http://www.opendemocracy.net/ecology-africa_democracy/article_1264.jsp

De Groote, H., Overholt, W., Ouma, J.O. and Mugo, S. (2003) Assessing the impact of Bt maize in Kenya using a GIS model. Paper presented at the International Agricultural Economics Conference, Durban, August 2003.

DeVries, J. and Toenniessen, G. (2001) *Securing the Harvest: Biotechnology, Breeding and Seed Systems for African Crops.* CAB International, Wallingford, UK.

Eicher, C.K., Maredia, K. and Sithole-Niang, I. (2006) Crop biotechnology and the African farmer Food Policy. *Food policy* 31, 504–527.

Environmental News Service Available at: http://www.ens-newswire.com/ens/jul2002/2002-07-29-01.asp

FABI 2005/2006 Biennial Report. Available at: http://src.fabinet.up.ac.za/main/pdf/FABI_Biennial_Report_2005_2006.pdf

FAO (2006) Food Prospects and Food Situation No1 April. Available at: ftp://ftp.fao.org/docrep/fao/009/J7511e/J7511e00.pdf

FAO (2007) Food prospects and Food Situation No 4 July. Available at: http://www.fao.org/docrep/010/ah868e/ah868e03.htm

Fleshman, M. (2006) Boosting African farm yields. Africa Renewal 20 No 2. Available at: www.un.org/ecosocdev/geninfo/afrec/vol20no2/202-boosting-farm-yields.html

Gepts, P. (2003) Ten thousands years of crop evolution. In: Chrispeels, M.J. and Sadava, D.E. (eds) *Plants, Genes and Crop Biotechnology.* Jones and Bartlett, Canada.

Gouse, M., Pray, C.E., Kirsten, J. and Schimmelpfenning, D. (2005) A GM subsistence crop in Africa: the case of Bt white maize in South Africa. *International Journal of Biotechnology* 7, 84–94.

Gressel, J. (2008) *Genetic Glass Ceilings: Transgenics for Crop Biodiversity.* Johns Hopkins University Press, Baltimore, Maryland.

Huang, J., Hu, R., Wang, Q., Keeley, J. and Zepeda, J.F. (2002) Agricultural Biotechnology Development, Policy and Impacts in China. Available at: http://www.ids.ac.uk/UserFiles/File/knots_team/China_Paper EPW.pdf

James, C. (2007) ISAAA Briefs No 35-2006: Global Status of Commercialised Biotech: GM Crops: 2006.

James, C. (2008) ISAAA Briefs No 37-2007: Global Status of Commercialised Biotech: GM Crops: 2007.

Kameri-Mbote (1991) Intellectual Property and Sustainable Industrial Development. A paper presented at the workshop on Sustainable Industrial Development in Africa: Policy perspectives for the 1990s, Nairobi, Kenya, 18–19 March.

Kasota, J. (1999) Recent Biotechnology Research and Development in Tanzania. Background Paper prepared for the regional Workshop on Biotechnology Assessment: Regimes and Experiences, organized by ACTS.

Madkour, M.A. (1999) Egypt: Biotechnology from the laboratory to the Marketplace: Challenges and Opportunities In: *Agricultural Biotechnology and the Poor.* Consultative Group on International Agricultural Research (GIAR) and the World Bank. Available at: http://www.cgiar.org/biotech/rep0100/Madkour.pdf

Mulder, M. and Henschel, T. (2003) National Biotech Survey 2003. Idea to Industry. Prepared for eGoli BIO and Department of Science and Technology, South Africa (eGoli BIO, Pinelands Office Park, Modderfontein, South Africa, 2003). Available at: http://www.oecd.org/dataoecd/7/37/36036991.pdf

Ndiritu, C.G. (1999) Kenya: Biotechnology in Africa: Why the Controversy? In: Agricultural Biotechnology and the poor. Consultative Group on International Agricultural Research (CGIAR) and the World Bank. Available at: http://www.cgiar.org/biotech/rep0100/Ndiritu.pdf

Njobe-Mbuli, B. (1999) South Africa: Biotechnology for Innovation and Development In: Agricultural Biotechnology and the poor. Consultative Group on International Agricultural Research (CGIAR) and the World Bank. Available at: http://www.cgiar.org/biotech/rep0100/Njobe.pdf

Persley, G.J. (2000) Agricultural biotechnology and the poor: promethean science. In: Parsley, G.J. and Lantin, M.M. (eds) *Agricultural Biotechnology and the Poor: Proceedings of an International Conference,* Washington, DC, 21–22 October 1999, Consultative Group on International Agricultural Research Washington, DC, pp. 3–21.

Raney, T. (2006) Economic impact of transgenic crops in developing countries. *Current Opinion in Biotechnology* 17, 174–178.

Romeis, J. *et al.* (2008) Assessment of risk of insect-resistant transgenic crops to nontarget arthropods. *Nature Biotechnology* 26, 203–208.

Rosegrant, M.W., Cai, X., Cline, S. and Nakagawa, N. (2002) The role of rainfed agriculture in the future of global food production. Environment and production technology division discussion paper no. 90. International Food Policy Research Institute. Washington, DC. Available at: http://ageconsearch.umn.edu/bitstream/123456789/19982/1/ep020090.pdf

Ruttan, V.W. (1999) Biotechnology and agriculture: a sceptical perspective. *AgBioForum2* (1999) 54–56, Available at: http://agbioforum.org/v2n1/v2n1a10-ruttan.htm

Science Daily (2002) Available at: http://www.sciencedaily.com/releases/2002/06/020619073727.htm

Thompson, J. (2004) The status of plant biotechnology in Africa. *AgBioForum* 7, 84–95.

Toenniessen, G.H., O'Tooley, J.C. and DeVriesz, J. (2003) Advances in plant biotechnology and its adoption in developing countries. *Current Opinion in Plant Biology* 6, 191–198.

UNESCO Science Report (2005) Available at: http://www.unesco.org/science/psd/publications/science_report2005.pdf

Vidal, J. (2006) Cost of water shortages: civil unrest, mass migration and economic collapse. Available at: http://www.guardian.co.uk/environment/2006/aug/17/water.internationalnews

Warwick, S.I. *et al.* (2008) Do escaped transgenes persist in nature? The case of an herbicide resistance transgene in a weedy Brassica rapa population. *Molecular Ecology* 17, 1387–1395.

IV The Future of Agriculture

18 Agriculture, Innovation and Environment

N. Ferry and A.M.R. Gatehouse

School of Biology, Institute for Research on Environment and Sustainability, Newcastle University, Newcastle upon Tyne, UK

Keywords: Agriculture, genetic modification (engineering), green revolution, gene revolution, environmental impact, GM crops

Summary

Approximately 10.3 million farmers in 22 countries grew biotech (genetically modified) crops in 2006. Yet this technology remains one of the most controversial agricultural issues of current times. Many consumer and environmental lobby groups believe that genetically modified (GM) crops will bring very little benefit to growers and to the general public and that they will have a deleterious effect on the environment. The human population is currently 6 billion and it is predicted to increase to 9–10 billion in the next 50 years. This is at a time when food and fuel are competing for land and climate change threatens to compromise current resources. It is, and will continue to be, a priority for agriculture to produce more crops on less land. From the dawn of agriculture, humans have modified their environment. Landscapes are shaped to suit our needs and the plants we grow as crops are engineered to our tastes and requirements. Throughout history food production has kept pace with population growth as a result of our innovative abilities, but it did so at a cost. Future agricultural production should not degrade the environment as it has in the past, it must become more sustainable. Will the adoption of biotech crops help to meet this challenge?

> *It seems that you cannot have a deep sympathy with both man and nature.*
> Henry David Thoreau (1854)

Whether new technology can resolve this dichotomy will be addressed in this chapter.

The Evolution of Agriculture

A lesson from prehistory

The end of the last Ice Age marks a fundamental turning point in human history. With the appearance of farming, there began a fundamental change in the relationship between humans and the natural world (Christian, 2005).

Early humans appeared in Africa about 250,000 years ago. Gradually they acquired new technologies and new ecological knowledge and became able to migrate and to exploit new lands. As humans spread, they began to have an impact on the environment. Even at this early stage in our history, humans were able to transform landscapes with fire-stick farming and with hunting that drove a large number of Pleistocene megafauna to extinction. However, small groups of humans hunting and gathering for subsistence had little power to degrade their environment on a scale that we are familiar with today. At the end of the Pleistocene era and the dawn of the Holocene era some 11,500 years ago most human populations were still considered to be hunter-gatherer communities; however, the end of the Ice Age signalled a change to a more sedentary way of life and this, together with increases in population led to the 'dawn of agriculture'.

When humans engaged in agriculture for the first time, they began to change the non-living environment (its soils, rivers and landscapes) to create new environments tailored specifically to meet their own needs. Agriculture by definition involves altering natural processes in ways that benefit humans, and in so doing interfering with natural ecological cycles. By removing unwanted species (weeding), agriculturalists deliberately create artificial landscapes in which processes of succession, which might have returned the land to its previous state, are prevented. The land is deliberately kept free of many species, and is therefore maintained below its natural productivity level. In return, the productivity of those species favoured by humans is increased, as they are given access to nutrients, water and sunlight. But reducing plant cover also increases the rate of erosion, because roots hold the soil together, create humus and reduce the kinetic energy of rainfall. Erosion, together with intense cultivation of a small number of crops, can accelerate nutrient cycles, forcing humans to maintain soil fertility by the addition of animal manures or crop rotation. Humans also managed to remake the organisms around them, by the genetic engineering of domestic crops and animals, but also by hunting down animals (such as wolves) that threatened survival of their domesticates. However, even in the early Holocene era, these changes affected only small parts of the world, and early agrarian technologies had a limited effect on the natural environment. Only when agricultural technologies began to spread widely did the human impact on the natural world become more significant.

The origins of genetic 'engineering'

At the dawn of agriculture, sedentary human populations, not yet practising agriculture as we recognize it, settled in regions with plentiful natural resources (e.g. along river banks) and so developed an increasing reliance upon a small number of abundant and easily harvested food sources. In doing so, they would have learned a great deal about the life cycles, growth pattern and diseases of a particular favoured species. The careful tending of these species encouraged genetic changes that favoured domestication, as poorer specimens were rejected. Over time careful selection and propagation of seeds with the desired characteristics led to the development of domesticated crops. In fact, some crops are now dependent on human intervention for reproduction. For example, maize/corn (*Zea mays*), first domesticated in the Americas, is now unable to drop and spread its own seeds due to human selection of plants with tougher rachis so that the crop is easily harvested (Galinat, 1975).

Founder crops and domestication

Plant domestication has been studied extensively in the Fertile Crescent using archaeological and palaeobotanical evidence. It is in this area that most cereal crops were domesticated (modern-day Turkey, Syria, Iraq, Iran, Lebanon, Israel). Wheat, a principal carbohydrate source worldwide today, has its origins in ancient Mesopotamia on the banks of the Tigris and Euphrates. The genomes of the wild *Triticum* species, the progenitors of modern einkorn and emmer wheat, have undergone extensive changes during their co-evolution with humans. First, wild species were selected by humans and went through normal evolution and chromosome divergence leading to different versions of the basic seven-chromosome set of wheat (labelled A, B, C, D, E, F, G; Feldman and Levy, 2005). Second, the *Triticum* species form a polyploidy series (which include: diploid, two sets of seven chromosomes; tetraploid, four sets; hexaploid, six sets) of species. These arose from rounds of intentional crossing between species with different chromosome sets, and although normally such hybrids would be infertile, occasional doubling led to fertile polyploidy progeny (Chrispeels and Sadava, 2003). These polyploidy progeny went on to become the modern day bread and pasta wheats. Einkhorn wheat (diploid AA) is not high yielding but hardy and was confined to mountainous regions. But two diploid AA and BB wild species were crossed and gave rise to emmer wheat which in turn gave rise to the modern pasta wheats. Following one further cross with a DD wild relative the hexaploid AABBDD bread wheat came into being. Wheat is not a crop that evolved independently in the wild. Simultaneous processes of domestication occurred globally outside of the Fertile Crescent, for example rice in Eastern Asia or maize and beans in the Americas, thus all our crops have been engineered to meet human needs.

The selection pressures on domesticated plants are very different from that which plants would experience in the natural environment. Thus, humans began the first genetic 'engineering' of crops from the very earliest domesticates. To return to the example of modern maize unable to shed its own seed – this is the result of generations of selection of plants that retain their seeds longer, crops that scatter their valuable resources before the farmer has been able to harvest have little value. Similarly seeds without long periods of dormancy were selected, as were plants with a compact growth habit, a favourable photoperiod, large harvestable organs and pest and disease resistance (Schlegel, 2007).

Genetic bottlenecks

Selection over prolonged periods of human history led to crops having remarkably different genomes from their wild predecessors, the first genetic bottleneck. Thus, an overall reduction in diversity has been in operation from the very first domesticates. After domestication, crops were disseminated from their centre of origin (often in the tropics) to other parts of the world, and often into very different environments. Small samples of seeds often served as the starter population which was then subjected to further selection for those best suited to the new environment, the second genetic bottleneck. Thus, many of the crops we depend on today have very little genetic diversity. Landraces of crops (those that continued to be grown in their centre of origin) tend to exhibit greater genetic heterogeneity, which can buffer crops from pathogens or pests. Crop breeding is dependent on the survival of crop lines with different traits; these are the basis for crosses between different plant lines, for example, desirable nutritional qualities or taste crossed with plants exhibiting pest or drought resistance. Such plant crosses constitute the basis of crop variety improvement from the dawn of agriculture to the present day. Remarkable increases in crop productivity have been achieved over time, particularly since the late 1950s, with modern crop breeding representing the third genetic bottleneck, and it may be argued that the major crops have now reached the end of their genetic potential for improvement (Gressel, 2008). While half of these changes stem from deliberate selection of more productive varieties, the remainder come from adopting crop production technologies that improve the environment in which crops grow.

The age of discovery; the agricultural revolution

In the west, the evolution of agriculture can be divided into four discrete periods, the Prehistoric, Roman, Feudal and Scientific Eras (Edwards and Gatehouse, 2007). While the Prehistoric is recognized as the era of crop domestication, the Roman Era (1000 BCE–500 CE) saw the introduction of metal tools, the use of animals for farm work and the manipulation of watercourses for irrigation, while the Feudal Era saw the beginning of international trade based on exportation of crops. Interestingly, the era known as the Scientific Era started as early as the 16th century and although there is documentary evidence for the use of pest control in ancient times, its adoption is primarily attributed to this era. Throughout human history agricultural innovation has led to increases in productivity. Irrigation, the use of secondary animal products including the animal powered plough, animal manures and the utilization of milk and wool all led to intensification. In 1798, Malthus predicted that population growth would outpace food production. He was wrong, in so far that our innovative abilities and technological advances have allowed us to keep pace. However, in the long term will he be proven to have been correct?

While productivity increased, agriculture still relied on labour and natural resources. Mechanization (starting in the 18th century) replaced manual labour and meant that fewer people could produce more food for more people. Recognition of the work of monk Gregor Medel on inheritance led to new varieties of plant grown that have been selected scientifically rather than by farmers saving seed. The advent of selective breeding increased yields dramatically. Much later (starting in the 1950s), the replacement of local varieties of rice and wheat with high-yielding hybrids constituted the so-called green revolution. About 40% of the increase in productivity in the past 50 years has stemmed from these new varieties (Chrispeels and Sadava, 2003). The rest of the improvements have come from changes in crop management;

inputs in the form of fertilizer and pesticides have dramatically increased productivity.

Energy inputs and the green revolution

Worldwide food production has been rising by 2.3% annually as a result of high-input agriculture (Chrispeels and Sadava, 2003). This started with the green revolution that was the product of alterations in plant architecture and physiological properties through breeding, not only in wheat (*T. aestivum*) and rice (*Oryza sativa*), but also maize (*Z. mays*), sorghum (*Sorghum bicolor*) and other crops. Semi-dwarf plants provided adequate nutrition with high productivity, without lodging, thus increasing the harvest index. Photoinsensitivity matched the crop seasons with appropriate rainfall availability (Swaminathan, 2006). This produced more, on less land, and conserved arable land and forests. The technology was, however, dependent upon purchased inputs. Excessive fertilizer and pesticide use (along with the growing of crops in large monocultures) created serious environmental problems, including the breakdown of resistance (to pests) and degradation of soil fertility. Changing energy inputs into agriculture over time is shown in Table 18.1. What is also evident from this table is that the green revolution helped developing countries (particularly India and China) feed their burgeoning populations. This would not have been possible without inputs in the form of agrochemicals.

Agrochemicals

Agrochemicals include two large groups of compounds: chemical fertilizers and pesticides. The use of fertilizers was in part responsible for the green revolution with the application of exogenous nitrogen, phosphorous and potassium along with irrigation providing huge leaps in yield.

Similarly, the use of pesticides, including insecticides, fungicides, herbicides and rodenticides to protect crops from pests significantly reduced losses, and improved yield, as well as protecting domestic livestock from arthropod-borne disease and humans from similar diseases such as malaria. World production of formulated pesticide has increased from *c*.1 million t in 1965 to *c*.6 million t in 2005 (Carvalho, 2006). Pesticides have been, and will be,

Table 18.1. Energy input and population density over time. (From Simmons, 1993).

	Energy input (GJ ha)	Food harvest (GJ ha)	Population density (persons km^{-2})
Foraging	0.001	0.003–0.006	0.01–0.9
Pastoralism	0.01	0.03–0.05	0.8–2.7
Shifting agriculture	0.04–1.5	10.0–25	10–60
Traditional farming	0.5–2	10–35	100–950
Modern agriculture	5–60	29–100	800–2000

a highly effective method to stop pests quickly when they threaten to destroy crops. Without pesticides crop damage globally ranges from 35–100%, whereas with pesticides damage is reduced to an estimated 0–20% (Chrispeels and Sadava, 2003). The chemical nature of pesticides has evolved over time. In early farming practices, inorganic chemicals were used for insect and disease control including sulfur and copper; however, with the advances in synthetic organic chemistry that followed two world wars the synthetic insecticides were born. In the 1940s, the neurotoxic organochlorine, DDT, was the pesticide of choice, but following its indiscriminate use it was suggested to bioaccumulate in the food chain were it affected the fertility of higher organisms, such as birds; this was first highlighted in the book *Silent Spring* published in 1962 (Carson, 1962), while this is now known to have been incorrect, it was nevertheless a signature event in the birth of the environmental movement. This pesticide was subsequently replaced by the comparatively safer organophosphate and carbmate-based pesticides (both acetylcholinesterase inhibitors) and many of these were replaced in turn by the even safer pyrethroid-based pesticides (axonic poisons). Synthetic phyrethroids continue to be used today despite the fact that they are broad spectrum. In parallel, the specific microbial toxins produced by the soil-dwelling bacterium *Bacillus thuringiensis* (*Bt*) are increasingly being adopted. In fact, microbial sprays of *Bt* are used in organic agriculture. This shift in chemistry is due in part to a demand for increased safety both for humans and for the environment. Other pesticides have followed similar trends. Herbicide use has seen a move towards more biodegradable chemicals that only need to be applied at a very low concentration of active ingredient. Thus, the new generation of herbicides is relatively environmentally benign. Many herbicides exploit the differences in plant physiology between the crop species and its weeds (usually the differences between monocots and dicots); they may be systemic or act on contact. The mode of action of common herbicides is varied (Naylor, 2002; Chapter 7, this volume). For example, inhibition of photosynthesis and light-dependent membrane destruction (acting on photosystems II and I, respectively) are the mode of action of the foliar acting non-selective herbicides like atrazine, paraquat and diquat. 2, 4-Dichlorophenoxyacetic acid induces abnormal plant growth by interfering with auxin regulation. The sulfonyl ureas, imidazolines and the environmentally benign, but non-selective glyphosate (acting on 5-enolpyruvyl-shikimate-3-phosphate synthase), inhibit amino acid synthesis. Others include inhibitors of lipid synthesis, inhibition of cell division and pigment synthesis. The advantages of herbicides are clear – they control multiple weed species, control perennial weeds, cause no injury to the crop plant and can readily be applied to large areas. Similarly, chemical strategies for disease control can be highly effective. Numerous compounds with antifungal or antibacterial activity have been discovered, most often applied as sprays, dusts or seed coatings. Many older compounds are broad spectrum and toxic with the newer chemistries acting systemically with a narrower target range. However, such chemicals tend to be expensive and norm-ally reserved for use on high-value fruit and vegetable crops. They are expensive to manufacture and require significant investment in terms of human and ecological safety testing before a product can be released (Chrispeels and Sadava, 2003).

While the application of external inputs led to higher yields and higher-quality crops, it did so in conjunction with significant advances in crop breeding.

Modern crop breeding, or 'unnatural' selection

Plants, pathogens and pests have co-evolved for millennia, and as a result plants possess endogenous defence mechanisms against pests and disease. In many cases, breeders have been able to exploit the genetic mechanism of resistance in the production of resistant plants. In the case of pathogen resistance, breeders have relied on single dominant characteristics controlled by R-genes that interact with pathogen *vir* (virulence) or *avr* (avirulence) gene products and produce a compatible or an incompatible response, respectively, in the plant. In classical plant breeding, this has relied on crosses between elite crops and wild relatives (that are more genetically diverse) to introduce new disease resistance traits into the crops. Extensive backcrossing of the elite line is then required to eliminate the undesirable traits in the wild relative and thus makes traditional breeding a time-consuming process with a time of c.15 years required before a new resistant variety is available for release to growers. Nevertheless, the development of F_1 hybrid crops (derived from crossing two pure inbred lines) resulted in vastly improved yields (Chrispeels and Sadava, 2003). In the USA, during the 1940s, virtually all the maize crop was such F_1 hybrids and yields increased fourfold. As any progeny of these crops would be heterozygous for the desired traits, seeds must be purchased each season and this provided the incentive for the private seed industry. Despite its enormous success in improving yield, the F_1 hybrids also serve as a warning as to the dangers of large-scale monoculture. In 1970, the uniform F_1 maize crops in the USA were left devastated by disease.

Plant breeding involves large genetic changes

While significant improvements in yield have been achieved through modern plant breeding, it has been described as a blunt tool rather than a precision implement (Gressel, 2008). Hybridization from two pure lines adds many genes from each line to the offspring. Even backcross breeding, where a donor plant (with, for example, a disease resistance gene) is crossed with a parent (an agronomically desirable crop plant), and the progeny selfed and selected for the desired trait, involves whole genome rearrangements, rather than the transfer of a single characteristic. In fact, it is a chromosome segment broken by recombination in meiosis that is transferred (Chrispeels and Sadava, 2003).

Quantitative traits

Unlike single traits transferred by backcrossing, many desirable crop qualities are in fact controlled in a more complex manner. So-called quantitative traits are controlled by multiple genes and require multiple crosses and phenotype screening. This process has been greatly accelerated with the use of marker-assisted breeding. Exploitation of these changes in short deoxyribonucleic acid

(DNA) sequences (such as in tandem repeats) has allowed the construction of genetic maps for a crop. If an interesting gene is tightly linked to a gene that is difficult to detect phenotypically, breeders can screen for the presence of the associated marker and thus rapidly identify progeny with the desirable trait. This has allowed for marker-assisted breeding, where the genetic control of quantitative trait locus (QTL) can be determined and thus the most desirable alleles bred into a single line. This technology revolutionized plant breeding in the 1980s and allowed identification of desirable alleles from wild species, such as those that regulate crop yield and nutritional qualities, to be transferred into elite lines (Collard and Mackill, 2008).

The crop breeding described so far relies upon natural variation within the plant species to improve a crop. However, this has not always been sufficient for human demands. As a result, many of our crops have been generated by mutation breeding. The most commonly used mutagens are radiation. X-ray and gamma rays (easily obtained from a radioactive source such as cobalt) have been used to bombard seeds, meristems, pollen and somatic cells in culture. This results in extensive damage to the plant's DNA, and the new variation generated is screened and selected for as in conventional breeding programmes. New characteristics generated include alterations in agronomic traits such as growth habit, disease resistance and nutritional content. In fact, many of the methods used to facilitate crop breeding, including tissue culture for propagation, embryo rescue (for interspecific crosses) and anther or pollen culture (haploid cells with many mutations) generate novel plants with characteristics that would not develop naturally.

There is no doubt that vast yield increases have been generated by scientific plant breeding and by the use of agrochemicals and that they have enabled food production to keep pace with population. However, modern crops and farming methods are not without significant drawbacks.

Problems with the agricultural revolution

In their long history of co-evolution, pests have continually adapted to natural plant defence mechanisms, such as higher levels of bioactive secondary metabolites. Thus, one may suggest that pests are preadapted to evolve resistance. Widespread use of a synthetic chemical that often targets a single enzyme creates strong selection pressure in the target pest population that will lead to the rapid development of resistance. This is true for the herbicides, insecticides and pathogen treatments and is a process that has led to the development of some of the worst agricultural weeds and pests (Ellstrand, 2003b).

Costs of the agricultural revolution

The development of agriculture irreversibly changed human lives, and while there is little doubt that advances in methods and technologies have allowed us to feed a burgeoning population, our dependence on agricultural production came at a high cost in how it shaped the natural world. Overall, since the dawn of agriculture, forests have declined by c.20%, from 5 to 4 billion ha (Christian, 2005). Until recently, the decline was more marked in temperate forests (32–35%), but today, deforestation is more rapid in regions of tropical forest, with an area the size of Wales (20,000 km^2) cleared from the Amazon every year (Lovelock, 2007).

From simple subsistence farming through larger agrarian civilizations, farming has intensified to become a global, multibillion-dollar enterprise controlled by a few large international companies. The human population has increased exponentially in recent years and so have agricultural yields. The change from hunter-gatherer to farmer, traded an unpredictable diet for a plentiful diet, but it also led to the development of urbanization and industrialization, thus creating environmental problems that are still growing.

Impact of Agriculture on the Environment

An agroecosystem is defined as 'semi-domesticated ecosystems that fall on a gradient between ecosystems that have experienced minimal human impact, and those under maximum human control, like cities'. Thus, agroecosystems are generally defined as novel ecosystems that produce food via farming under human guidance (Hecht, 1995). Thus, by definition, no form of agriculture is natural. Natural ecosystems are defined as 'ecosystems free of human activities' and composed of 'native biodiversity', though it is doubtful if many such environments exist today. Rather seminatural ecosystems (in which human activity is limited) are typical (Amman, 2008). The modern failure to appreciate the reality of the agroecosystem often leads to unreasonable expectations of what is a novel crop or what is reasonable environmental impact. That said, it is not unreasonable to expect that having learnt from past mistakes we must manage our agricultural production in as sustainable a manner as possible in the future and limit environmental degradation resultant from past agricultural practices.

Land use

As natural habitat was and is converted for agricultural use, complex, species-rich ecosystems are replaced by simple, species-poor agroecosystems. To date humans are using just over a third of the total land area for growing crops and in doing so have cleared an estimated 25% of the worlds grassland and 30% of the worlds forest (McGavin, 2006). Tropical forests now cover *c.*6% of the total land surface area, down from 14% in prehistoric times. Several studies have shown that these habitats harbour 60–80% of the earth's total biodiversity. Tropical forests are under immense pressure not only from remaining subsistence farmers, but also from timber extraction, ranching and large-scale agriculture. Malaysia, for example, has already lost 60% of its forest cover (McGavin, 2006). In the Amazon, the number of cattle increased from 26 million to over 55 million in the years 1990–2000 and recent demand for soybean has resulted in huge areas being taken over for plantations. A similar situation has arisen in South-east Asia with oil palm. Agricultural expansion and colonization, fuelled by cheap land and high market prices is encroaching on the forests year by year.

When natural vegetation is cleared for agriculture, it becomes vulnerable to drought and soil erosion with billions of tonnes of topsoil blowing away in the wind or being washed from the land. Over time, the fertility of the land decreases and overgrazing and intensive cultivation leads to fall in yields and, ultimately desertification. Of major concern is the fact that agriculture now takes 70% of the available freshwater and while irrigated land can be very productive, it eventually becomes useless due to salination of the soil. It is estimated that around 6 million ha of land are lost every year to desertification and one-sixth of the world's population will be directly affected by it (McGavin, 2006).

However, with suitable temperature and rainfall, people can successfully use land to grow crops and provide food, but agricultural land is not evenly distributed across the globe. Eleven per cent is used for crops and 24% for pasture; these resources are concentrated in the USA and Canada, Europe, India, China and South-east Asia, with the rest of the land surface being too cold or dry for plant growth. The most productive land came from the clearance of grassland. The process of converting natural ecosystems to agricultural land is ongoing and encroaching upon the tropical forests as discussed above. Obviously, land is a finite resource and greater productivity must be achieved on the land currently under cultivation. This has been achieved to date through innovation and the use of purchased inputs, but at a cost.

The impact of agriculture on the landscape

During the 20th century there was a common trend in most European countries away from 'traditional landscapes' towards 'modern agricultural landscapes' (Glebe, 2007). This change was induced by the industrial revolution, but accelerated by economic boom in the post-war era following World War II. Intensification of highly productive land has led to a reduction in the number of farms, with less diversified, larger-scaled landscapes and with fewer boundaries maintained.

The impact of agriculture on biodiversity

Agricultural landscape changes have an impact on ecosystem biodiversity, since they affect wild plant and animal habitats. Adverse effects of farming on natural ecosystems are particularly found on arable land. Arable farming enhances a few domesticated species, thus the overall species diversity is lower. This is particularly true when monoculture is practised (Green et al., 1994) as bird and insect species rely on certain environmental essentials for food, breeding sites and shelter, which they often cannot find in simplified environments (see Chapter 12, this volume). In addition, the seed population declines substantially under arable farming, since ploughing brings buried seeds up to the surface where they germinate and are lost (Pywell et al., 1997).

The decline of species diversity on arable land is further exacerbated when traditional extensive farming systems are converted to high-input systems. Diverse crop rotations and set-aside will promote plant, insect and bird populations; however; species richness declines with monoculture and with high levels of pesticide and fertilizer input (Crabb *et al.*, 1998; Hansson and Fogelfors, 1998; Henderson *et al.*, 2000). Spraying with pesticides reduces abundance and diversity of weeds and insects, and may have side effects in the food chain (De Snoo, 1999). Similarly, nitrogen and phosphorous fertilizer application reduces species richness, both on the field and in the boundary vegetation of arable fields (Kleijn and Verbeek, 2000).

The impact of agriculture on water quality and the atmosphere
The majority of the negative environmental effects of farming are associated with the use of pesticide and fertilizer (nitrogen, phosphorous, potassium) applications (Glebe, 2007). Phosphorous and pesticides pollute lakes and rivers with nitrates and pesticides effecting groundwater. Water pollution stimulates algae and plant growth, which may lead to eutrophication due to high respiration rates, reducing the abundance of other aquatic organisms (Ginting *et al.*, 1998). Nitrate leaching into ground-water also poses a health risk if the water is a primary source of drinking water (Brandi-Dohrn *et al.*, 1997). While such leaching is a natural phenomenon, the rate of nitrate leaching is heavily affected by agricultural management (Ready and Henken, 1999).

Intensively managed agriculture also has a negative effect on the atmosphere (Glebe, 2007). Emissions of nitrogen oxides and ammonia from soil fertilization with both manure and inorganic fertilizer may lead to acidification of soils (ammonia) and increases in greenhouse gases (Lewandrowski *et al.*, 1997).

The impact of agriculture on soils
Soil formation is a long and complex process (Chrispeels and Sadava, 2003). Parent rock is broken down into mineral articles which are chemically modified, organic residues from plants, microbes and animals decay and are continuously added to it. It can be a very long process of accumulation, some of the agriculturally important soils today were formed 10,000 years ago. Soil may be lost by erosion and plants, which bind the soil together with their roots, play a major role in preventing this. But soil is dynamic and living. It is an ecosystem supporting thousands of living species. Only one-third of the earth's land surface is used for cultivation of crops or animal husbandry, equal portions are forested or unsuitable for agriculture. Thus, arable soils represent only 11% of the total land surface and of this 75% has poor fertility.

If the soil is deficient in nutrients (minerals), plant growth will be retarded. Treatment with animal manures was common place, but the green revolution saw widespread adoption of inorganic fertilizer use. These release nutrients rapidly and as such are available to growing plants when they are most needed. However, the application of inorganic nitrogen, phosphorous and potassium has led to problems with leaching into watercourses. Poor management of natural resources, deforestation and misuse of agricultural lands has led to extensive soil degradation worldwide; 67% of agricultural soils have been, or are being, degraded by erosion, salinization, compaction, nutrient losses, pollution and biological deterioration. It is estimated that this has reduced world crop productivity by *c.* 16% (Chrispeels and Sadava, 2003).

The Millennium Ecosystem Assessment

An ecosystem is a natural unit consisting of all plants, animals and microorganisms (biotic factors) in an area functioning together with all of the non-living physical (abiotic) factors of the environment. In 2005, the largest ever assessment of the earth's ecosystems was conducted by a research team of over 1000 scientists. The findings of the assessment were published in the multivolume *Millennium Ecosystem Assessment*, which concluded that in the past 50 years humans have altered the earth's ecosystems more than any other time in our history (Millennium Ecosystem Assessment (MA), 2003). Ecosystem services are a view on the relationship between society and nature and are defined as the benefits that people obtain from ecosystems. Ecosystem services are the 'fundamental life-support services upon which human civilization depends', and can be direct or indirect. Examples of direct ecosystem services are pollination, wood and erosion prevention. Indirect services could be considered climate moderation, nutrient cycles, detoxification of natural substances among many more. Almost two-thirds of ecosystem services are found to be in decline worldwide (MA, 2003). The capacity of ecosystems to provide services is determined by many human-induced factors that result in change. These drivers may act directly or indirectly (Alcamo *et al.*, 2005). The direct drivers taken into account in the MA analysis include greenhouse gas emissions, air pollution emissions, risk of acidification and excess nitrogen emissions, climate change, sea level rise, changes in land use and land cover, the use of nitrogen fertilizers, and nitrogen loading to rivers and coastal marine systems. In turn, indirect drivers will influence these direct drivers. The main indirect drivers include population development, economic development, technology development, energy, agricultural demand and production, and human behaviour. Food is one of the most important life-supporting services provided to humans by the ecosystem.

Productivity

As illustrated by the patterns of land use in the wheat-growing capital of the USA, the landscape is dominated by irrigated fields of a single crop (Fig. 18.1). Intensive agriculture such as this has had a dramatic impact on the environment but has sustained a global population of 6 billion people.

At present there are still abundant food resources (shortages arise from problems with distribution, not production) although 80% of the world's population is now dependent on just three crops: wheat, rice and maize (Chrispeels and Sadava, 2003).

The Future of Agriculture

Population

Early agrarian communities were subject to Malthusian cycles. Population pressure was greater than rates of innovation required to sustain growth. Hence,

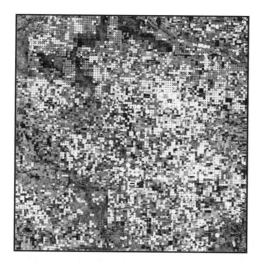

Fig. 18.1. Patterns of agriculture: Garden City, Kansas, USA, 2000. Agriculture is the most widespread use of land by humans consuming huge amounts of water. Kansas is the wheat capital of the USA with hundreds of irrigated fields. Crops appear in white. (Photograph courtesy of 183 USGS/EROS Data Centre.)

there were periodic famines that shaped the basic rhythms of early human history. The most striking feature of the past two centuries is that innovation (for a time at least) was so rapid and so sustained that levels of productivity have kept pace with, and even outstripped, population growth. In fact, humans stepped up into overproduction. Although there have been many devastating regional famines, on a global scale, food production has kept pace with population growth, which is precisely why populations have risen so fast. The human population was about 200 million people in 6000 BCE, and in the following 7000 years there was an increase of just over 100 million. However, with the advent of the agricultural revolution in the 18th century the population began to increase rapidly and in the 1800s, during the industrial revolution, the world entered an exponential growth phase. In the past 50 years, the population has doubled to over 6 billion people, with conservative estimates by the United Nations Population Division predicting 7.4–10.6 billion people by 2050 (http://www.un.org/esa/population).

Boom and Bust (St Matthew Island, 1944): a brief digression
In 1944, 24 female and five male reindeer were taken to St Matthew Island, in the middle of the Bering Sea, and were released by members of the American Coastguard for recreational hunting. Once the coastguard station was shut down, the reindeer were left to their own devices with no predators on the island and bountiful resources. Birth rate was correspondingly high and death rate low. As a consequence the small population grew rapidly until in 1963 when it reached a peak of 6000 (in 19 years). However, a year later there were only 42 surviving animals and by 1966 the reindeer were declared extinct on St Matthew Island. Ecologists studying the population noted that as food became

scarce, the reindeer were losing condition. Lichen, the preferred winter food source, can grow in dense mats of up to 5 in. (1 in. = 2.5 cm) in depth; however, it is very slow growing and under heavy grazing pressure and trampling could not survive. The reindeer on St Mathew Island faced their last winter (which was particularly harsh) underfed and died of starvation. The case of St Matthew Island is a classic example of what happens when the ability of a habitat to support a species (the carrying capacity) is exceeded (McGavin, 2006). The comparison of human population growth to that of the reindeer on St Matthew Island (Fig. 18.2) while not intended as an apocalyptic vision should serve as a warning.

The human population is ever expanding and the ability to provide enough food is now becoming increasingly difficult (Chrispeels and Sadava, 2003). The planet has a finite quantity of land available to agriculture and the need for increasing global food production. It is clear that a priority for agriculture will be to produce enough food in a manner that does not further degrade the environment.

Global climate change

One of the greatest dangers to agriculture is its vulnerability to global climate change. The expected impacts are for more frequent and severe drought and flooding, and shorter growing seasons (http://www.cgiar.org/impact/global/climate.html). The performance of crops, wild species, livestock and aquatic resources under stress will depend on their inherent genetic capacity and on the whole agroecosystems in which they are managed. It is for this reason that any efforts to increase the resilience of agriculture to climate change must

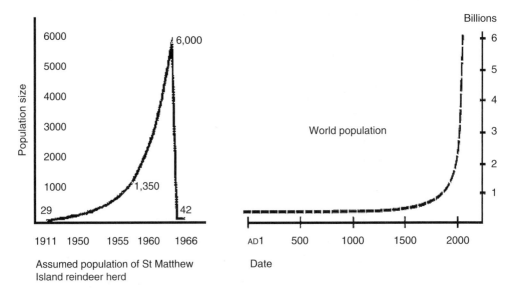

Fig. 18.2. Comparison of the population of reindeer on St Matthew Island to human population growth.

involve the adoption of stress-tolerant varieties as well as more prudent management of crops, animals and the natural resources that sustain their production, while at the same time providing vital services for both people and the environment. A recent report of the United Nations Intergovernmental Panel on Climate Change (IPCC) states that the average temperature of the Earth's surface is likely to increase by about 3°C, on average, over the next century, assuming greenhouse gas emissions continue to rise at current rates. The scientific evidence behind those predictions leaves 'no doubt as to the dangers mankind is facing' (Yvo de Boer, Executive Secretary of the UN's Framework Convention on Climate Change).

Biotechnology

To feed the world population under the threat of climate change as it heads towards 10 billion in the next 25 years will require a further massive increase in food production.

Today, we understand a great deal more about our environmental impact than at any previous point in human history. Future advances in technology have the power to damage the environment still further, but also to mitigate the effects.

Can biotechnology or 'The Gene Revolution' bridge the gap? 1996–2008, more than a decade of commercial GM plantings

The year 2008 marks the 13th year of the commercialization of genetically modified (GM) crops (James, 2007). Adoption rates of GM crops between 1996 and 2006 were unprecedented by recent agricultural industry standards (Fig. 18.3). There is a growing body of consistent evidence across years,

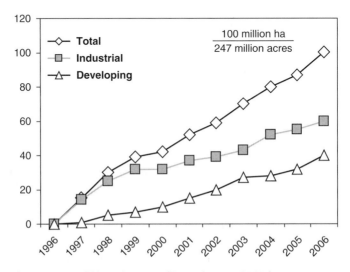

Fig. 18.3. Global area of biotech crops. (From James, 2007.)

countries, crops and traits generated by public sector institutes that demonstrate that GM crops have delivered substantial agronomic, environmental, economic, health and social benefits to farmers in this time frame.

Approximately 10.3 million farmers in 22 countries grew biotech crops in 2006. Yet this technology remains controversial and reportedly unwelcome in many countries, particularly within the European Union. Possibly the most compelling case for agricultural biotechnology is their capability to contribute to:

- Increase crop productivity, and hence food, fibre and feed security.
- Conserve biodiversity; GM crops can be a land-saving technology capable of higher productivity on currently available land, thereby reducing loss of other natural ecosystems and their biodiversity.
- Reduce the environmental footprint of agriculture by contributing to more efficient use of external inputs.
- Increase the stability of production by shoring up resistance to biotic and abiotic stress.
- Improve the livelihoods of those farmers dependent on agriculture in developing countries.
- Increase the cost-effective production of biofuels, thus reducing dependency on fossil fuels.

Three crops account for 95% of the land under GM cultivation: soybean (51%), maize (31%) and cotton (13%). The traits primarily grown are herbicide tolerance (63%) and insect pest resistance (18%) or a combination of both in the same crop (19%; James, 2007).

UN millennium development goals

The UN has eight millennium development goals (MDGs) all with a target date of 2015:

1. *Eradicate extreme poverty and hunger.*
2. Achieve universal primary education.
3. Promote gender equality and empower women.
4. Reduce child mortality.
5. Improve maternal health.
6. Combat HIV/AIDS, malaria and other diseases.
7. *Ensure environmental sustainability.*
8. Develop a global partnership for development.

Changes in agricultural practice will play a part either directly (goals 1 and 7) or indirectly, for example in improving maternal health and reduction of child mortality.

Throughout history human innovation has allowed us to produce food at a level that has kept pace with population increases. However, it is a fallacy that we can meet all our population's food needs on the current area of land available for agriculture. Further intensification of current farming methods would require more land which, in turn, would have an even greater impact on the environment; this is compounded by the fact that global warming predictions will result in less land available, and more unpredictable weather conditions with loss of land already in cultivation. Will biotechnology enable us to increase yield, be it by minimizing crop losses to pests (Chapters 5 and 6, this volume) or by reducing competition for resources with weeds (Chapter 7, this volume), as seen with the current commercial GM crops, or by the adoption of varieties resistant to abiotic stress?

Genetically modified crops

Insect resistance (IR)

B. thuringiensis (*Bt*) is a soil-dwelling bacterium of major agronomic and scientific interest (see Chapters 2, 5, 6, and 15–17, this volume). The subspecies of this bacterium colonize and kill a large variety of host insects, but each strain does so with a high degree of specificity. This is mainly determined by the crystal proteins that the bacterium produces during sporulation, which form an extensive range of *Bt* δ-endotoxins (deMaagd *et al.*, 2001).

Different plasmids encode toxins of different sequence and different specificity of action against insects. The *cry* gene encodes a crystalline protoxin protein, with individual Cry toxins having a defined spectrum of activity, usually restricted to a few species within one particular order of insects. To date, toxins for insects in the orders Lepidoptera (butterflies and moths), Diptera (flies and mosquitoes), Coleoptera (beetles and weevils) and Hymenoptera (wasps and bees) have been identified (deMaagd *et al.*, 2001). A minority of toxins showing activity against nematodes have also been identified (Table 18.2; Gatehouse *et al.*, 2002).

The toxin exerts its pathological effect by forming lytic pores in the cell membrane of the insect gut. Upon ingestion toxins are solubilized in the midgut and proteolytically cleaved by insect digestive proteases to an N-terminal, 65–70 kDa truncated form (active form); and the active toxin molecule binds to a specific high-affinity receptor in the insect midgut epithelial cells, inserts into the membrane, and forms pores that kill the epithelial cells (and thus the insect) by osmotic lysis (Schnepf *et al.*, 1998; deMaagd *et al.*, 2001). The economic need for an effective insecticide, the availability of *Bt* genes and the proven safety of *Bt* biopesticide sprays made *Bt*-expressing plants obvious candidates

Table 18.2. Insecticidal properties of *Bt* toxins.

Insect order	Cry protein
Lepidoptera	Cry1A, Cry1B, Cry1C, Cry1E, Cry1F, Cry1I, Cry1J, Cry1K, Cry2A, Cry9A, Cry9I, Cry15A
Coleoptera	Cry1I, Cry3A, Cry3B, Cry3C, Cry7A, Cry8A, Cry8B, Cry8C Cry14A, Cry23A
Diptera	Cry2A, Cry4A, Cry10A, Cry11A, Cry11B, Cry16A, Cry19A, Cry20A, Cry21A
Hymenoptera	Cry22A
Nematodes	Cry5A, Cry6A, Cry6B, Cry12A, Cry13A, Cry14A
Liver fluke	Cry5A

for commercial exploitation in plant biotechnology. However, bacterial *cry* genes are rich in A/T content compared to plant genes. As a result the *cry* genes have undergone considerable modification of codon usage and removal of polyadenylation sites with both full-length and truncated versions of *cry* genes expressed in plants (deMaagd *et al.*, 1999).

At present 20.3 million ha of land is planted with *Bt* cotton and maize (James, 2007), with economic benefits from *Bt* cotton estimated at US$9.6 billion and maize US$3.6 billion (James, 2007). Significantly, Phipps and Park (2002) showed that on a global basis GM technology has reduced pesticide use. It estimated that GM soybean, oilseed rape, cotton and maize varieties modified for herbicide tolerance, and insect-protected varieties of cotton, reduced pesticide use by a total of 22.3 million kg of formulated product in 2000.

Herbicide-tolerant (HT) crops

Effective weed control is a prerequisite for high-yielding quality crops. The preferred herbicides at present are those with low environmental persistence. Unfortunately the current generation of highly effective, low persistence herbicides are broad spectrum. Furthermore, the few remaining effective highly specific herbicides are speeding up the development of resistance (Mulwa and Mwanza, 2006). In the 1940s, only *c.*500 compounds needed to be screened to select a potential herbicide (Gressel, 2002). By 1989, it was estimated that 30,000 compounds needed to be screened and then further modified to improve their toxicity (Parry, 1989). It is thus becoming harder to identify new herbicides with novel modes of action, and of course this incurs greater costs.

Glyphosate (Round-up) is a highly effective broad-spectrum herbicide that inhibits 5-enolpyruvyl-shikimate-3-phosphate (EPSP) synthase, a branch point enzyme in aromatic amino acid biosynthesis. A naturally occurring EPSP synthase gene (*cp4*) was identified from *Agrobacterium* sp. Strain CP4, whose protein product provided glyphosate tolerance in plants (Padgette *et al.*, 1995). Furthermore, glyphosate detoxification pathways are known in microbes involving glyphosate oxidoreductase (*gox*) genes (Jacob *et al.*, 1988). These two genes, in combination, confer glyphosate resistance in selected crops. While resistance to other herbicides has been engineered, glyphosate-resistant crops demand the major market share.

To date 72.2 million ha of HT soybean, maize, canola, cotton and lucerne are grown globally. The economic benefits to the farmers are estimated at US$17.5 billion. While HT crops reduce the amounts of active ingredient required for weed control, they also promote the usage of no/low-till farming and thus lower fuel consumption with direct benefits to both soil structure and carbon emissions (James, 2007). (For detailed information on HT crops, refer to Chapter 7, this volume.)

Stacked traits

Nineteen per cent of the global area of transgenic crops expresses a combination of IR and HT traits. Increasingly multiple traits are being stacked for more multiple resistance. For example in 2007 in the USA, almost two-thirds of the maize grown expressed a double or triple construct of *Bt* and herbicide traits. One *Bt* to control the European corn borer and the other the corn rootworm, both major economic pests costing US farmers up to US$1 billion each year in losses (James, 2007). Ultimately the stacking of transgenes in a crop with different targets and modes of action may help to increase durability of the technology by decreasing the risk of resistance (Ferry *et al.*, 2006).

While IR and HR crops have been major successes for the companies that developed them and do have potential to increase yield, it will be the next generation of transgenic crops that may bring highly significant environmental benefits as the focus shifts towards improving drought tolerance, nitrogen-use efficiency and intrinsic yield.

The Environmental Impact of Transgenic Crops

Almost from the beginning of the production of transgenic crops there have been concerns over their use and introduction into the environment. There is international agreement that GM crops should be evaluated for their safety, including their environmental impact (Dale, 2002). During the past 15–20 years, there have been extensive research programmes of risk assessment, with several areas of major concern identified.

Insect resistance

Perhaps one of the most important issues relates to the development of target pest resistance which would limit the lifespan of the technology. In the case of *Bt* toxins, this is a major concern for the organic farming community, since the potential for insect populations to evolve resistance to *Bt* will not only limit the effectiveness of *Bt*-expressing crops but also *Bt*-based biopesticides. *Bt* resistance in insect pests has already been reported to develop in four to five generations in the laboratory (Stone *et al.*, 1989). Currently transgenic plants express the *Bt* toxin constitutively in all tissues and through all life stages of the plant, so although it took 40 years for resistance to biopesticide to appear, the process may be accelerated due to high selection pressure exerted by transgenic plants (de Maagd *et al.*, 2001). Considerable effort has been devoted to delaying the evolution of resistance, e.g. the use of refugia has been recommended and adopted in most regions growing *Bt*-crops (Betz *et al.*, 2000; Chapters 5 and 6, this volume), but when one considers the ability of insects to evolve resistance to chemical pesticides (ffrench-Constant *et al.*, 2004) the development of resistance is inevitable and has in fact already occurred (Chapter 5, this volume). This concern is equally valid for the appearance of herbicide-resistant weeds (discussed below; see also Chapter 7, this volume).

Weeds, gene flow, invasiveness and biodiversity

There have been significant concerns regarding the potential of GM crops to become 'superweeds' (by invasion, volunteerism) or to create 'superweeds' (by cross-pollination; Ellstrand, 2003a).

GM crops do have the potential to cross-pollinate other crops and wild relatives, but there are four basic elements determining the likelihood and consequences of gene flow: first, the distance of pollen movement from the GM crop; second, the synchrony of flowering between crop and pollen recipient; third, sexual compatibility between crop and recipient; and fourth, ecology of the recipient species (Dale *et al.*, 2002). Research has shown that pollination declines sharply with distance from the pollen source (Lutman, 1999) and one could reduce the chances of GM pollen reaching other crops through the use of buffer zones, although it may travel further if insect-pollinated. Ellstrand *et al.* (1999) review the sexual compatibility of crops with weeds and feral species. For example, oilseed rape (canola), barley, wheat and beans can hybridize with weeds in some countries; however, in the UK the probability of hybridization with weeds is considered minimal for wheat, low for oilseed rape and barley and high for sugarbeet. Although sugarbeet can readily hybridize, in the case of herbicide-tolerant varieties of sugarbeet the crop is harvested before flowering and hence shed no pollen. Indeed, methods have been developed to block expression in the pollen of transgenic plants, including engineering of the chloroplast genome (Heifetz, 2000) as well as transgene mitigation strategies (Gressel, 2008). Also, the potential exists for GM crops to become invasive; there has been a great deal of concern that such crops could persist in the wild and disperse from their cultivated habitat. However, studies have indicated that their ability to invade and persist was no better than their conventional counterparts (Crawley *et al.*, 2001). Finally, GM crops persisting in fields after harvest thus becoming a weed in a different crop may be dealt with in two ways; simple treatment with an appropriate herbicide or technologies that prevent the transgene being carried over to the next generation (Gressel, 2008).

In order to put these concerns into perspective, one must understand that flow from the agroecosystem to natural ecosystems has always occurred. Gene flow is a continuing process and is the source of biological diversity (Thies and Devare, 2007). There has always been gene flow from commercial crops to relatives living in near proximity. In several regions of Mexico, maize is cultivated in close proximity to teosinte, its wild progenitor. Under these circumstances gene flow between the plants and the formation of hybrids is frequent (Chrispeels and Sadava, 2003) and leads to the steady improvement of the landraces (Wisniewski *et al.*, 2002). There should be no greater concern with this occurring with *Bt* maize. In reality the vast majority of the major cultivated crops have no wild or weedy relatives outside of their centres of origin (Gressel, 2008); however, some crops are grown in areas where gene flow may occur, but the farmers select the landraces they wish to cultivate – thus, if *Bt* maize were to cross with a landrace, unless the local farmers showed a significant preference for it then they would, as they always have, select against it (Gressel, 2008). Which leads one to ask – is this a problem if the farmers derive a benefit from it?

Breeding has selected for traits in our crops that are not found in wild or weedy relatives such as responsiveness to nitrogen fertilizer. One must consider if a trait transferred from crop to weed would actually confer a selective advantage to the weed outside of the agroecosystem as this would limit its invasiveness into natural ecosystems. Nevertheless, crops do cross with wild relatives and have historically produced some of the worst agricultural weeds (i.e. invading the agroecosystem). For example, the weedy beets of Europe have evolved from crosses between conventional (non-transgenic) sugarbeet and its progenitor, as well as from de-domestication of the crop back to feral forms (Gressel, 2008). Oilseed rape is a crop that may present a major volunteer problem in agriculture; its seed pods may shatter prior to harvest and give rise to weeds in the next crop. Conventional oilseed rape can form volunteer weeds and become feral from where they can then mix with the crop. In fact, some older varieties of non-transgenic oilseed rape high in erucic acid and glucosinolates significantly affect the value of a modern crop through contamination (Diepenbrock and Leon, 1988).

Genes have always moved between the natural and agroecosystem, and to date – despite the formation of hybrids between HT canola and wild relatives in Canada (Gressel, 2008) – the technology has proven safe and effective (Cerdeira and Duke, 2006; James, 2007; Darmency *et al.*, 2007; Gressel, 2008).

Impact on non-target organisms

Assessing the consequences of pest control on non-target organisms is an important precursor to their becoming adopted in agriculture. The expression of transgenes that confer enhanced levels of resistance to insect pests is of particular significance since it is aimed at manipulating the biology of organisms in a different trophic level to that of the plant. Potential risks to beneficial non-target arthropods exist. Those groups most at risk include: non-target Lepidoptera, beneficial insects (pollinators, natural enemies) and soil organisms.

Exposure of non-target Lepidoptera to insecticidal transgene products may occur through both direct consumption of transgenic plant tissues or via consumption of transgenic pollen, many non-target Lepidoptera are rare butterflies having great conservation value. The case of the Monarch butterfly (*Danaus plexippus*), a conservation flagship species in the USA, highlighted the need for ecological impact research. In a letter to *Nature*, Losey *et al.* (1999) claimed that both survival and consumption rates of Monarch larvae fed milkweed leaves (natural host) dusted with *Bt* pollen were significantly reduced, and that this would have profound implications for the conservation of this species. However, a series of ecologically based studies rigorously evaluated the impact of pollen from such crops on Monarchs and demonstrated that the commercial wide-scale growing of *Bt* maize did not pose a significant risk to the Monarch population (Hellmich *et al.*, 2001; Gatehouse *et al.*, 2002). In fact, the initial experiments did not quantify the dose of pollen used, or indeed, if this was a realistic level likely to be encountered in the field, nevertheless, this work highlighted the importance of studying non-target effects. In a separate field study, Wraight *et al.* (2000) showed that *Papilio polyxenes* (black swallowtail) larvae were unaffected by pollen from *Bt* maize event Mon810 at 0.5, 1, 2, 4 and 7 m from the transgenic field edge, highlighting the need for a case-by-case study of organisms considered to be at risk. In addition to the potential direct impacts of *Bt* toxins on susceptible target insects, as in the case of the Morarch butterfly, some Lepidoptera have been shown to have a reduced sensitivity to the lepidopteran-specific *Bt* toxins. For example, *Spodoptera littoralis* can survive on maize expressing Cry1Ab (Hilbeck *et al.*, 1998) and thus present a route of exposure to the next trophic level. In the case of the coleopteron specific *Bt* Cry3Aa- or Cry3Bb-expressing potatoes or maize, some Lepidoptera may represent non-target secondary pests, and while not directly affected by the transgene product themselves may again present a route of exposure to the next trophic level, as do other non-target herbivores. Organisms such as those belonging to the orders Homoptera, Hemiptera, Thysanoptera and Tetranychidae are not targeted by *Bt* toxins expressed in transgenic plants; however, they do utilize the *Bt* crop (Groot and Dicke, 2002). The direct effect that this may have on these insects is dependent on the presence of *Bt* receptors in the first instance, and it is so far unclear whether such receptors are present in non-target organisms (de Maagd *et al.*, 2001). In addition, the fate of the toxin ingested by non-target herbivores is unclear, since if it retains toxicity then this may have implications at the next trophic level.

The impacts of insect-resistant transgenic crops at higher trophic levels have also been considered, where there are concerns over the risks to beneficial arthropod biodiversity (Schuler *et al.*, 1999; Bell *et al.*, 2001), in particular, predators and parasitoids, which play an important role in suppressing insect pest populations both in the field and under specialized cultivation systems (glasshouses). Natural enemies may ingest transgene products via feeding on herbivorous insects that have themselves ingested the toxin from the plant; such tritrophic interactions will be influenced by the susceptibility of the herbivore to the plant protection product. If, as in the case with *Bt* toxins, the prey item is susceptible to the toxin, then the predator will not come into contact with the toxin as the pest will effectively be controlled, and in target insects the toxin is bound to receptors in the midgut epithelium that are structurally rearranged and may lose their entomotoxicity (de Maagd, 2001). In non-target insects (and resistant insects), the toxins do not bind and may thus retain biological activity. However, the overwhelming weight of evidence from independent laboratory and field studies show that *Bt* toxins have a limited ability to affect the next trophic level (reviewed in Sanvido *et al.*, 2007; Romeis *et al.*, 2008; Chapter 8, this volume).

Pollinators represent another group of non-target organisms highlighted as at risk from *Bt* toxins in GM crops. The current generation of transgenic crops produce *Bt* toxin in the pollen as well as in the vegetative tissues. Several studies have been conducted to determine toxicity of *Bt* toxins to pollinators (Vandenberg, 1990; Sims, 1995, 1997; Arpaia, 1997; Malone and Pham-Delegue, 2001); generally they all conclude that neither the adults nor larvae of bees were affected by *Bt* toxins (see Chapter 9, this volume).

Finally, non-target species may come into contact with *Bt* toxins via the environment. Several studies have shown that *Bt* toxins released from transgenic plants bind to soil particles (Palm *et al.*, 1996; Crecchio and Stotzky, 1998; Saxena *et al.*, 1999). Soil-dwelling and epigeic insects such as Collembola and Carabidae may thus be exposed to the toxins. Several studies (Saxena and

Stotzky, 2001; Ferry *et al.*, 2007) show no differences in mortality or body mass of bacteria, fungi, protozoa, nematodes and earthworms or carabid beetles exposed to *Bt*, but as with non-target herbivores, some of these organisms could mediate exposure to predators.

Exposure to the transgene products, however, does not necessarily imply a negative impact. Most studies to date have demonstrated that crops transformed for enhanced pest resistance have no deleterious effects on beneficial insects (reviewed in Ferry *et al.*, 2003; Romeis *et al.*, 2008).

Ultimately one must consider the impact of *Bt* toxins in comparison to other pest control strategies, e.g. conventional crop protection using insecticides. While pesticides have no doubt brought vast yield improvements, they have well documented undesirable non-target effects (Devine and Furlong, 2007). It is worth remembering that while potential risks do exist to the environment from the cultivation of GM crops, their current (IR and HT) and future potential to decrease reliance on external inputs and to increase the availability of genetic resources available to breeders is great.

Future

The challenges that face 21st century agriculture are to increase yield while limiting the environmental impact of agriculture, this will necessitate not only a reduction in pesticide usage, but also improvement in stress responses in crops, and improvement in nutritional content. Currently, we may be at the limit of the existing genetic resources available in our major crops (Gressel, 2008). Thus, new genetic resources must be found and new technologies will enable this.

Agriculture must focus on: global food security, farming in a sustainable manner and increasingly plants as biomass. As food and fuel begin to compete for land, this brings the need to improve yields on land already under cultivation into even sharper focus.

Conclusion

Modern farming requires significantly less labour than at any point through our history, particularly in the more developed regions of the world. Consequently an ever-increasing number of people are becoming detached from farming and food production, thus many consumers have lost touch with the complexities of farming. Activist groups, and some sections of the media, increasingly advocate banning synthetic chemicals (including approved pesticides) and GM crops from the market place. Unfortunately in many cases the facts are distorted.

Increasingly, in Europe, agricultural technology is perceived as inherently 'bad' by the public. Decades of agricultural disasters in the UK (bovine spongiform encephalopathy (BSE), salmonella, foot-and-mouth, the poisoning of non-target large animal species with pesticides, the eutrophication of watercourses) have led to a deep-seated mistrust of agricultural companies, scientists and policy makers and politicians. Paradoxically this is at the same time when

other new technologies have seen a rapid period of growth and wide acceptance (PCs, digital media, iPods), despite having their own environmental cost,
particularly in terms of their eventual disposal. However, despite the perceived
risks of new biotechnology, it may now be time to rely on our innovative abilities to produce more food globally in a changing climate, and to farm in as sustainable a manner as possible while preserving the surrounding environment.
Agriculture is an inherently unnatural situation and once this is fully understood
by the broader community, we may be able to advance towards a rational
debate on the role of biotechnology in food production.

References

Alcamo, J., van Vuuren, D., Ringler, C.,
Cramer, W., Masui, T., Alder, J. and
Schulze, K. (2005) Changes in nature's
balance sheet: model-based estimates of
future worldwide ecosystem services.
Ecology and Society 10(2), 19.

Amman, K. (2008) The needs for plant biodiversity: the general case. In: Gressel, J.
(ed.) *Genetic Glass Ceilings. Transgenics
for Crop Biodiversity*. The John Hopkins
University Press, Baltimore, Maryland.

Arpaia, S. (1997) Ecological impact of Bt-
transgenic plants: 1. Asessing possible
effects of CryIIIB toxin on honeybee (*Apis
mellifera*) colonies. *Journal of Genetic
Breeding* 50, 315–319.

Bell, H.A., Fitches, E.C., Marris, G.C., Bell, J.,
Edwards, J.P., Gatehouse, J.A. and
Gatehouse, A.M.R. (2001) Transgenic crop
enhances beneficial biocontrol agent performance. *Transgenic Research* 10, 35–42.

Betz, F.S., Hammond, B.G., Fuchs, R.L. (2000).
Safety and advantages of *Bacillus thuringiensis*-protected plants to control insect
pests. *Regulatory Toxicology and
Pharmacology*, 32(2), 655–666.

Brandi-Dohrn, F.M., Dick, R.P., Hess, M.,
Kauffman, S.M., Hemphill, D.D. and Selker,
J.S. (1997) Nitrate leaching under a cereal
rye cover crop. *Journal of Environmental
Quality* 26, 181–188.

Carson, R. (1962) *Silent Spring*. Houghton
Mifflin, Boston, Massachusetts.

Carvalho, F.P. (2006) Agriculture, pesticides
food security and food safety. *Environmental
Science and Policy* 9, 685–692.

Cerdeira, A.L. and Duke, S.O. (2006) The current status and environmental impacts of
glyphosate-resistant crops: a review.
Journal of Environmental Quality 35,
1633–1658.

Chrispeels, M. and Sadava, D. (2003) *Plants,
Genes and Crop Biotechnology*. ASPB./
Jones and Bartlett, Boston, Massachusetts.

Christian, D. (2005) *Maps of Time. An
Introduction to Big History*. University of
California Press, California.

Collard, B.C.Y. and Mackill, D.J. (2008)
Marker-assisted selection: an approach for
precision plant breeding in the twenty-first
century. *Philosophical Transactions of the
Royal Society B-Biological Sciences* 363,
557–572.

Crabb, J., Firbank, L., Winter, M., Parham, C.
and Dauven, A. (1998) Set-aside landscapes: farmer perceptions and practices
in England. *Landscape Research* 23,
237–254.

Crawley, M.J., Brown, S.L., Hails, R.S., Kohn,
D.D. and Rees, M. (2001) Biotechnology –
transgenic crops in natural habitats. *Nature*
409, 682–683.

Crecchio, C. and Stotzky, G. (1998) Insecticidal
activity and biodegradation of the toxin
from *Bacillus thuringiensis* subspecies
kurstaki bound to humic acids from soil.
Soil Biology and Biochemistry 30,
463–470.

Dale, P.J. (2002) The environmental impact of
genetically modified (GM) crops: a review.
Journal of Agricultural Science 138,
245–248.

Dale, P.J., Clarke, B. and Fontes, E.M.G. (2002) Potential for the environmental impact of transgenic crops. *Nature Biotechnology* 20, 567–574.

Darmency, H., Vigouroux, Y., Gestat De Garambe, T., Richard-Molard, M. and Muchembled, C. (2007) Transgene escape in sugar beet production fields: data from six years farm scale monitoring. *Environmental Biosafety Research* 6, 197–206.

de Maagd, R.A., Bosch, D. and Stiekema, W. (1999) *Bacillus thuringiensis* toxin-mediated insect resistance in plants. *Trends in Plant Science* 4, 9–13.

de Maagd, R.A., Bravo, A. and Crickmore, N. (2001) How *Bacillus thuringiensis* has evolved specific toxins to colonize the insect world. *Trends in Genetics* 17, 1993–1999.

De Snoo, G.R. (1999) Unsprayed field margins: effects on environment, biodiverstiy and agricultural practise. *Landscape and Urban Planning* 46, 151–160.

Devine, G.J. and Furlong, M.J. (2007) Insecticide use: contexts and ecological consequences. *Agriculture and Human Values* 24, 281–306.

Diepenbrock, W. and Leon, J. (1988) Quantitative effects of volunteer plants on glucosinolate content in double low oilseed rape (Brassica napus L.) – a theoretical approach. *Agronomie* 8, 373–377.

Edwards, M.G. and Gatehouse, A.M.R. (2007) Biotechnology in crop protection: towards sustainable insect control. In: Vurro, M. and Gresel, J. (eds) *Novel Biotechnologies for Biocontrol Agent Enhancement and Management.* Springer, Dordrecht, The Netherlands, pp. 1–23.

Ellstrand, N.C. (2003a) Current knowledge of gene flow in plants. *Philosophical Transactions of the Royal Society of London B Biological Sciences* 358, 1163–1170.

Ellstrand, N.C. (2003b) *Dangerous Liasons – When Cultivated Plants Mate with Their Wild Relatives.* John Hopkins University Press, Baltimore, Maryland.

Ellstrand, N.C., Prentice, H.C. and Hancock, J.F. (1999) Gene flow and introgression from domesticated plants into their wild relatives. *Annual Review of Ecology and Systematics* 30, 539–563.

Feldman, M. and Levy, A.A. (2005) Allopolyploidy – a shaping force in the evolution of wheat genomes. *Cytogenetic and Genome Research* 109, 250–258.

Ferry, N., Edwards, M.G., Mulligan, E.A., Emami, K., Petrova, A., Frantescu, M., Davison, G.M. and Gatehouse, A.M.R. (2003) Engineering resistance to insect pests. In: Christou, P. and Klee, H. (eds) *Handbook of Plant Biotechnology.* Wiley, Hoboken, New Jersey.

Ferry, N., Edwards, M.G., Gatehouse, J.A., Capell, T., Christou, P. and Gatehouse, A.M.R. (2006) Transgenic plants for insect pest control: a forward looking scientific perspective. *Transgenic Research* 15, 3–19.

Ferry, N., Mulligan, E.A., Majerus, M.E.N. and Gatehouse, A.M.R. (2007) Bitrophic and tritrophic effects of Bt Cry3A transgenic potato on beneficial, non-target, beetles. *Transgenic Research* 16, 795–812.

ffrench-Constant, R.H., Daborn, P.J. and Le Goff, G. (2004) The genetics and genomics of insecticide resistance. *Trends in Genetics* 20, 163–170.

Galinat, W.C. (1975) The evolutionary emergence of maize. *Bulletin of the Torrey Botanical Club* 102, 313–324.

Gatehouse, A.M.R., Ferry, N. and Raemaekers, R.J.M. (2002) The case of the Monarch butterfly; a verdict is returned. *Trends in Genetics* 18, 249–251.

Ginting, D., Moncrief, J.F., Gupta, S.C. and Evans, S.D. (1998) Interaction between manure and tillage systems on phosphorous uptake and runoff losses. *Journal of Environmental Quality* 27, 1403–1410.

Glebe, T.W. (2007) The environmental impact of european farming: how legitimate are agri-environmental payments? *Review of Agricultural Economics* 29, 87–102.

Green, R.E., Osborne, P.E. and Sears, E.J. (1994) The distribution of passerine birds in Hedgerows during the breeding season in relation to characteristics of the Hedgerow and adjacent farmland. *Journal of Applied Ecology* 31, 677–692.

Gressel, J. (2002) *Molecular Biology of Weed Control.* Taylor & Francis, London.

Gressel, J. (2008) *Genetic Glass Ceilings. Transgenics for Crop Biodiversity.* The

John Hopkins University Press, Baltimore, Maryland.

Groot, A.T. and Dicke, M. (2002) Insect-resistant transgenic plants in a multi-trophic context. *Plant Journal* 31, 387–406.

Hansson, M. and Fogelfors, H. (1998) Management of permanent set-aside on Arable Land in Sweden. *Journal of Applied Ecology* 35, 758–771.

Hecht, S.B. (1995) The evolution of agroecological thought. In: Altieri, M.A. (ed.) *Agroecology: the Scientific Basis of Alternative Agriculture*. Westview Press, Boulder, Colorado, p. 4.

Heifetz, P.B. (2000) Genetic Engineering of the Chloroplast. *Biochimie*, 82(6-7), 655–666

Hellmich, R.L., Siegfried, B.D., Sears, M.K., Stanley-Horn, D.E., Daniels, M.J., Mattila, H.R. and Spencer, T. (2001) Monarch larvae sensitivity to *Bacillus thuringiensis*-purified proteins and pollen. *Proceedings of the National Academy of Sciences of the USA* 98, 11925–11930.

Henderson, I.G., Cooper, J., Fuller, R.J. and Vickery, J. (2000) The relative abundance of birds on set-aside and neighbouring fields in summer. *Journal of Applied Ecology* 37, 335–347.

Hilbeck, A., Baumgartner, M., Fried, P.M. and Bigler, F. (1998) Effects of transgenic *Bacillus thuringiensis* corn-fed prey on mortality and development of immature *Chrysoperla carnea* (Neuroptera: Chrysopidae). *Environmental Entomology* 27, 480–487.

Jacob, G.S., Garbow, J.R., Hallas, L.E., Kimack, N.M., Kishore, G.M. and Schaefer, J. (1988) Metabolism of glyphosate in Pseudomonas sp. Strain LBr. *Applied Environmental Microbiology* 54, 2953.

James, C.A. (2007) Global Status of Commercialized Biotech/GM Crops: 2007. ISAAA Briefs. Brief 37.

Kleijn, D. and Verbeek, M. (2000) Factors affecting the species composition of Arable field vegetation. *Journal of Applied Ecology* 37, 256–266.

Lewandrowski, J., Tobey, J. and Cook, Z. (1997) The interface between agricultural assistance and the environment: chemical fertilizer consumption and area expansion. *Land Economics* 73, 404–427.

Losey, J.E., Rayor, L.S. and Carter, M.E. (1999) Transgenic pollen harms monarch larvae. *Nature* 399, 214–214.

Lovelock, J. (2007) *The Revenge of Gaia*. Allen Lane, London.

Lutman, P. (ed.) (1999) *Gene Flow and Agriculture: Relevance for Transgenic Crops (BCPC Symposium Proceedings no.72, Keele Proceedings)*. British Crop Protection Council, London.

Malone, L.A. and Pham-Delegue, M-H. (2001) Effects of transgene products on honeybees (*Apis mellifera*) and bumblebees (*Bombus sp.*). *Adidologie* 32, 287–304.

McGavin, G.C. (2006) *Endangered: Wildlife on the Brink of Extinction*. Cassell Illustrated, London.

Millenium Ecosystem Assessment (MA) (2003) Ecosystems and human well-being. Current state and human well-being. Current state and trends. Island Press, Washington, DC. Available at: http://www.millenniumassessment.org/en/framework.aspx (accessed 09/10/08)

Mulwa, R.M.S. and Mwanza, L.M. (2006) Biotechnology approaches to developing herbicide tolerance/selectivity in crops. *African Journal Biotechnology* 5, 396.

Naylor, R.E.L. (2002) *Weed Management Handbook*, 9th edn. Blackwell, Oxford.

Padgette, S.R., Kolacz, K.H., Delannay, X., Re, D.B., La Vallee, B.J., Peschke, V.M., Nida, D.L., Taylor, N.B. and Kishore, G.M. (1995) Development, identification and characterization of a glyphosate tolerant soybean line. *Crop Science* 35, 1451–1461.

Palm, C.J., Schaller, D.L., Donegan, K.K. and Seidler, R.J. (1996) Persistence in soil of transgenic plant produced *Bacillus thuringiensis* var. *kurstaki* δ-endotoxin. *Canadian Journal of Microbiology* 42, 1258–1262.

Parry, K.P. (1989) Herbicide use and invention. In: Dodge, A.D. (ed.) *Herbicides and Plant Metabolism*. Cambridge University Press, Cambridge, pp. 1–20.

Phipps, R.H. and Park, J.R. (2002) Environmental benefits of genetically modified crops: global and European perspectives on their ability to reduce pesticide use. *Journal of Animal and Feed Sciences* 11, 1–8.

Pywell, R.F., Putwain, P.D. and Webb, N.R. (1997) The decline of heathland seed populations following the conversion to agriculture. *Journal of Applied Ecology* 34, 949–960.

Ready, R.C. and Henken, K. (1999) Optimal self-protection from nitrate contaminated groundwater. *American Journal of Agricultural Economics* 81, 321–334.

Romeis, J., Bartsch, D., Bigler, F., Candolfi, M.P., Gielkens, M.M.C., Hartley, S.E., Hellmich, R.L., Huesing, J.E., Jepson, P.C., Layton, R., Quemada, H., Raybould, A., Rose, R.I., Schiemann, J., Sears, M.K., Shelton, A.M., Sweet, J., Vaituzis, Z. and Wolt, J.D. (2008) Assessment of risk of insect-resistant transgenic crops to nontarget arthropods. *Nature Biotechnology* 26, 203–208.

Sanvido, O., Romeis, J. and Bigler, F. (2007) Ecological impacts of genetically modified crops: ten years of field research and commercial cultivation. Green Gene Technology. *Research in an Area of Social Conflict* 107, 235–278.

Saxena, D. and Stotzky, G. (2001) *Bacillus thuringiensis* (*Bt*) toxin released from root exudates and biomass of *Bt* corn has no apparent effect on earthworms, nematodes, protozoa, bacteria, and fungi in soil. *Soil Biology and Biochemistry* 33, 1225–1230.

Saxena, D., Flores, S. and Stotzky, G. (1999) Insecticidal toxin in root exudates from *Bacillus thuringiensis* corn. *Nature* 402, 480.

Schlegel, R.H.J. (2007) *Introduction to the History of Crop Development: Theories, Methods, Achievements, Institutions, and Persons*. Haworth Press, New York, London, Oxford.

Schnepf, E., Crickmore, N., Van Rie, J., Lereclus, D., Baum, J., Feitelson, J., Zeigler, D.R. and Dean, D.H. (1998) *Bacillus thuringiensis* and its pesticidal crystal proteins. *Microbiology and Molecular Biology Reviews* 62, 775–806.

Schuler, T.H., Potting, R.P.J., Denholm, I. and Poppy, G.M. (1999) Parasitoid behaviour and *Bacillus thuringiensis* plants. *Nature* 400, 825–826.

Simmons, I.G. (1993) *Environmental History: A Concise Introduction*. Blackwell, Oxford, p.37.

Sims, S.R. (1995) *Bacillus thuringiensis* var. *kurstaki* [Cry1A(c)] protein expressed in transgenic cotton: effects on beneficial and other non-target insects. *Southwestern Entomologist* 20, 493–500.

Sims, S.R. (1997) Host activity spectrum of the CryIIA *Bacillus thuringiensis* subsp. *kurstaki* protein: effects on Lepidoptera, Diptera, and non-target arthropods. *Southwestern Entomologist* 22, 395–404.

Stone, T.B., Sims, S.R. and Marrone, P.G. (1989) Selection of tobacco budworm for resistance to a genetically engineered *Pseudomonas fluorescens* containing the δ-endotoxin of *Bacillus thuringiensis* subsp. Kurstaki. *Journal of Invertebrate Pathology* 53, 228–234.

Swaminathan, M.S. (2006) An evergreen revolution. *Crop Science* 46(5), 2293–2303.

Thies, J.E. and Devare, M.H. (2007) An ecological assessment of transgenic crops. *Journal of Development Studies* 43, 97–129.

Vandenberg, J.D. (1990) Safety of four entomopathogens for cages adult honeybees (Hymenoptera: Apidae). *Journal of Economic Entomology* 83, 755–759.

Wisniewski, J.P., Fragne, N., Massonneau, A. and Dumas, C. (2002) Between myth and reality: genetically modified maize, an example of a sizeable scientific controversy. *Biochimie* 84, 1095–1103.

Wraight, C.L., Zangerl, A.R., Carroll, M.J. and Berenbaum, M.R. (2000) Absence of toxicity of *Bacillus thuringiensis* pollen to black swallowtails under field conditions. *Proceedings of the National Acadamy of Science of the USA* 14, 7700–7703.

Index